This is the most comprehensive book to date on the production, structure, properties and applications of metal matrix composites, an important new class of materials that is making a major impact in many diverse areas of industry.

The emergence of metal matrix composites is partly a consequence of an improved understanding of their potential and limitations, based on principles of physical metallurgy, interfacial chemistry, stress analysis and processing science. This book is intended as an introduction to the microstructure, behaviour and usage of these materials. In each chapter, a simple outline is given of the underlying principles, followed by an assessment of the current state of research knowledge in the area. At a more detailed level, the mathematical background to the analytical treatments involved, including the Eshelby method, has been incorporated into the book, although emphasis is placed throughout on the concepts and mechanisms involved.

The book encompasses particulate, short fibre and long fibre reinforcement. The accent is on mechanical performance, describing how the presence of reinforcement in a metallic matrix influences the stiffening, strengthening and failure characteristics. This involves a critical examination of both load transfer and microstructural modification effects brought about by the presence of the fibres or particles. Comprehensive coverage is also given of other properties, including thermal/electrical conductivity and resistance to thermal shock, wear and corrosive environments. Fabrication and processing are also treated in some detail. The final two chapters provide a source of useful practical information, covering a range of specialist techniques for the study of these materials and detailed examples of commercial applications.

This book is aimed primarily at scientists, engineers, production managers and all those involved in research on new materials in general and metal matrix composites in particular, but it is also suitable for use as a text in graduate and undergraduate courses.

620.118 ENGINEERING
(S.C. Use only)

microstructure
shear moduli
plastic deformation
fracture processes
jet cutting
diesel engine piston
motor drive shaft
brake disc
bicycle frame
engine block
ashby charts

AN INTRODUCTION TO METAL MATRIX COMPOSITES

Leeds

Titles in print in this series

Polymer Surfaces
B. W. Cherry

An Introduction to Composite Materials
D. Hull

Thermoluminescence of Solids
S. W. S. McKeever

Modern Techniques of Surface Science
D. P. Woodruff and T. A. Delchar

New Directions of Solid State Chemistry
C. N. R. Rao and J. Gopalakrishnan

The Electrical Resistivity of Metals and Alloys
P. L. Rossiter

The Vibrational Spectroscopy of Polymers
D. I. Bower and W. F. Maddams

Fatigue of Materials
S. Suresh

Glasses and the Vitreous State
J. Zarzycki

Hydrogenated Amorphous Silicon
R. A. Street

Microstructural Design of Fiber Composites
T-W. Chou

Liquid Crystalline Polymers
A. M. Donald and A. H. Windle

Fracture of Brittle Solids, 2nd Edition
B. R. Lawn

An Introduction to Metal Matrix Composites
T. W. Clyne and P. J. Withers

AN INTRODUCTION TO METAL MATRIX COMPOSITES

T. W. CLYNE
Department of Materials Science and Metallurgy, University of Cambridge

P. J. WITHERS
Department of Materials Science and Metallurgy, University of Cambridge

Published by the Press Syndicate of the University of Cambridge
The Pitt Building, Trumpington Street, Cambridge CB2 1RP
40 West 20th Street, New York, NY 10011-4211, USA
10 Stamford Road, Oakleigh, Melbourne 3166, Australia

First published 1993
First paperback edition 1995

A catalogue record for this book is available from the British Library

Library of Congress cataloguing in publication data

Clyne, T. W.
An introduction to metal matrix composites / T. W. Clyne, P. J. Withers
p. cm. – (Cambridge solid state science series)
Includes bibliographical references.
ISBN 0-521-41808-9
1. Metallic composites. I. Withers, P. J. II. Title. III. Series.
TA481.C55 1993
620.1′6 – dc20 92–24679 CIP

ISBN 0 521 41808 9 hardback
ISBN 0 521 48357 3 paperback

Transferred to digital printing 2003

KW

Contents

Preface

Over the last 30 years or so, metal matrix composites have emerged as an important class of materials. During this period, a very substantial research effort has been directed towards an improved understanding of their potential and limitations, invoking principles of physical metallurgy, stress analysis and processing science. This book is intended as an introduction to the field, covering various aspects of the structure, behaviour and usage of these materials. It is designed primarily for scientists and technologists, but the content is also suitable for final year degree course students of materials science or engineering and for postgraduate students in these disciplines.

The structure of the book is designed to allow several different modes of usage. Chapters 2 and 3 provide a background to stress analysis techniques used to describe the mechanical behaviour of MMCs. In these chapters we have aimed to introduce the concepts pictorially, while the details are discussed in the main text. The finer points of these treatments are relevant to those with a keen interest in composite mechanics, but they are not essential for use of the rest of the book. The following four chapters then form a core description of the load-bearing behaviour. Chapters 4 and 5 cover the basic deformation mechanisms and characteristics, over a range of temperature. A chapter is then devoted to various aspects of the interface between matrix and reinforcement. This is relevant to several areas, particularly the fracture behaviour outlined in Chapter 7. A chapter follows dealing with various transport properties (such as thermal conductivity) and environmental performance (such as the resistance to wear and corrosion). The final four chapters then deal with a variety of special topics. Chapter 9 covers fabrication and processing in some detail, followed by a short chapter on the evolution of matrix microstructure during fabrication and in service. The last two chapters are largely independent of the rest of the book. The first introduces a number of experimental techniques and examines their potential for the study of composite materials, while the second highlights, through a series of illustrative applications, how useful property combinations offered by MMCs can be exploited in practice. To aid in the use of the book, a compendium of materials properties and a nomenclature can be found in the Appendices.

There is a strong research effort in metal matrix composites at Cambridge University and we have both been involved in this over the past several years. We decided to cooperate on this book late in 1989 and the writing was done during 1990 and 1991. Each chapter is very much a joint effort. We hope that this is reflected in a homogeneity of style and nomenclature throughout.

We are grateful for numerous discussions and contacts with colleagues and visitors here in Cambridge. In particular, we would like to thank M. F. Ashby, C. Y. Barlow, W. J. Clegg, I. M. Hutchings, D. Juul Jensen, J. E. King, O. B. Pedersen and W. M. Stobbs. We would also like to acknowledge the substantial contributions from past and present research students in the Materials Science and Metallurgy Department, especially those within our own groups. In addition, we are indebted to all those who have furnished us with micrographs and unpublished information, notably A. Ardekani, A. Begg, A. W. Bowen, G. S. Daehn, D. Double, J. F. Dolowy, T. J. Downes, D. C. Dunand, F. H. Gordon, E. A. Feest, J. A. G. Furness, W. R. Hoover, S. J. Howard, F. J. Humphreys, A. R. Kennedy, R. R. Kieschke, J. J. Lewandowski, J. Lindbo, Y. L. Liu, D. J. Lloyd, J. F. Mason, V. Massardier, A. Mortensen, A. J. Phillips, P. B. Prangnell, D. A. J. Ramm, A. J. Reeves, K. A. Roberts, C. A. Stanford-Beale, R. A. Shahani, R. A. Ricks, A. Tarrant, T. J. Warner, C. M. Ward-Close, C. M. Warwick, M. C. Watson, W. Wei, J. White, A. F. Whitehouse and F. Zok. We would also like to acknowledge the financial and moral support we have received over recent years for our own research work on metal matrix composites, in particular from the Science and Engineering Research Council, Alcan International, British Petroleum, Imperial Chemical Industries, National Physical Laboratory, Pechiney, RAE Farnborough, Risø National Laboratory and Rolls Royce. We have had extensive scientific contact with various people from these and other organisations, which has been of considerable benefit to us.

Finally, we would like to thank our wives, Gail and Lindsey, for a degree of forbearance well beyond any reasonable call of marital duty. Without their support, this book would never have been finished.

T. W. Clyne
P. J. Withers
Cambridge, January 1992

1
General introduction

After more than a quarter of a century of active research, composites based on metals are now beginning to make a significant contribution to industrial and engineering practice. This is partly a consequence of developments in processing methods. However, equally important have been advances in the understanding of various structure–property relationships, assisting in the identification of cost effective solutions and highlighting important objectives in the control of microstructure and the design of components. In this first chapter, a brief overview is given of the nature of MMCs and the background to their development.

1.1 Types of MMC and general microstructural features

The term metal matrix composite (MMC) encompasses a wide range of scales and microstructures. Common to them all is a contiguous metallic matrix[†]. The reinforcing constituent is normally a ceramic, although occasionally a refractory metal is preferred. The composite microstructures may be subdivided, as depicted in Fig. 1.1, according to whether the reinforcement is in the form of continuous fibres, short fibres or particles. Further distinctions may be drawn on the basis of fibre diameter and orientation distribution. Before looking at particular systems in detail, it is helpful to identify issues relating to the microstructure of the final product. A simplified overview is given in Table 1.1 of the implications for composite performance of the main microstructural features. Whereas some of these microstructural parameters are readily pre-specified, others can be very difficult to control. Necessarily, the

[†] A further sub-class is that in which the matrix is an intermetallic compound, but such materials are not wholly within the scope of this book.

Table 1.1 *Overview of structure/property relationships for MMCs. The symbols indicate whether an increase in the microstructural parameter in the left-hand column will raise, lower or leave unaffected the properties listed. Also shown are the section numbers within the book where the trends concerned are covered in detail.*

Microstructural feature	Composite property					
	α_{axial}	E_{axial}	Tensile YS (0.2% PS)	Work hardening rate	Creep resistance	Toughness (ductility)
Ceramic content f	⇓ § 5.1.2	⇑ §3.6	⇑ §4.1.1	⇑ §4.3.1	⇑ §5.2.3	⇓ §7.3.3
Fibre aspect ratio s	⇓ §5.1.2	⇑ §3.6	⇑ or ⇓ §4.1.1	⇑ §4.3.1	⇑ §5.2.3	⇑ or ⇓ §7.4.2
Misalignment $g(\theta)$	⇑ §5.1.2	⇓ §3.6	⇑ or ⇓ §7.1.2	⇓ §4.3.1	⇓ §5.2.3	⇑ or ⇓ §7.4.2
Fibre diameter d	–	–	⇓ §4.2	⇓ §4.3.2	⇑ §5.2.3	⇑ §7.4.3
Inhomogeneity of f	–	–	–	–	⇓ §5.2.3	⇓ §7.4.5
Bond strength τ_i	⇓	⇑ §6.1.4	⇑	⇑ §4.4.2	⇑ §5.2.3	⇑ or ⇓ §6.1.4
Reaction layer t	⇑ or ⇓	⇓ §6.3.2	⇓ §6.1.4	⇓ §6.1.4	⇓ §5.2.3	⇓ §6.1.4
$\Delta\alpha\,\Delta T$ stresses	–	–	⇓ §4.1.2	⇓ §4.1.2	⇓ §5.2.3	⇓ §7.2.2
Matrix porosity	⇓ §5.1.1	⇓ §3.6	⇓	⇓	⇓	⇓ §7.3.3
Matrix YS	–	–	⇑ §4.2	⇑ §4.3.2	⇑ §5.2.3	⇓ §7.4.6

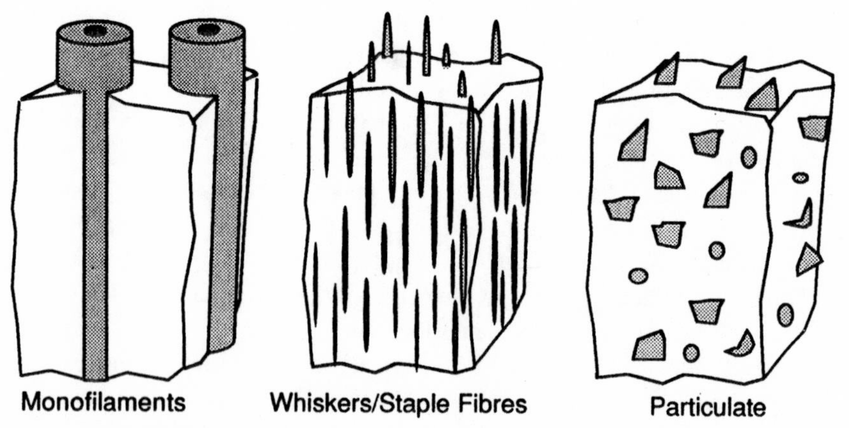

Fig. 1.1 Schematic depiction of the three types of MMC, classified according to the type of reinforcement.

interplay between microstructure and properties is much more complex than the representation shown, which also conceals various uncertainties. Nevertheless, it is important to identify simple microstructural objectives and methods for attaining them before embarking on the design and fabrication of any specific component. One of the aims of this book is to help with that task.

An important lesson learnt in connection with polymeric composites (PMCs) is that the material specification and the component design must be integrated into a single operation. For MMCs, the high formability of the matrix (particularly when compared with polymer resins) means that this need not be the case, or that it is only partially true. Several types of discontinuously reinforced MMC material are available in stock form, such as billet, rod and tube, suitable for various secondary fabrication operations. Even long fibre MMC stock material can sometimes be subjected to certain shaping and forming operations. This highlights the potential versatility of MMCs, but it should nevertheless be recognised that, in common with PMCs, they are essentially anisotropic, potentially to a much greater extent than unreinforced metals. Unlike aligned polymer-based composites, however, excellent axial performance can be combined with transverse properties which are more than satisfactory. The freedom to separate material preparation and component design/production procedures, in much the same way as with unreinforced metals, presents both an opportunity and a potential pitfall. Certainly, there will be many instances where the integration of these operations is likely to assist in optimisation of the use of MMCs.

1.2 Historical background

Examples of metal matrix composites stretch back to the ancient civilisations. Copper awls from Cayonu (Turkey) date back to about 7000 BC and were made by a repeated lamination and hammering process, which gave rise to high levels of elongated non-metallic inclusions[1]. Among the first composite materials to attract scientific as well as practical attention were the dispersion hardened metal systems. These developed from work in 1924 by Schmidt[2] on consolidated mixtures of aluminium/alumina powders and led to extensive research in the 1950s and 1960s[3]. The principles of precipitation hardening in metals date from the 1930s[4,5] and were developed in the following decade[6,7]. Recent collected papers[8,9], celebrating such landmarks in the use of metals, give a fascinating insight into the major metallurgical advances during this period.

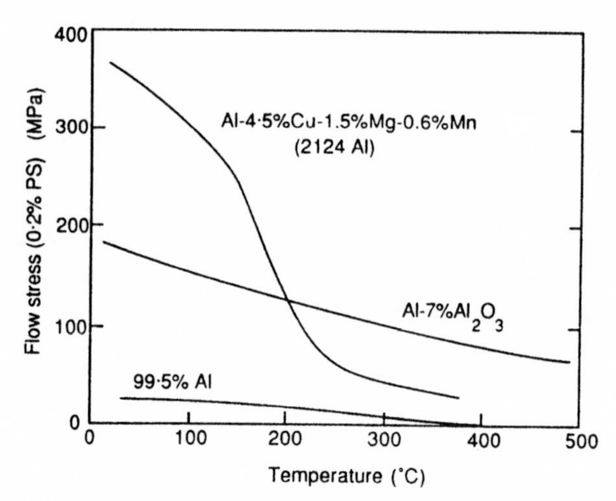

Fig. 1.2 Variation with temperature of the flow stress of three aluminium-based materials[3], illustrating the retention of dispersion strengthening to temperatures above which hardening precipitates coarsen or dissolve.

For both dispersion hardening and precipitation hardening, the basis of the strengthening mechanism is to impede dislocation motion with small particles. This is achieved by the incorporation of either fine oxide particles or non-shearable precipitates within a metallic matrix. Of prime importance in this context is the minimisation of the spacing between the inclusions. Since it is generally possible to achieve finer distributions in precipitation hardened systems, these normally exhibit higher strengths at room temperature. However, dispersion strengthened systems show advantages at elevated temperature, because of the high thermal stability of the oxide particles (Fig. 1.2). While extremely low volume fractions are desirable in terms of ductility, dispersoid contents as high as 15 vol% have been used in applications requiring good high temperature strength and creep resistance. Creep is effectively suppressed because the dislocations must climb over the dispersoids by diffusive processes and this results in creep rates decreasing with increasing dispersoid size.

More recent developments have brought the concept of metal matrix composites closer to engineering practice. An interesting example is provided by the so-called 'dual phase' steels, which evolved in the 1970s[10]. These are produced by annealing fairly low carbon steels in the $\alpha + \gamma$ phase field and then quenching so as to convert the γ phase to martensite. The result is a product very close to what is now referred to as a particulate

MMC, with about 20% of very hard, relatively coarse martensite particles distributed in a soft ferrite matrix. This is a strong, tough and formable material, now used extensively in important applications such as car bodywork. Its success could be interpreted as confirming the viability of the particulate MMC concept, although its properties have not often been considered in terms of composite theory.

Interest in fibrous metal matrix composites mushroomed in the 1960s, with effort directed mainly at aluminium and copper matrix systems reinforced with tungsten and boron fibres. In such composites the primary role of the matrix is to transmit and distribute the applied load to the fibres. High reinforcement volume fractions are typical (~40–80%), giving rise to excellent axial performance. Consequently, matrix microstructure and strength are of secondary importance. Research on continuously reinforced composites waned during the 1970s, largely for reasons of high cost and production limitations. The continuing need for high temperature, high performance materials for various components in turbine engines has triggered a resurgence of interest, mainly directed towards titanium matrices. As mentioned in §12.2.9, attention has focused on hoop-wound unidirectional systems. Factors such as thermal fatigue (§5.3.3), interfacial chemical stability (§6.3.1) and thermal shock resistance (§8.1.5) have become of equal importance to those of straightforward mechanical performance.

Discontinuously reinforced composites fall somewhere between the dispersion strengthened and fibre strengthened extremes, in that both matrix and reinforcement bear substantial proportions of the load. They have been rapidly developed during the 1980s, with attention focused on Al-based composites reinforced with SiC particles and Al_2O_3 particles and short fibres. The combination of good transverse properties, low cost, high workability and significant increases in performance over unreinforced alloys has made them the most commercially attractive system for many applications (Chapter 12). They are distinguished from the dispersion hardened systems, of which they are a natural extension, by the fact that, because the reinforcement is large (~1–100 μm), it makes a negligible contribution to strengthening by Orowan inhibition of dislocation motion (Fig. 1.3). In addition, since the volume fraction is relatively high (5–40%), load transfer from the matrix is no longer insignificant (Chapters 3 and 4). However, unlike continuously reinforced systems, the matrix strength, as affected by precipitation and dislocation strengthening (§4.2, §4.3.2, §10.1 and §10.2), also plays an important role.

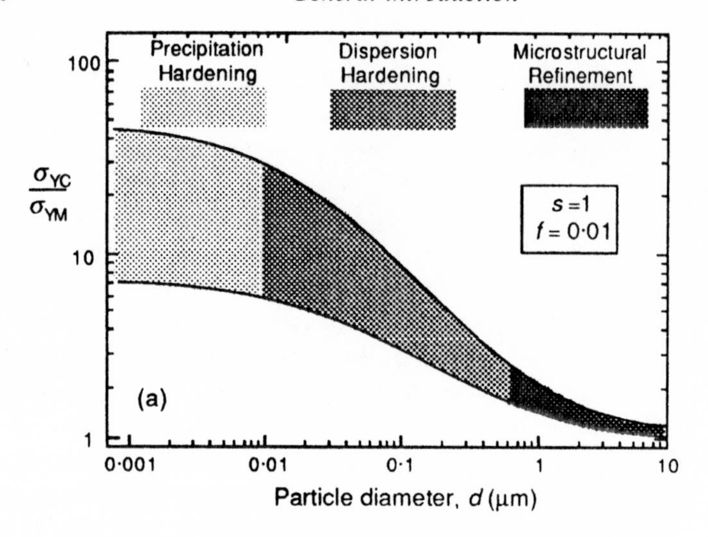

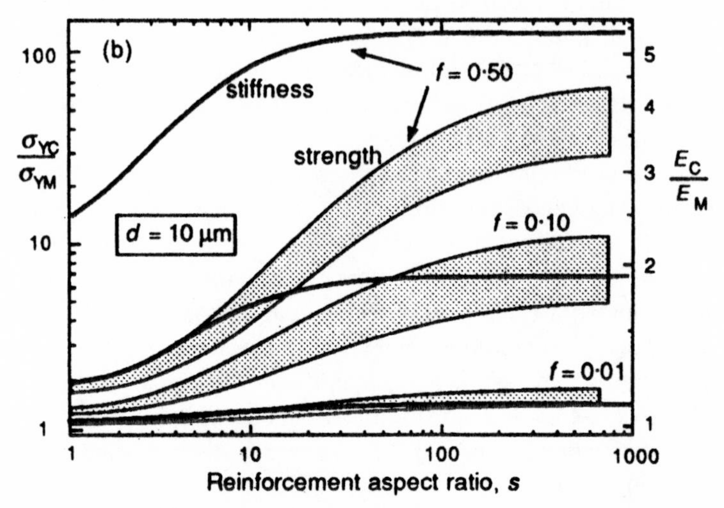

Fig. 1.3 Schematic illustration of the magnitude and range of operation of the primary composite strengthening mechanisms as a function of inclusion shape, size and volume fraction. In (a), matrix strengthening dominates, with the inclusions constituting too low a volume fraction to carry a significant proportion of the load. In (b), strengthening and stiffening are primarily a consequence of load transfer to the reinforcement. In some cases, both types of mechanism may be significant, as with particle reinforcement of age-hardening alloys. Furthermore, the reinforcement may itself give rise to both types of strengthening, directly by load transfer and indirectly by stimulating changes in matrix microstructure.

1.3 Interactions between constituents and the concept of load transfer

One definition (Longman's dictionary) of the word 'composite' is simply: 'something combining the typical or essential characteristics of individuals making up a group'. Central to the philosophy behind the use of any composite material is the extent to which the qualities of two distinct constituents can be combined, without seriously accentuating their short-comings. In the context of MMCs, the objective might be to combine the excellent ductility and formability of the matrix with the stiffness and load-bearing capacity of the reinforcement, or perhaps to unite the high thermal conductivity of the matrix with the very low thermal expansion of the reinforcement.

In attempting to identify attractive matrix/reinforcement combinations, it is often illuminating to derive a 'merit index' for the performance required, in the form of a specified combination of properties. Appropriate models can then be used to place upper and lower bounds on the composite properties involved in the merit index, for a given volume fraction of reinforcement. The framework for such predictions has been clearly set out by Ashby[11], and an example of a composite property map is shown in Fig. 1.4. This graph refers to a merit index for the minimisation of thermal distortion during heating or cooling. Plotted on a field of expansivity against conductivity, this merit index will be high at bottom right and low at top left. The data shown therefore indicate that the resistance of aluminium to thermal distortion can be improved by the incorporation of silicon carbide, but will be impaired if boron nitride is added.

In practice, there is often interest in establishing composite properties to a greater precision than is possible by the use of bounds, which in many cases are widely separated. This can be relatively complex, particularly for MMCs – in which certain matrix properties may be significantly affected by the presence of the reinforcement. Much of this book is thus devoted to understanding, through an examination of the relevant thermophysical interactions, the principles underlying effective hybridisation of this type. Rules will be developed governing mechanical loadings (Chapters 2, 3, 4 and 7), thermal expansion (§5.1), thermal/electrical conductivity (§8.1) and other more complex properties such as wear resistance (§8.2) and damping capacity (§8.3). In many of these areas, the role of the interface is critical and Chapter 6 is devoted to interfacial phenomena and their effects on composite behaviour.

Central to an understanding of the mechanical behaviour of a

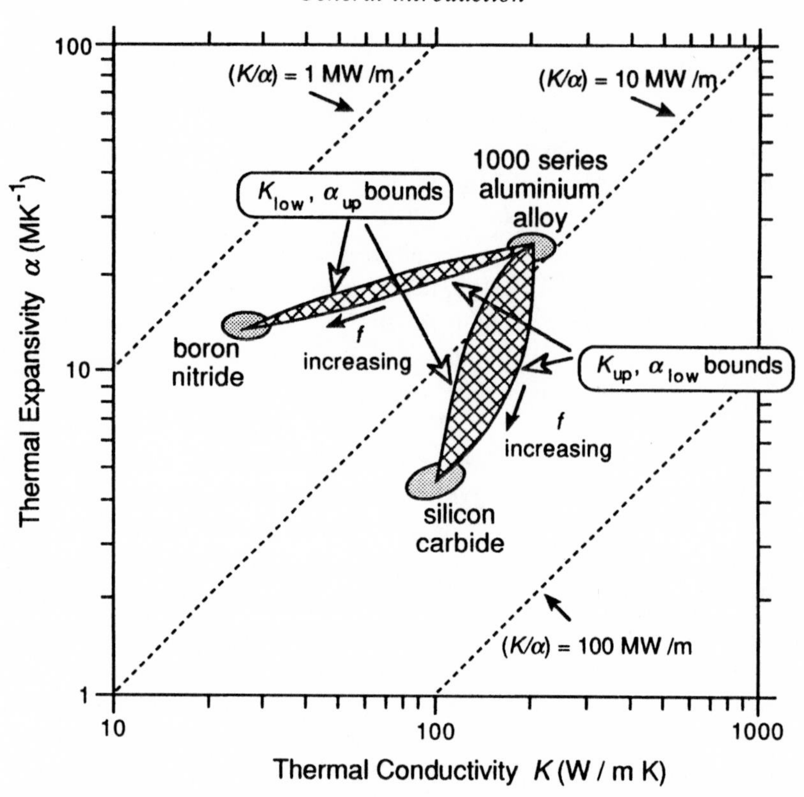

Fig. 1.4 Predicted map[11] of thermal expansivity, α, against thermal conductivity, K, for composites made up of either silicon carbide or boron nitride in an aluminium matrix. The diagonal dotted lines represent constant values of a merit index, given by K/α, taken as indicative of the resistance to thermal distortion. The shaded areas, defined by upper and lower bounds of the two parameters (obtained from appropriate equations), indicate the possible combinations of K and α expected for Al–SiC and Al–BN composites, depending on the volume fraction, shape and orientation distribution of the reinforcement.

composite is the concept of load sharing between the matrix and the reinforcing phase. The stress can vary sharply from point to point, but the proportion of the external load borne by each of the individual constituents can be gauged by volume-averaging the load within them. Of course, at equilibrium, the external load must equal the sum of the volume-averaged loads borne by the constituents† (e.g. the matrix and

† In the absence of an externally applied load, the individual constituents may still be stressed (due to the presence of residual stresses), but these must balance each other according to eqn (1.1).

the reinforcement). This gives rise to the condition

$$(1 - f)\bar{\sigma}_M + f\bar{\sigma}_I = \sigma^A \tag{1.1}$$

governing the volume-averaged matrix and inclusion (fibre or particle) stresses ($\bar{\sigma}_M$, $\bar{\sigma}_I$) in a composite under an external stress σ^A, containing a volume fraction f of reinforcement. (In the most general form of eqn (1.1), $\bar{\sigma}_M$, $\bar{\sigma}_I$ and σ^A are tensors.) Thus, for a simple two-constituent MMC under a given applied load, a certain proportion of that load will be carried by the reinforcement and the remainder by the matrix. Provided the response of the composite remains elastic, this proportion will be independent of the applied load and it represents an important characteristic of the material. As is examined in later chapters, it depends on the volume fraction, shape and orientation of the reinforcement and on the elastic properties of both constituents. The reinforcement may be regarded as acting efficiently if it carries a relatively high proportion of the externally applied load. This can result in higher strength, as well as greater stiffness, because the reinforcement is usually stronger, as well as stiffer, than the matrix. The concept of elastic load transfer has been entirely familiar to those working with (polymer-based) fibre composites over the past 30–40 years. It is readily translated to the elastic behaviour of MMCs, although calculation of the load partitioning is often more complex as a result of the greater interest in discontinuous (short fibre and particulate) reinforcement, as opposed to continuous fibres (Chapters 2 and 3).

While stiffening is well understood, confusion occasionally arises when considering the origin of strengthening, in the sense that the strengthening contribution arising from load transfer is sometimes attributed to enhanced *in situ* matrix strengthening and vice versa (Fig. 1.3). Both types of strengthening are discussed in Chapter 4. As with unreinforced metals, MMCs are expected to be able to sustain a certain amount of plastic deformation in normal service. It is particularly important, therefore, that their stress–strain response beyond the elastic limit be well understood. In practice, at least for discontinuously reinforced MMCs, both load transfer and the effect of the presence of the reinforcement on the *in situ* properties of the matrix need to be invoked in order to explain the observed behaviour. The latter can often be predicted using well-established laws and correlations drawn from dislocation theory and metallurgical experience. Treatment of the load transfer, however, is a little more problematical – particularly when the matrix starts to undergo plastic deformation. In the present book, considerable use is made of an analytical method of predicting load transfer – the Eshelby technique (see Chapter 3). Central to this approach is the concept of the ***misfit strain*** – i.e. the difference between the 'natural' shape of the reinforcement and the 'natural' shape of the hole it occupies in the matrix†. Equations can be derived to predict the load transfer for any misfit strain – such as can arise from the difference in stiffness of the two constituents (Chapter 3), from differential thermal contraction (Chapter 5) or from plastic deformation of the matrix (Chapter 4). (The same phenomena can also be simulated using numerical methods – see Chapter 2, but the accessibility of an analytical technique facilitates the study of general trends.)

Finally, to complicate the issue a little further, even a rigorous treatment of load transfer and an appreciation of the *in situ* yielding and work hardening behaviour of the matrix does not give the complete picture. Frequently, an applied stress can cause the reinforcement to start to carry a very high load indeed. For example, even a small amount of matrix plasticity creates a relatively large misfit strain and hence transfers load very strongly to the reinforcement. Such load transfer may not be sustainable. For example, with discontinuous reinforcement it would require very high local stresses in certain regions of the matrix and interface. Under these circumstances, stress relaxation phenomena (such as interfacial sliding or diffusion) will tend to occur (Chapter 4), which act to reduce the load borne by the reinforcement and hence to impair the load-bearing capacity of the composite as a whole. These stress relaxation processes are difficult to model accurately, but at least their effect can be understood and predicted for well-characterised material under specific conditions.

References

1. J. D. Muhly (1988) The Beginnings of Metallurgy in the Old World, in *The Beginning of the Use of Metals and Alloys*, R. Maddin (ed.), MIT Press, Cambridge, pp. 2–20.
2. E. Schmidt (1924) German Patents Nos 425451, 425452 and 427370.
3. N. Hansen (1971) *Dispersion Strengthened Aluminium Products – Manufacture, Structure and Mechanical Properties*, Risø National Laboratory, Denmark.
4. G. I. Taylor (1934) The Mechanism of Plastic Deformation of Crystals, I – Theoretical, *Proc. Roy. Soc.*, **A145**, pp. 362–404.
5. E. Orowan (1934) Zur Kristallplastizität. III. Über den Mechanismus des Gleitvorganges, *Z. Phys.*, **89**, pp. 634–59.

† 'Natural' in this context means the shape that the constituent would have if the other constituent(s) were not present.

6. E. Orowan (1948) in *Internal Stresses in Metals and Alloys*, Inst. of Metals, London, pp. 451–65.
7. N. F. Mott and F. R. N. Nabarro (1948) in *Strength of Solids*, Physical Society, London, pp. 1–19.
8. M. H. Loretto (ed.) (1985) *Dislocations and Properties of Real Materials*, Inst. of Metals, London.
9. J. A. Charles and G. C. Smith (eds.) (1991) *Advances in Physical Metallurgy*, Inst. of Metals, London.
10. M. D. Maheshwari, A. Chatterjee, T. Mukherjee and J. J. Irani (1983) Experience in the Heat Treatment of Dual Phase Steels, in *Heat Treatment '91*, London, R. H. Johnson (ed.), Inst. of Metals, London, pp. 67–77.
11. M. F. Ashby (1993) Material Selection for Composite Design, *Acta Met. et Mat.*, in press.

2
Basic composite mechanics

Composite materials are inherently inhomogeneous, in terms of both elastic and inelastic properties. One consequence of this is that, on applying a load, a non-uniform distribution of stress is set up within the composite. Much effort has been devoted to understanding and predicting this distribution, as it determines how the material will behave and can be used to explain the superior properties of composites over conventional materials. In this chapter, a brief survey is given of the methods used for modelling stress distributions in composites. These techniques range widely in nature and complexity. Some are more suited to certain types of composite and attention is drawn to areas of particular relevance to metal matrix composites. No treatment is presented in this chapter of the Eshelby method, which is particularly useful for MMCs, since it is considered in detail in Chapters 3 and 4.

2.1 The slab model

The simplest way to model the behaviour of a composite containing continuous, aligned fibres is to treat it as if it were composed of two slabs bonded together, one of the matrix and the other of the reinforcement, with the relative thickness of the latter in proportion to the volume fraction of the fibres (designated as f). The response of this 'composite slab' to external loads can be predicted quite easily, but its behaviour will closely mirror that of the real composite only under certain conditions (Fig. 2.1).

2.1.1 Axial stiffness

The model is most useful for the case of a normal stress being applied parallel to the fibre axis (the '3' direction). The two components of the

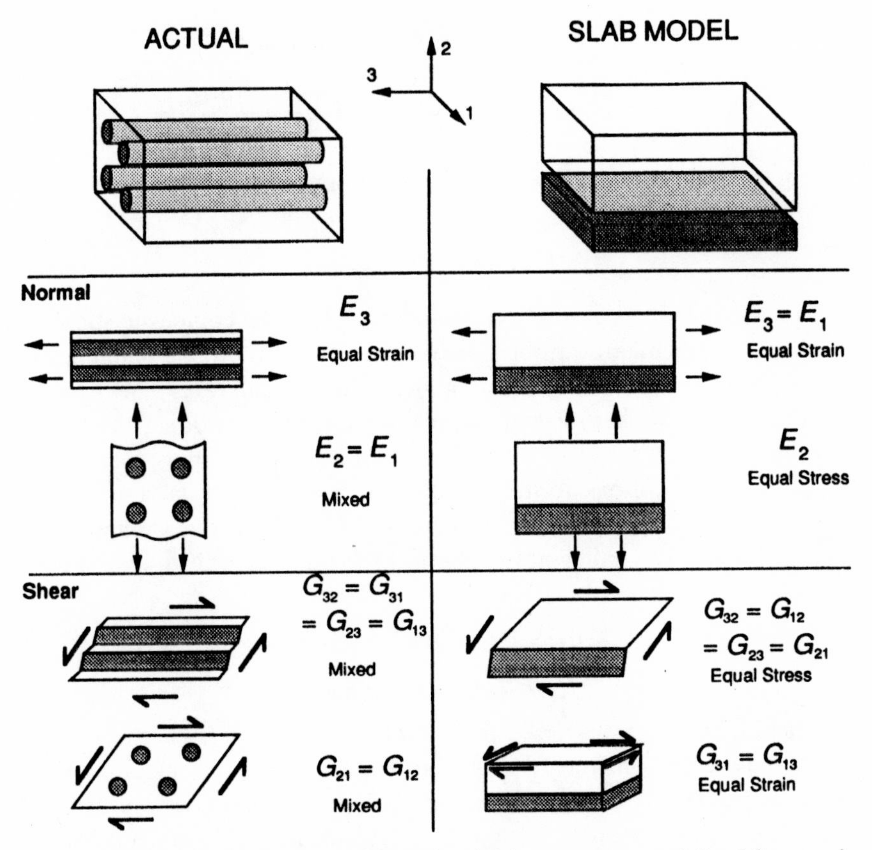

Fig. 2.1 Schematic illustrations of how the elastic constants are defined for a real fibre composite and for the slab model representation. The partitioning of stress and strain between the constituents is noted in each case.

slab composite must have the same strain in this direction, ε_{3C}, equal to the ratio between the stress and the Young's modulus for each of the two components

$$\varepsilon_{3C} = \varepsilon_{3I} = \frac{\sigma_{3I}}{E_I} = \varepsilon_{3M} = \frac{\sigma_{3M}}{E_M} \tag{2.1}$$

where the subscript I refers to the inclusion (fibre) and M to the matrix. To establish the composite stiffness, the overall composite stress σ_{3C} must be expressed in terms of the contribution from each phase

$$\sigma_{3C} = (1 - f)\sigma_{3M} + f\sigma_{3I} \tag{2.2}$$

The Young's modulus of the composite can now be written

$$E_{3C} = \frac{\sigma_{3C}}{\varepsilon_{3C}} = \frac{[(1-f)\sigma_{3M} + f\sigma_{3I}]}{(\sigma_{3I}/E_I)} = E_I\left[\frac{(1-f)\sigma_{3M}}{\sigma_{3I}} + f\right] \tag{2.3}$$

Using the ratio between the stresses in the components given by eqn (2.1), this may be simplified to

$$E_{3C} = (1-f)E_M + fE_I \tag{2.4}$$

This well-known '***Rule of Mixtures***' indicates that the composite stiffness is simply a weighted mean between the moduli of the two components, depending on the volume fraction of reinforcement. This equation is expected to be valid to a high degree of precision, providing the fibres are long enough for the equal strain assumption to apply (see §2.2). Very minor deviations from the equation are expected as a result of stresses which arise when the Poisson's ratios of the two components are not equal. The equal strain treatment is often described as a '***Voigt model***'.

2.1.2 *Transverse stiffness*

Prediction of the transverse stiffness of a composite from the elastic properties of the constituents is far more difficult than the corresponding axial value. This is because, in a real composite, the phases are neither equally stressed nor equally strained, as depicted in Fig. 2.1. Furthermore, the symmetry of the composite is different from that of the slab model. The 'equal stress' bound is obtained by considering an applied stress in the '2' direction

$$\sigma_{2C} = \sigma_{2I} = \varepsilon_{2I}E_I = \sigma_{2M} = \varepsilon_{2M}E_M \tag{2.5}$$

so that the component strains can be expressed in terms of the applied stress. The overall net strain can be written as

$$\varepsilon_{2C} = f\varepsilon_{2I} + (1-f)\varepsilon_{2M} \tag{2.6}$$

Substituting expressions for the component strains, the composite modulus is thus given by

$$E_{2C} = \left[\frac{f}{E_I} + \frac{(1-f)}{E_M}\right]^{-1} \tag{2.7}$$

The equal stress treatment is often described as a '***Reuss model***'.

Although this treatment is simple and convenient, it gives only a rather crude approximation for E_{2C}. A variety of other empirical or semi-empirical expressions have been suggested, with that of Halpin and Tsai[1] being widely used. This analysis is not based on rigorous elasticity theory, but broadly takes account of enhanced fibre load bearing, relative to the equal stress assumption. The predictions behave correctly in the limits of small and large fibre volume fractions. The expression for the transverse stiffness is

$$E_{2C} = \frac{E_M(1 + \xi\eta f)}{(1 - \eta f)} \quad \text{with} \quad \eta = \frac{\left(\dfrac{E_I}{E_M} - 1\right)}{\left(\dfrac{E_I}{E_M} + \xi\right)} \tag{2.8}$$

The value of ξ is taken as an adjustable parameter, but its magnitude is generally of the order of unity. A comparison is shown in Fig. 2.2(a) between predicted transverse moduli for a Ti–SiC fibre composite according to equal stress and Halpin–Tsai models. Also shown for comparison is the prediction according to the (more rigorous) Eshelby model (see Chapter 3). It is clear that the equal stress curve represents an underestimate, while the Halpin–Tsai curve gives a very good approximation. Other expressions have also been proposed. For example, an analytical equation for E_{2C} has been derived by Spencer[2] for a square array of fibres, although it is based on the relatively crude assumption that the matrix strain is uniform within any slice taken parallel to the loading direction. Its predictions become inaccurate at high fibre volume fractions.

2.1.3 Shear moduli

The shear moduli can also be predicted using the slab model, by evaluating the net shear strain induced when a shear stress is applied to the composite, in terms of the individual displacement contributions from the two constituents. It is important to understand the nomenclature convention which is used. A shear stress designated $\tau_{ij} (i \neq j)$ refers to a stress acting in the i direction on the plane with a normal in the j direction. Similarly, a shear strain γ_{ij} is a rotation towards the i direction of the j axis. The shear modulus G_{ij} is the ratio of τ_{ij} and γ_{ij}. As the composite body is assumed not to be rotating, the condition $\tau_{ij} = \tau_{ji}$ must hold. In addition, $G_{ij} = G_{ji}$, so that $\gamma_{ij} = \gamma_{ji}$. As the '1' and '2' directions are equivalent in a fibre composite, it follows that there are two shear moduli, because $G_{32C} = G_{23C} = G_{13C} = G_{31C} \neq G_{21C} = G_{12C}$.

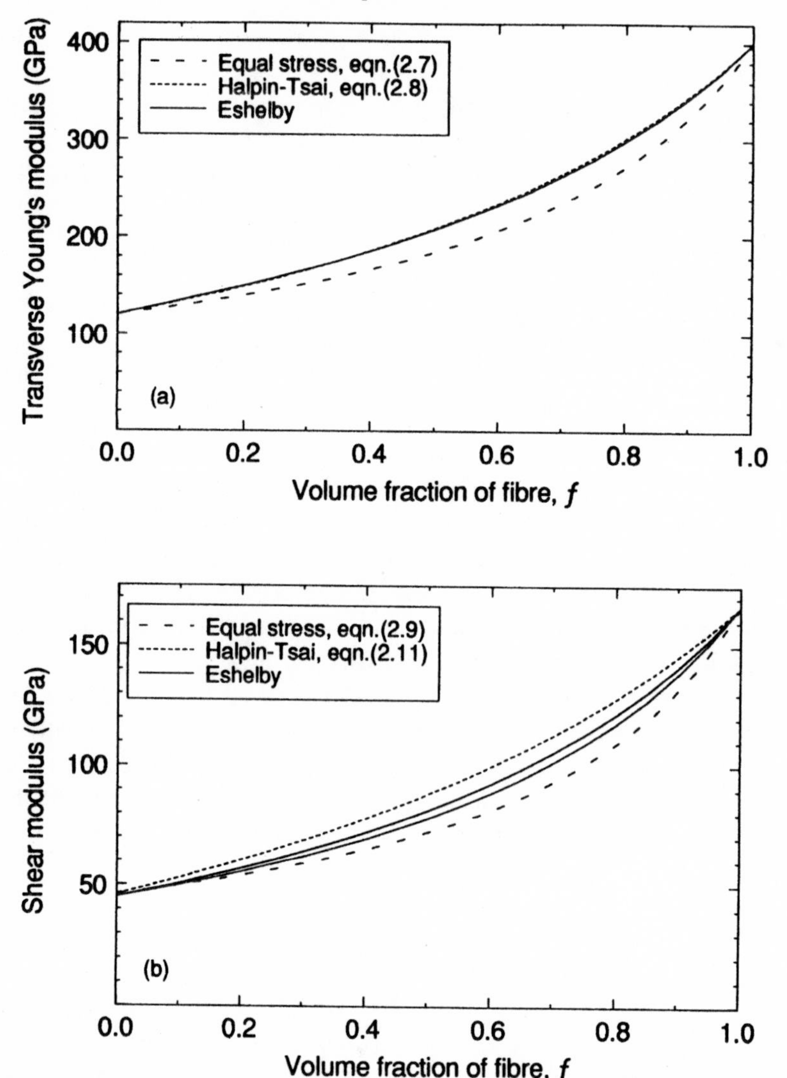

Fig. 2.2 Predicted dependence on fibre volume fraction of (a) the transverse Young's modulus and (b) shear modulus of continuous composites, according to several models, for SiC fibres in titanium with $\xi = 1$. In (b), the upper Eshelby prediction corresponds to G_{32}, the lower to G_{21}.

There are also two shear moduli for the slab model (Fig. 2.1), but these are not expected to correspond closely with the values for the fibre composite. The stresses τ_{32} and τ_{23} are assumed to operate equally within both of the constituents. The derivation is similar to the equal stress treatment leading to eqn (2.7) for transverse stiffness

$$G_{32C} = \frac{\tau_{32}}{\gamma_{32}} = \frac{\tau_{32I}}{f\gamma_{32I} + (1-f)\gamma_{32M}} = \left[\frac{f}{G_I} + \frac{(1-f)\gamma_{32M}}{\tau_{32I}}\right]^{-1}$$

i.e.

$$G_{32C} = \left[\frac{f}{G_I} + \frac{(1-f)}{G_M}\right]^{-1} \qquad (2.9)$$

The other shear modulus shown by the slab model, $G_{31} = G_{13}$ in Fig. 2.1, corresponds to an equal shear strain condition and is analogous to the axial tensile modulus case. It is readily demonstrated that

$$G_{31C} = fG_I + (1-f)G_M \qquad (2.10)$$

which is similar to eqn (2.4). It can be shown that the actual values of G_{32} and G_{21} for a typical fibre composite are generally rather close to each other, with G_{32} slightly larger in magnitude. In comparison with these, eqn (2.9) gives a significant underestimate relative to both of them, while eqn (2.10) is a gross overestimate. In view of this, the semi-empirical expressions of Halpin and Tsai[1], are frequently employed. In this case, the appropriate equation is

$$G_{32C} = \frac{G_M(1+\xi\eta f)}{(1-\eta f)} \quad \text{with} \quad \eta = \frac{\left(\frac{G_I}{G_M} - 1\right)}{\left(\frac{G_I}{G_M} + \xi\right)} \qquad (2.11)$$

and the parameter ξ is often taken to have a value of unity. This has been done for the curves produced in Fig. 2.2(b), which compares predictions of eqn (2.10) with those of the equal stress/equal strain (eqn (2.8)) and Eshelby models. It can be seen that the Halpin–Tsai expression represents a fair approximation to the Eshelby axial shear modulus (G_{32}).

2.1.4 Poisson contractions

Finally, certain Poisson contractions can be predicted using the slab model. The Poisson's ratio ν_{ij} describes the contraction in the j direction on applying a stress in the i direction and is defined by the equation

$$\nu_{ij} = -\frac{\varepsilon_j}{\varepsilon_i} \qquad (2.12)$$

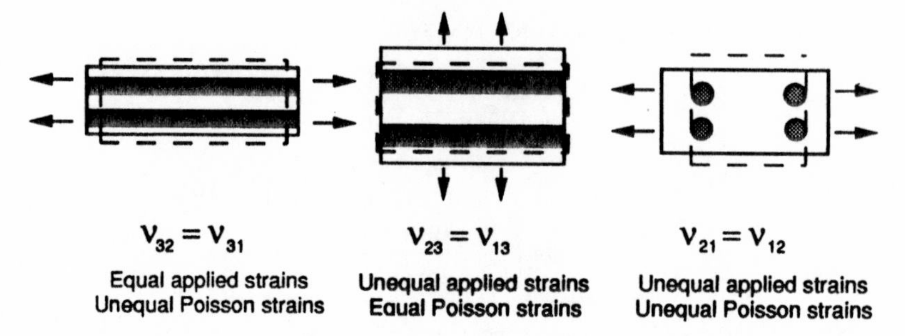

Fig. 2.3 Schematic illustration of how the Poisson's ratios are defined for a fibre composite.

For an aligned fibre composite, there are three different Poisson's ratios, as illustrated in Fig. 2.3. This brings the total number of elastic constants identified for the material to seven. However, these constants are interrelated, such that only five independent values are needed to describe the behaviour of such a transversely isotropic material (e.g. see Nye[3]). The following two relationships account for this

$$\frac{\nu_{32}}{E_{3C}} = \frac{\nu_{23}}{E_{2C}} \tag{2.13}$$

$$G_{21C} = \frac{E_{2C}}{2(1 + \nu_{21})} \tag{2.14}$$

The slab model is only useful for calculating ν_{32}, for which the Poisson contractions for the two constituents can be evaluated independently and summed. Thus

$$\varepsilon_{2I} = -\nu_I \varepsilon_{3I} = -\nu_I \frac{\sigma_{3I}}{E_I} \qquad \varepsilon_{2M} = -\nu_M \varepsilon_{3M} = -\nu_M \frac{\sigma_{3M}}{E_M}$$

so that

$$\varepsilon_{2C} = -\left[\frac{f\nu_I \sigma_{3I}}{E_I} + \frac{(1-f)\nu_M \sigma_{3M}}{E_M}\right]$$

$$= -[f\nu_I \varepsilon_{3C} + (1-f)\nu_M \varepsilon_{3C}]$$

and

$$\nu_{32C} = -\frac{\varepsilon_2}{\varepsilon_3} = f\nu_I + (1-f)\nu_M \tag{2.15}$$

A simple rule of mixtures is therefore applicable and, because the equal strain assumption is accurate for axial stressing of the composite, this is expected to be a fairly reliable prediction.

The ratio of the axial contraction to the transverse extension on stressing transversely, ν_{23C}, can be obtained from the reciprocal relationship given as eqn (2.13), so that

$$\nu_{23C} = [f\nu_I + (1 - f)\nu_M] \frac{E_{2C}}{E_{3C}} \tag{2.16}$$

This will be lower than ν_{32C} because, on stressing transversely, the fibres will offer strong resistance to axial contraction. This leads to pronounced contraction in the other transverse direction, so that ν_{21C} is expected to be high. An expression for ν_{21C} may be obtained by considering[4] the overall volume change experienced by the material, leading to

$$\nu_{21C} = 1 - \nu_{23C} - \frac{E_{2C}}{3K_C} \tag{2.17}$$

in which the bulk modulus of the composite K_C can be estimated via an equal stress assumption

$$K_C = \left[\frac{f}{K_I} + \frac{(1 - f)}{K_M}\right]^{-1} \tag{2.18}$$

and the bulk moduli of the constituents are related to other elastic constants by expressions such as

$$K_I = \frac{E_I}{3(1 - 2\nu_I)} \tag{2.19}$$

The accuracy of eqn (2.17) is determined largely by the error in E_{2C}. This is large for the equal stress model, so that, in the comparisons shown in Fig. 2.4, the Halpin and Tsai values of E_{2C} predicted by eqn (2.8), were used to obtain values of ν_{23C} and ν_{21C}. It can be seen that agreement with the Eshelby predictions is good. These plots convey an idea of the pronounced tendency under transverse loading for the composite to contract in the other transverse direction in preference to the axial direction.

In summary, slab model-based equations can be useful for predicting the elastic constants of long fibre composites, providing the shortcomings of the model are understood and empirical approximations are used where appropriate. The model does not in general lead to useful predictions of the internal stresses, except for the simple instance of the axial stresses during axial loading. This, combined with its inapplicability to

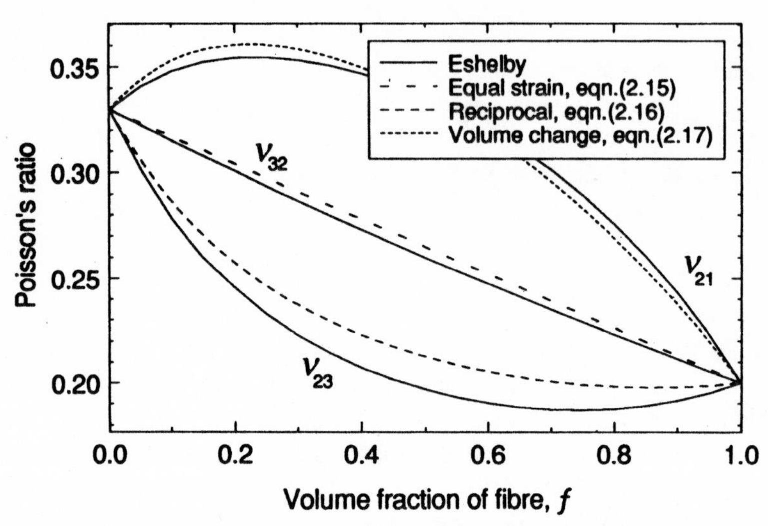

Fig. 2.4 Predicted Poisson's ratios as a function of fibre volume fraction for SiC fibres in titanium.

discontinuously reinforced composites, means that its usefulness for MMCs is rather limited.

2.2 The shear lag model

The most widely used model describing the effect of loading an aligned short fibre composite is the so-called shear lag model, originally proposed by Cox[5] and subsequently developed by others[6–8], which centres on the transfer of tensile stress from matrix to fibre by means of interfacial shear stresses. The basis of the calculations is shown schematically in Fig. 2.5. The external loading is applied parallel to the fibre axis. The model is based on considering the radial variation of shear stress in the matrix and at the interface.

2.2.1 Stress and strain distributions

The radial variation of shear stress in the matrix, τ (at a given axial distance z from the fibre mid-point), is obtained by equating the shear forces on neighbouring annuli (with radii r_1, r_2) of length dz (Fig. 2.5(c))

$$2\pi r_1 \tau_1 \, \mathrm{d}z = 2\pi r_2 \tau_2 \, \mathrm{d}z$$

$$\therefore \frac{\tau_1}{\tau_2} = \frac{r_2}{r_1} \tag{2.20}$$

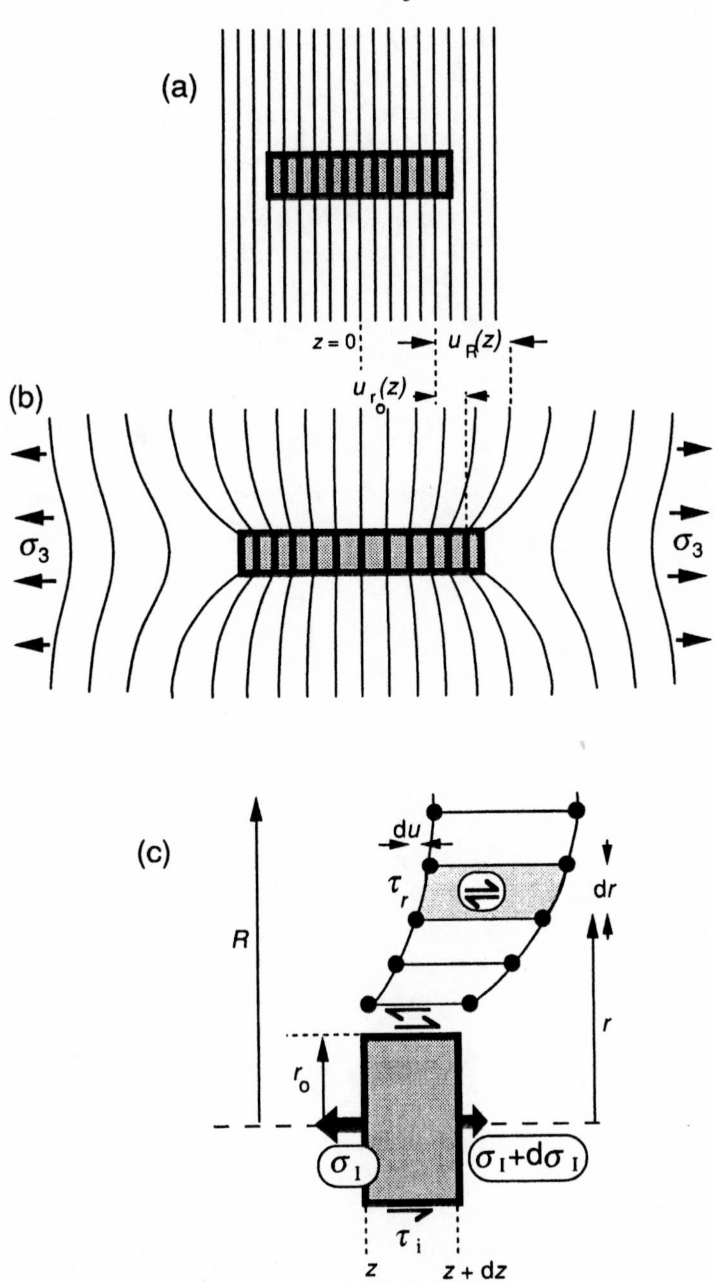

Fig. 2.5 Schematic illustration of the basis of the shear lag model, showing (a) the unstressed system, (b) the axial displacements, u, introduced on applying tension parallel to the fibre and (c) the variation with radial location of the shear stress and strain in the matrix.

It follows that the shear stress τ at any particular radius r can be related to that at the interface (radius r_0) written as τ_i

$$\tau_r = \tau_i\left(\frac{r_0}{r}\right) \tag{2.21}$$

The concept is now introduced of the ***displacement*** $\boldsymbol{u_r(z)}$ of the matrix in the z direction, relative to the position with no applied stress. The increment of this displacement, du, on moving out from the fibre axis by dr, is then determined by the shear strain γ, and hence by the shear modulus G_M

$$\frac{du}{dr} = \gamma = \frac{\tau_r}{G_M} = \frac{\tau_i}{G_M}\left(\frac{r_0}{r}\right) \tag{2.22}$$

It follows that, for a given z, the difference between the displacement of the matrix at a radius R and that of the interface is given by a simple integration

$$\int_{u_{r_0}}^{u_R} du = \frac{\tau_i r_0}{G_M}\int_{r_0}^{R}\frac{dr}{r}$$

$$(u_R - u_{r_0}) = \frac{\tau_i r_0}{G_M}\ln(R/r_0) \tag{2.23}$$

The assumption is made that the matrix strain is uniform remote from the immediate vicinity of the fibre. The radius R represents some far field location where this condition becomes operative. The appropriate value of (R/r_0) is related to the proximity of neighbouring fibres and hence to the fibre arrangement and volume fraction, f: however, because (R/r_0) appears in a logarithmic term, the final result is relatively insensitive to the details of the geometry. A hexagonal array leads to

$$\left(\frac{R}{r_0}\right)^2 = \frac{\pi}{2f\sqrt{3}} \sim \frac{1}{f} \tag{2.24}$$

and in view of the relatively crude nature of the model, this approximation is suitable for all cases. The build-up of tensile stress in the fibre $\sigma_I(z)$ is now related to the distribution of interfacial shear stress. Referring to Fig. 2.5(c), force balance requires

$$2\pi r_0 \tau_i\, dz = -\pi r_0^2\, d\sigma_I$$

$$\frac{d\sigma_I}{dz} = -\frac{2\tau_i}{r_0} \tag{2.25}$$

Now, the variation of τ_i with z is unknown a priori, but eqn (2.23) can be used to relate it to displacements and hence to axial strains. Assuming perfect interfacial adhesion and no shear strain in the fibre (so that within the fibre $u_r = u_{r0}$, the displacement of the fibre surface), and substituting for the shear modulus in terms of the Young's modulus and Poisson's ratio leads to the result

$$\frac{\mathrm{d}\sigma_I}{\mathrm{d}z} = -\frac{2E_M(u_R - u_{r0})}{(1 + \nu_M)r_0^2 \ln(1/f)} \tag{2.26}$$

The displacements themselves are unknown, but their differentials are related to identifiable strains. That of the fibre is simply

$$\left.\frac{\mathrm{d}u}{\mathrm{d}z}\right|_{r=r0} = \varepsilon_I = \frac{\sigma_I}{E_I}$$

but the corresponding expression for the matrix is less well-defined in view of uncertainties about the stress distribution. The differential of u_R is expected to approximate to the far field matrix strain, at least over most of the length of the fibre, and this in turn should be close to the overall composite strain ε_{3C}

$$\left.\frac{\mathrm{d}u}{\mathrm{d}z}\right|_{r=R} \approx \varepsilon_M \sim \varepsilon_{3C}$$

This is clearly not rigorous, but it is expected to represent a fairly good approximation – particularly for relatively long fibres. This view of the far field matrix strain being uniform along (and beyond) the fibre length is broadly reflected in Fig. 2.5(b), while the fibre strain (and stress) evidently builds up with distance from the ends.

The solution is now obtained in a straightforward manner. Differentiation of eqn (2.26) and substitution leads to

$$\frac{\mathrm{d}^2\sigma_I}{\mathrm{d}z^2} = \frac{n^2}{r_0^2}(\sigma_I - E_I\varepsilon_{3C}) \tag{2.27}$$

in which n is a dimensionless constant given by

$$n = \left[\frac{2E_M}{E_I(1 + \nu_M)\ln(1/f)}\right]^{1/2}$$

Eqn (2.27) is a standard second order linear differential equation, with the solution

$$\sigma_I = E_I\varepsilon_{3C} + B\sinh\left(\frac{nz}{r_0}\right) + D\cosh\left(\frac{nz}{r_0}\right)$$

On applying the boundary conditions

$$\sigma_{\mathrm{I}} = 0 \qquad \text{at} \qquad z = \pm L$$

where L is the half-length, and writing the fibre aspect ratio (L/r_0) as s, this gives the solution

$$\sigma_{\mathrm{I}} = E_{\mathrm{I}}\varepsilon_3[1 - \cosh(nz/r_0)\,\mathrm{sech}(ns)] \tag{2.28}$$

From this, the variation of interfacial shear stress along the fibre length is readily derived, using eqn (2.25), to give

$$\tau_{\mathrm{i}} = \frac{n\varepsilon_{3\mathrm{C}}}{2} E_{\mathrm{I}} \sinh\left(\frac{nz}{r_0}\right) \mathrm{sech}(ns) \tag{2.29}$$

These two equations can be used to explore several aspects of composite behaviour. However, before examining these, it is worth noting that the main source of error in the basic model lies in the neglect of load transfer across the fibre end. Corrections have been developed to account for this.

2.2.2 *Transfer of normal stress across fibre ends*

Several attempts have been made[9–11] to introduce corrections for the neglect of stress transfer across the fibre ends. This becomes increasingly necessary for lower fibre aspect ratios and smaller fibre/matrix stiffness ratios and is thus particularly significant for MMCs. Any attempt to account for the effect, while retaining the attractive simplicity of the shear lag approach, must involve postulating an analytical expression for the fibre end stress σ_{e}. Evidently, this must be an arbitrary postulate, as there is no scope within the shear lag framework for any rigorous description of stresses beyond the fibre end. An example is provided by the suggestion[11] that σ_{e} be set equal to the average of the peak fibre stress and the remote matrix stress values predicted by the standard shear lag model

$$\sigma_{\mathrm{e}} = \frac{\sigma_{\mathrm{I}0} + \sigma_{\mathrm{M}0}}{2}$$

in which $\sigma_{\mathrm{I}0}$ is given by substituting $z = 0$ in eqn (2.28) and $\sigma_{\mathrm{M}0}$ is taken as $E_{\mathrm{M}}\varepsilon_{3\mathrm{C}}$ (the average matrix stress – see below). This leads to an expression for σ_{e}

$$\sigma_{\mathrm{e}} = \frac{\varepsilon_{3\mathrm{C}}\{E_{\mathrm{I}}[1 - \mathrm{sech}(ns)] + E_{\mathrm{M}}\}}{2} = \varepsilon_{3\mathrm{C}} E_{\mathrm{M}}^{*} \tag{2.30}$$

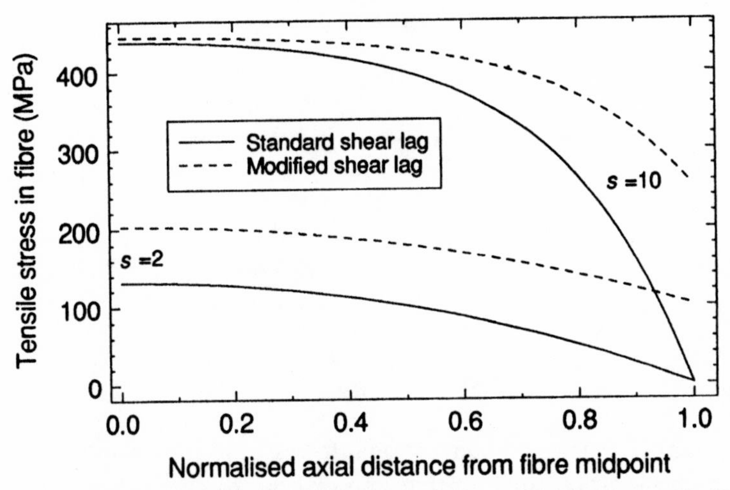

Fig. 2.6 Predicted variations in axial stress within a SiC fibre in an Al–30 vol% SiC composite subject to a strain of 10^{-3}, for two fibre aspect ratios, according to the standard[5] and modified[11] shear lag models.

and hence, using the new boundary conditions $\sigma_{\text{I}} = \sigma_{\text{e}}$ at $z = \pm L$, to a new expression for σ_{I}, analogous to eqn (2.28)

$$\sigma_{\text{I}} = \varepsilon_{3\text{C}}\left[E_{\text{I}} - (E_{\text{I}} - E_{\text{M}}^{*})\cosh\left(\frac{nz}{r_0}\right)\text{sech}(ns)\right] \tag{2.31}$$

Predictions from eqn (2.31) are compared with those from the standard shear lag model (eqn (2.28)) in Fig. 2.6, which shows the build-up of tensile stress along fibres having two different fibre aspect ratios, both in Al–30 vol% SiC composites under a composite strain of 0.1%. Naturally, taking account of fibre end stress transfer leads to the fibres carrying more load, particularly for short fibres, which will result in an increase in the predicted stiffness of the composite.

2.2.3 Composite mechanics

The basic results of the shear lag treatment can be used to predict the elastic deformation of the composite. Consider a section of area A taken normal to the loading direction (in which all the fibres are aligned). This will intersect individual fibres at random positions along their length. Now, the applied load can be expressed in terms of the contributions from the two components

$$\sigma_3^A A = fA\bar{\sigma}_I + (1 - f)A\bar{\sigma}_M$$

$$\therefore \sigma_3^A = f\bar{\sigma}_I + (1 - f)\bar{\sigma}_M \tag{2.32}$$

in which $\bar{\sigma}_I$ and $\bar{\sigma}_M$ are the volume-averaged stresses carried by fibre and matrix. This equation is often termed the '***Rule of Averages***'. The average fibre stress is easily evaluated from eqn (2.28)

$$\bar{\sigma}_I = \frac{E_I \varepsilon_{3C}}{L} \int_0^L \left[1 - \frac{\cosh(nz/r_0)}{\cosh(ns)}\right] dz$$

$$\bar{\sigma}_I = E_I \varepsilon_{3C}\left[1 - \frac{\tanh(ns)}{ns}\right] \tag{2.33}$$

For the matrix, it is again conventional to resort to the assumption of a uniform tensile strain equal to that imposed on the composite

$$\bar{\sigma}_M \approx E_M \varepsilon_{3C}$$

Combining these equations gives the axial Young's modulus of the composite

$$E_{3C} = \left\{fE_I\left[1 - \frac{\tanh(ns)}{ns}\right] + (1 - f)E_M\right\} \tag{2.34}$$

The same procedure for the modified model, taking account of fibre end stress transfer, leads[11] to

$$E_{3C} = \left\{fE_I\left[1 - \frac{(E_I - E_M^*)\tanh(ns)}{E_I ns}\right] + (1 - f)E_M\right\} \tag{2.35}$$

These equations can be tested by making comparisons with predictions from the (more rigorous) Eshelby model. For example, Fig. 2.7 shows the variation in composite Young's modulus with fibre/matrix modulus ratio for two fibre aspect ratio values. It can be seen that the standard shear lag model is inaccurate for low fibre aspect ratios, particularly if the fibre/matrix stiffness ratio is relatively small. It is also clear that the end stress modification is a useful approximation for (short fibre) metal matrix composites. To illustrate this, Fig. 2.8 compares predictions from the three models with experimental data for a particulate MMC. The standard shear lag model is clearly quite unsuitable for application to such materials.

It will be noted from eqns (2.34) and (2.35) that the stiffness approaches the limiting (Rules of Mixtures) value as s becomes large enough for

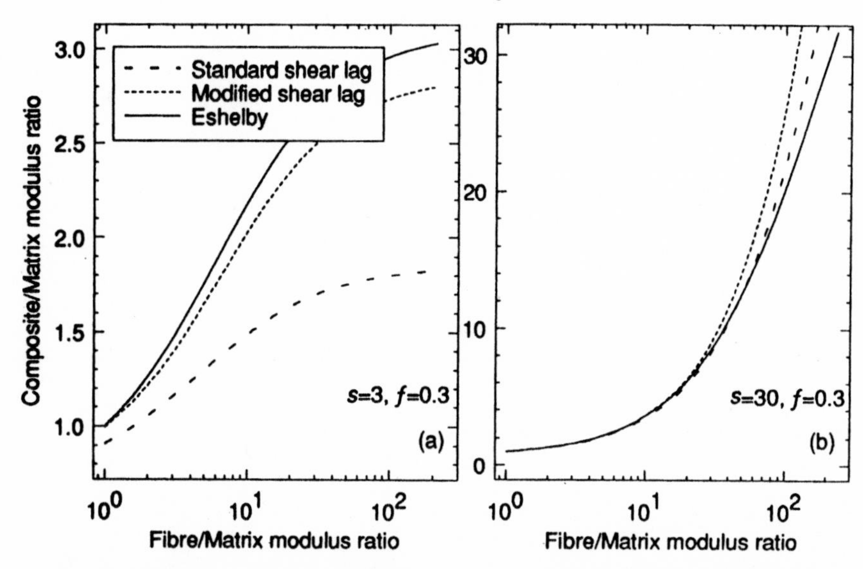

Fig. 2.7 Predicted composite/matrix modulus ratio as a function of the fibre/matrix modulus ratio, for composites with 30 vol% reinforcement and two different fibre aspect ratios. (Poisson's ratio values of fibre and matrix were taken as 0.2 and 0.3 respectively.)

$\tanh(ns)/ns$ to become negligible. Since $\tanh(ns) \sim 1$ for $ns \gtrsim 3$, and taking 0.1 as a suitably small value for $\tanh(ns)/ns$, we could define

$$s_{\mathrm{RM}} \approx \frac{10}{n} \tag{2.36}$$

in which s_{RM} is the fibre aspect ratio needed for the modulus to approach its maximum value. The magnitude of n obviously depends on the composite concerned, but for most MMCs it will be around 0.4 (cf. $\sim$0.1 for a typical PMC). This suggests a value for s_{RM} of about 25 for MMCs (100 for PMCs). These values could be regarded as typical target aspect ratios in aligned composites for which the main objective is to maximise the load transfer and hence the stiffness.

2.2.4 *Onset of inelastic behaviour*

The shear lag model can be used to predict the onset of matrix plasticity or interfacial sliding. This should occur at the fibre ends, where the matrix shear stress is a maximum. Specifying a critical shear stress τ_{i*} and substituting this into eqn (2.29) with $z = L$ gives the composite strain at

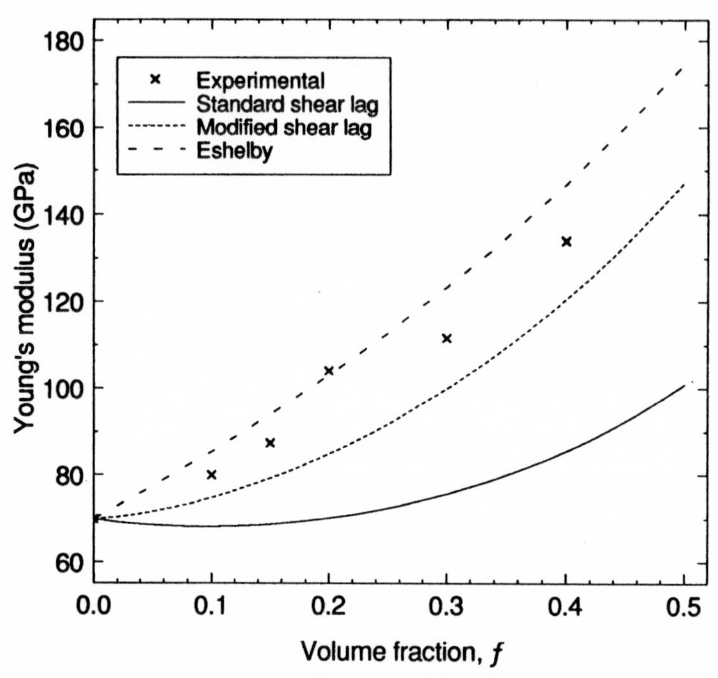

Fig. 2.8 Comparison between experimental data[12] and model predictions for the stiffness of Al–SiC (particulate) composites produced by extrusion. As the particles are not truly equiaxed, and tend to become aligned during processing, an aspect ratio of 2 was used in the predictions.

this point

$$\varepsilon_{3C*} = \frac{2\tau_{i*} \coth(ns)}{nE_I} \tag{2.37}$$

which can be converted to an applied stress using eqn (2.34). This gives the expression

$$\sigma_{3*}^{A} = \frac{2\tau_{i*}}{nE_I}\left\{[fE_I + (1-f)E_M]\coth(ns) - \frac{fE_I}{ns}\right\}$$

It should be emphasised that this does not correspond to a clearly identifiable composite yield stress, as yielding (or sliding) will only be taking place in a small localised region. It will, however, be at this point that the stress–strain curve starts to depart from a linear plot. As an illustration of the use of eqn (2.37), the predicted composite strain at the onset of local plastic flow for Al–30 vol% SiC with $s = 5$, assuming $\tau_{i*} = 100$ MPa, would be about 0.1%. (In practice, it often occurs even

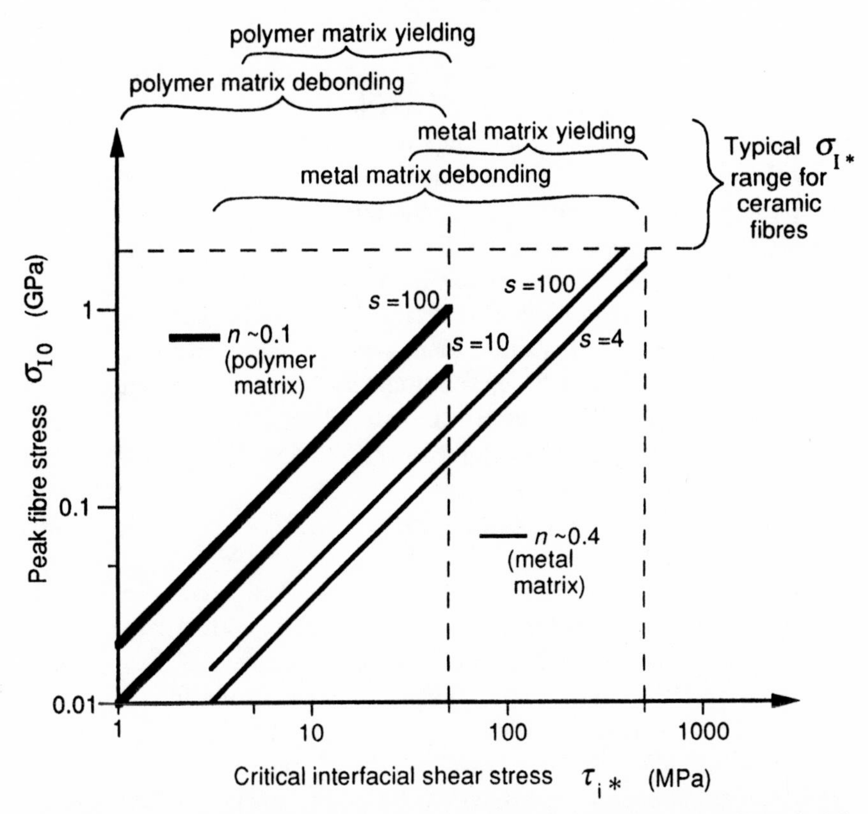

Fig. 2.9 Plots of the dependence of peak fibre stress, σ_{I0} (at the onset of interfacial sliding or matrix yielding), as a function of critical shear stress for the onset of these phenomena, according to eqn (2.38). Plots are shown for different aspect ratios, with n values typical of polymer- and metal-based composites. Also indicated are typical ranges for fracture of ceramic fibres and matrix yielding and interfacial debonding.

earlier than this with MMCs as a result of the presence of residual stresses from differential thermal contraction after fabrication – see §4.1.2.)

A question of interest here concerns the likelihood of fibre fracture taking place before matrix yielding or interfacial sliding. The peak stress in the fibre at this point is found by evaluating eqn (2.28) for $z = 0$ at the composite strain given by eqn (2.37). This leads to

$$\sigma_{I0} = \frac{2\tau_{i*}}{n}(\coth(ns) - \operatorname{cosech}(ns)) \tag{2.38}$$

Schematic plots of this relationship are shown in Fig. 2.9, which also gives an indication of the range of values expected for τ_{i*} in metallic and polymeric matrices and for σ_{I*} exhibited by ceramic fibres. It is clear from this plot that, on increasing the load applied to both types of composite, some type of yielding or sliding process at the fibre ends will almost certainly take place before fibres start to fracture.

As the composite strain is increased, yielding (or sliding) will spread along the length of the fibre, raising the tensile stress in the fibre as the interfacial shear stresses increase. Fracture of fibres may then become possible and a simple treatment can be used to explore the limit of this effect. If it is assumed that the interfacial shear stress becomes uniform at τ_{i*} along the length of the fibre, then a ***critical aspect ratio*** s_* can be identified, below which the fibre cannot undergo any further fracture. This corresponds to the peak (central) fibre stress just attaining its ultimate strength σ_{I*}, so that, by integrating eqn (2.25) along the fibre half-length

$$s_* = \frac{\sigma_{I*}}{2\tau_{i*}} \tag{2.39}$$

It follows from this that a distribution of aspect ratios between s_* and $s_*/2$ would be expected if the composite were subjected to a large strain. For an MMC, the value of s_* might range from, say, 2–3 for a strong matrix such as a Ti alloy to ~20–30 or so for a strong fibre in a soft metal (e.g. SiC whiskers in Al at elevated temperature).

In summary, the shear lag model can be of some value in exploring basic features of composite behaviour. Simple estimates can be made of the stiffness of short fibre and particulate composites and some insights obtained into the internal distribution of stress. However, the model is based on crude simplifications and the nature of these must be appreciated if misleading deductions are to be avoided. In general, realistic calculations of matrix yielding, thermal stresses, etc., require more complex models in which the multiaxial nature of the stress state is recognised.

2.3 Continuous coaxial cylinder models

For continuous fibre composites, a potentially useful modelling approach involves treating the composite as a set of coaxial cylinders. Provided all of the materials are transversely isotropic, then analytical solutions exist for the elastic stress state within such a system when it is subjected to a uniform external load (axial or radial), or to some temperature change. Several papers have been published outlining these solutions[13–16], which are obtained by the imposition of various boundary conditions for stress and strain compatibility, giving a set of linear simultaneous equations

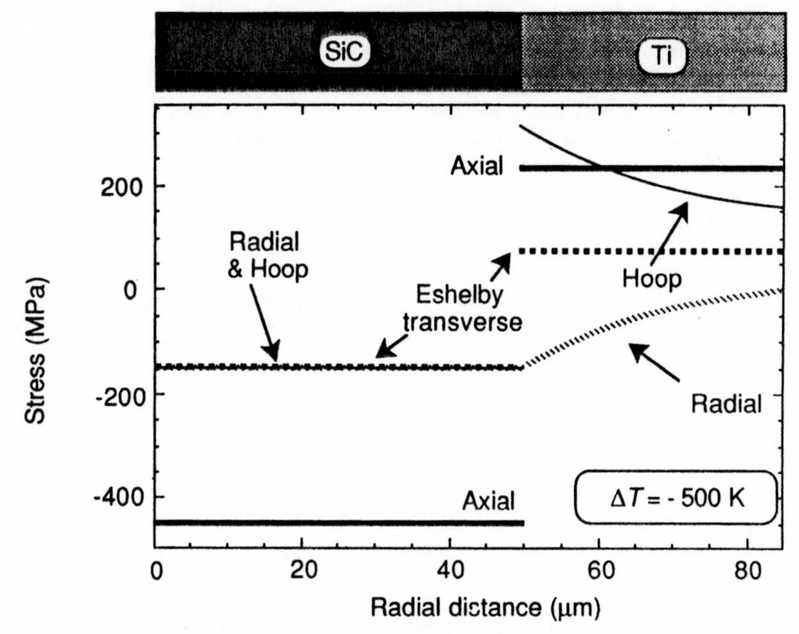

Fig. 2.10 Stress distributions for a Ti–35 vol% SiC fibre composite, after cooling through 500 K. The variation in axial, radial and hoop stresses has been calculated according to the coaxial cylinder model, with Ti thickness chosen to give the correct fibre volume fraction. The volume-averaged stresses have also been calculated using the mean field modified Eshelby model; in the axial case, no radial variation in axial stress is predicted so that the coaxial model prediction and the volume-averaged stress coincide.

which can be solved by standard methods. The procedure involves converting the five independent engineering constants exhibited by a transversely isotropic material to a stiffness tensor, using the matrix representation. The necessary set of equations has been presented by Mikata and Taya[15] for a four-layer structure composed of fibre, coating, matrix and (infinite) surrounding composite and by Warwick and Clyne[16] for any number of layers, with a free outer surface.

Use of a two-layer model with a free outer surface can be used to explore internal stresses in fibre composites, with the relative thickness of the two layers chosen to reflect the fibre content. This is illustrated in Fig. 2.10, which refers to a Ti–35 vol% SiC fibre composite. Fig. 2.10 shows the radial distribution of the three principal stresses arising from a temperature decrease of 500 K; this can be compared with the volume-averaged axial and transverse stresses shown for the same case, obtained

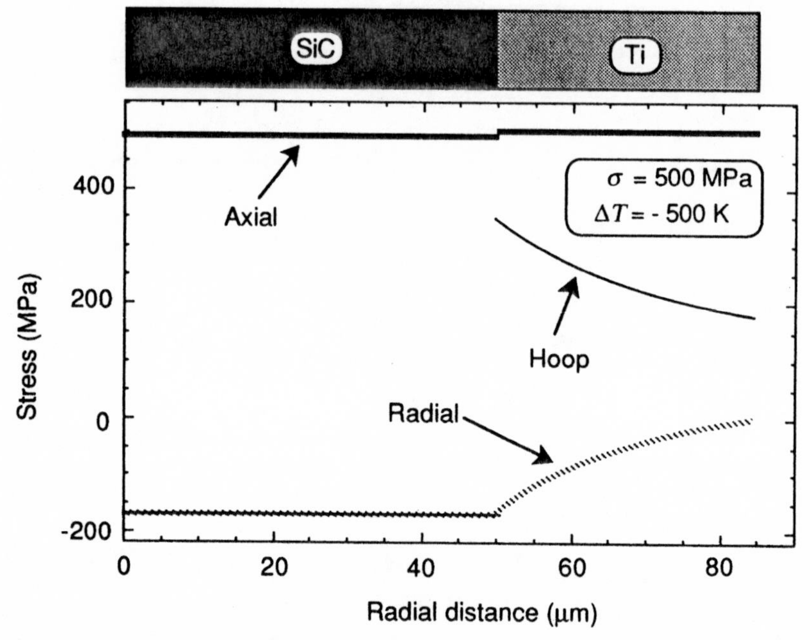

Fig. 2.11 Predicted stress distribution for a Ti–35 vol% SiC composite subjected to a 500 MPa applied tensile stress along the fibre axis, together with residual stresses resulting from a 500 K temperature decrease.

using the Eshelby model (Chapter 3). It can be seen that these two sets of predictions are entirely consistent, although the approaches involved are very different.

Of course, stresses predicted by the coaxial cylinder model will not be exactly the same as those within and around a fibre in a real composite, as the boundary conditions are incorrect at the free surface. Only by using numerical methods (see §2.4) can the stress state be predicted for some arbitrary local arrangement of fibres. Nevertheless, the coaxial cylinder model can give a useful quantitative insight into the stress distributions within long fibre composites. For example, the curves in Fig. 2.10 illustrate how large deviatoric stresses are created in the matrix adjacent to the fibre as a result of temperature change, making this region a likely site for plastic flow. It is also possible to examine the effects of combined thermal and mechanical loading. Fig. 2.11 shows the stress state predicted to result from an applied axial tensile load of 500 MPa, superimposed on the effect of the temperature drop shown in Fig. 2.10. Note that preferential load bearing by the fibre, expected as a result of its greater

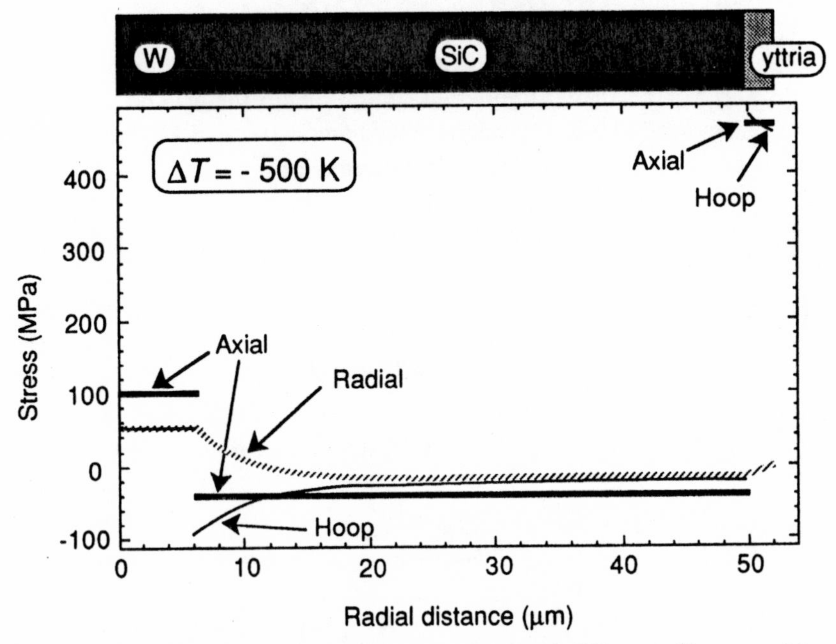

Fig. 2.12 Predicted stress distribution for a W-cored SiC monofilament with a 2 μm thick yttria coating, after cooling through 500 K.

stiffness, is countered by the thermal stresses, so that (if the system remained elastic), the fibres would have a negligible reinforcing effect at 500 MPa and a very large deviatoric stress would exist in the matrix near the interface, where it is in radial compression.

Finally, there are some situations of interest which are ideally suited to the coaxial composite model. For example, stresses in coated fibres (before incorporation into a composite) can be predicted with some confidence, provided the necessary thermophysical properties are known. Fig. 2.12 shows the stress distribution in a (W-cored) SiC monofilament with a thin coating of yttria, after cooling through 500 K. Tungsten and SiC have rather similar thermal expansivities (α), so that the core–fibre thermal contraction stresses are quite small, but $\alpha_{Y_2O_3}$ is relatively high and it can be seen that very high tensile stresses (axial and hoop) consequently build up in the coating. This is of practical significance, as such coatings might be deposited at high temperatures and these stresses could then cause them to crack on cooling to room temperature[17].

In general, exact solutions for a set of continuous coaxial cylinders can be of some value in exploring stresses in fibre composites. Although a computer program is needed to obtain these solutions, this need not be highly complex and there are no conceptual difficulties in using the method. However, there are obvious limitations to the use of the model, particularly for MMCs. Only long fibre composites can be treated and there is no scope for examining inelastic behaviour (beyond predicting its onset). Finally, most elastic constants cannot be predicted because axial symmetry must be maintained, so that, for example, an applied shear stress is not admissible.

2.4 Finite difference and finite element models

Most mathematical problems in materials science reduce to a search for particular solutions to the general two-dimensional second order equation:

$$a\frac{\partial^2\phi}{\partial x^2} + b\frac{\partial^2\phi}{\partial x\,\partial y} + c\frac{\partial^2\phi}{\partial y^2} + d\frac{\partial\phi}{\partial x} + e\frac{\partial\phi}{\partial y} + f\phi + g + h\frac{\partial\phi}{\partial t} = 0 \quad (2.40)$$

where the coefficients a–h may in general be functions of the independent variables x, y (space) and t (time) and of the dependent variable ϕ (which might be temperature, concentration, potential, momentum, some stress function, etc.). In fact the equations of Laplace, Poisson, Gauss, Fick, Fourier, Hooke, Cauchy–Riemann and Navier–Stokes are all of this form. Except in very simple situations, either the coefficients a–h or the boundary conditions are too complicated for a functional form $\phi(x, y, t)$ to be found analytically. Recourse is then necessary either to analogue physical devices (e.g. photoelastic stress measurement), or to a search for a numerical solution in the form of discrete ϕ values at a finite set of (x, y, t) coordinates. For MMCs, there is particular interest in determining internal stresses and strains, with and without matrix plastic flow, for given fibre/matrix configurations under external loading and temperature changes.

Methods for obtaining such solutions are subdivided into finite difference methods (FDM) and finite element methods (FEM). Both techniques involve ***spatial discretisation***, i.e. dividing the component of interest into a number of small domains or ***volume elements***. For time-dependent problems, time is also discretised, so that successive solutions are found after a series of time steps. While the mathematical procedures involved in FDM and FEM described below (§2.4.1 and 2.4.2) are complex, they are now well-established and routines for their implementation are widely available. However, the modeller must decide how to specify the mesh of volume elements and nodal points and this can be difficult, particularly for a spatially extensive system such as a large array of bodies within a continuum, as in MMCs. Ideally, one would like to be able to model composites of variable aspect ratio, and both spatial and orientational distributions of reinforcement. However, this would require a unit cell equal to the size of the composite, or at least big enough to contain dozens of fibres in a pseudo-random array. At the present time, this is quite impractical computationally and a much simpler matrix/fibre distribution must be used. In practice, the complexity of the problem is often reduced by the use of a repeated representative domain, which can be meshed (§2.4.2) and then modelled (§2.4.3) subject to carefully chosen boundary conditions, in a similar way to that conventionally carried out for whole objects or components treated as continua.

2.4.1 Mathematical basis

In FDMs[18,19], the dependent variable ϕ is assumed to have a uniform value within each volume element, in the centre of which is located a ***nodal point***. Derivatives in the governing equation (2.40) are then approximated by difference quotients (e.g. $\delta\phi/\delta x$) between pairs or small groups of volume elements. (For second derivatives, these approximations are commonly made using ***truncated Taylor series expansions***.) A solution is then obtained from a set of n algebraic equations involving the value of ϕ in n volume elements and the imposed boundary conditions. The set of simultaneous equations is solved by a standard method, such as ***Gaussian elimination***. For time-dependent problems, successive solutions are obtained from preceding ones by evaluating $(\delta\phi/\delta t)$ for each volume element, using either an ***explicit*** scheme (which gives the solution immediately, but may require a very short time step) or an ***implicit*** (e.g. ***Crank–Nicolson***[18]) method (which requires an iterative procedure but permits a longer time step).

In general, FEMs[20,21] are better suited than FDMs to (steady state) stress analysis problems and to complex geometries. Hence, they are commonly used for modelling of MMCs. For stress analysis, the basic equation is of the form

$$F = Ka \tag{2.41}$$

where F is a 'force' vector, K is a 'stiffness' matrix and a is a vector of unknowns (often displacements). If plastic flow takes place, then K depends on displacement, making the equation non-linear. For unsteady problems, eqn (2.41) becomes

$$F(t) = Ka + C\dot{a} + M\ddot{a} \tag{2.42}$$

where C and M represent damping effects and mass respectively. The basic FEM steps (for stress analysis) are as follows

(1) Identification of the partial differential equation (involving some stress function).
(2) Spatial discretisation, with the volume elements commonly triangular or quadrilateral in shape. Values of the stress function, ϕ, are associated with nodal points at element boundaries (and sometimes within elements).
(3) Evaluation of K and F for each volume element. This is the core of the FEM and several alternative procedures are available. These subdivide into ***variational*** (e.g. ***Rayleigh–Ritz***) and ***weighted residual*** (e.g. ***Galerkin***) methods. Variational methods depend on finding stationary values for a ***functional***, often the energy of the system, while weighted residual methods involve minimisation of the sum of the terms which would be zero if the governing equation were exactly satisfied.
(4) Assembly of a set of simultaneous equations, using K and F for each volume element and equating the net effect to that with a global K and a global F. This is sometimes referred to as use of the principle of virtual work, balancing internal and external energies.
(5) Solution of the set of equations (e.g. by Gaussian elimination) to give the unknown nodal variables a (e.g. displacements and hence strain fields).

Coupled thermo-mechanical problems can also be solved, by ensuring compatibility between thermal and displacement fields; this has been done[22] for residual thermal stresses in MMCs, using a Rayleigh–Ritz method for stress and a Galerkin method for heat flow.

2.4.2 Mesh generation

The simplest meshing arrangement is the orthogonal mesh, which is often preferred for FDM formulation, and would be suitable for modelling a single fibre in a matrix. Fig. 2.13 shows the elastic strain field for such an isolated fibre with an imposed far field matrix strain. Such modelling[23,24] can reveal interesting features, but it is not in general suitable for realistic simulation of MMC behaviour; this requires a

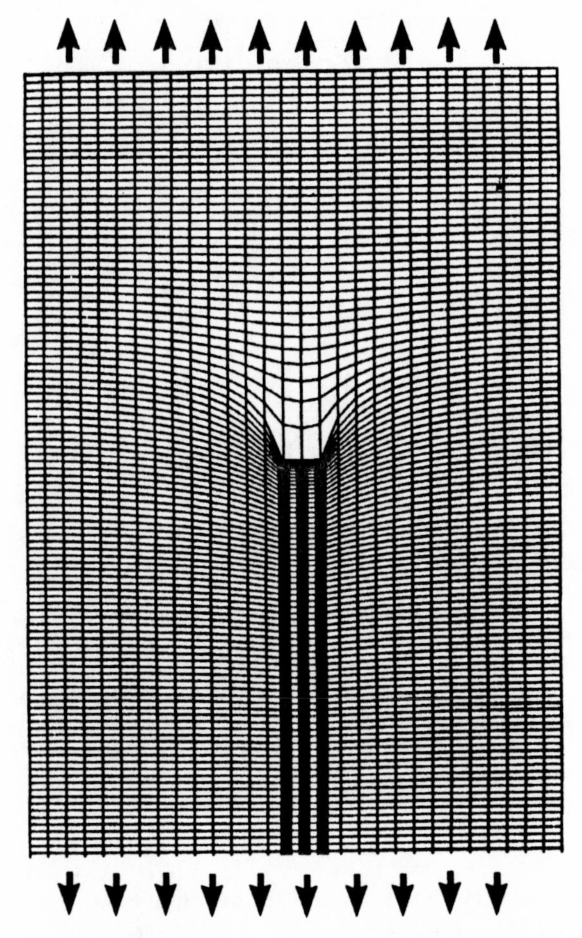

Fig. 2.13 A deformed FDM grid[23] showing half of a fibre embedded in a matrix, with an imposed far field matrix strain, for a fibre/matrix stiffness ratio, E_I/E_M, of 40. The grid was originally a uniform mesh. The scale along the fibre axis has been compressed, in order to see the distortions more easily. The deformation is purely elastic.

versatile meshing scheme (to give good precision in regions of high strain gradient), a space-filling fibre/matrix assembly (to explore load-bearing capacity as a function of fibre content) and the capacity to treat plastic as well as elastic deformation. FEMs are preferable in terms of capacity for complex geometry and software is now widely available for the generation of FEM meshes, some examples of which are shown in Fig. 2.14(b). These meshes are normally two-dimensional, often with cylindrical symmetry.

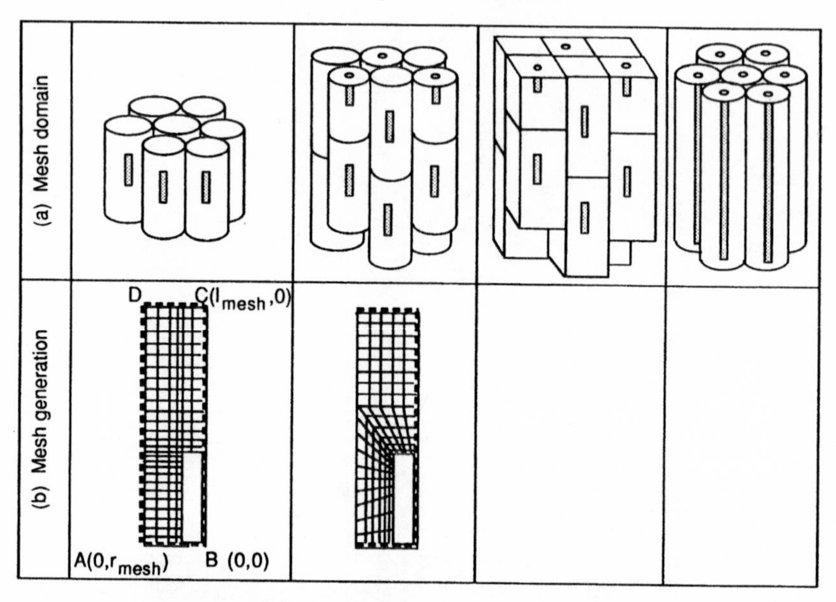

Fig. 2.14 FEM mesh generation and tessellation schemes[25–27]. (a) Tessellation schemes for filling space with modelled domains and hence for imposing the surface boundary conditions, (b) some possible meshes for the matrix around a fibre, chosen to give greater precision in regions of high stress. In all cases, lines of symmetry constrain the movement of the domain boundaries (dashed domain walls). The expected behaviour can be bracketed by taking extreme boundary condition cases (Fig. 2.16).

A further point to note about mesh specification concerns the problems that arise with large (plastic) strains. These can so distort the original mesh that it is no longer a suitable framework for describing further deformation. The solution, developed mainly for metalworking problems[28], is to carry out an ***adaptive remeshing*** procedure periodically. Such a meshing procedure often referred to as ***convected coordinate*** or ***Lagrangian***, is commonly used for elastoplastic MMC simulation. (A static mesh is termed ***Eulerian***.)

2.4.3 Tessellation of the modelled domain

Often more problematical than generation of the mesh is the specification of boundary conditions at the surface of the meshed domain, i.e. the question of how the domains are assumed to ***tessellate*** (fit together in space). Some possible assembly schemes[25–27] are shown in Fig. 2.14(a).

Cylindrically symmetrical domains do not fill space, so that approximations are necessary[29] in imposing the boundary conditions.

The finite element method is ideally suited to the modelling of highly ordered composite systems, such as an hexagonal array of continuous fibres, because the choice of the mesh domain and its repeat arrangement are unambiguous. The best choice of domain tessellation for disordered systems is less clear. Unfortunately, in many cases it turns out that the predictions are very sensitive to the arrangement chosen, because it affects the severity of the constraint imposed on different regions of the matrix due to the presence of neighbouring fibres or particles, which in turn influences matrix plasticity[25,27,30] (and damage formation[31,32]). In cases where only one fibre is contained within the domain, the differences between predictions based on different tessellation schemes can be quite dramatic, particularly for effects sensitive to constraint on matrix plasticity. (Fig. 2.16 illustrates this point for straining transverse to long fibres[33] and for short fibres.) This problem is not especially serious for the modelling of long fibre systems, because it is computationally feasible to use a domain which includes a substantial number of fibres and is thus fairly representative of the composite of interest. An example of this is given in Fig. 2.15, which illustrates the local variation in matrix hydrostatic stress, σ_H, in an FEM domain containing 30 aligned fibres in a matrix under transverse loading.

At the present time, a multi-fibre approach is not really applicable to short fibre or randomly oriented continuous fibre composites, because of the three-dimensional nature of the stress state and consequent excessive computing time requirements. In the short fibre case, the domain can in practice contain no more than one or two fibres. The simplest option is to adopt a configuration similar to that used in the shear lag analysis, i.e. a cylindrically symmetric domain containing a single fibre. This has the advantage that cylindrical symmetry gives rise to a two-dimensional mesh. The main difficulty is that of invoking physically reasonable boundary conditions. The two simplest choices are a free cylinder containing a single fibre and the same cylinder constrained by neighbours so that the side-walls remain vertical throughout. These conditions clearly represent two extremes; in the former case the neighbouring domains offer no interaction at all, while in the latter they allow no deformation across the side-walls[†30]. This is equivalent to tensile deformation of a single

[†] This is topographically impossible, although it could be achieved if the domain were hexagonal[30].

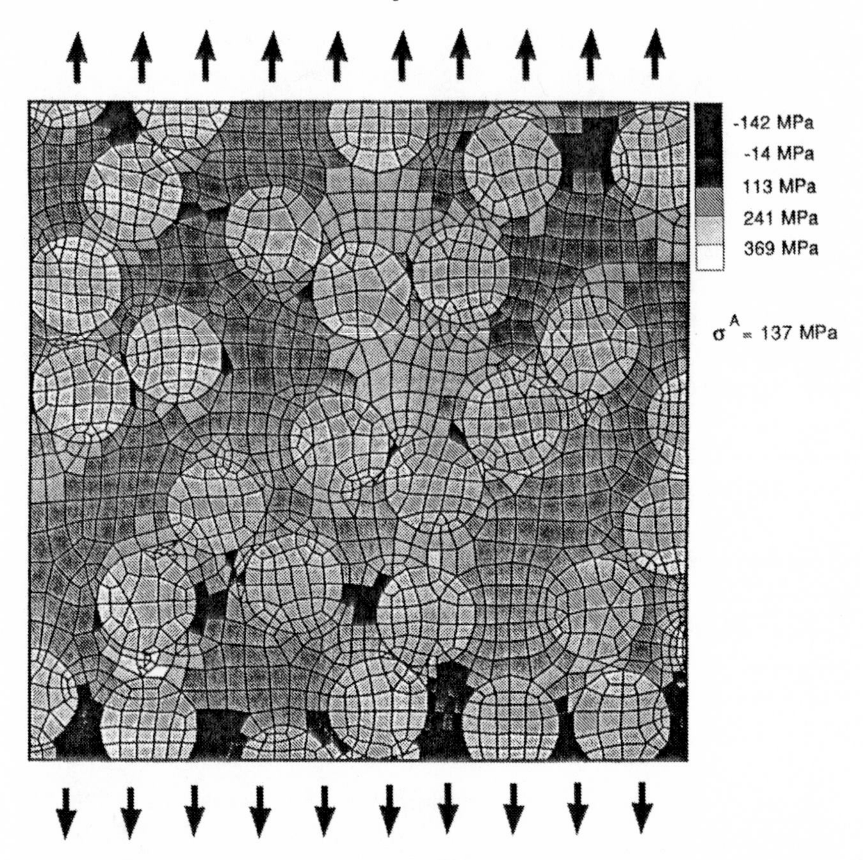

Fig. 2.15 Results from FEM calculations[33] for 50 vol% of transverse boron fibres in an Al (6061) matrix subjected to a tensile stress of 137 MPa. The mesh is for a random fibre distribution, mapping the variation in matrix hydrostatic stress. (The shading of the fibres is not indicative of the stresses within them.)

fibre system within a straight-sided rigid can, overestimating the lateral constraint caused by the neighbouring material. As shown in Fig. 2.14, other packing arrangements have been proposed, including regularly staggered fibre arrangements and hexagonal cells[27]. As one would expect, in these cases the constraint is intermediate between the two examples discussed above, but even then predictions[25,26,33] (Fig. 2.16) vary markedly. It is thus very important to assess the extent to which the symmetry of the tesselation will overestimate or underestimate the interaction between neighbouring regions.

In summary, FE methods are versatile and powerful and can be used to reveal useful information about both the local and global deformation

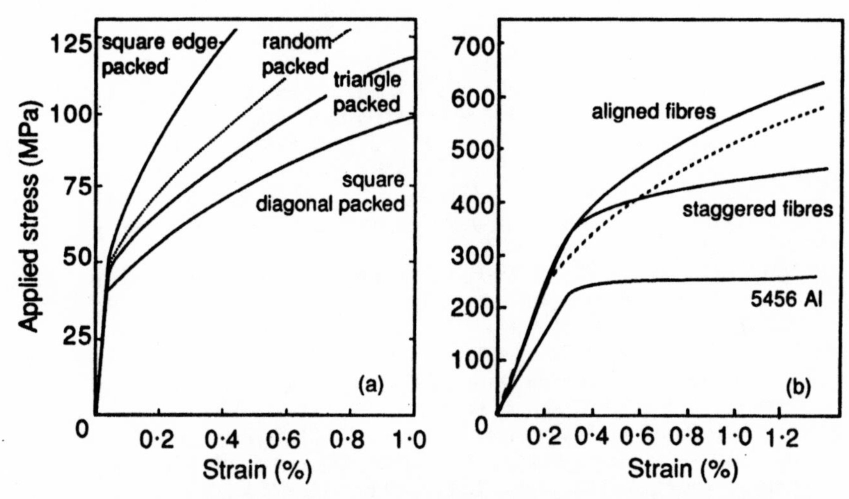

Fig. 2.16 Predicted stress–strain responses for (a) Al–50 vol% B long fibre composite under transverse loading[33], and (b) 5456 Al–20 vol% aspect ratio 4 SiC[27] for different fibre packing distributions. Note that large differences in the work-hardening behaviour result from these changes, even though there is no change in fibre volume fraction or aspect ratio, or in any physical property of either constituent. The dashed line in (b) represents experimental values.

characteristics of MMCs. However, a sound understanding of the basis of numerical techniques, and an appreciation of the significance of the boundary conditions being used, are essential if misleading deductions are to be avoided. It may be noted that in the Eshelby method (introduced in the next chapter) volume-averaged matrix stresses can be explored without it being necessary to specify the local spatial distribution of fibres, so that complications of this sort are largely avoided.

References

1. J. C. Halpin and S. W. Tsai (1967) *Environmental Factors in Composite Design*, Air Force Materials Laboratory, AFML-TR-67-423.
2. A. Spencer (1986) The Transverse Moduli of Fibre Composite Material, *Comp. Sci. & Techn.*, **27**, pp. 93–109.
3. J. F. Nye (1985) *Physical Properties of Crystals – Their Representation by Tensors and Matrices*, Clarendon, Oxford.
4. T. W. Clyne (1990) A Compressibility-based Derivation of Simple Expressions for the Transverse Poissons Ratio and Shear Modulus of an Aligned Long Fibre Composite, *J. Mat. Sci. Letts.*, **9**, pp. 336–9.
5. H. L. Cox (1952) The Elasticity and Strength of Paper and other Fibrous Materials, *Brit. J. Appl. Phys.* **3**, pp. 73–9.

6. J. O. Outwater (1956) The Mechanics of Plastics Reinforced in Tension, *Modern Plastics*, **33**, pp. 56–65.
7. B. W. Rosen (1960) Mechanics of Fibre Strengthening, in *Fibre Composite Materials*, B. W. Rosen (ed.), ASM, Metals Park, Ohio.
8. N. F. Dow (1963) *Study of Stresses near Discontinuity in a Filament-Reinforced Composite Metal*, GE Co., Missile and Space Div., R635D61.
9. H. Fukuda and T. W. Chou (1981) An Advanced Shear Lag Model Applicable to Discontinuous Fiber Composites, *J. Comp. Mat.*, **15**, pp. 79–91.
10. V. C. Nardone and K. M. Prewo (1986) On the Strength of Discontinuous Silicon Carbide-Reinforced Aluminium Composites, *Scripta Met.*, **20**, pp. 43–8.
11. T. W. Clyne (1989) A Simple Development of the Shear Lag Theory appropriate for Composites with a Relatively Small Modulus Mismatch, *Mat. Sci. & Eng.*, **A122**, pp. 183–92.
12. D. L. McDanels (1985) Analysis of Stress–Strain, Fracture, and Ductility of Aluminium Matrix Composites Containing Discontinuous Silicon Carbide Reinforcement, *Metall. Trans.*, **16A**, pp. 1105–15.
13. T. Ishikawa, K. Koyoma and S. Kobayashi (1978) Thermal Expansion Coefficients of Unidirectional Composites, *J. Comp. Mat.*, **12**, pp. 153–68.
14. D. Iesan (1980) Thermal Stresses in Composite Cylinders, *J. Thermal Stresses*, **3**, pp. 496–508.
15. Y. Mikata and M. Taya (1985) Stress Field in a Coated Continuous Fibre Composite Subjected to Thermo-Mechanical Loadings, *J. Comp. Mat.*, **19**, pp. 554–79.
16. C. M. Warwick and T. W. Clyne (1991) Development of a Composite Coaxial Cylinder Stress Analysis Model and its Application to SiC Monofilament Systems, *J. Mat. Sci.*, **26**, pp. 3817–27.
17. C. M. Warwick, R. R. Kieschke and T. W. Clyne (1991) Sputter Deposited Barrier Coatings on SiC Monofilaments for Use in Reactive Metallic Matrices – Part II. System Stress State, *Acta. Met. et Mat.*, **39**, pp. 437–44.
18. G. D. Smith (1985) *Numerical Solution of Partial Differential Equations: Finite Difference Methods*, Clarendon, Oxford.
19. S. V. Patankar (1971) *Numerical Heat Transfer and Fluid Flow*, McGraw-Hill, New York.
20. D. R. J. Owen and E. Hinton (1980) *A Simple Guide to Finite Elements*, Pineridge, Swansea.
21. O. C. Zienkiewicz (1977) *The Finite Element Method*, McGraw-Hill, London.
22. G. L. Porvik, A. Needleman and S. R. Nutt (1990) An Analysis of Residual Stress Formation in Whisker Reinforced Al–SiC Composites, *Mat. Sci. & Eng.*, **A125**, pp. 129–40.
23. Y. Termonia (1987) Theoretical Study of the Stress Transfer in Single Fibre Composites, *J. Mat. Sci.*, **22**, pp. 504–8.
24. A. S. Carrara and F. J. McGarry (1968) Matrix and Interfacial Stresses in a Discontinuous Fibre Composite Model, *J. Comp. Mat.*, **2**, pp. 222–43.
25. T. Christman, A. Needleman and S. Suresh (1989) An Experimental and Numerical Study of Deformation in MMCs, *Acta Metall.*, **37**, pp. 3029–50.
26. V. Tvergaard (1990) Analysis of Tensile Properties for a Whisker Reinforced MMC, *Acta Met. et Mat.*, **38**, pp. 185–94.
27. A. Levy and J. M. Papazian (1990) Tensile Properties of Short Fibre-Reinforced SiC/Al Composites, part II. Finite Element Analysis, *Metall Trans.*, **21A**, pp. 411–20.

28. O. C. Zienkiewicz and G. C. Huang (1989) Adaptive Modelling of Transient Coupled Metal Forming Processes, in *Numerical Methods in Industrial Forming Processes* (*Proc. Numiform 89*), E. G. Thompson (ed.), A. A. Balkema, Rotterdam, pp. 3–10.
29. V. Tvergaard (1976) Effect of Thickness Inhomogeneities in Internally Pressurised Elasto-plastic Spherical Shells, *J. Mech. Phys. Solids*, **24**, pp. 291–304.
30. T. Christman, A. Needleman, S. R. Nutt and S. Suresh (1989) On Microstructural Evolution and Micromechanical Modelling of Deformation of a Whisker-Reinforced MMC, *Mat. Sci. & Eng.*, **A107**, pp. 49–61.
31. S. R. Nutt and A. Needleman (1987) Void Nucleation at Fibre Ends in Al-SiC Composites, *Scripta Met.*, **21**, pp. 705–10.
32. A. Needleman and S. R. Nutt (1990) Void Nucleation in Short Fiber Composites, in *Advances in Fracture Research* (*Proc. ICF7*), K. Salama (ed.), Pergamon, Oxford, pp. 2211–18.
33. J. R. Brockenbrough, S. Suresh and H. A. Wienecke (1991) Deformation of MMCs with Continuous Fibers: Geometrical Effects of Fiber Distribution and Shape, *Acta Met. et Mat.*, **39**, pp. 725–52.

3
The Eshelby approach to modelling composites

In the previous chapter a number of models were presented for estimating the partitioning of loads between the constituents of composites subjected to external loads. These models involve mathematical approximations ranging from the good to the very poor. Some are rather limited in terms of the properties which can be predicted, while others are computationally daunting. For an isolated inclusion (reinforcing constituent) having an ellipsoidal shape, the approach presented in this chapter is mathematically rigorous. Later we shall see that it is also a good model at higher inclusion volume fractions and for other inclusion shapes. This analysis, commonly named the ***Eshelby method****, turns out to be useful for predicting a wide range of composite properties. On a practical level, the standard equations highlighted by boxes in the text can be used to predict many composite properties quickly and fairly accurately.*

Internal stresses are commonplace in almost any material which is mechanically inhomogeneous. Typically, their magnitude varies according to the degree of inhomogeneity: for an externally loaded polycrystalline cubic metal, differently oriented crystallites will be stressed to different extents, but these differences are usually quite small. For a composite, consisting of two distinct constituents with different stiffnesses, these disparities in stress will commonly be much larger. Internal stresses arise as a result of some kind of ***misfit*** between the shapes of the constituents (matrix and reinforcement, i.e. fibre, whisker or particle). Such a misfit could arise from a temperature change, but a closely related situation is created during mechanical loading – when a stiff inclusion tends to deform less than the surrounding matrix. Analysis of the stresses required to mate up the inclusion and matrix across the interface allows the prediction of properties such as thermal expansivity and stiffness. For an arbitrary

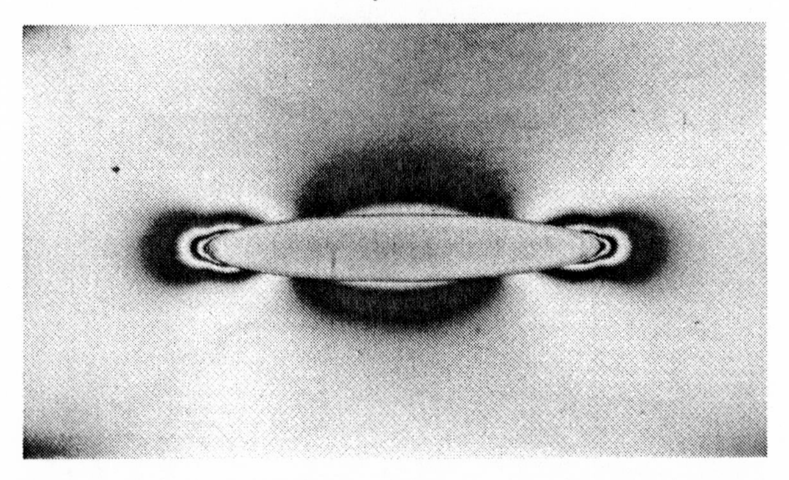

Fig. 3.1 Slice from an axisymmetric three-dimensional photoelastic model containing a prolate ellipsoidal inclusion of stiffness ~2.5 times that of the matrix, viewed between crossed polars. The model was loaded in axial compression. The pattern of fringes (contours of equal principal stress difference) shows that the stress within the matrix fluctuates in a complicated manner, the stress being largest at the inclusion ends and smaller than the applied stress about the equator. The stress within the inclusion, on the other hand, is larger than the applied stress and is uniform throughout.

inclusion shape (or, indeed, for a cylinder), this analysis can only be carried out numerically, but for the special case of an ellipsoid an analytical technique founded on the pioneering work in the 1950s by J. D. Eshelby[1,2] can be employed. The key point here is that the ellipsoid, which can have any aspect ratio, has a uniform stress at all points within it (Fig. 3.1). The technique is based on representing the actual inclusion by one made of matrix material (an '***equivalent homogeneous inclusion***') which has an appropriate misfit strain (an ***equivalent transformation strain***), such that the stress field is the same as for the actual inclusion.

3.1 Misfit stresses

Suppose a region within a homogeneous medium were suddenly to transform in shape, so that it no longer fitted freely into the hole in the matrix from which it came; what would the stress field look like? The answer to this question would at first sight appear to have little to do with calculating the stresses within MMCs, but Eshelby[1] showed that there is an elegant solution to this problem, which can be applied to a wide variety of other situations.

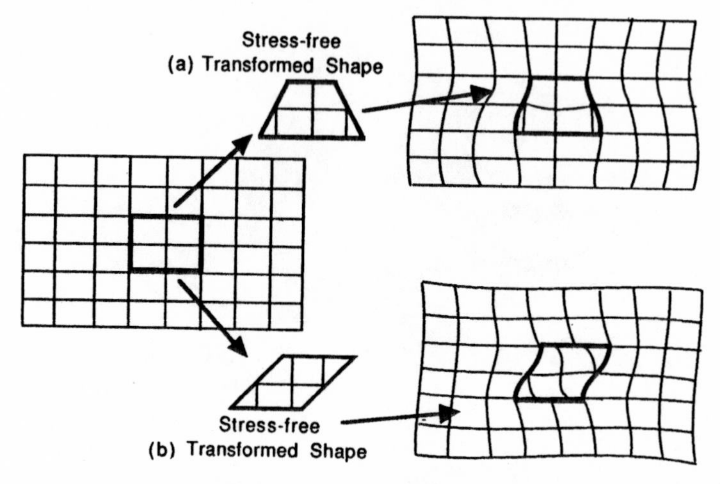

Fig. 3.2 A schematic illustration of how a transformation (i.e. a shape change) imposed on a region within a matrix tends to cause complex distortions in both the transformed region and the surrounding matrix. This is so for (a) non-uniform strains (in this case a linear change in width with height) and (b) uniform strains (in this case a simple shear). Consequently, even simple transformations are usually associated with complicated strain fields which are not easily calculated mathematically.

The consequences of a spontaneous transformation of the type discussed above can best be visualised in terms of ***displacement maps*** (Fig. 3.2). (In these diagrams, the grid lines represent the displacement of an originally square mesh, while the thickness of the lines represents the stiffness.) An example of such a spontaneous shape change is provided by a martensitic transformation. It is clear from Fig. 3.2 that the elastic strain field is very complicated, both inside and outside the constrained transformed region, and for this reason an analytic solution is not usually possible[1,2]. When the transformed region is ellipsoidal in shape and the shape change is a uniform one (i.e. ellipsoid → ellipsoid), however, the mathematics become tractable. This is because, under these conditions, the stress and strain within the enclosed phase are uniform (see Fig. 3.1)†.

Eshelby[1] approached the problem by visualising a series of cutting and welding exercises, as illustrated in Fig. 3.3. A region (the inclusion) is cut from the unstressed elastically homogeneous material, and is then imagined to undergo a shape change (the transformation strain ε^{T}) free

† A similar effect is observed for the field and potential lines about ellipsoidal regions of different dielectric constants in electrostatics[3].

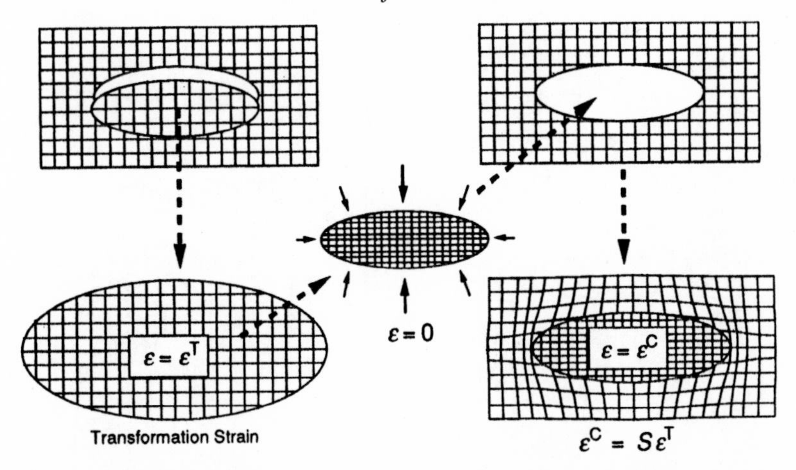

Fig. 3.3 Eshelby's cutting and welding exercises for the uniform stress-free transformation of an ellipsoidal region. Eshelby related the natural shape mismatch between the hole and the inclusion (ε^T) to the final constrained inclusion shape (ε^C). This relationship is described by the Eshelby tensor, values of which are tabulated for different inclusion shapes (Appendix III).

from the constraining matrix (so that the strain is 'stress-free'). The inclusion cannot now be replaced directly back into the hole from whence it came. Instead surface tractions are first applied in order to return it to its original shape. Once back in position, the two regions are then welded together once more, i.e. there is no movement or sliding along the interface, and the surface tractions are then removed. Equilibrium is then reached between the matrix and the inclusion at a ***constrained*** strain (ε^C) of the inclusion relative to its *initial* shape before removal.

Since the inclusion is strained uniformly throughout, the stress within it can be calculated using Hooke's Law in terms of the *elastic* strain ($\varepsilon^C - \varepsilon^T$) and the stiffness tensor of the material (C_M).

$$\sigma_I = C_M(\varepsilon^C - \varepsilon^T) \tag{3.1}$$

For a specified shape change ε^T, all that is now required in order to calculate the inclusion stress is a knowledge of the final constrained strain ε^C. Eshelby found that ε^C can be obtained from ε^T by means of a tensor, termed the Eshelby 'S' tensor, which can be calculated in terms of the inclusion aspect ratio and the Poisson's ratio of the material

$$\varepsilon^C = S\varepsilon^T \tag{3.2}$$

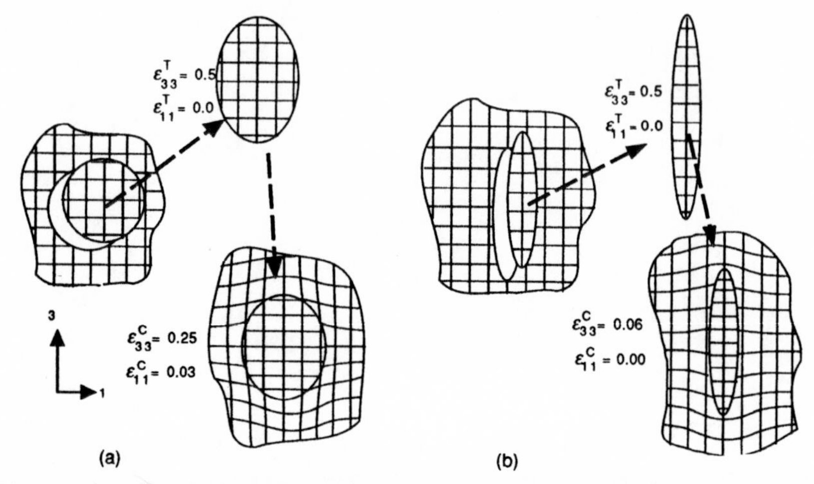

Fig. 3.4 A scaled diagram illustrating how the S tensor relates ε^{T} to ε^{C} for two inclusion shapes. In (a) a sphere and in (b) an aspect ratio 5 prolate ellipsoid (cigar shape) have undergone a 50% stress-free expansion in the '3'-direction. Note that in (b) the constrained axial strain is very small; this means that the compressive axial inclusion stress ($C_{\mathrm{M}}(\varepsilon^{\mathrm{C}} - \varepsilon^{\mathrm{T}})$) will be much larger than for the sphere.

The tensor S thus expresses the relationship between the final constrained inclusion shape and the natural shape mismatch (see Fig. 3.4). The form of the S tensor for some common inclusion shapes is given in Appendix III. (Care must be taken to avoid confusion between the Eshelby tensor and the compliance, also commonly represented by the symbol S.)

Now that the constrained shape is known, the inclusion stress can be evaluated in terms of the stress-free shape misfit by combining eqns (3.1) and (3.2) to give

$$\sigma_{\mathrm{I}} = C_{\mathrm{M}}(S - I)\varepsilon^{\mathrm{T}} \tag{3.3}$$

where I is the identity matrix.

An important feature of the S tensor is that it allows calculation of the (uniform) stress and strain within the inclusion, without having to look in detail at the complicated form of the matrix stress field.

3.2 An example of a misfit strain – differential thermal contraction

When the temperature of a composite is changed, a misfit is generated between the natural shapes of the reinforcing inclusion and the corresponding hole in the matrix in which it sits (assuming that they have different expansion coefficients). In many ways this is very similar to the misfit for which the inclusion stress was calculated in the last section. The results presented above can thus be used to deduce the thermal residual stresses within a single reinforcing inclusion.

By subtracting the strain of the matrix ($\alpha_M \Delta T$) which takes place on lowering the temperature by $|\Delta T|$, one can consider the differential thermal contraction misfit as a shape change (transformation strain) similar to that used above ($\varepsilon^{T^*} = (\alpha_I - \alpha_M)\Delta T$, where ΔT would be negative in this example). Similar cutting and welding arguments (see Fig. 3.5) then allow one to write the inclusion stress in terms of the stiffness (C_I) and the elastic strain of the inclusion:

$$\sigma_I = C_I(\varepsilon^C - \varepsilon^{T^*}) \tag{3.4}$$

Although the constrained strain will still be uniform, it will not now be related to the stress-free misfit by eqn (3.2). This is because, for MMCs, the matrix and inclusions normally have different stiffnesses†. Given that the inclusion is stiffer than the matrix, one would expect

$$\varepsilon^C > S\varepsilon^{T^*} \tag{3.5}$$

i.e. the stiff inclusion would resist being forced back towards the natural shape of the hole more strongly than did the inclusion matched in stiffness to the matrix (the homogeneous inclusion). However, since transformed ellipsoids always give rise to constrained shapes which are ellipsoidal, it must be possible to imagine a '***ghost***' inclusion (made of matrix material) with a shape such that, when surface tractions are applied to give it the *constrained* shape of the real inclusion, it will attain the same uniform stress state (Fig. 3.5). In such a case, the actual inclusion and the equivalent inclusion could be interchanged without disturbing the matrix. The inclusion which satisfies this condition is termed the ***equivalent homogeneous inclusion***. This 'equivalent' elastically homogeneous problem was solved in §3.1, so that, by calculating the ***equivalent transformation strain***‡ ε^T required to imitate the constrained stress state in the actual inclusion, we have also calculated the stress state of the inhomogeneous composite. The trick therefore is to find the shape of the 'ghost' inclusion (matched in stiffness to the matrix) which is 'equivalent' to the reinforcing inhomogeneous inclusion of interest.

† Such composites are often referred to as ***elastically inhomogeneous***[4]. In the text, the inclusion is taken to be stiffer and as having a lower thermal expansivity than the matrix. The equations derived are, however, applicable to other cases.

‡ Sometimes called the ***eigenstrain***[5].

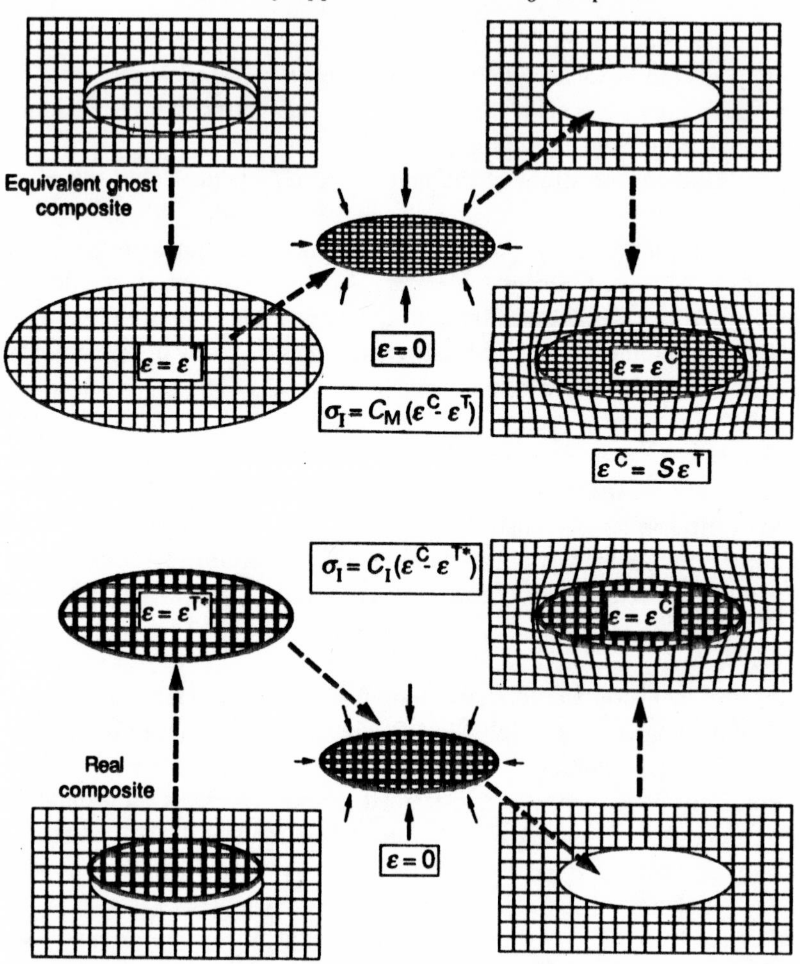

Fig. 3.5 Eshelby's cutting and welding exercises for a stiff misfitting ellipsoidal inclusion and a 'ghost' inclusion, with the properties of the matrix, which has been chosen so that the final stress states in the two composites are identical everywhere. Note that although the elastic *strain* in the inclusion is different in the two cases (it must be larger in the equivalent elastically homogeneous case in order to generate the same stress) the matrix and inclusion *stress* fields are identical.

The constrained strain (ε^C) for the elastically homogeneous problem has already been calculated in the last section in terms of the homogeneous transformation strain (ε^T). If it is to be truly equivalent, it must be the same as the constrained shape of the real inclusion. Furthermore, in order to be able to interchange the two inclusions, the stress in the equivalent 'ghost' inclusion must be identical to that given by eqn (3.4) for the real inclusion

$$\sigma_I = C_M(\varepsilon^C - \varepsilon^T) \tag{3.1}$$

$$= C_I(\varepsilon^C - \varepsilon^{T*}) \tag{3.4}$$

$$\therefore C_I(S\varepsilon^T - \varepsilon^{T*}) = C_M(S - I)\varepsilon^T \tag{3.6}$$

from which the appropriate transformation strain ε^T can be deduced for any shape change (ε^{T*}) and stiffness mismatch ($C_I - C_M$) between the two phases

$$\varepsilon^T = [(C_I - C_M)S + C_M]^{-1}C_I\varepsilon^{T*} \tag{3.7}$$

Now that the elastically equivalent 'ghost' problem has been identified, the results from §3.1 can be used directly. The inclusion stress is thus

$$\sigma_I = C_M(S - I)\varepsilon^T \tag{3.3}$$

$$\sigma_I = C_M(S - I)[(C_I - C_M)S + C_M]^{-1}C_I\varepsilon^{T*} \tag{3.8}$$

This equation allows us to calculate the internal stresses (and strains) generated within a single inclusion as a result of a (thermal expansion) misfit ε^{T*} (Fig. 3.6).

3.3 Internal stresses in loaded composites

In many situations it is the ability of an MMC to bear a large external load which is important. So far we have only calculated the stress in an inclusion arising from a shape misfit between the two regions. However, our approach is also fruitful when studying external loading, because it is the difference in natural shapes of inclusion and hole when both phases bear the applied stress that is responsible for load transfer to the reinforcement.

In all cases the loading of an elastically homogeneous composite is straightforward. It is an elastically homogeneous medium, so that the strain arising from the applied load is the same throughout (ε^A). Consequently, the stress (and strain) at each point is simply the sum of the applied stress (strain) and the internal stress (strain) already calculated in §3.1 for the unloaded system (see Fig. 3.7). Accordingly, in the homogeneous inclusion

$$\sigma_I + \sigma^A = C_M(\varepsilon^C - \varepsilon^T) + C_M\varepsilon^A \tag{3.9}$$

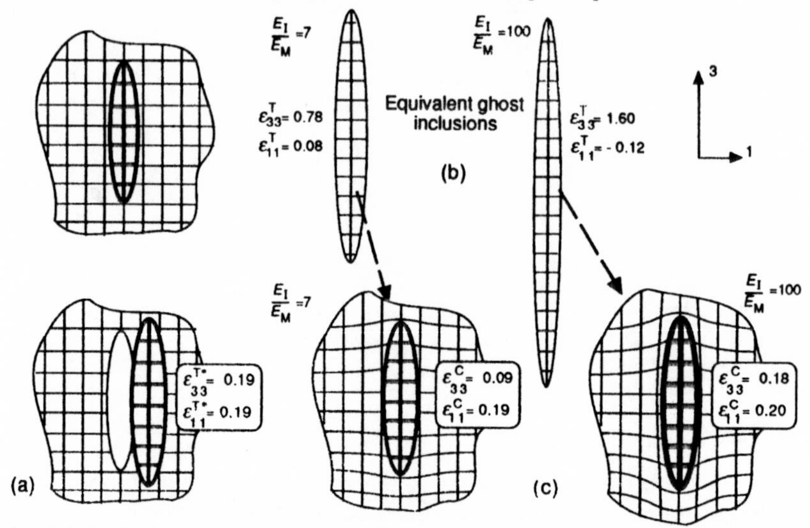

Fig. 3.6 Scale drawings illustrating how the constrained strain (ε^C) is calculated from a misfit strain (ε^{T*}). The diagram shows (a) an isotropic misfit strain ε^{T*} of 0.19 (as would arise from a 10 000 K temperature drop for the Al–SiC system), (b) the strain of the corresponding 'ghost' inclusion ε^T (obtained from ε^{T*} using eqn (3.7)), and (c) values of ε^C (obtained from ε^T using eqn (3.2)). Data are shown for ellipsoidal inclusions of aspect ratio 5 and stiffnesses 7 (~SiC/Al) and 100 times that of the matrix. Notice how a stiffer inclusion requires a more exaggerated stress-free shape change of the corresponding 'ghost' inclusion in order to end up with the same constrained shape and stress. This requires ε^T to become highly anisotropic, even though the natural strain misfit ε^{T*} it represents is purely dilatational (isotropic).

where σ_I, arising from the misfit in natural shapes of inclusion and hole, is given by eqn (3.3), and

$$\sigma^A = C_M \varepsilon^A \tag{3.10}$$

We can now look at the real composite. Clearly the response of the reinforced composite will be very different from that of the elastically homogeneous one. However, if one is given complete freedom over the choice of the original shape of the 'ghost' inclusion, it must still be possible to select a stress-free misfit (ε^T) so as to duplicate the stress in the real inclusion when the composite is under a specific load. In such a case, the real and ghost inclusions can then be interchanged as before without disturbing the matrix at all (see Figs. 3.7 and 3.8).

The shape and stress states of the ghost inclusion are

$$\textit{constrained shape change} = \varepsilon^C + \varepsilon^A \qquad \textit{– ghost inclusion} \quad (3.11)$$

$$\textit{inclusion stress} = \sigma_I + \sigma^A = C_M(\varepsilon^C + \varepsilon^A - \varepsilon^T) \qquad \textit{– ghost inclusion} \quad (3.9)$$

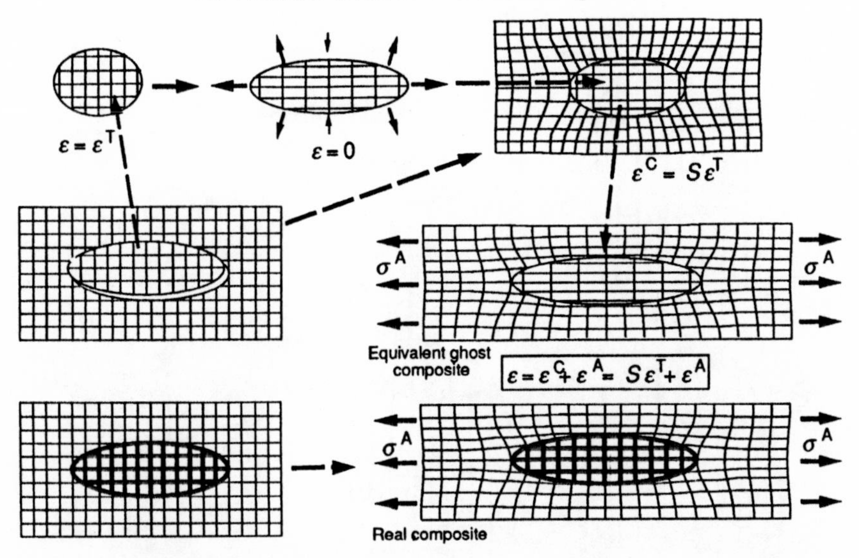

Fig. 3.7 Modelling the response of a reinforced composite to an applied load, and the cutting and welding exercises required to select the equivalent transformation strain ε^{T} of the 'ghost' composite appropriate to the applied stress σ^{A}. Note that again the strain within the real inclusion is uniform; it is this which enables one to match the inclusion stress and shape with that of a 'ghost' inclusion having the same stiffness as the matrix. To do this the elastic strain of the ghost inclusion must be larger than that of the real inclusion. A particular ghost inclusion misfit is only correct for a specific applied load. In fact, because the composite is initially stress-free and because larger and larger misfits are required of the ghost inclusion as the load is increased, ε^{T} is directly proportional to σ^{A}.

where ε^{C} is the *constrained strain* of the inclusion *before* external loading. The shape and stress state of the real inclusion must also be given by eqns (3.9) and (3.11)

$$\textit{constrained shape change} = \varepsilon^{C} + \varepsilon^{A} \quad \textit{– real inclusion} \tag{3.11}$$

However, the stress can also be written in terms of the elastic strain of the inclusion using Hooke's Law

$$\textit{stress} = C_{I}(\varepsilon^{C} + \varepsilon^{A}) \quad \textit{– real inclusion} \tag{3.12}$$

Combining the two equivalent statements of the stress in the inclusion with the expression for ε^{C} given in eqn (3.2), the shape misfit required for the ghost inclusion can be found

$$\varepsilon^{T} = -[(C_{I} - C_{M})S + C_{M}]^{-1}(C_{I} - C_{M})\varepsilon^{A} \tag{3.13}$$

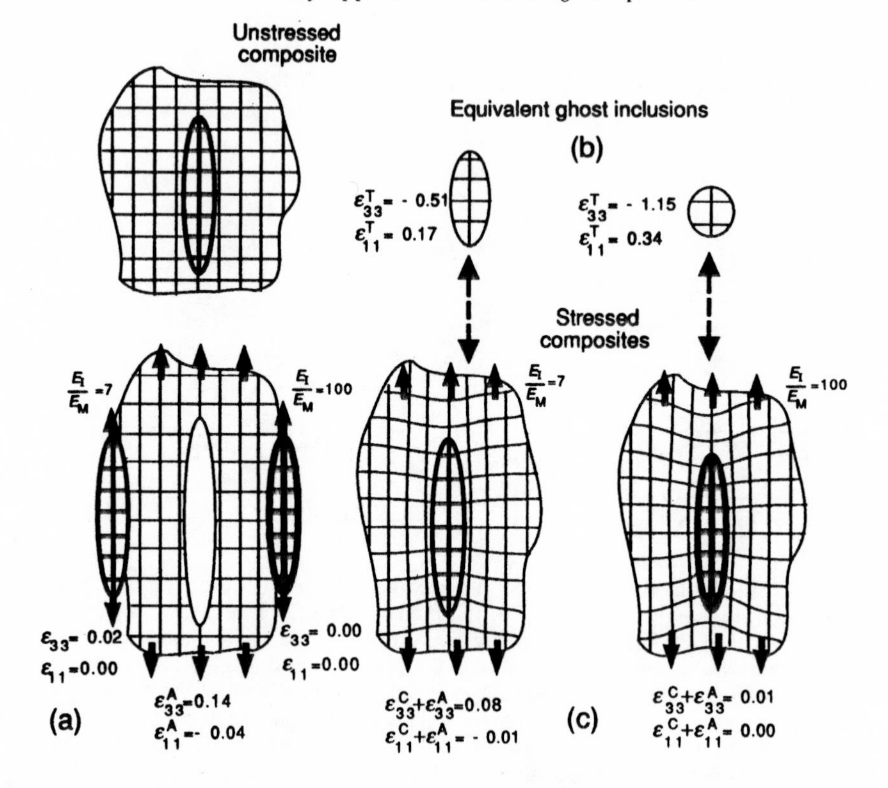

Fig. 3.8 Scale drawings illustrating how the final constrained strain is calculated for a composite under an applied stress σ^A. By comparison with the 'ghost' composite (Fig. 3.7) the inclusion strain is shown to be equal to $(\varepsilon^C + \varepsilon^A)$, where ε^C has already been calculated (eqns (3.2) and (3.13)). Data are shown for ellipsoidal inclusions of aspect ratio 5 and stiffnesses 7 ($\sim$SiC/Al) and 100 times that of the matrix. The diagram shows (a) the natural strains of the inclusion and hole (as would arise for Al if σ^A were 10 000 MPa), (b) the transformation strains of the corresponding 'ghost' inclusions ε^T (obtained from ε^A using eqn (3.13)), and (c) values of ε^C (obtained from ε^T using eqn (3.2)). Note how the 'ghost' inclusion must be made shorter for the stiffer inclusion, so that it is more heavily loaded when inserted in the hole.

This can be regarded as the equivalent inclusion transformation strain associated with the misfit between the shapes the reinforcing inclusion and the hole would adopt were the two phases to be subjected to the applied stress independently (see Fig. 3.7). As one would expect, it is proportional to the applied stress and increases with the disparity in stiffness between the phases (Fig. 3.8).

As before, now that the transformation strain of the equivalent 'ghost' inclusion has been identified, the inclusion stress can be determined, in this case from eqn (3.12), or more simply from eqn (3.9)

$$\sigma_I + \sigma^A = -C_M(S - I)[(C_I - C_M)S + C_M]^{-1}(C_I - C_M)\varepsilon^A + C_M\varepsilon^A \quad (3.14)$$

In summary, it is *always* possible to imagine a parallel experiment involving an inclusion of the same elastic constants as the matrix, suitably transformed in a stress-free manner, such that, under the relevant loading conditions, it is 'equivalent' to the reinforcing inclusion of interest, in that the two can be interchanged without changing the stress distribution anywhere. As the applied load is increased, the shape of the ghost inclusion will vary so that the internal stress state shadows that of the actual reinforcing system. It is from this parallel 'ghost' experiment that the stress and strain common to both situations can be calculated.

3.4 The matrix stress field

In §3.1 we saw that, even for simple inclusion/matrix misfits, the matrix stress fields are a complicated function of position as well as shape mismatch and elastic properties (Figs. 3.1 and 3.2). For the transformation of an ellipsoidal region, an analytic matrix strain field solution is possible and has been discussed by Eshelby[1] and Mura[5]. The result is expressed in terms of a tensor ($D(x)$) similar to the Eshelby S tensor, but now relating the constrained strain at a point x within the matrix to the stress-free transformation strain ε^T of the 'ghost' inclusion

$$\varepsilon^C(x) = D(x)\varepsilon^T \quad (3.15)$$

The form of $D(x)$ is given explicitly by Mura[5]; it is rather complicated, but it can be evaluated quite easily using a microcomputer. From eqn (3.15) the matrix and inclusion stresses can then be calculated

$$\sigma_I = C_M(\varepsilon^C - \varepsilon^T) \qquad (= C_I(\varepsilon^C - \varepsilon^{T*})) \quad (3.1)\ \&\ (3.4)$$

$$\sigma_M(x) = C_M\varepsilon^C(x) \quad (3.16)$$

While such expressions can give a useful indication of the overall form of the stress field around an inclusion (e.g. Fig. 4.17), practical composites do not contain isolated ellipsoidal inclusions. It is therefore important to ask, when considering the stress field about reinforcing fragments and 'cylindrical' whiskers in real MMCs, 'How useful is a model based upon such an idealised inclusion shape?'

To investigate the sensitivity of the predictions to inclusion shape,

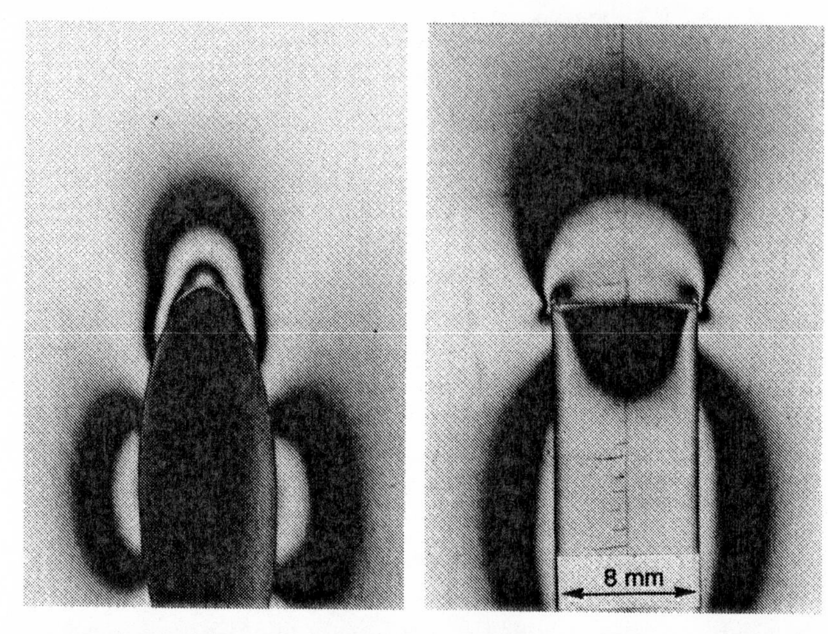

Fig. 3.9 Photoelastic fringe patterns for ellipsoidal and cylindrical inclusions[9], both having an equatorial diameter of 8 mm and compressively loaded along the fibre axis by 0·16 MPa. There are large differences in the matrix stress fields, but (in broad agreement with Saint-Venant's principle[8]) they are confined to the immediate vicinity of the inclusion 'corners'. Whereas the stress within the ellipsoid is uniform, that within the cylinder rises to a maximum at the centre. Such differences local to the fibre affect predictions of the peak stresses, limiting the usefulness of the Eshelby technique for matrix failure analysis. Overall, the stress fields are comparable, so that Eshelby predictions should be reliable for properties dependent on volume-averaged stresses.

three-dimensional frozen stress photoelastic studies[6,7] have been undertaken of the stress fields about cylindrical and ellipsoidal inclusions under an applied load. These results and their implications for composite modelling are summarised in Figs. 3.9 and 3.10. Broadly speaking, the results indicate that the exact details of the local stress fields are sensitive to the precise reinforcement shape, but that in most of the matrix the fields are strikingly similar. Whilst this restricts the usefulness of Eshelby-based calculations of the local stress field (finite element methods being preferable), the model provides reliable estimates of volume-averaged stress fields, and thus of many composite properties. Furthermore, as is shown in the following section, these average matrix stresses can be determined without recourse to the rather complicated expressions describing the local variations of stress.

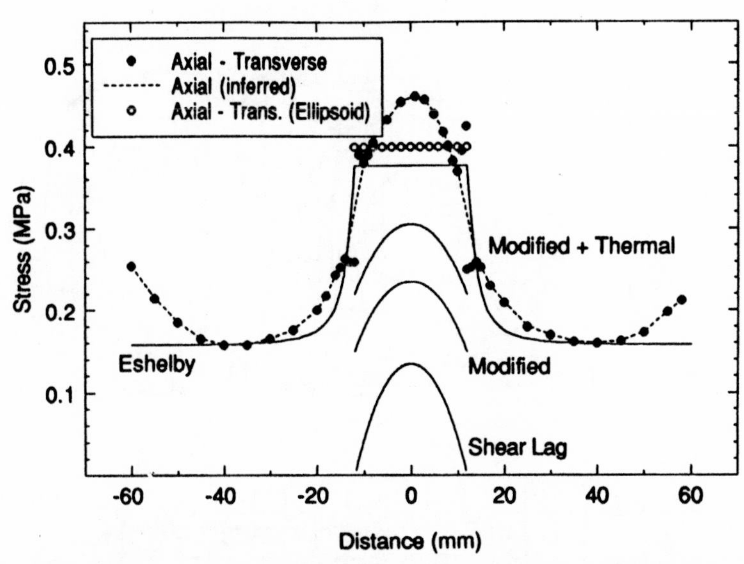

Fig. 3.10. Predicted and experimental stress difference variations along the symmetry axes of the composites shown in Fig. 3.9. In the cylindrical inclusion, the observed results are similar in form to predictions from the modified shear lag (§2.2) and FEM models, in that the stress rises towards the centre. However, shear lag models underestimate the average fibre loading, which is close to the uniform value predicted by the Eshelby model (solid line)[9]. This suggests that the Eshelby approach is an adequate model for the evaluation of inclusion loading. Furthermore, the balance of internal stress (eqn (3.18)) suggests that the average matrix stress must also be similar in the two models, given the similarity of the average fibre stresses.

3.5 Modelling non-dilute systems

The previous sections have been based on a single inclusion embedded within an infinite matrix. The results are therefore applicable only to 'dilute' composites, in which the reinforcement volume fraction is less than a few percent. In this section, the scope of the model is extended to higher reinforcement contents. It is clear from the displacement lines in Fig. 3.11 that the matrix stress field approaches zero asymptotically with distance from the inclusion. Thus, upon cutting the matrix near the inclusion, the boundary moves, causing a readjustment of the internal stresses. One way of examining this redistribution is to look at the average stress in each phase[1,10,11] (see Table 3.1).

In Fig. 3.11 it can be seen that, to maintain the balance of stress, the composite distorts upon cutting, so as to provide an average matrix stress to oppose the inclusion stress. For an oversized inclusion, both matrix

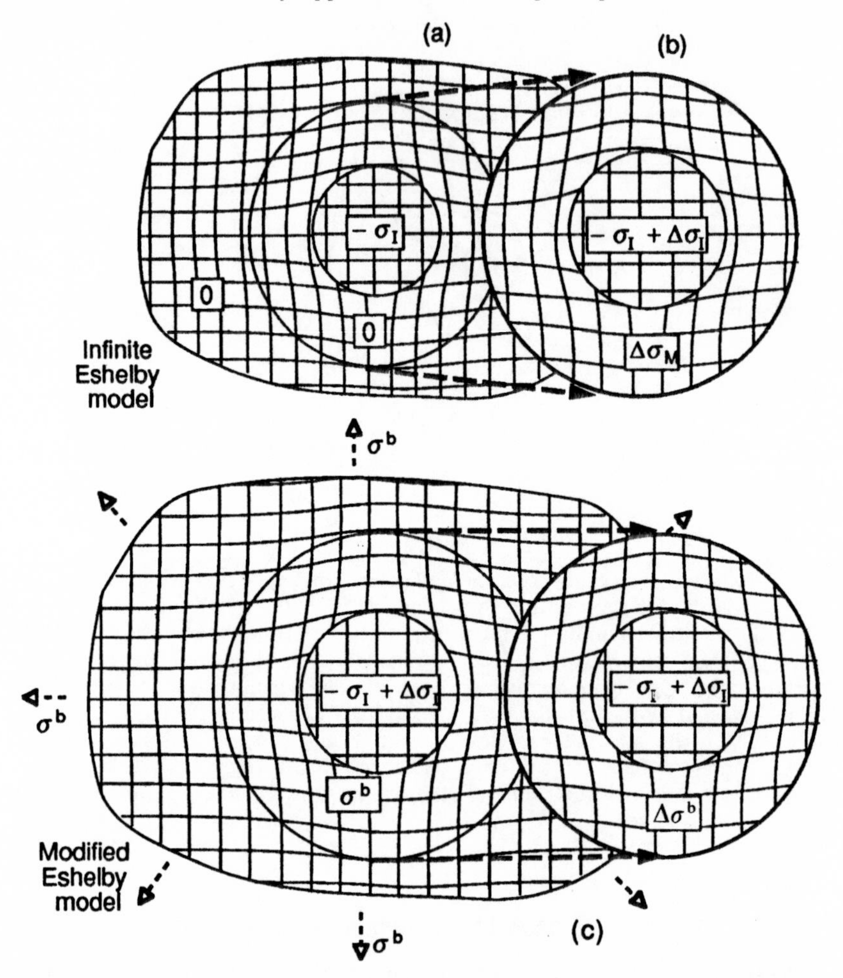

Fig. 3.11 (a) An infinite matrix containing an oversized inclusion of the same elastic constants (the boxes show the average stresses in the different regions). When a finite composite is cut (such as in (b)) from the infinite, the position of the new external boundary moves outward (the more so the nearer the cut to the inclusion). This is because the matrix so removed was helping to constrain the oversized inclusion. This movement generates a tensile average matrix stress to balance the compressive (but lowered) inclusion stress, so that the internal stresses within the new finite composite sum to zero. Rather than solve this new problem afresh, this distortion of the composite is modelled (c) simply by superimposing a tensile background stress (σ^b) upon the infinite solution as if it were externally applied; this satisfies the stress balance and causes the matrix and inclusion to relax outwards uniformly.

Table 3.1 *Expressions for the volume-averaged stresses in the two constituents.*

Region over which average taken	Mean stress		
	Matrix	Inclusion	Composite
Infinite composite (inclusion vol. fract. $f_{\text{single}} \to 0$)	0	σ_{I}	$(1 - f_{\text{single}})0 + f_{\text{single}}\sigma_{\text{I}} = 0$
Inside region to be cut[a]	0	σ_{I}	$(1 - f_{\text{single}})0 + f_{\text{single}}\sigma_{\text{I}} \neq 0$

[a] Using Eshelby's analysis, it is found that the volume-averaged matrix stress is zero within any ellipsoidal region cut from the infinite body which contains the reinforcing inclusion[11,12].

and inclusion will expand outwards. One way of incorporating this into the model is to think of it occurring by the superposition of a (in this case tensile) ***background stress*** (σ^{b})† which partitions between the phases as if it were externally applied[10]. The balance of stress then becomes

$$(1 - f_{\text{single}})\sigma^{\text{b}} + f_{\text{single}}(\sigma_{\text{I}} + \sigma^{\text{b}}) = 0 \tag{3.17}$$

where f_{single} represents the fraction of the total volume which is occupied by the single inclusion. It is far less clear how the stresses should redistribute for a reinforcing system. Pedersen[12,13] proposed that, just as in the homogeneous case, one could also regard the correction as occurring through a background stress acting as though externally applied (see Fig. 3.11). In these circumstances, eqn (3.17) still holds, except that now σ_{I} is also a function of σ^{b}. This 'mean field' approach gives rise to lower bound estimates of the composite properties[4,14,15].

The concept of the background stress also facilitates the extension of the model to composites containing more than one inclusion. Through its dependence on the volume fraction of reinforcement, it communicates to one inclusion the background stress arising from all the others. Consequently, Eshelby's solution can be used to model multiple inclusion composites simply by modifying eqn (3.17) to include the volume fraction of all the inclusions (f, as against the proportion of the composite occupied by the single inclusion f_{single} above)

$$(1 - f)\langle\sigma\rangle_{\text{M}} + f\langle\sigma\rangle_{\text{I}} = 0 \qquad \langle\sigma\rangle_{\text{M}} = -\frac{f}{1-f}\langle\sigma\rangle_{\text{I}} \tag{3.18}$$

† This is commonly called the ***image stress***, or ***mean field stress***; since it is equal to the mean internal stress in the matrix $\langle\sigma\rangle_{\text{M}}$, a separate symbol is rarely needed.

This procedure does not take into account individual inclusion/inclusion interactions, but only the average effect of all the other inclusions, and hence models a random spatial distribution of reinforcing particles. As it happens, such a distribution is commonly exhibited by many types of MMC. Eqn (3.18) is extremely useful, because it relates the volume-averaged internal matrix stress to the inclusion stress – which can be calculated easily, without requiring details of the form of the matrix stress field itself.

For the thermal misfit case, the background stress modifies the equations of §3.2. The 'ghost' inclusion in the equivalent composite is now stressed by $\sigma_I + \sigma^b$ or, equivalently, by $\sigma_I + \langle\sigma\rangle_M$ and has a constrained shape of $\varepsilon^C + \langle\varepsilon\rangle_M$. Consequently eqns (3.1), (3.6) and (3.7) lead to

$$\begin{aligned} \sigma_I + \langle\sigma\rangle_M &= C_M(\varepsilon^C + \langle\varepsilon\rangle_M - \varepsilon^T) \quad \text{– ghost inclusion} \\ \sigma_I + \langle\sigma\rangle_M &= C_I(\varepsilon^C + \langle\varepsilon\rangle_M - \varepsilon^{T*}) \quad \text{– real inclusion} \end{aligned} \tag{3.19}$$

where

$$\langle\sigma\rangle_M = C_M\langle\varepsilon\rangle_M$$

The internal stress balance, eqn (3.18), can then be used to calculate the level of the background or mean matrix stress ($\langle\sigma\rangle_M$)

$$\begin{aligned} (1-f)\langle\sigma\rangle_M + fC_M(\varepsilon^C + \langle\varepsilon\rangle_M - \varepsilon^T) &= 0 \\ \langle\sigma\rangle_M &= -fC_M(\varepsilon^C - \varepsilon^T) \end{aligned} \tag{3.20}$$

This, in combination with the two equivalent statements of the inclusion stress, eqn (3.19), allows the deduction of the equivalent transformation strain

$$\boxed{\varepsilon^T = -\{(C_M - C_I)[S - f(S - I)] - C_M\}^{-1}C_I\varepsilon^{T*}} \tag{3.21}$$

As $f \to 0$, this expression for the transformation strain becomes equal to the dilute composite version (eqn (3.7)).

The case of an external load (§3.3) also requires modification for non-dilute composites. In this case the 'ghost' inclusion is stressed by $\sigma_I + \langle\sigma\rangle_M + \sigma^A = \langle\sigma\rangle_I + \sigma^A$ and has a constrained shape corresponding to a strain of $\varepsilon^C + \varepsilon^A + \langle\varepsilon\rangle_M$. Consequently, eqns (3.9), (3.11) and (3.13) become

$$\begin{aligned} \langle\sigma\rangle_I + \sigma^A &= C_I(\varepsilon^C + \varepsilon^A + \langle\varepsilon\rangle_M) \quad \text{– real inclusion} \\ &= C_M(\varepsilon^C + \varepsilon^A + \langle\varepsilon\rangle_M - \varepsilon^T) \quad \text{– ghost inclusion} \end{aligned} \tag{3.22}$$

$$\boxed{\varepsilon^T = -\{(C_M - C_I)[S - f(S - I)] - C_M\}^{-1}(C_M - C_I)\varepsilon^A} \tag{3.23}$$

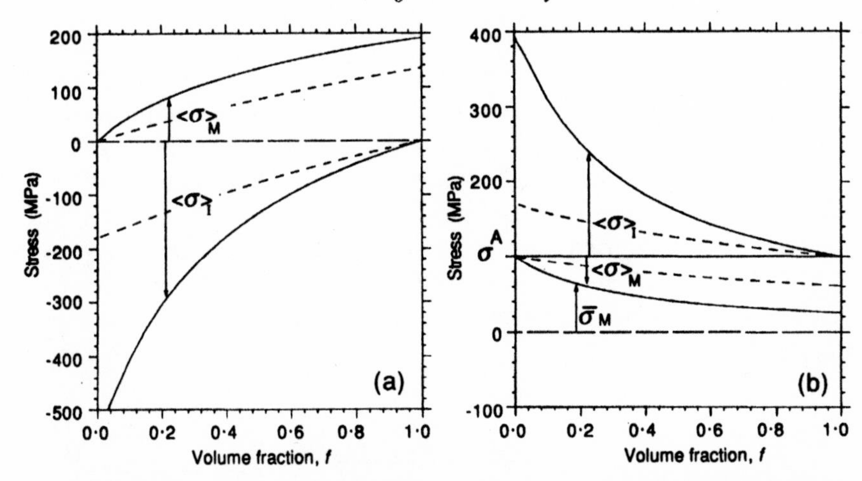

Fig. 3.12 Matrix and inclusion stresses for an Al–SiC system containing aspect ratio 5 'fibres' (solid lines) and spherical inclusions (dashed lines). Plots are shown for (a) thermal loading (100 K drop) and (b) applied loading (100 MPa). Under applied loading, the mean internal matrix stress ($\langle\sigma\rangle_M^A$) is a useful parameter, because it expresses the extent of load transfer to or from the matrix and can be easily calculated using eqns (3.23) and (3.24).

Once ε^T is established, the mean stresses can be calculated

$$\boxed{\langle\sigma\rangle_M = -fC_M(S-I)\varepsilon^T} \tag{3.24}$$

$$\boxed{\langle\sigma\rangle_I = (1-f)C_M(S-I)\varepsilon^T} \tag{3.25}$$

These are very important results and show the matrix stress to be proportional to an applied load, eqn (3.23), or to a misfit strain, eqn (3.21), or to their sum†. Now we can calculate the average stresses and strains in each phase, from which many materials properties can be predicted.

Fig. 3.12 shows the predicted variation of matrix and inclusion stresses with inclusion volume fraction under thermal and applied loading, for particulate and short fibre Al–SiC systems. From these curves, it is clear that there are two stress parameters of interest, namely the ***average stress*** in each constituent ($\bar{\sigma}_{M,I}$), and the ***mean internal stress*** ($\langle\sigma\rangle_{M,I}$) defined as the difference between the average phase stress and the external stress.

† In the elastic regime the transformation strains (ε^T) of different origins can be added together.

Under thermal loading, the two quantities are equal in both matrix and reinforcement. Under applied mechanical loading, the balance of internal stress means that the mean stresses in the two constituents are related by eqn (3.18) – see Fig. 3.12(b). The internal mean stress is a useful parameter, since it expresses the sense and effectiveness of the load transverse taking place. A large negative matrix mean stress is desirable and indicates that load has been transferred elastically from the matrix to the reinforcement. Furthermore, given the form of eqns (3.24) and (3.23), the mean matrix stress is approximately proportional to the reinforcement content at low contents.

3.6 Composite stiffness

The expressions for the average phase stresses derived in the previous section can be used to make various predictions about composite behaviour. In this section we shall use them to derive the elastic constants. By definition, the composite stiffness (C_C) is related to the volume-averaged composite strain under an applied load ($\bar{\varepsilon}_C^A$)

$$\sigma^A(=C_M\varepsilon^A) = C_C\bar{\varepsilon}_C^A = C_C(\varepsilon^A + \langle\varepsilon\rangle_C) \tag{3.26}$$

where $\langle\varepsilon\rangle_C$ is the mean composite strain arising from the internal stresses caused by the stiffness mismatch. This can be evaluated by recognising that, in the matrix, the local internal strains arising from the inclusion average to zero (see §3.4). Therefore, the overall strain is made up of the background or mean matrix strain (present in both constituents) and the constrained strain (present in the inclusions only).

$$\langle\varepsilon\rangle_C^A = \langle\varepsilon\rangle_M + f\varepsilon^C \tag{3.27}$$

The mean matrix strain can be found by considering the ghost composite, for which the internal stress balance of eqn (3.18) may be written

$$\begin{aligned}(1-f)C_M\langle\varepsilon\rangle_M + fC_M(\langle\varepsilon\rangle_M + \varepsilon^C - \varepsilon^T) &= 0\\ \therefore\ \langle\varepsilon\rangle_M = -f(\varepsilon^C - \varepsilon^T) &= 0\end{aligned} \tag{3.28}$$

so that

$$\langle\varepsilon\rangle_C^A = f\varepsilon^T \tag{3.29}$$

It follows that the overall average composite strain is given by

$$\bar{\varepsilon}_C^A = \varepsilon^A + f\varepsilon^T \tag{3.30}$$

so that, substituting for ε^{T} from eqn (3.23),

$$\bar{\varepsilon}_{\mathrm{C}}^{\mathrm{A}} = \varepsilon^{\mathrm{A}} - f\{(C_{\mathrm{M}} - C_{\mathrm{I}})[S - f(S - I)] - C_{\mathrm{M}}\}^{-1}(C_{\mathrm{M}} - C_{\mathrm{I}})\varepsilon^{\mathrm{A}} \quad (3.31)$$

Finally, substitution of this into eqn (3.26) gives

$$\sigma^{\mathrm{A}} = C_{\mathrm{C}}[\![C_{\mathrm{M}}^{-1}\sigma^{\mathrm{A}} - f\{(C_{\mathrm{M}} - C_{\mathrm{I}})[S - f(S - I)] - C_{\mathrm{M}}\}^{-1} \times (C_{\mathrm{M}} - C_{\mathrm{I}})C_{\mathrm{M}}^{-1}\sigma^{\mathrm{A}}]\!]$$

leading to the equation for the composite stiffness tensor

$$\boxed{C_{\mathrm{C}} = [\![C_{\mathrm{M}}^{-1} - f\{(C_{\mathrm{I}} - C_{\mathrm{M}})[S - f(S - I)] + C_{\mathrm{M}}\}^{-1}(C_{\mathrm{I}} - C_{\mathrm{M}})C_{\mathrm{M}}^{-1}]\!]^{-1}} \quad (3.32)$$

The engineering constants of the composite can be derived from this tensor. Some reduce to a rather simple form. For example, the axial Young's modulus is given by

$$\boxed{E_{3\mathrm{C}} = \frac{1}{C_{3\mathrm{C}}^{-1}} = \frac{\sigma^{\mathrm{A}}}{\varepsilon_3^{\mathrm{A}} + f\varepsilon_3^{\mathrm{T}}}} \quad (3.33)$$

Equation (3.32), and the composite engineering constants, are in general best evaluated using a simple computer program. A listing of a program designed to do this is given in Appendix IV. Some predictions obtained in this way are shown in Fig. 3.13, which illustrates the effect of inclusion stiffness and aspect ratio on composite stiffness.

3.6.1 Experimental values of Young's moduli

Researchers have found it a much more difficult task to undertake the measurement of Young's modulus on discontinuous MMCs than on unreinforced alloys (see §11.1). This is largely because the proportional regime is very short (see §4.1). This arises for two reasons; firstly, within the matrix the (average) thermal stresses are normally tensile, which tends to encourage early tensile matrix flow. Secondly, the matrix stress state is very inhomogeneous, so that highly stressed regions (usually in the vicinity of the fibre ends) become critically stressed much earlier than the matrix as a whole. The associated microplasticity causes a premature departure from the linear elastic line (see §4.1). Furthermore, for short fibre or particulate composites, the more 'effective' the reinforcing phase, the more inhomogeneous the matrix stress field, so the more pronounced the effect. To counter these difficulties, a number of methods have been proposed for the measurement of composite stiffness and these are outlined in §11.1.

3.6.2 Particulate systems

Experimental stiffness data for SiC particulate- and whisker-reinforced Al systems are shown in Fig. 3.14. Since the matrix stiffness is approximately the same for all these data, the observed scatter requires explanation. This scatter is doubtless partly due to the measurement difficulties mentioned above, but it is also a consequence of different fabrication routes and microstructures. Given that the Eshelby method provides a lower bound[14] estimate of stiffness, it is reassuring to find that nearly all the particulate data are confined to a band marginally above the spherical particle prediction. However, in the light of the agreement between Eshelby and FEM methods (§2.4) in the elastic regime, it seems probable that this slight underestimation of the results is significant. The most likely explanation for this lies in the fact that the SiC fragments are not truly equiaxed and tend to become partially aligned during processing. In fact, an effective aspect ratio of only 1.4 is required to give a good fit to most of the data. This correlates well with measurements[26] on extruded particulate material which, for an estimated average particle aspect ratio of 2.3, found an effective (aligned) aspect ratio of around 1.6.

3.6.3 Whisker systems

The experimental data for whisker-reinforced aluminium are even more scattered than those for the corresponding particulate materials (Fig.3.14). This is not really surprising, because whisker composites tend to be more sensitive to processing variables than particulate material (Chapter 9). Whisker alignment and aspect ratio are the most important parameters. Both differ markedly with fabrication route and, unfortunately, gains in one are usually made at the expense of the other. Fibre bunching, breakage and alignment generally become worse as the whisker volume fraction is increased. This would seem to be reflected in the much slower increase in stiffness with increasing whisker content in Fig. 3.14 than is either observed for the particulate composites or predicted theoretically. It should be noted that FEM calculations (§2.4) suggest that, while average reinforcement shape and volume fraction are important, stiff-

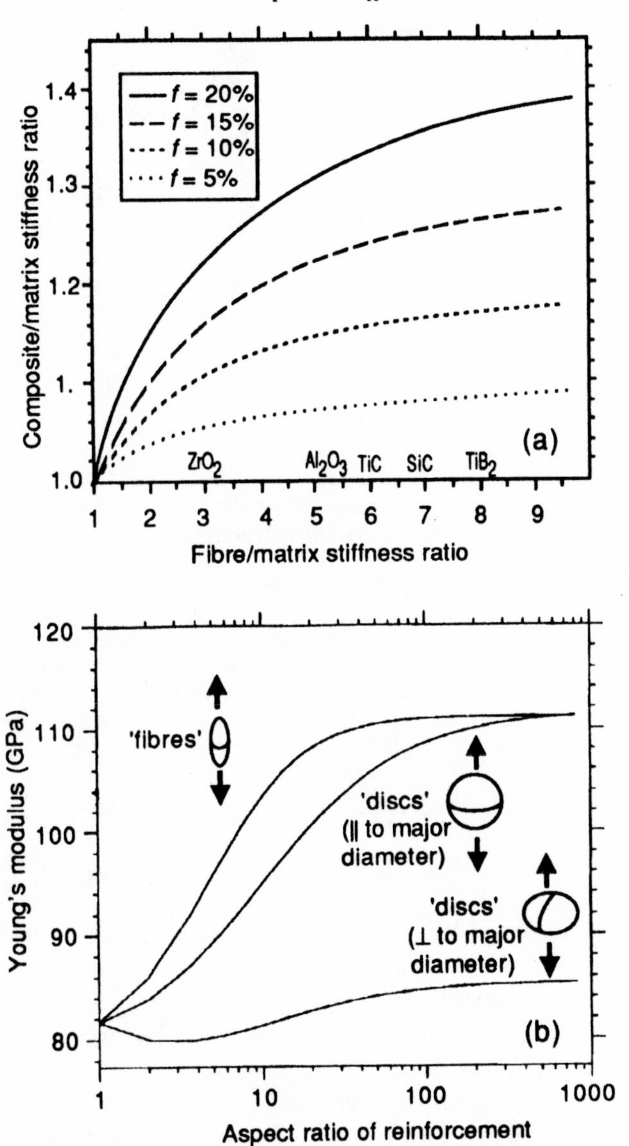

Fig. 3.13 The variation in composite stiffness with (a) increasing inclusion stiffness and volume fraction for spherical reinforcement and (b) increasing aspect ratio for 10 vol% SiC fibres and discs within Al. Note that, as one would expect, composite stiffness increases with aspect ratio, stiffness and volume fraction of the reinforcement. But it is also clear that the rewards for incorporating increasingly stiff inclusions tails off, especially at low volume fractions. In (b) the aligned fibre and plate predictions tend to the rule of mixtures value (111 GPa) at high aspect ratio, while for plates normal to the tensile axis the predictions lie above those of the slab model of §2.1 (76 GPa) because of Poisson contraction effects.

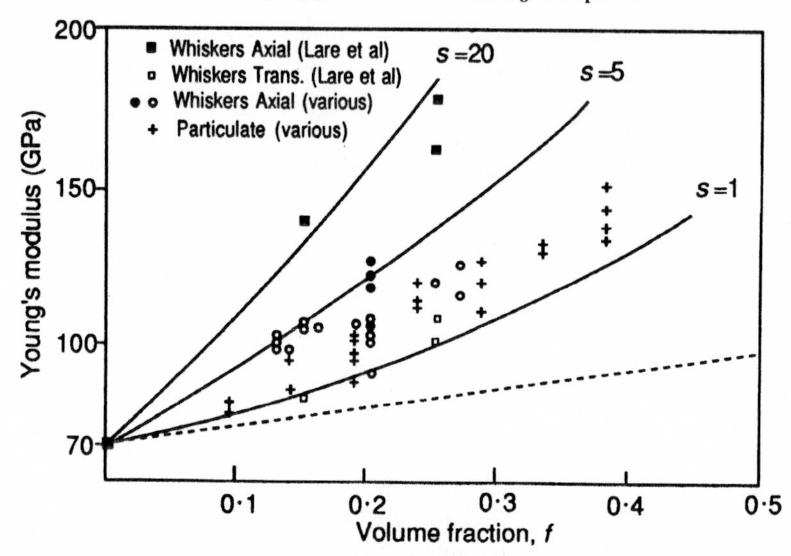

Fig. 3.14 Experimental data for Al–SiC particle[16–19] (+) and whisker[18,20–24] (■ □ ● and ○) composites plotted alongside predictions of the Eshelby model. An aspect ratio of around 5 is typical of most whisker systems and much of the scatter in the data derives from the degree of alignment achieved by different processing routes. The data of Lare *et al.*[25] were obtained using material with good whisker alignment and high aspect ratio. The solid lines correspond to predictions made using eqn (3.33) and the dashed line to the simple expression given by eqn (3.40) for spherical particles.

ness is largely insensitive to local variations in these parameters (e.g. clustering).

The results of Fig. 3.14 exemplify the large differences between even nominally similar whisker-reinforced systems. Whisker composites must therefore be characterised carefully before one can hope to predict the mechanical properties with any precision. Manufacturing has not yet reached the stage where the observed properties are sufficiently reproducible for precise comparisons with theory to be made. Consequently, at the present time fine distinctions between FEM and Eshelby-based approaches in the elastic regime are not really justified and either method is equally satisfactory.

3.7 Simple representations of Eshelby's equations

A number of simple analytical equations describing the internal stress state have been developed for particular reinforcement and loading geometries. Some of these are rigorous, whilst others are only approximations. If used with care, these expressions can provide a fast and simple means of estimating composite behaviour.

3.7.1 The elastic properties of spherical particle systems

Several analytic expressions for the stress state surrounding a spherical particle predate Eshelby's approach by a considerable margin. Lamé[27] (1852) was able to express the radial (r) and hoop or tangential (θ) stress field within a matrix containing a spherical bubble (radius a) with an internal pressure of P

$$\sigma_{\mathrm{M}r}(r) = -\frac{Pa^3}{r^3} \qquad \sigma_{\mathrm{M}\theta}(r) = \frac{Pa^3}{2r^3} \tag{3.34}$$

and the hydrostatic (H) stress field is given by

$$\sigma_{\mathrm{MH}}(r) = [\sigma_{\mathrm{M}r}(r) + 2\sigma_{\mathrm{M}\theta}(r)]/3 = 0$$

Of course this is also the solution for the field around an oversized inclusion for which the inclusion hydrostatic stress is equal to $-P$. If the matrix and inclusion have the same elastic constants, such a stress field might be generated by a change of temperature (ΔT). The strain field is given by[28]

$$\textit{Within the inclusion:} \quad \varepsilon_r = \varepsilon_\theta = \frac{1}{3}\frac{(1+\nu)}{(1-\nu)}(\alpha_\mathrm{I} - \alpha_\mathrm{M})\,\Delta T$$

$$\textit{Outside the inclusion:} \quad \varepsilon_r = -\frac{2}{3}\frac{(1+\nu)}{(1-\nu)}\frac{a^3}{r^3}(\alpha_\mathrm{I} - \alpha_\mathrm{M})\,\Delta T; \tag{3.35}$$

$$\varepsilon_\theta = \frac{1}{3}\frac{(1+\nu)}{(1-\nu)}\frac{a^3}{r^3}(\alpha_\mathrm{I} - \alpha_\mathrm{M})\,\Delta T$$

Eshelby's approach treats the stress state in terms of the difference in shape of the inclusion when constrained within the matrix ($\varepsilon_{ij}^{\mathrm{C}}$) and unconstrained ($\varepsilon_{ij}^{\mathrm{T}}$). To simplify the notation, the strain tensors (e.g. $\varepsilon_{ij}^{\mathrm{T}}$) are split into dilatational (Δ^{T}) and deviatoric (${}'\varepsilon_{ij}^{\mathrm{T}}$) components

$$\varepsilon_{ij}^{\mathrm{T}} = \tfrac{1}{3}\Delta^{\mathrm{T}}\delta_{ij} + {}'\varepsilon_{ij}^{\mathrm{T}} \qquad \text{where} \qquad \Delta^{\mathrm{T}} = (\varepsilon_{11}^{\mathrm{T}} + \varepsilon_{22}^{\mathrm{T}} + \varepsilon_{33}^{\mathrm{T}}) \tag{3.36}$$

When the matrix and reinforcement have the same elastic constants[1], the constrained strain can be written

$$\varepsilon_{ij}^{C} = \frac{1}{3}\frac{(1+\nu)}{(1-\nu)}\Delta^{T}\delta_{ij} + \frac{2}{15}\frac{(4-5\nu)}{(1-\nu)}{}'\varepsilon_{ij}^{T} \\ = A\,\Delta^{T}\delta_{ij} + B\,{}'\varepsilon_{ij}^{T} \tag{3.37}$$

Note that this equation gives the same result as eqn (3.35) when the misfit is purely dilatational $\frac{1}{3}\Delta^{T} = \varepsilon_{i}^{T} = [(\alpha_{I} - \alpha_{M})\,\Delta T]$. When the matrix and reinforcement have different elastic constants[1], the misfit strain of the ghost inclusion can be written[1,29] in terms of the bulk modulus K and shear modulus G

$$\varepsilon_{ij}^{T} = \frac{1}{3}\frac{K_{I} - K_{M}}{(K_{M} - K_{I})A - K_{M}}\Delta^{T*} + \frac{K_{M} - K_{I}}{3K_{I}}\Delta^{A} \\ + \frac{G_{I} - G_{M}}{(G_{M} - G_{I})B - G_{M}}{}'\varepsilon_{ij}^{T*} + \frac{G_{M} - G_{I}}{G_{I}}{}'\varepsilon_{ij}^{A} \tag{3.38}$$

To take an example, consider axial loading with $\nu_{M} = \frac{1}{3}$

$$\Delta^{T*} = {}'\varepsilon_{ij}^{T*} = 0 \qquad \Delta^{A} = \frac{\sigma^{A}}{3E_{M}} \qquad {}'\varepsilon_{33}^{A} = \frac{8\sigma^{A}}{9E_{M}} \qquad \varepsilon_{33}^{A} = \frac{\sigma^{A}}{E_{M}}$$

from which ε_{3}^{T} $(=\varepsilon_{33}^{T})$ can be calculated

$$\varepsilon_{3}^{T} = \frac{K_{M} - K_{I}}{3K_{I}}\frac{\sigma^{A}}{8G_{M}} + \frac{G_{M} - G_{I}}{G_{I}}\frac{\sigma^{A}}{3G_{M}} \tag{3.39}$$

Inserting the materials properties relating to the Al–SiC system, this expression facilitates the prediction of composite stiffness via eqn (3.33)

$$E_{3C} = \frac{\sigma^{A}}{\varepsilon_{3}^{A} + f\varepsilon_{3}^{T}} \approx \frac{1}{0.0143 - f(1 \times 10^{-3} + 0.011)}\,\text{GPa} \tag{3.40}$$

This relation gives rise to predictions according to the dashed line in Fig. 3.14 and predicts a stiffness increment of ~6 GPa for 10% spherical reinforcement. As is shown by the discrepancy between this and the more complete treatment, the expressions given in this section are only suitable at low inclusion volume fractions ($f < 10\%$), because they refer to isolated particles.

References

1. J. D. Eshelby (1957) The Determination of the Elastic Field of an Ellipsoidal Inclusion, and Related Problems, *Proc. Roy. Soc.*, **A241**, pp. 376–96.
2. J. D. Eshelby (1961) Elastic Inclusions and Inhomogeneities, in *Prog. Solid Mech.*, I. N. Sneddon and R. Hill (eds.), pp. 89–140.
3. I. S. Grant and W. R. Phillips (1982) *Electromagnetism*, Wiley, New York.
4. O. B. Pedersen (1983) Thermoelasticity and Plasticity of Composites – I. Mean Field Theory, *Acta Metall.*, **31**, pp. 1795–808.
5. T. Mura (1987) *Micromechanics of Defects in Solids*, Nijhoff, The Hague.
6. P. J. Withers, A. N. Smith, T. W. Clyne and W. M. Stobbs (1989) A Photoelastic Examination of the Validity of the Eshelby Approach to the Modelling of MMCs, in *Fundamental Relationships between Microstructure and Mechanical Properties of MMCs*, Indianapolis, P. K. Liaw and M. N. Gungor (eds.), TMS, pp. 225–40.
7. P. J. Withers, E. M. Chorley and T. W. Clyne (1991) Use of the Frozen Stress Photoelastic Method to Explore Load Partitioning in Short Fibre Composites, *Mat. Sci. & Eng.*, **A135**, pp. 173–8.
8. B. D. Saint-Venant (1855) *Memoirs des Etrangers*, Paris.
9. P. J. Withers, G. Cecil and T. W. Clyne (1992) Frozen Stress Photoelastic Determination of the Extent of Fibre Stressing in Short Fibre Composites, in *Proc. 2nd European Conf. on Adv. Mats. and Processes* (*Euromat '91*), Cambridge, UK, T. W. Clyne and P. J. Withers (eds.), Inst. of Metals, pp. 134–40.
10. L. M. Brown and W. M. Stobbs (1971) The Work-Hardening of Cu–SiO_2 – I. A Model Based on Internal Stresses, with no Plastic Relaxation, *Phil. Mag.*, **23**, pp. 1185–99.
11. K. Tanaka and T. Mori (1972) Note on Volume Integrals around an Ellipsoidal Inclusion, *J. Elasticity*, **2**, pp. 199–200.
12. O. B. Pedersen (1978) Transformation Theory for Composites, *Z. ang. Math. Mech.*, **58**, pp. 227–8.
13. O. B. Pedersen (1978) Thermoelasticity and Plasticity of Composite Materials, in *Proc. ICM–3*, Cambridge, England, K. J. Miller and R. F. Smith (eds.), Pergamon, pp. 263–73.
14. O. B. Pedersen and P. J. Withers (1991) Iterative Estimates of Internal Stresses in Short-Fibre MMCs, *Phil. Mag.*, In Press.
15. T. Mori and K. Tanaka (1973) Average Stress in a Matrix and Average Elastic Energy of Materials with Misfitting Inclusions, *Acta Metall.*, **23**, pp. 571–4.
16. Y. Flom and R. J. Arsenault (1987) Fracture of SiC/Al Composites, in *Proc. ICCM VI/ECCM 2*, London, F. L. Matthews, N. C. R. Buskell, J. M. Hodgkinson and J. Morton (eds.), Elsevier, pp. 189–98.
17. M. W. Mahoney, A. K. Ghosh and C. C. Bampton (1987) in *Proc. ICCM VI/ECCM 2*, London, F. L. Matthews, N. C. R. Buskell, J. M. Hodgkinson and J. Morton (eds.), Elsevier, p. 372.
18. C. R. Crowe, R. A. Gray and D. F. Hasson (1985) Microstructure Controlled Fracture Toughness of SiC/Al Metal Matrix Composites, in *Proc. 5th Int. Conf. Comp. Mats.* (*ICCM V*), San Diego, W. C. Harrigan, J. Strife and A. K. Dhingra (eds.), TMS-AIME, pp. 843–66.
19. D. L. McDanels (1985) Analysis of Stress–Strain, Fracture, and Ductility of

Aluminium Matrix Composites Containing Discontinuous Silicon Carbide Reinforcement, *Metall. Trans.*, **16A**, pp. 1105–15.
20. P. Jarry, W. Loué and J. Bouvaist (1987) Rheological Behaviour of SiC/Al Composites, in *Proc. ICCM VI/ECCM 2*, London, F. L. Matthews, N. C. R. Buskell, J. M. Hodgkinson and J. Morton (eds.), Elsevier, pp. 2.350–2.361.
21. T. G. Nieh (1984) Creep Rupture of a Silicon Carbide Reinforced Aluminium Composite, *Metall. Trans.*, **15A**, pp. 139–46.
22. T. G. Nieh and R. F. Karlak (1983) Hot-Rolled SiC–Al Composites, *J. Mat. Sci. Letts.*, **2**, pp. 119–22.
23. V. C. Nardone and K. M. Prewo (1986) On the Strength of Discontinuous Silicon Carbide-Reinforced Aluminium Composites, *Scripta Met.*, **20**, pp. 43–48.
24. D. Webster (1984) Properties and Microstructures of MMCs, in *Proc. 3rd Int. Conf. on Comp. Mats.* (*ICCM III*), pp. 1165–76.
25. P. J. Lare, F. Ordway and H. Hahn (1971) *Research on Whisker Reinforced Metal Composites*, Naval Air Systems Command, contract N00019 70 C 0204.
26. T. J. Warner and W. M. Stobbs (1989) Modulus and Yield Stress Anisotropy of Short Fibre MMCs, *Acta Metall.*, **37**, pp. 2873–81.
27. G. Lamé (1852) *Leçons sur la Théorie de l'Élasticité*, Gauthier-Villars, Paris.
28. S. P. Timoshenko and J. N. Goodier (1982) *Theory of Elasticity*, McGraw-Hill, New York.
29. O. B. Pedersen (1985) Mean Field Theory and the Bauschinger Effect in Composites, in *Proc. IUTAM Eshelby Mem. Symp.*, Sheffield, UK, K. J. Miller, B. Bilby and J. R. Willis (eds.), Cambridge University Press, pp. 263–73.

4
Plastic deformation

Why do many MMCs often behave asymmetrically in tension and compression? Why do they frequently have higher ultimate tensile strengths, yet lower proportional limits than unreinforced alloys? Since the reinforcement usually remains elastic as the composite is loaded, the answers to these and other questions concerning the mechanical behaviour of MMCs lie with the factors which govern matrix plasticity. These can be broadly divided into two areas; those which affect the stress state of the matrix, and those which alter the flow properties of the matrix via changes in microstructure induced by incorporation of the reinforcement. This chapter illustrates, with the aid of relatively simple models, how these two aspects interact and combine to determine the behaviour, from the onset of flow to the development of large plastic strains.

A considerable body of mechanical test data for discontinuously reinforced MMCs is now available, although some of these results have been obtained with rather poor quality material. However, study of data such as those for Al/SiC summarised in Table 4.1, reveals some systematic trends:

- the incorporation of reinforcement improves both yield stress (0.2% proof stress) and ultimate tensile stress (UTS)
- whiskers provide more effective reinforcement than particles
- yield stress rises with increasing volume fraction; UTS is not always similarly affected
- for whisker-reinforced composites, increases in yield strength are often much greater in compression than in tension
- for whisker-reinforced composites, increases in tensile yield strength are greater transverse to the whisker alignment than parallel to it
- there is a wide scatter in experimental results

Table 4.1 *Published mechanical test data on Al/SiC systems: improvements with respect to the unreinforced alloys are shown in parentheses.*

Particle-Reinforced

Al Alloy	Axis	f (%)	YS (MPa) 0.2%	UTS (MPa)
A356 (T6)[1]		15	324 (+124)	331 (+55)
6061[1]		15	317 (+40)	359 (+60)
LM13 (T6)[3]		15	348 (+65)	348 (+40)
6061 (T6)[2]	Lo	15	280 (+10)	340 (+60)
	Tr		310 (+40)	310 (+30)
6061 (T6)[2]	Lo	20	370 (+100)	410 (+130)
	Tr		360 (+90)	370 (+90)
6061 (T6)[2]	Lo	30	370 (+100)	380 (+100)
	Tr		410 (+140)	480 (+200)
2124[5]		20	430 (−10)	615 (+125)

Whisker-Reinforced

Al Alloy	Ten/ Com	Axis	f (%)	YS (MPa) 0.2%	UTS (MPa)
5083 (T6)[2]	T	Lo	10	200 (−15)	260 (−60)
		Tr		245 (+30)	320 (0)
5083 (T6)[2]	T	Lo	20	280 (+65)	320 (0)
		Tr		370 (+155)	390 (+70)
5083 (T6)[2]	T	Lo	30	380 (+165)	270 (−50)
		Tr		390 (+175)	420 (+100)
2024[4]	T	Lo	20	386 (+45)	532 (+30)
		Tr		419 (+75)	550 (+35)
2024[4]	C	Lo	20	565 (+220)	—
		Tr		475 (+110)	—
2024[4]	T	Lo	15	476 (+140)	648 (+150)
		Tr		545 (+205)	683 (+185)
6061[6]	T	Lo	20	414 (+135)	545 (+245)
		Tr		413 (+135)	439 (+140)

Symbols: Lo – longitudinal and Tr – transverse to an extrusion axis; C – compressive and T – tensile testing.

Through the following sections these, and other, experimental observations will be discussed in terms of the influence of matrix microstructure and the mechanics of load transfer.

4.1 Onset of yielding: the effect of internal stress

For an MMC, the onset of composite yielding is governed by the onset of matrix yielding and it is important to ask the question 'At what point in space and time does matrix yielding occur?' In this section we will examine the role of internal stresses and we shall see that many of the features noted above can be explained on the basis of quite simple models. In this respect it will be assumed that the matrix yield stress is invariant with strain – i.e. we will assume that the matrix exhibits *no inherent* work-hardening.

Even without plastic flow, applied loads always partition very unevenly between the phases of an MMC, because of the higher stiffness of the ceramic phase (Chapter 3). As a natural consequence of this, the stress state in the matrix varies greatly from point to point (see Fig. 3.9), and intuitively one would expect flow to initiate near the fibre end, where the stress is the greatest. This expectation is supported by finite element calculations[7] which show that, with increased straining, plastic flow initiates near the fibre end and expands progressively into the matrix (Fig. 4.1). But at what point does plastic flow become significant? In this section we shall compare the predicted internal stress state with observed stress–strain behaviour in order to assess the influence of local and average internal stresses on yielding.

By way of example, consider the elastic loading of an Al–15 vol% SiC whisker composite. For a given whisker aspect ratio (e.g. 5), the average stress in each constituent can be calculated using the Eshelby model (Chapter 3), as shown in Fig. 4.2. The lowering of the matrix stress line, relative to the unreinforced alloy, is the result of load transfer to the reinforcement as expressed by eqns (3.23) and (3.24)

$$\langle\sigma\rangle_{\mathrm{M}}^{\mathrm{A}} = -fC_{\mathrm{M}}(S-I)\{(C_{\mathrm{M}}-C_{\mathrm{I}})[S-f(S-I)]-C_{\mathrm{M}}\}^{-1}(C_{\mathrm{I}}-C_{\mathrm{M}})\varepsilon^{\mathrm{A}} \tag{4.1}$$

in this case

$$\langle\sigma_3\rangle_{\mathrm{M}}^{\mathrm{A}} = -0.30\sigma^{\mathrm{A}} \qquad \text{so that} \qquad \bar{\sigma}_{3\mathrm{M}}^{\mathrm{A}} = 0.70\sigma^{\mathrm{A}}$$

Note that, while the average stress in the matrix is lower than that in the unreinforced alloy, the peak stresses (near the whisker ends) are much

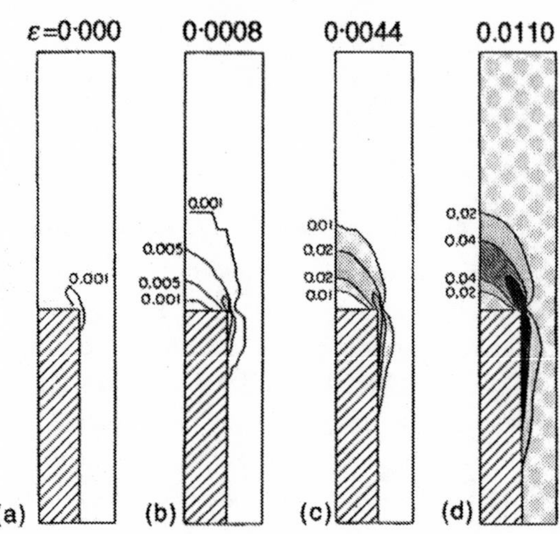

Fig. 4.1 Contours of equivalent plastic strain around a fibre end with progressive straining of the composite, as predicted by FEM modelling[7]. Plastic flow starts near the fibre corner, but the plastic zone becomes contiguous at low overall strains. Note the similarity between the plastic flow contours and the elastic fringe pattern of Fig. 3.9. Interestingly, for equiaxed particles Levy and Papazian[8] predict that the point of initial yielding moves from the fibre corner to the midpoint between fibre ends.

higher than this. This is because stress continuity of normal stress across the inclusion/matrix interface requires that the peak matrix stresses be approximately equal to the average inclusion stresses (which are uniform throughout the inclusion, if it is ellipsoidal). As Fig. 4.2 illustrates, this means that, for a matrix with a yield stress of, say, 200 MPa, the stress in regions of the matrix near a whisker end exceeds the yield stress at very low applied stresses (~70 MPa). On the other hand, the *average* stress in the matrix does not exceed the matrix yield stress until the applied load reaches about 290 MPa. Experimental observation[9,10] accords well with the occurrence of micro-yielding at low loads (in §4.2 we shall see that thermal stresses often accentuate premature yielding), and this has hampered the accurate measurement of the Young's modulus of MMCs (§11.1). However, as evidenced by the data of Table 4.1, micro-yielding does not seem to control the onset of global yielding, as measured by a conventional 0.2% proof stress, which is typically well in excess of that for the unreinforced alloy. This observation suggests that the overall yield stress is governed, not so much by premature localised yielding, but rather by the attainment of an average matrix value sufficient for global yielding.

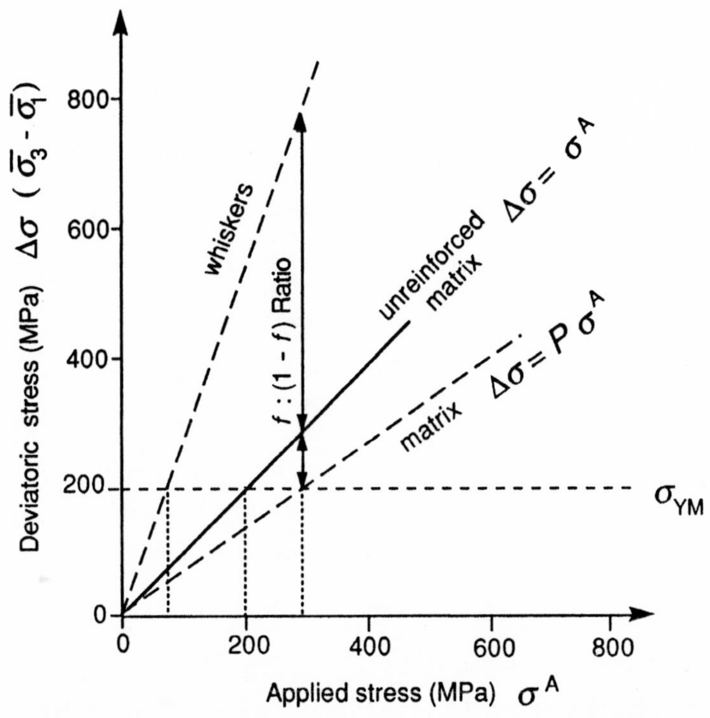

Fig. 4.2 The predicted variation (see Chapter 3) in the average stresses within fibre and matrix for an aligned Al–15 vol% SiC ($s = 5$) composite under elastic loading. The whisker stresses are clearly much greater than the matrix stresses, but the whisker strength is such that plastic flow occurs within the matrix before whisker fracture.

This can be explained at a microscopic level as being consistent with the stress required for the repeated *long-range* glide of dislocations throughout the matrix, while at a continuum level it can be argued that extensive macroscopic flow requires that regions of plastic flow be interconnected.

4.1.1 A yield criterion for matrix flow

In common with criteria derived for monolithic materials which reflect the triaxial state of stress, it would seem natural to expect plastic flow within an MMC to be dependent upon the transverse stresses as well as those in the axial direction. A number of models use a Tresca-type yield criterion[11–13], modified by assuming that bulk flow occurs when the average matrix stress state reaches that which would give rise to flow were it experienced by the matrix† free from the reinforcement (σ_{YM}). For an aligned composite, the average axial and transverse matrix stresses are given by

$$\bar{\sigma}_{3M} = \sigma^A + \langle\sigma_3\rangle_M^A \qquad \bar{\sigma}_{1M} = \bar{\sigma}_{2M} = \langle\sigma_1\rangle_M^A = \langle\sigma_2\rangle_M^A \tag{4.2}$$

and the Tresca yield criterion may be written as

$$\sigma_{YM} = \bar{\sigma}_{3M} - \bar{\sigma}_{1M} \, (=\Delta\sigma) \tag{4.3a}$$

so that flow occurs at an applied stress (σ^A) equal to

$$\boxed{\sigma^A = \sigma_{YM} - (\langle\sigma_3\rangle_M^A - \langle\sigma_1\rangle_M^A)} \tag{4.3b}$$

Now, the mean matrix stress resulting from an applied load can be evaluated immediately using eqns (3.23) and (3.24) to give

$$\langle\sigma\rangle_M^A = -fC_M(S - I)\{(C_M - C_I)[S - f(S - I)] - C_M\}^{-1} \times (C_I - C_M)C_M^{-1}\sigma^A \tag{4.4}$$

This expression can be used to find $\langle\sigma_3\rangle_M^A$ and $\langle\sigma_1\rangle_M^A$. Since both quantities are proportional to σ^A, we can use a simple scalar form, $\Delta\sigma$

$$\Delta\sigma = (\bar{\sigma}_{3M} - \bar{\sigma}_{1M}) = \sigma^A + (\langle\sigma_3\rangle_M^A - \langle\sigma_1\rangle_M^A) = P\sigma^A \tag{4.5}$$

in which P is a dimensionless constant. The magnitude of P should be as small as possible for efficient reinforcement; it quantifies the degree to which the load is borne by the matrix after the elastic transfer of load to the reinforcement. For the present example (Al–15 vol% SiC ($s = 5$))

$$\langle\sigma_3\rangle_M^A = -0.30\sigma^A \qquad \langle\sigma_1\rangle_M^A = +0.005\sigma^A$$

so that P has the value 0.695‡, from which

$$\sigma_{YC}^A = \frac{\sigma_{YM}}{P} = 1.43\sigma_{YM} \tag{4.6}$$

i.e. the yield stress has been raised by about 43% by the presence of the reinforcement (Fig. 4.2).

† The *in situ* yield stress of the matrix may in fact differ from its value when unreinforced (§4.2 & Chapter 10).

‡ For comparison, P would equal 0.55 for a similar long fibre composite.

4.1.2 Differential thermal contraction stresses

Thermal residual stresses can have a marked influence on mechanical behaviour, especially during the early stages of deformation. Unfortunately, accurate measurement of the initial stages of the yielding response is difficult (§11.2). This difficulty, compounded by the use of non-standard loading configurations (often triggered by constraints on materials supply), has no doubt contributed to the large scatter in published data. Nevertheless, a number of general features seem to be emerging and these are discussed below.

The tensile–compressive yielding asymmetry

The asymmetry in the tensile/compressive yielding behaviour illustrated in Fig. 4.3 and Table 4.1 is well documented for whisker-reinforced[8,14,15], long fibre[16] and eutectic[17] composites; the established explanation lies in the state of residual stress prior to loading. As discussed in §5.1.1, the large disparity in the coefficients of thermal expansion (CTE) of matrix

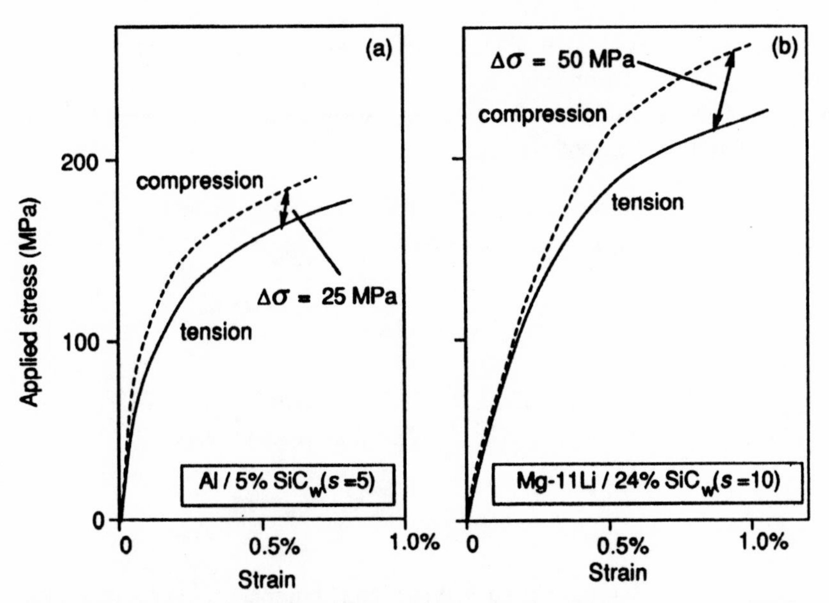

Fig. 4.3 Experimentally determined stress–strain curves for two aligned fibre composites, in tension and compression[14,18]. Yielding and flow take place at a lower applied stress in tension because the internal thermal stresses act to reinforce those from the applied load.

and reinforcement can lead to the build up of substantial residual stresses on cooling from the fabrication temperature. The matrix, typically having the greater CTE, is left in net residual *tension*, the reinforcement in residual *compression*. These internal stresses effectively pre-stress the matrix and are of a magnitude which can be estimated using the Eshelby method (eqn. (3.21)) and setting the transformation strain ε^{T*} equal to the thermal misfit $(\alpha_M - \alpha_I)\,\Delta T$

$$\langle\sigma\rangle_M^{\Delta T} = -fC_M(S - I)\{(C_M - C_I)[S - f(S - I)] - C_M\}^{-1} \times C_I(\alpha_M - \alpha_I)\,\Delta T \qquad (4.7a)$$

From which it follows, in an analogous way to the case of the applied stress, that we can write

$$\Delta\sigma_M^{\Delta T} = (\bar{\sigma}_{3M} - \bar{\sigma}_{1M}) = Q\,\Delta T \qquad (4.7b)$$

where Q (+ve for Al–SiC), which is expressed in MPa K^{-1} (per degree decrease), represents the difference between axial and transverse matrix stresses caused by cooling the composite through 1 deg K. Fig. 4.4 illustrates how P and Q vary with volume fraction. It can be seen that, for our reference example (Al–15%SiC, $s = 5$), $Q \approx 0.43$ MPa K^{-1}. This matrix deviatoric stress ($\Delta\sigma_M^{\Delta T}$) is simply added to that from the applied stress, as is shown in Fig. 4.5. This means that the applied stress at which the Tresca criterion eqn (4.3) becomes satisfied is reduced in tension and increased in compression

tensile stressing | compressive stressing

$$\sigma_{YM} = P\sigma_{ten}^A + Q\,\Delta T \qquad -\sigma_{YM} = -P\sigma_{com}^A + Q\,\Delta T$$

$$\sigma_{ten}^A = \frac{\sigma_{YM} - Q\,\Delta T}{P} \qquad \sigma_{com}^A = \frac{\sigma_{YM} + Q\,\Delta T}{P} \qquad (4.8)$$

from which it is clear that the extent of the tensile/compressive loading asymmetry is independent of the (*in situ*) matrix yield stress

$$\sigma_{Ycom}^A - \sigma_{Yten}^A = \frac{2Q\,\Delta T}{P} \qquad (4.9)$$

It is now possible to evaluate ΔT_{esf}, the difference between the test temperature and that at which the specimen would be free of any deviatoric stress. This can be done from the values of P, Q and the experimentally observed difference between the tensile and compressive

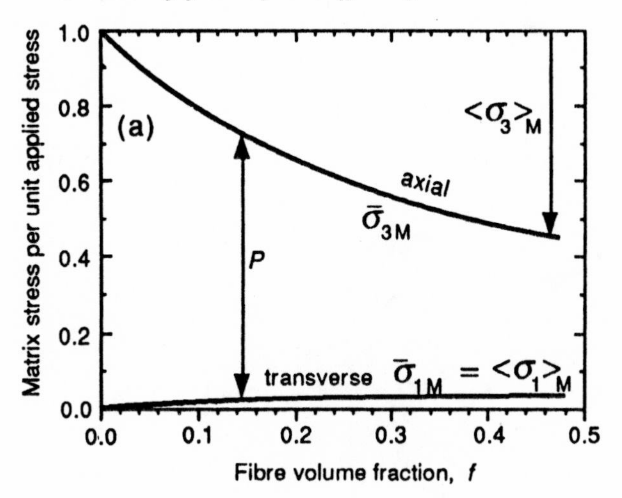

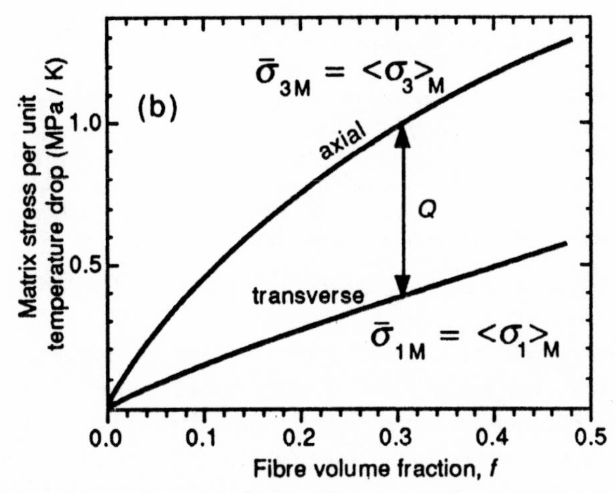

Fig. 4.4 The volume-averaged deviatoric matrix stresses for the Al–SiC ($s = 5$) system (a) calculated using eqn (4.4) for applied loading, and (b) calculated using eqn (4.7a) for an effective temperature *drop*. P and Q are of interest because, under the Tresca yield criterion (eqn (4.3)), it is these quantities which determine when flow will occur.

yield stresses. In the example case

$$\Delta T_{\mathrm{esf}} = \frac{P(\sigma^{\mathrm{A}}_{\mathrm{Ycom}} - \sigma^{\mathrm{A}}_{\mathrm{Yten}})}{2Q}$$

$$\approx 3.3(\sigma^{\mathrm{A}}_{\mathrm{Ycom}} - \sigma^{\mathrm{A}}_{\mathrm{Yten}}) \tag{4.10}$$

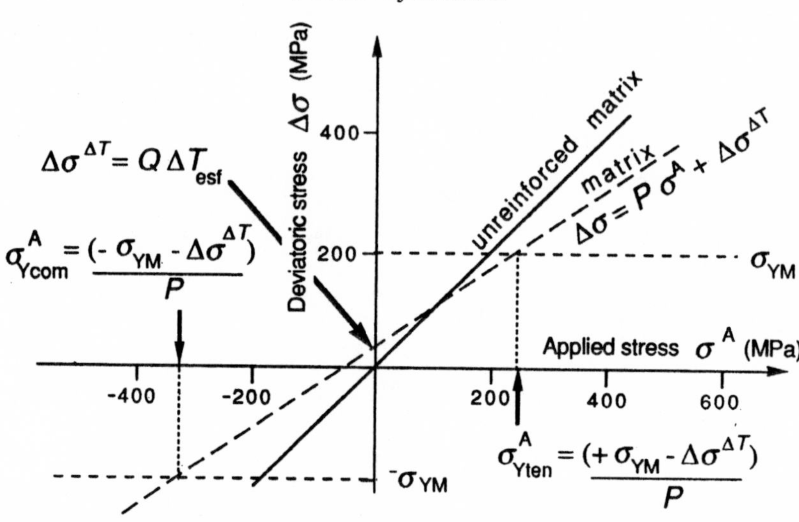

Fig. 4.5 The effect of thermal residual stresses on the yielding behaviour of a fibre composite is best considered in terms of the difference in average matrix stress between the axial and transverse directions ($\Delta\sigma$). As shown in Fig. 4.4 for a temperature *drop*, the stress difference ($\Delta\sigma_M^{\Delta T}$) is *positive* for fibrous reinforcement – corresponding to the axial stress being more strongly tensile than the transverse stress. This raises the overall axial/transverse stress difference $\Delta\sigma$, hastening tensile yielding and retarding compressive yielding. In some cases, the tensile yield stress can be below that of the unreinforced alloy (i.e. the effect of Q (thermal stress) more than offsets the effect of P (elastic load transfer)).

This should be quoted as an 'effective' temperature change, because stress relieving processes can act to reduce its values to substantially below the actual temperature change. Consequently, measurements of ΔT_{esf} can be used to investigate the extent of stress relaxation. Table 4.2 shows values of ΔT_{esf} derived for the two whisker composites of Fig. 4.3. Measurements of thermal residual stress made using the tensile/ compressive asymmetry have been compared with those made using diffraction methods[15,20] and have been found to be somewhat smaller (50–80%). This might be at least partly due to some relaxation of the thermal stresses during yielding.

It is interesting to note that, in contrast to experimental observations and the simple arguments presented above, many finite element models predict no substantial yield stress asymmetry (Fig. 4.6). This may be a consequence of the low work-hardening rates of unreinforced metals on which matrix behaviour has usually been modelled. This leads to a rapid expansion of the plastic zone at very low strains, so that a

Table 4.2 *Average differential thermal contraction stresses in the matrix for two aligned fibre composites*[14,18,19].

Matrix	SiC reinforcement	Theoretical P	Measured asymmetry (MPa)	$\Delta\sigma_M^{\Delta T}$ (MPa)	Theoretical Q (MPa K^{-1})	ΔT_{esf} (K)
Al	3.5% (s = 5)	0.91	25	11.4	0.135	84.4
Mg–11 wt% Li	24% (s = 10)	0.41	50	10.2	0.90	11.4

Note that the yield stress asymmetry is greater for the higher volume fraction composite, but not proportionately so. This is because the Mg–Li matrix is much more prone to stress relaxation[19] and consequently has a much lower effective temperature drop. While this parameter is a useful benchmark for comparing thermal stresses, it may be very difficult to relate it to the actual thermomechanical history, unless the stress relaxation characteristics of the composite are fully established.

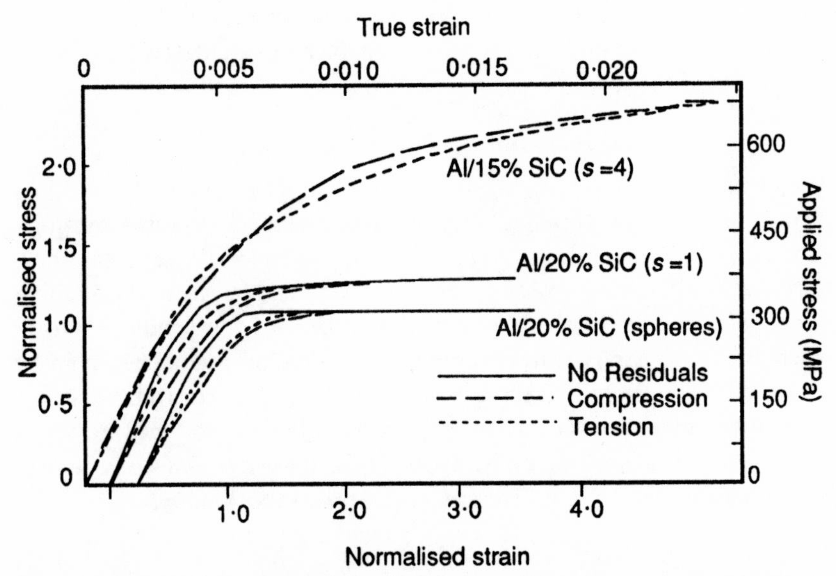

Fig. 4.6 Most finite element models do not predict the tensile–compressive macroyielding asymmetry reported experimentally for whisker composites, and this is well-illustrated by the whisker composite (Al/Sic$_W$ (s = 4), T_{esf} = 350 °C) response predicted by Porvik *et al.*[21] However, differences in microyielding are predicted, because of differences in local flow behaviour; these are even more evident in the aspect ratio 1 cylinder and sphere curves calculated by Zahl and McMeeking[22] (T_{esf} = 256 °C) which, perhaps surprisingly, also show a substantially higher yield stress for unit cylinders than for spheres, despite them having the same length to diameter ratio.

'memory' of the initial thermal residual stress state is quickly lost[21]. While the tensile/compressive yielding asymmetry is difficult to characterise accurately (§11.2), and is sensitive to microvoiding in tension as well as thermal stresses[23], this discrepancy might provide a useful insight into the future development of finite element models. In this vein, the model recently proposed by Levy and Papazian[24] *does* predict a compression/tension asymmetry, at least over the published strain range (0–1%), in the presence of substantial thermal stresses.

Particulate composites

If the reinforcement is equiaxed, then no tensile/compressive asymmetry would be expected under the yield criterion of eqn (4.3), because it is based on the anisotropy in the average matrix stresses (i.e. $Q \rightarrow 0$). Experimental data would seem to confirm this hypothesis. However, difficulties in carrying out accurate plastic strain tests (see §11.2) make detailed comparisons hazardous. Bearing in mind the form of the thermal residual stresses local to the particles, one might expect differences in local plasticity under tension and compression to manifest themselves in the early parts of the loading curve (Fig. 4.6).

The axial–transverse yielding asymmetry

The data in Table 4.1 highlight a further asymmetry in the yielding behaviour, although this has attracted less attention than the tension–compression characteristics. It appears that, during tensile testing of short fibre composites, yielding often occurs more readily when the loading direction is parallel to the fibre axis than when it is perpendicular to it, whereas the reverse is true in compression. This behaviour can also be explained in terms of the thermal residual stresses. For transverse loading, the reinforcement is less effective than for axial loading so that the matrix bears a greater proportion of the applied load (larger P). This can be seen by comparing the stress differences under transverse loading,

$$\Delta\sigma_{M}^{A} = (\bar{\sigma}_{1M} - \bar{\sigma}_{3M}) = P_{trans(1-3)}\sigma^{A} \tag{4.5a}$$

$$\Delta\sigma_{M}^{A} = (\bar{\sigma}_{1M} - \bar{\sigma}_{2M}) = P_{trans(1-2)}\sigma^{A} \tag{4.5b}$$

which are shown in Fig. 4.7, with that for the axial case (Fig. 4.4).

However, as before, thermal stresses must also be considered. For an aligned fibre composite, the matrix is under greater tension parallel to the fibres than normal to them and these stresses must be added to those from the applied load. Returning to our Al–15vol% SiC ($s = 5$) example

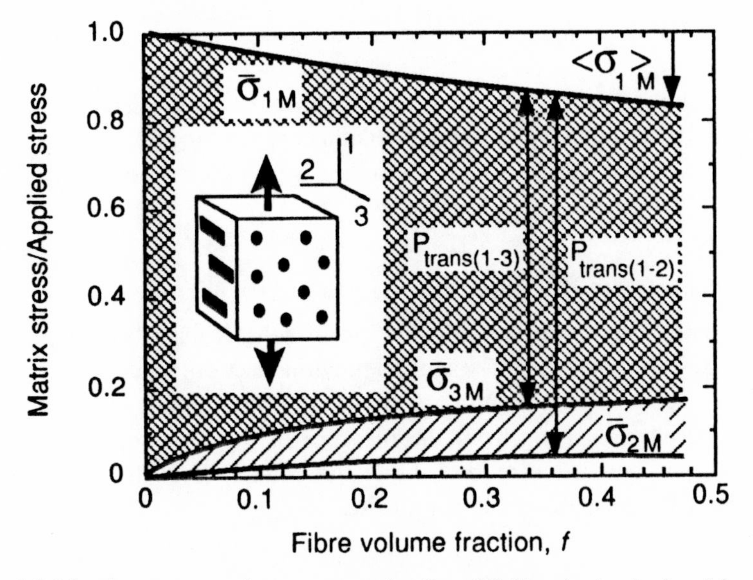

Fig. 4.7 Matrix stresses under transverse loading ('1' direction) calculated for the Al–SiC$_W$ ($s = 5$) system. Unlike the axially loaded case (Fig. 4.4(a)), the two directions normal to the applied load are not equivalent, so that there are two alternative P values (stress difference factors) to consider. Commonly, the stress difference will be greater in the (1–2) plane, since the stress in the '3' direction is raised by inhibition of the matrix Poisson contraction by the fibres.

composite, with $\sigma_{YM} = 200$ MPa and $\Delta T_{esf} = 100$ K, and using the relevant values of P_{axial}, $P_{trans(1-3)}$, $P_{trans(1-2)}$ and Q (0.69, 0.87, 0.94 and 0.43 MPa/K respectively), the applied stresses for the onset of yielding become

$$\begin{array}{llll} & \text{axial testing} & \text{transverse testing} & \\ \sigma^A_{\text{Yten}}: & \dfrac{\sigma_{YM} - Q\,\Delta T_{esf}}{P_{axial}}, & \dfrac{\sigma_{YM} + Q\,\Delta T_{esf}}{P_{trans(1-3)}} \quad \text{or} & \dfrac{\sigma_{YM}}{P_{trans(1-2)}} \\ & = 226\text{ MPa} & = 273\text{ MPa} & = 210\text{ MPa} \end{array} \tag{4.11}$$

These equations show that the yield stress could be higher under axial or transverse loading, depending on the precise level of the thermal stress. The onset of yielding in compression, on the other hand, is given by

$$\begin{array}{llll} & \text{axial testing} & \text{transverse testing} & \\ \sigma^A_{\text{Ycom}}: & \dfrac{\sigma_{YM} + Q\,\Delta T_{esf}}{P_{axial}}, & \dfrac{\sigma_{YM} - Q\,\Delta T_{esf}}{P_{trans(1-3)}} \quad \text{or} & \dfrac{\sigma_{YM}}{P_{trans(1-2)}} \\ & = 350\text{ MPa} & = 180\text{ MPa} & = 210\text{ MPa} \end{array} \tag{4.12}$$

indicating that a higher yield stress is always expected under axial loading.

Even though microstructural factors have not yet been taken into account, these simple yield stress predictions are in broad agreement with the experimental observations of Lederich and Sastry summarised in Table 4.1.

4.2 Onset of yielding: the effect of matrix microstructure

While many aspects of the deformation of MMCs have been explained on the basis of internal stresses in the previous section, assuming a simple elastic–plastic matrix, a more complete picture requires that microstructural effects arising from the presence of the fibres or particles be considered. These often modify the *in situ* behaviour of the matrix, and thereby the properties of the composite.

4.2.1 Dislocation strengthening

The link between dislocation density and strengthening has long been established. In a composite, dislocations can be formed either as a result of applied straining, or through the relaxation of thermal residual stresses. In the latter case, Arsenault and coworkers[25,26], among others, have estimated the strengthening effect by assuming that the misfit strain is relaxed by the punching of dislocation loops (see Fig. 10.1). Arsenault and Shi[25] predict an increase in dislocation density given by

$$\Delta\rho = \frac{\Delta\alpha\,\Delta TNA}{b} \tag{4.13}$$

where $\Delta\alpha\,\Delta T$ is the thermal misfit strain, N the number of particles, b the Burgers vector and A the total surface area of each particle. Miller and Humphreys[27] have obtained a similar expression assuming cube-shaped particles

$$\Delta\rho = 12\frac{\Delta\alpha\,\Delta Tf}{bd} \tag{4.14}$$

where d is the particle size. These models predict that the dislocation density increases with decreasing particle size. This is predicted to be especially marked for particle sizes below about 1 μm. (For a 300 °C temperature drop: $\Delta\rho \sim 10^{13}\,\mathrm{m}^{-2}$ for 5% of 2 μm SiC_p in Al, rising to $\sim 10^{14}\,\mathrm{m}^{-2}$ for 0.2 μm particles). The former model also predicts a decrease in dislocation density as the particles become more spherical[25].

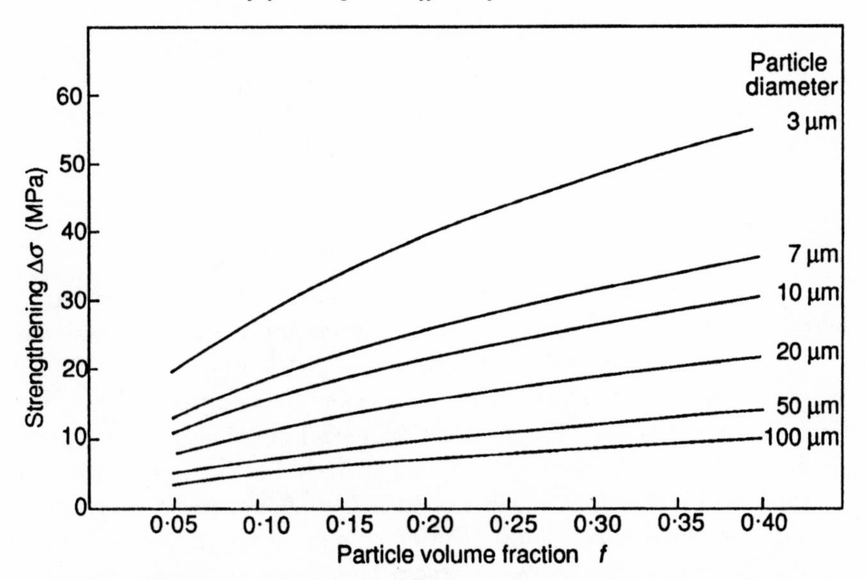

Fig. 4.8 The effect of dispersoid volume fraction and size on the strength increment arising from thermally induced dislocations. The plots were obtained[27] using eqns (4.14) and (4.15), with a temperature drop of 100 °C.

In order to predict the influence of dislocation density on matrix strength, a relation of the type[28]

$$\Delta\sigma_{\mathrm{YM}} \sim Gb\sqrt{\rho} \tag{4.15}$$

is conventionally used. In Fig. 4.8 the predicted dependence of this strengthening increment on inclusion size and volume fraction is shown for Al–SiC; it is clear that, particularly for high volume fractions of fine reinforcement, a significant enhancement of the *in situ* yield stress could arise.

4.2.2 Strengthening from grain size refinement

Discontinuously reinforced MMCs typically have a very fine grain size, much smaller than their unreinforced matrices. Perhaps the simplest way of estimating the strengthening from this effect is to assume that each particle nucleates a single grain (see §10.3). The contribution to the yield stress of this refinement can then be estimated using the Hall–Petch relation[29,30], which relates yield stress enhancement to grain size D

$$\Delta\sigma_{\mathrm{YM}} \approx \beta D^{-1/2} \approx \beta d^{-1/2}\left(\frac{1-f}{f}\right)^{1/6} \tag{4.16}$$

The value of β depends on a number of factors, but is typically[31] around 0.1 MPa $\sqrt{m}$. Given that D might be decreased to around 1 μm for fine reinforcements, this suggests that grain refinement can have a large strengthening effect (~100 MPa). In practice, increases might typically be a few tens of MPa.

4.2.3 Orowan and dispersion strengthening

Orowan strengthening caused by the resistance of closely spaced hard particles to the passing of dislocations is important in many aluminium alloys. It is widely acknowledged, however, that this is not significant for MMCs, because the reinforcement is coarse and the interparticle spacing large[25,32]. For example, 17% of 3 μm SiC_p reinforcement has been predicted[32] to increase the yield stress by only 6 MPa. Furthermore, since the reinforcement is often found to lie on matrix grain boundaries, it is unclear whether the Orowan mechanism can operate at all under these circumstances. Of course, while Orowan strengthening by the reinforcing particles or fibres is small, it is important for age-hardened matrices to consider the effect the reinforcement may have upon the distribution and size of the fine precipitates. For example, increased dislocation density (caused by the accommodation of thermal misfit strain) may accelerate the nucleation of precipitates, reducing the incubation period for the attainment of peak strength (§10.2). This effect is similar to that caused by an increased dislocation density brought about by extensive cold working of the unreinforced metal[33].

4.3 Modelling of bulk plastic flow

4.3.1 The potential for internal stress work-hardening

As Brown[34] points out, models of plastic deformation can be roughly divided into two groups; those with a continuum viewpoint, and those based on dislocation theories. Finite element, Eshelby and other continuum models have the advantage over dislocation-based approaches that they are conceptually straightforward, though the mathematics may be complex. However, it is not possible to include considerations such as dislocation strengthening and dispersion strengthening with an exclusively continuum-based approach. Rather than revert to the complexities of dislocation mechanics, the simplest approach is to include dislocation considerations somewhat empirically within the framework of a continuum model. In order to do this, it is important to understand the relationship between the two viewpoints.

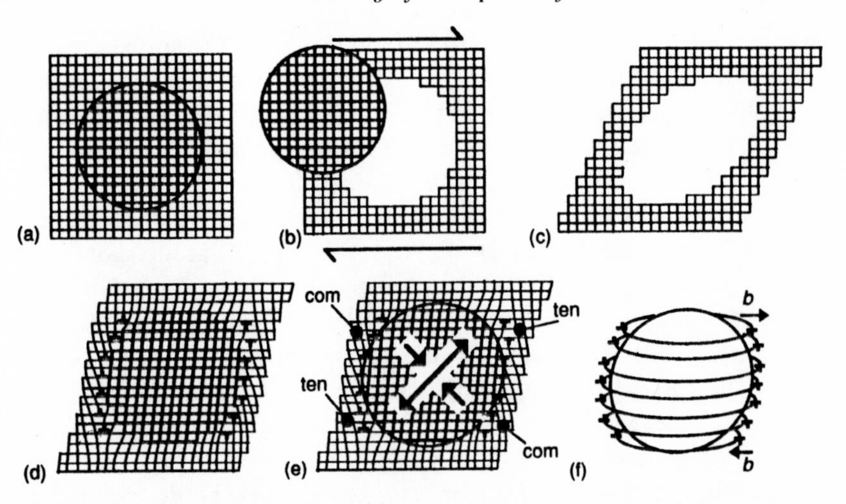

Fig. 4.9 A schematic depiction of the way in which the Eshelby method models the internal stress pattern generated by plastic flow. Here the lines represent atomic planes, although in reality many more lines should be drawn. A uniform plastic deformation of the matrix is assumed, in this case a simple shear (c), to give a new (stress-free) shape. This uniformity of plastic strain is an oversimplification[36], but it allows one to use a simple misfit strain similar to that considered in Fig. 3.3. In order to mate up the two constituents, dislocations must be created near the interface. These contribute to the highly inhomogeneous matrix stress field. This is also expected from the continuum viewpoint, because the shear requires that the spherical particle be fitted into an ellipsoidal hole. This procedure generates a tensile stress along the ellipsoid axis in the particle as well as the nearby matrix.

Provided the reinforcing phase is large enough (≥ 20 nm)[35] for the atomic nature of the two phases to be ignored, then continuum models are more than adequate for the description of completely elastic behaviour. With the onset of plastic flow, dislocation theories require a precise knowledge of the dislocation arrangement in order to calculate the associated stress field. Continuum methods on the other hand, require no such knowledge and yet allow a wide range of materials properties to be analysed solely in terms of matrix flow criteria and the transfer of load via the plastic misfit between matrix and inclusion. To see how the two approaches interrelate, it is helpful to consider an example and to explain the resultant stress field from the two viewpoints.

Consider a composite containing spherical particles before and after being subjected to a shear stress. Prior to loading, the composite is unstressed, so there are no shape misfits or dislocations (Fig. 4.9(a)). If the particle could be dissolved away after plastic shear loading, then the shape of the cavity within the matrix would be different from the original shape, Fig. 4.9(c). Eshelby-type[37,38] continuum approaches model this shape change by adopting the concept of an effectively homogeneous plastic strain in the matrix, i.e. the matrix is assumed to deform uniformly throughout in a manner similar to that of a matrix containing no reinforcing particle. In view of the highly variable state of stress around the particle prior to the onset of flow, this simplification should be regarded with suspicion, but this approach has been justified experimentally for dispersion strengthened materials[36,39] and continuous fibre composites[12]. This assumption has the advantage that the internal stress field can be derived in terms of the shape misfit between the spherical particle and the ellipsoidal cavity, using the approach of §3.2. Since this is clearly the largest misfit that could be generated, it corresponds to an upper bound for load transfer and hence work hardening. Other continuum approaches (e.g. FEM) assume or predict a different pattern of flow around the fibre, but in each case the stress field is essentially a result of a shape misfit.

From a dislocation viewpoint, the uniform shear depicted in Fig. 4.9(c) is achieved by the passing of nine dislocations through the matrix from one side to the other. Because the particle is non-shearable, seven of these dislocations are held up and the only way they can reach the other side of the composite is by leaving an Orowan loop around the particle (Fig. 4.9(d),(e),(f)). The resulting dislocation array forms a series of 'shear dipoles'[34], which has a local component which averages to zero and a long-range component which can be thought of as arising from the introduction of a kink (low angle grain boundary). Since the angle of this kink is independent of the precise arrangement of the dislocations, i.e. the number on each slip line, the long-range matrix stress will be largely independent of the number of dislocations per pile-up. Thus, provided a reasonable number of slip lines intersect the particles, the plastic deformation can be smeared out and the number of loops related to the plastic distortion in the Eshelby approach[40].

As with the continuum viewpoint, this initial dislocation structure gives rise to the maximum possible matrix/inclusion misfit. The possibility of dislocation rearrangement so as to reduce the misfit, and hence the work-hardening capacity, will be discussed in the next section. Finite element models take these processes together in that they do not distinguish between the primary dislocation structures caused by the external constraint and those which are internally generated so as to bring about a reduction in internal stress. Both occur simultaneously under the chosen yield criterion. Consequently, they predict a reduced misfit (work-hardening rate) brought about by a complex local flow pattern driven by the high local stresses that would otherwise occur (Fig. 4.9(e)). In the following, the Eshelby approach is used.

An Eshelby model of plastically induced stresses

As depicted in Fig. 4.9, provided plastic flow is assumed to occur uniformly throughout the matrix, the resulting misfit between the hole and reinforcing particle can be represented by a misfit strain similar to that used in §3.2. This transfers load strongly to the reinforcement – see Fig. 4.10. Neutron diffraction evidence[41] suggests that very large inclusion stresses are indeed generated under plastic loading, and these must inevitably result in very large matrix stresses local to the whisker ends. One reason why such high local stresses are to a large extent sustained is the large hydrostatic component generated at the fibre end, but in any case, we have already seen that the highest stresses can be relaxed locally without greatly affecting the onset of global yielding. Neglecting these local effects, the plastic misfit can be treated as a transformation strain, similar to that arising from thermal expansion[11]

$$\varepsilon^{T*} = -\varepsilon_M^P(-\tfrac{1}{2}, -\tfrac{1}{2}, 1, 0, 0, 0) \tag{4.17}$$

where ε_M^P is the global plastic strain undergone by the matrix in the direction of loading† ('3'-axis). Inserting this matrix/inclusion misfit into eqns (3.21) and (3.24) gives the mean stress component arising from plastic flow

$$\langle\sigma\rangle_M^P = -fC_M(S - I)\{(C_M - C_I)[S - f(S - I)] - C_M\}^{-1}C_I\varepsilon_M^P \tag{4.18}$$

This equation indicates that the transfer of load to the reinforcement with plastic straining is a linear function of the matrix plastic strain, and that the rate of load transfer increases with increasing volume fraction and inclusion aspect ratio. This gives rise to a high rate of work-hardening, even when the matrix itself exhibits no inherent work-hardening capacity. To illustrate this, consider the example Al–15 vol% SiC ($s = 5$) composite of §4.1. Analogous to the factors P and Q used to quantify the stiffness and thermal load transfer terms, a parameter W (internal stress difference per unit plastic strain) representative of the plastic component can be defined

$$\Delta\sigma_M^P = \langle\sigma_3\rangle_M^P - \langle\sigma_1\rangle_M^P = -W\varepsilon_C^P \tag{4.19}$$

† Note that this differs from that for thermal expansion in that the tensor is no longer isotropic. Here ε_3 is equal to $-\varepsilon_M^P$, while ε_1 and ε_2 are equal to $\tfrac{1}{2}\varepsilon_M^P$, as required to conserve matrix volume.

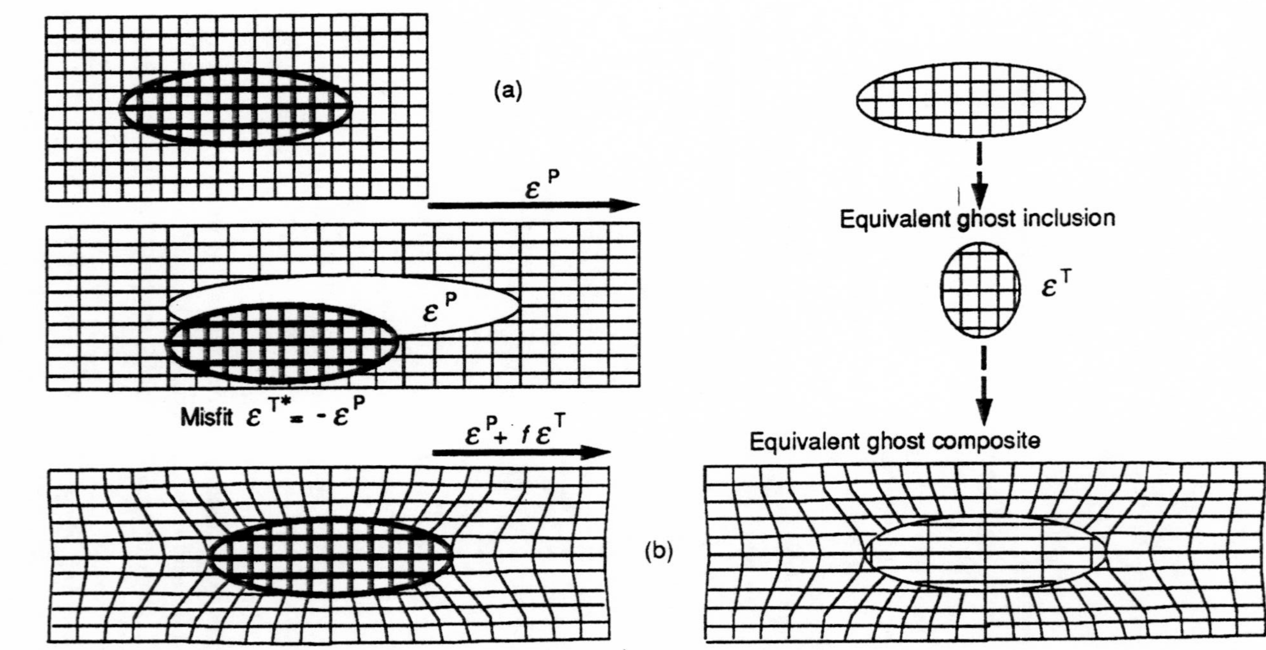

Fig. 4.10 A displacement map showing the inclusion and matrix shapes (a) prior and (b) subsequent to tensile plastic deformation of the composite. After plastic flow of the matrix, the inclusion is residually in axial tension and the matrix in overall compression. As both phases are in tension during tensile loading, plastic flow sets up a hole/inclusion misfit which reduces the average matrix stress and increases the stress borne by the reinforcement. This gives rise to an increase in the composite yield stress as plastic flow progresses, i.e. to work-hardening, even when the matrix has no inherent capacity to work-harden. The plastic strain in the matrix (ε_{M}^{P}) is greater than in the composite as a whole ($\varepsilon_{M}^{P} + f\varepsilon^{T}$).

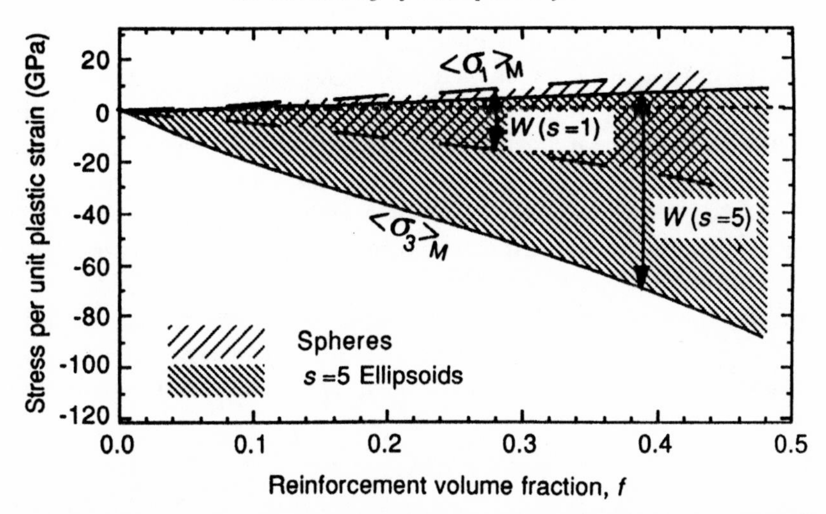

Fig. 4.11 The variation of average internal matrix stress with fibre volume fraction arising from matrix plasticity. The stresses shown correspond to an Al–SiC system. Because the Tresca yield criterion depends on the average matrix stress difference, W is the important parameter. The sense of this internal stress parameter is to reduce the matrix stress so that the applied load can be increased before further yielding occurs under the yield criterion of eqn (4.2). Increases in W with volume fraction lead to enhanced load transfer and work-hardening rate. The rate at which W increases with volume fraction is greater at higher reinforcement contents since the matrix strain, which determines the misfit (eqn (4.17)), increases at a faster rate relative to the overall composite strain as the volume fraction is increased.

with the composite plastic strain given by

$$\varepsilon_{\mathrm{C}}^{\mathrm{P}} = \varepsilon_{\mathrm{M}}^{\mathrm{P}} + f\varepsilon^{\mathrm{T}} \; (\approx (1-f)\varepsilon_{\mathrm{M}}^{\mathrm{P}} \; \textit{at low volume fractions})$$

The dependence of $\langle\sigma_3\rangle_{\mathrm{M}}^{\mathrm{P}}$ and $\langle\sigma_1\rangle_{\mathrm{M}}^{\mathrm{P}}$, and hence W, on fibre volume fraction is shown in Fig. 4.11 for Al–SiC composites. It follows that, in the absence of thermal stresses, matrix yielding will continue under conditions such that, (eqn (4.3)),

$$P\sigma^{\mathrm{A}} - W\varepsilon_{\mathrm{C}}^{\mathrm{P}} = \sigma_{\mathrm{YM}} \tag{4.20}$$

so that the work-hardening rate can be written as

$$\frac{\partial \sigma^{\mathrm{A}}}{\partial \varepsilon_{\mathrm{C}}} = \left(E_{\mathrm{C}}^{-1} + \frac{P}{W}\right)^{-1} \tag{4.21a}$$

and

$$\frac{\partial \sigma^{\mathrm{A}}}{\partial \varepsilon_{\mathrm{C}}^{\mathrm{P}}} = \frac{W}{P} \tag{4.21b}$$

with respect to the total, and solely plastic, composite strains. It can be inferred from Fig. 4.11 that the value of W for our example Al–15 vol% SiC ($s = 5$) system is about 30 GPa. Given that the corresponding value of P was 0.695, the theoretical work-hardening rate according to eqn (4.21) is about 43 GPa (i.e. about one-third of the elastic modulus). Comparing this with a typical maximum value for a high strength unreinforced Al alloy of around a few GPa, it can be seen that this is potentially a highly significant effect. Clearly any inherent work-hardening exhibited by the matrix as a result of microstructural changes (§4.3.2), will supplement that of eqn (4.21). This is readily incorporated into this analysis, provided the relationship between matrix strengthening, $\Delta\sigma_{YM}$, and ε_M^P is known. In this case the matrix plasticity condition becomes

$$P\sigma^A - W\varepsilon_C^P = \sigma_{YM}(\varepsilon_M^P) \tag{4.22}$$

which, for low reinforcement volume fractions ($\varepsilon_C^P \approx (1-f)\varepsilon_M^P$ for $f < 10\%$), gives a work-hardening rate

$$\frac{\partial\sigma^A}{\partial\varepsilon_C^P} = \frac{W}{P} + \frac{1}{P(1-f)}\frac{\partial\sigma_{YM}}{\partial\varepsilon_M^P} \tag{4.23}$$

Of course, if the *in situ* matrix work-hardening rate changes with plastic strain, then so will the overall rate. In any event, since the first term (load transfer) dominates in all but very dilute composites, *unrelaxed* plastic flow should give rise to approximately linear work-hardening. Humphreys[42] suggests that the unrelaxed linear work-hardening regime discussed above should typically extend to a plastic strain of only 0.05%, beyond which a great number of secondary dislocations (§4.3.2) would be produced. Experimental evidence also suggests that, while long fibre composites exhibit a prolonged linear work-hardening regime[43], any linear work-hardening in short fibre or particulate systems is extremely short-lived[44]. This is a consequence of pronounced stress relaxation limiting the load transfer and reducing the extent to which the full work-hardening capacity is realised (see §4.4).

4.3.2 Microstructural work-hardening

In unreinforced alloys, the stimulation of long-range dislocation glide during plastic straining is usually accompanied by a gradual increase in the yield stress. In view of the interruption of flow caused by the reinforcement in a composite, one might expect dislocation generation to be accentuated and for the composite matrix to show increased work-hardening. A picture of dislocation structure development during composite straining is outlined in §10.1. Essentially, with increasing plastic strain the dislocation density around the reinforcement increases, first ($\varepsilon_M^P < 3\%$) in a localised manner[45] (see Fig. 10.3), and then extending further into the matrix. This, and the associated development of subgrain structure, can be understood, first in terms of the primary dislocation structures formed about the particles, and then through the secondary dislocation structures formed upon their relaxation. There are a number of ways in which the resulting dislocations and dislocation structures can increase the matrix yield strength (reviewed by Kelly[46]) and these are discussed below.

Forest hardening

Forest hardening arises from the inhibiting effect of forests of secondary dislocations about the reinforcement. These tangles work-harden the matrix locally and inhibit further relaxation of the inclusion stress by dislocation motion. A rough estimate of this effect can be made by assuming the tangle volume to be approximately independent of the plastic strain[39]; the dislocation density, expressed as the number of dislocations per unit volume, N, is then proportional to the plastic strain

$$N \sim \frac{f\varepsilon_M^P}{bd^2} \tag{4.24}$$

where b is the Burgers vector and d the particle size. This relation is insensitive to details of the model, i.e. the interaction of a dislocation with this grouping is more or less the same whether the dislocation structure around the particles comprises simple dislocation loops or a dislocation forest. Relating the flow stress to this dislocation density, the hardening increment is approximately[39]

$$\Delta\sigma_{YM}^{fh} \cong 0.2G_M\sqrt{\frac{fb\varepsilon_M^P}{d}} \tag{4.25}$$

where G_M is the matrix shear modulus. For 20 vol% of 3 μm particles, this mechanism raises the yield stress by ~2 MPa at a strain of 1%.

Source shortening

Naturally, the matrix stress is highly non-uniform, and the stress field in the vicinity of the reinforcements is such that dislocations are repelled away from them, towards the intermediate regions[47]. Brown and Stobbs[39] have estimated the degree to which the field near the inclusions causes an approaching dislocation to stand off, so as to reduce their effective spacing and thus the matrix volume in which flow can occur. Because of the large interparticle spacing typical of MMCs, one would not expect the source shortening contribution $\Delta\sigma^{ss}$ from the reinforcing particles to be large. However, Pedersen[43] has suggested that this mechanism is responsible for the high rate of linear *matrix* work-hardening observed for the Cu–W and Al–W long fibre systems.

Combining the strengthening terms

Several schemes for the addition of the various strengthening terms have been proposed, the most commonly accepted being that reviewed by Lilholt, which differentiates between the range of the strengthening contributions and their relative strengths[48]. Strictly, computer models are required to sample statistically the different contributions[49,50]. Clearly, stress contributions which act more or less uniformly throughout the matrix must be superposed linearly, whereas mechanisms of similar strengthening ability, which act unevenly throughout the matrix (e.g. the source shortening), are most suitably combined as the root of the sum of the squares

$$\Delta\sigma = \sqrt{\Delta\sigma_1^2 + \Delta\sigma_2^2} \qquad \text{for } \Delta\sigma_1 \sim \Delta\sigma_2$$

Bearing these statements in mind perhaps the most suitable expression is of the form[51,52]:

$$\Delta\sigma_{YM} = \sqrt{[\Delta\sigma^{ss}(\varepsilon_M^P) + \Delta\sigma^{Orow}]^2 + \Delta\sigma^{fh}(\varepsilon_M^P)^2} \tag{4.26}$$

In any event, it is worth noting that the net contribution from these effects is expected to be small compared with the capacity for work-hardening via the plastically induced misfit. Consequently, it is the capability of *stress relaxation processes* to reduce that misfit which is crucial in determining the work-hardening behaviour of discontinuous MMCs.

4.4 Internal stress relaxation

As we have seen in the previous section, global plastic flow causes the reinforcement to become highly stressed, giving rise to a high rate of work-hardening. Eventually this build-up becomes limited by processes which act to reduce the stress within and around the reinforcement. These processes are driven by a reduction in energy of the system as a whole. The actual mechanisms involved range from diffusional mass transport to interfacial sliding and cavitation. They can be important, not only in changing the deformation behaviour, but also in influencing the onset of fracture. In this section, the nature of stress relaxation is first examined, before going on to discuss the individual mechanisms. The first part of this treatment is conceptually a little abstruse, but the consequences are straightforward and the physical processes readily understood when the driving force is considered in terms of a misfit strain.

4.4.1 Energy minimisation: the driving force for relaxation

During plastic flow, the internal stresses between the matrix and reinforcement increase. Relaxation processes lower the internal strain energy and cause the work-hardening rate to depart significantly from a linear response (Fig. 4.12). In order to understand these processes, it is helpful to consider the energy changes which occur when relaxation takes place. To do this we shall use Eshelby's model, not so much to produce quantitative predictions, but rather to obtain a simple pictorial insight into the phenomena concerned.

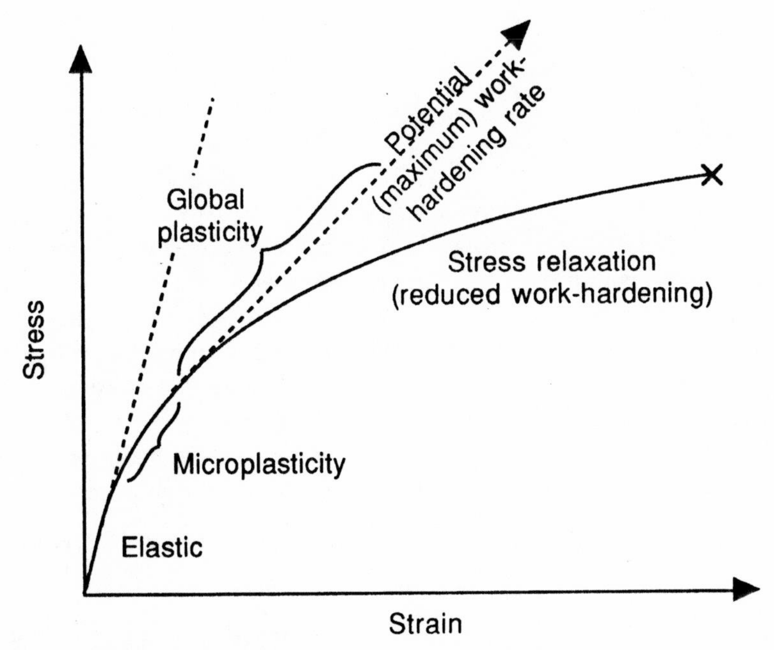

Fig. 4.12 Schematic showing the different regimes of the stress–strain curve. Note that the unrelaxed work-hardening regime is indistinct, because of the onset of relaxation mechanisms.

Consider first the stressing by σ^A of a single phase medium, in terms of a reduction in potential energy of the load as it descends during extension and the increase in elastic energy stored in the material

$$\begin{aligned} \textit{Energy} &= \textit{Potential Energy Change} + \textit{Elastic Strain Energy} \\ U &= \quad -\sigma^A\varepsilon^A \quad + \quad \sigma^A\varepsilon^A/2 \\ &= \quad -\sigma^A\varepsilon^A/2 \end{aligned} \tag{4.27}$$

Since it is rather difficult to calculate directly the elastic strain energy of a composite comprising two constituents of different stiffness, consider the energy of the 'ghost' composite used in §3.2. Eshelby[53] has calculated the internal strain energy of a misfitting (by ε^T) inclusion within a matrix of the same elastic constants

$$U_{Ghost}(\textit{Unloaded}) = -f\langle\sigma\rangle_I\varepsilon^T/2 \tag{4.28}$$

where $\langle\sigma\rangle_I$ is the internal stress in the inclusion. This corresponds to our 'ghost' composite, and since it behaves as a single phase medium under the application of a load, the increase in strain energy must be given by eqn (4.27). The potential energy change is slightly different, because the load is suspended at a different height ($-\varepsilon^A - f\varepsilon^T$, rather than $-\varepsilon^A$)

$$\begin{aligned} \textit{Energy} &= \textit{Potential Energy Change} + \textit{Elastic Strain Energy} \\ U_{Ghost}(\textit{Loaded}) &= \quad -(\varepsilon^A + f\varepsilon^T)\sigma^A \quad + \sigma^A\varepsilon^A/2 - f\langle\sigma\rangle_I\varepsilon^T/2 \\ &= -\sigma^A\varepsilon^A/2 - f\varepsilon^T\sigma^A - f\langle\sigma\rangle_I\varepsilon^T/2 \end{aligned} \tag{4.29}$$

This expression for the energy of the 'ghost' composite agrees with that derived by Pedersen and Brown[54]. Since the displacements of the load and the strains within the matrix are the same as for the real composite, the only difference between the two systems arises from the disparity between the strain energies of the inclusions. This occurs because in §3.3 and §3.4 the 'ghost' composite was chosen to imitate the stress state of the real composite. However, in order to generate the same *stresses*, the 'ghost' inclusion must be *strained* to a larger extent than the real one. This means that the strain energies of the inclusions are different

$$\begin{aligned} U_{Real}(\textit{Loaded}) &= U_{Ghost}(\textit{Loaded}) + \textit{Strain Energy of Real Inclusion} \\ &\quad - \textit{Strain Energy of Ghost Inclusion} \end{aligned}$$

$$U_{Real}(\textit{Loaded}) = -\sigma^A\varepsilon^A/2 - f\langle\sigma\rangle_I\varepsilon^{T*}/2 - f\sigma^A(\varepsilon^T + \varepsilon^{T*})/2 \tag{4.30}$$

Now the potential for relaxation, driven by an overall reduction in energy (i.e. a negative ΔU), arises from the possibility of changing (or generating) the misfit between the two phases ($\Delta\varepsilon^{T*}$). By considering changes in the misfit it can be shown that[14]

$$\Delta U_{\text{Real}} = -f(\sigma^A + \langle\sigma\rangle_I + \Delta\langle\sigma\rangle_I/2)\,\Delta\varepsilon^{T*} \tag{4.31}$$

While the form of eqn (4.31) seems rather daunting, it can be understood in quite simple terms. The first term is dominated by changes in the potential energy (which occur when the misfit between the matrix and inclusion is changed), while the second and third terms are concerned predominantly with changes in the strain energy. In order to get a feel for how relaxation might occur, it is helpful to consider a number of examples.

Relaxation in the absence of applied loading

In this case $\sigma^A = 0$ and so the system can only decrease in energy by a reduction in internal strain energy. This is shown schematically in Fig. 4.13.

Relaxation under applied loading

Whilst in theory it is possible for the composite to increase its potential energy (e.g. to contract under a tensile load) or to increase its strain energy,

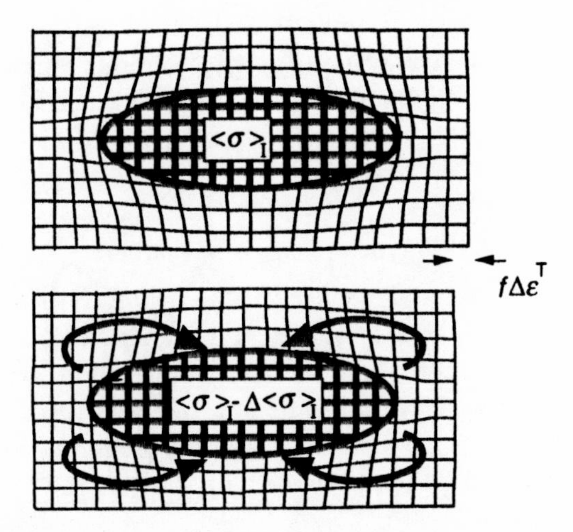

Fig. 4.13 Relaxation (with no applied load) takes place so as to decrease the internal strain energy. Here this is achieved by the local movement of matrix material (perhaps by local dislocation movement or by (bulk or interfacial) diffusion) so as to reduce the misfit. Because the net tension in the more compliant phase is reduced, this process is also accompanied by an overall contraction of the composite.

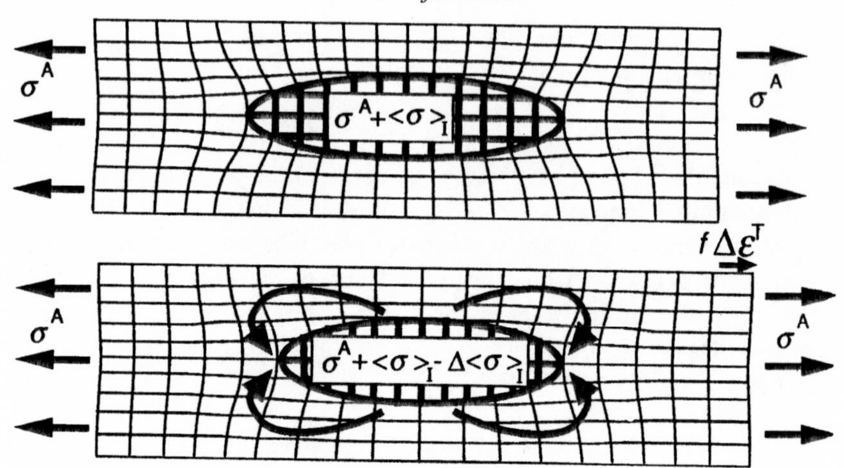

Fig. 4.14 Relaxation within a loaded composite can take place in two ways. Here the strain energy is reduced by a reduction of the internal stress-related misfit and the potential energy is reduced by extension of the composite. This extension occurs because, subsequent to relaxation, the stiffer phase bears a lower proportion of the load.

provided the overall change in energy is negative, the most common situation is that shown in Fig. 4.14, where both the internal strain energy and the potential energy decrease. Usually relaxation occurs by mass transport within the matrix phase (Fig. 4.14), although processes such as fibre fracture and cavitation must also be regarded as relaxation processes since they bring about a reduction in the energy of the composite.

4.4.2 Micromechanisms of relaxation

Whilst relaxation is thermodynamically possible only if there is an overall reduction in free energy, it is important to remember that alone this does not mean that relaxation will take place. Relaxation also requires a suitable mechanism which can operate at a sufficient rate. The following mechanisms can reduce the load carried by the reinforcement, and hence can be classified as relaxation processes:

- o inclusion fracture
- o interfacial debonding/sliding
- o matrix cavitation

- • dislocation motion and rearrangement
- • diffusion
- • structural transformation
- • recrystallisation

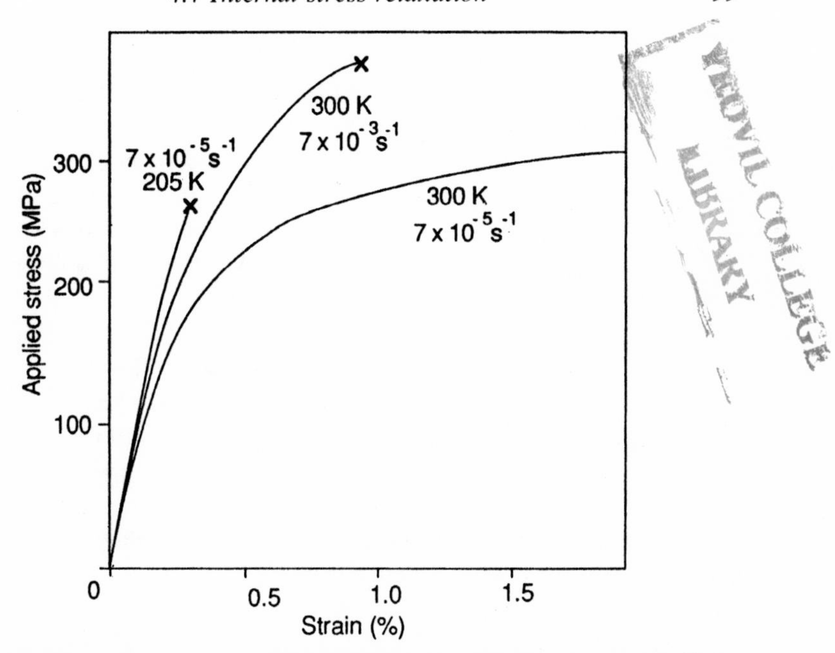

Fig. 4.15 Stress–strain curves for Mg–11 wt% Li/24 vol% SiC_W composite[55]. At room temperature this matrix is very prone to diffusion-assisted relaxation. This relaxation process is limited by lower temperatures and faster strain rates. The concomitant increase in load transfer to the reinforcement gives rise to higher rates of work-hardening. Because diffusive relaxation reduces the high local matrix stresses non-catastrophically, reduced diffusive relaxation results in lower ductility through the earlier activation of more detrimental processes.

The mechanisms represented by open symbols are catastrophic relaxation processes and as such are discussed in §7.2 in connection with composite failure, while the remainder are discussed below. For any matrix/reinforcement system, the relative incidence of all these processes is a complex function of variables such as temperature, inclusion aspect ratio, the applied load and the rate at which it is applied. As these processes affect both load-bearing capacity at a given strain and the onset of component failure, it is important to identify the individual mechanisms so as to be able to exert some degree of control over their relative predominance. For example, diffusion-related relaxation can be limited by low temperatures and high strain rates and, as Fig. 4.15 shows, this increases the work-hardening rate, but decreases ductility. There is also considerable scope for controlling certain relaxation processes by restructuring the interface.

Dislocation motion and rearrangement

In order to avoid confusion between plastic relaxation and unrelaxed bulk plastic flow, it is helpful to draw a distinction between repeated long range dislocation glide in the matrix, which *increases* the matrix/inclusion misfit via the deposition of Orowan loops around the reinforcement (Fig. 4.9), and local rearrangements of dislocation structures, which *decrease* the misfit. For example, the punching of dislocation loops, used in §4.2.1 and §10.1 to explain the thermal stress induced stimulation of dislocations, reduces the misfit by transporting discs of vacancies or extra atoms to or from the inclusion[52,56] (Fig. 4.16 and Fig. 10.1). Other secondary dislocation structures are possible, such as secondary shear loops which can subsequently cross-slip[58]. From a continuum viewpoint, these structures reduce energy by reducing the internal strain energy, while at a dislocation level this can be explained in terms of a decrease in the long-range stress because the new configurations have shorter-range stress fields. Such processes are energetically favoured provided the reduction in

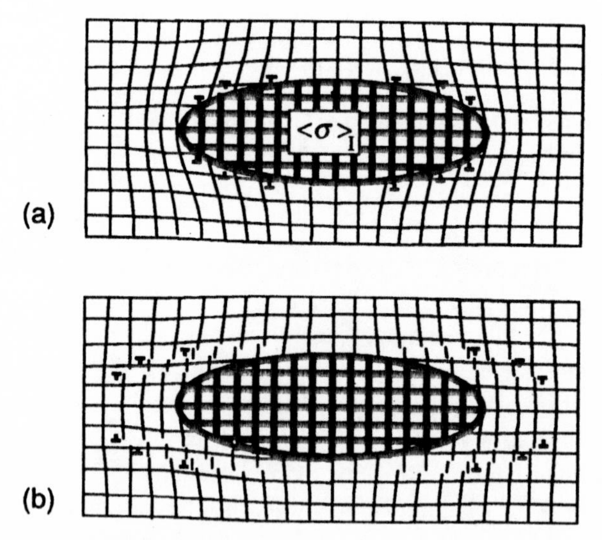

Fig. 4.16 Provided the energetics are favourable, a thermal strain misfit represented as surface dislocations (a) can be punched out into the matrix (b). This reduces the elastic stress field, but at the expense of increased work done by moving the dislocations. In this illustration the prismatic dislocation loops have cut the grid lines in the displacement map as they have been punched out, giving rise to a region of reduced strain relative to (a). The array is similar to the ellipsoidal array envisaged in the Taya and Mori model[57].

elastic energy is greater than the effects of the increased dislocation line length.

Taya and Mori[57] have looked at the energetics of prismatic punching as a means of relaxing the thermal stress. This was done by representing the thermal misfit strain as a series of surface dislocations (see Fig. 4.16), which (provided it is energetically favourable) are punched into the matrix, thus smearing out the thermal misfit. This model[57,59] predicts that at large fibre aspect ratios this process is energetically inhibited (i.e. the punching distance tends to zero), while the shear lag stress-based analysis of Dunand and Mortensen[60] suggests that at high aspect ratios dislocation loops are produced, but that their number reaches a plateau.

It may be noted that, in the context of relaxation by dislocation motion, the ability to cross-slip is often important and hence stacking fault energy (SFE) is often a relevant parameter (particularly at relatively low temperature where climb is inhibited). Aluminium has a high SFE and cross-slips easily, unlike copper, for example. This may be a factor favouring the observed tendency of aluminium alloy MMCs to exhibit relatively pronounced relaxation effects, even at low temperatures.

Diffusion

A number of diffusion-related processes lead to a lowering of energy. Three will be discussed here:

- *Interfacial diffusion* – for most systems the metal/ceramic boundary is disordered and as such acts as a low energy channel for the movement of atoms. Gradients of hydrostatic stress provide a sustained driving force for vacancy transport (Fig. 4.17). Because only local rearrangements can be brought about by interfacial movement, this process cannot completely relax a dilatational misfit, but it can reduce the especially high stresses which might arise in certain directions at the expense of others. It is quite simple to show[61] that when interfacial diffusion is complete only hydrostatic stresses remain within the reinforcement. In other words, only deviatoric misfits can be completely relaxed by interfacial movement.
- *Volume diffusion* – at higher temperatures, or when interfacial diffusion is impossible or complete, volume diffusion can become important (Fig. 4.17). For spherical particles, it is the only diffusive mechanism available for the relaxation of thermal stresses. In addition to the larger activation energy required for volume diffusion, the greater

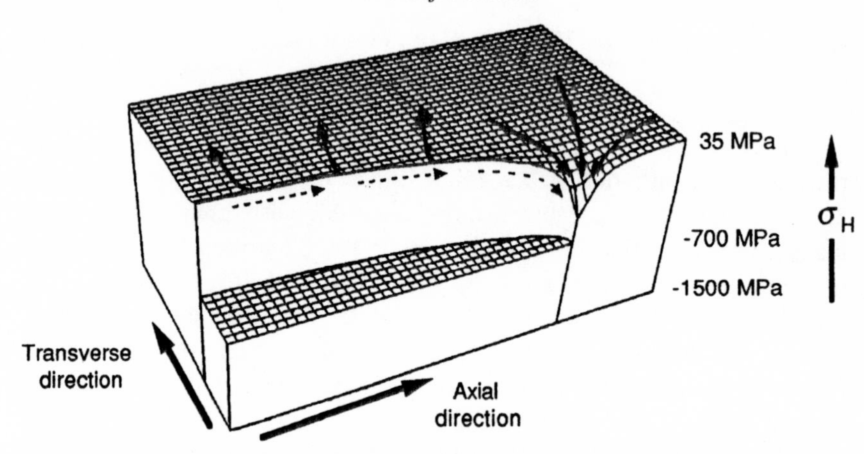

Fig. 4.17 The variation in the hydrostatic component of the stress field upon cooling an Al–SiC ($s = 5$) whisker reinforced composite by 300 °C. Note how variations around the inclusion circumference drive vacancies (atoms) along the interface towards (away from) the fibre end. Local diffusion through the volume of the matrix is also indicated. To completely relax the compressive fibre stress requires the long-range diffusion of vacancies (atoms) towards (away from) the inclusion.

distance required for the relaxation of hydrostatic stresses means that relaxation times are expected to be considerably longer.

- *Diffusion-assisted dislocation climb* – interaction between diffusive processes and dislocation motion covers a wide range of possibilities. For example, dislocation rearrangement (plastic relaxation) can be assisted in this way. This is especially important in connection with elevated temperature creep because it limits reinforcement stressing via removal of piled-up dislocations. Conversely, pipe diffusion along dislocation cores can contribute to mass transport processes. As both pipe and volume diffusion can produce climb, it may be difficult to identify single activation energy values if this is a dominant relaxation mechanism.

Structural transformation

The majority of the relaxation phenomena discussed above have focused on lowering the inclusion stress by changes taking place in the matrix. If the energetics are favourable, it is possible for the reinforcement to transform structurally to another phase with an associated change in shape. Though not yet widely explored for metal-based composites[62,63], structural transformations are widely used in the toughening of ceramic matrix composites.

Transforming particles could be exploited for MMCs in two ways: either they could be incorporated as a highly dispersed secondary phase so as to reduce stresses in areas of the matrix where they had reached the transformation threshold, or they could function as the primary reinforcement, the transformation being triggered when the stress field they themselves had created became sufficiently large. In this case the energy expression of eqn (4.31) becomes

$$\Delta U_{\mathrm{Real}} = -f(\sigma^{\mathrm{A}} + \langle\sigma\rangle_{\mathrm{I}} + \Delta\langle\sigma\rangle_{\mathrm{I}}/2)\,\Delta\varepsilon^{\mathrm{T}^*} + \Delta U_{\mathrm{Trans}} \qquad (4.32)$$

Recrystallisation

Under some circumstances, the region around a reinforcing inclusion may recrystallise during thermomechanical loading. Since a reconstructive process is taking place at the recrystallisation front, there is considerable scope for a change in the natural misfit during this process. Of course, at the same time the hardness of the matrix itself undergoes a sharp decrease as the dislocation density falls. There has been relatively little attention paid to recrystallisation as a stress relaxation process in MMCs, but it is well-established[64] that in some circumstances it can occur very readily. This depends on metal purity, the presence of fine (Zener pinning) particles, the temperature and the strain history. Strain localisation in the matrix around the reinforcement facilitates the formation of recrystallisation nuclei, giving a marked dependence on volume fraction and distribution homogeneity of the ceramic[65]. While it would be unusual for recrystallisation to take place during room temperature mechanical testing, it can be important during elevated temperature loading and during fabrication and forming procedures (see Chapters 5, 9 and 10).

4.5 Reduced internal stress work-hardening

While there is a wide variety of mechanisms by which relaxation can occur, these have the common feature of reducing the load borne by the reinforcing phase. In this section we shall look at their influence on the shape of the stress–strain curve. As was pointed out in §4.4.2, relaxation is generally achieved by a reduction in the misfit between the 'natural' shapes of the hole and inclusion, whether it be gradual (diffusive, dislocation movement, etc.) or catastrophic (inclusion fracture or transformation, matrix cavitation, etc.). The original work on plastic relaxation

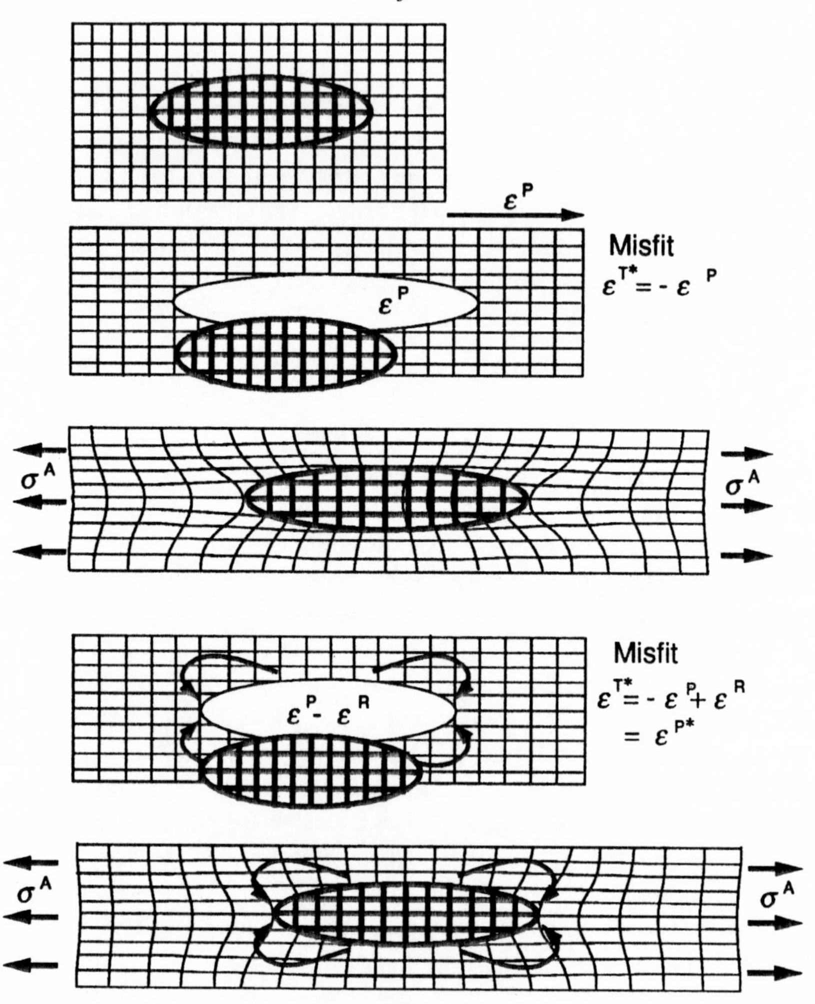

Fig. 4.18 Schematic showing the partial inelastic accommodation of the plastically induced shape misfit between matrix and inclusion.

in two-phase systems[58] used the Eshelby description to work through the change in load partitioning that takes place. We will use this method because of the simple manner in which it highlights a number of features common to MMC stress–strain curves.

As depicted in Fig. 4.18, relaxation usually reduces the shape misfit between the matrix and inclusion. At large plastic strains, the unrelaxed misfit is mainly of plastic origin and has the form given by eqn (4.17) and

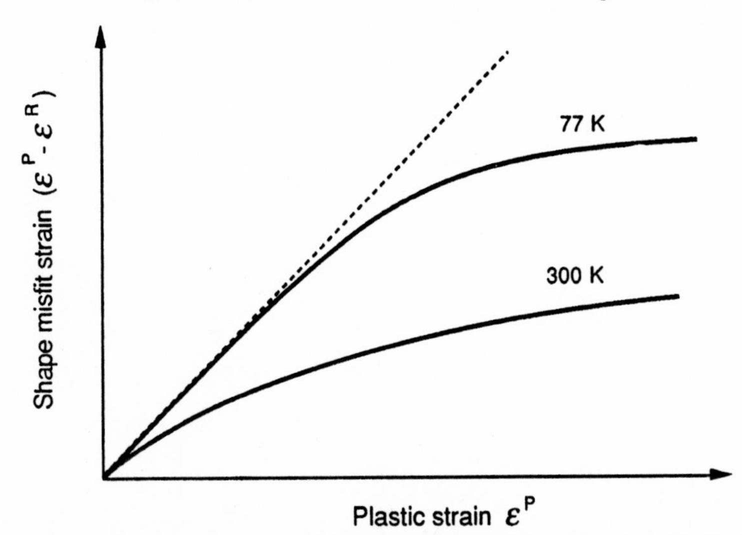

Fig. 4.19 The experimentally determined relationship between the total plastic strain and the internal matrix/inclusion shape misfit for Cu strengthened with a dispersion of SiO_2[39]. Only at 77 K is there a significant linear regime. This would suggest that for Al ($T_{mp} \sim 930$ K) reinforced with a high volume fraction of SiC particles, one would expect the misfit to increase with straining but at a much reduced rate.

the net misfit after relaxed plastic flow is given by

$$\varepsilon_M^{P*} = \varepsilon_M^P - \varepsilon_M^R \tag{4.33}$$

Experimentally it has been found that, for dispersion hardened copper, the relation between the unrelaxed and the relaxed misfit has the form shown in Fig. 4.19. If the net misfit for Al–SiC had a similar form, then empirically

$$\varepsilon_M^{P*} \sim \sqrt{\varepsilon_M^P} \tag{4.34}$$

giving the plastic component of the internal stress (cf. eqn (4.18))

$$\langle\sigma\rangle_M^P = -fC_M(S - I)\{(C_M - C_I)[S - f(S - I)] - C_M\}^{-1}C_I\varepsilon_M^{P*} \tag{4.35}$$

Such a relation agrees quite well with the results shown in Fig. 4.20 (and Fig. 4.24(a)).

Assuming that the misfit develops according to eqn (4.34), it is now possible to sketch out the predicted behaviour on the basis of the Tresca

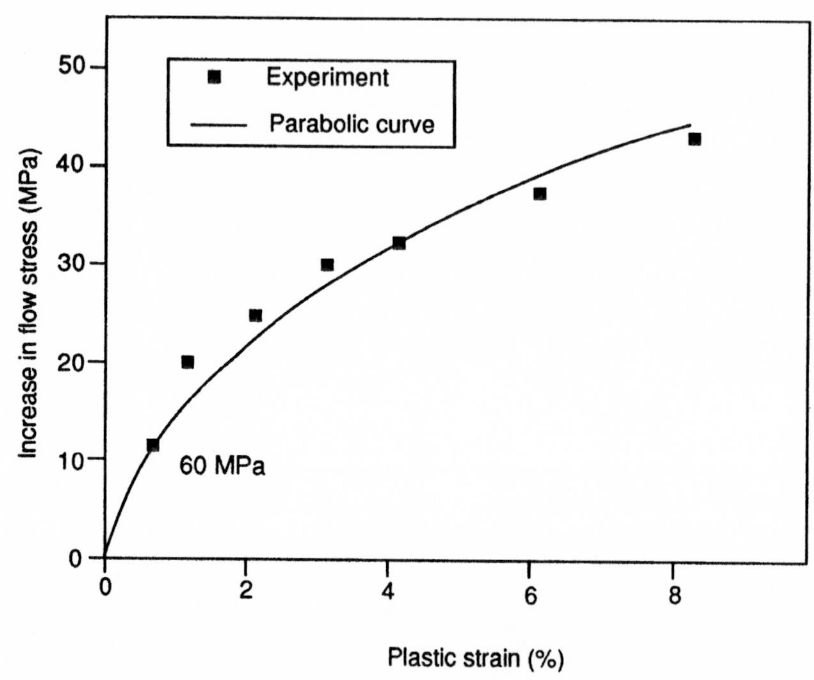

Fig. 4.20 Experimental data[45] for the increase in flow stress on incorporating 2 vol% SiC_W in commercially pure Al–0.8% Al_2O_3. The form of the increase as a function of plastic strain is very similar to the parabolic dependence of the misfit in Fig. 4.19. (A parabolic curve is also plotted for comparison.) Of course the situation is complicated by the effect of matrix hardening terms, but this does suggest a progressive increase in the yield stress via the matrix/inclusion misfit.

yield criterion, taking into account the stiffness, thermally and plastically induced load transfer terms. This has been done in Fig. 4.21, for 5 and 20 vol% SiC whiskers in Al using P and Q values taken from Fig. 4.4. It has been shown[66] that good agreement with experimental data can be obtained in this way. This very simple model explains a number of features characteristic of MMC stress–strain curves.

- *MMCs typically have a steep but short-lived linear work-hardening regime* – this is because load transfer is very effective; this would lead to a very rapid increase in the inclusion stress were it not for the early activation of relaxation processes involving secondary dislocations, diffusion, fibre fracture, etc., which lower the growth of the inclusion/matrix misfit resulting in a reduced work-hardening rate.

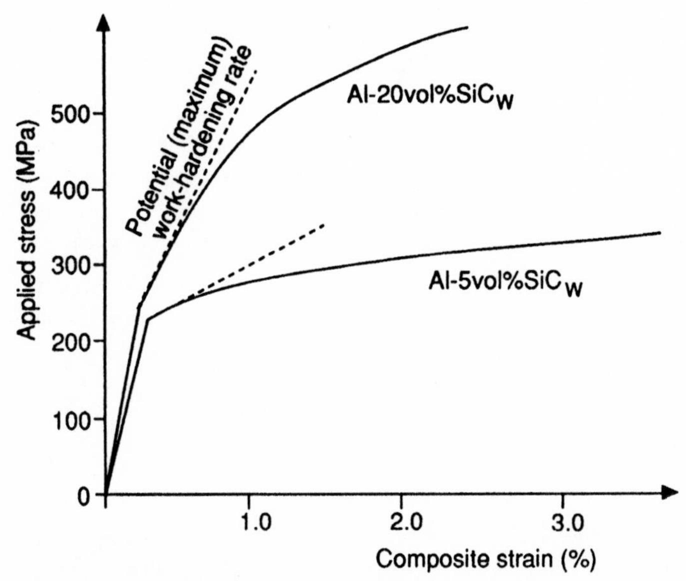

Fig. 4.21 Predicted stress–strain curves plotted assuming that relaxation occurs according to eqn (4.34). An effective temperature drop of ~100 K has been assumed (Table 4.2) (although it may be more realistic to assume that the composites maintain the same thermal matrix stress – i.e. T_{esf} would be lower for the higher volume fraction composite). This graph is not expected to represent composite behaviour accurately, partly because matrix strengthening has not been taken into account and also because the extent of relaxation has been estimated in a rather arbitrary way, but it does show how the internal stress terms combine to give rise to a high initial work-hardening rate, which decreases with increasing strain.

- *The incorporation of SiC reinforcement improves the UTS* – provided premature failure is not instigated at low plastic strains, the higher work-hardening rate means that the yield stress normally surpasses that of the unreinforced alloy.
- *Fibrous MMCs often have higher rates of work-hardening than particulate MMCs* – this is because the transfer of load to the reinforcement is more effective for higher aspect ratio inclusions.
- *Though the work-hardening rate decreases rapidly, it is always greater than that for the unreinforced matrix* – this arises for two reasons: firstly, although the misfit grows at an ever decreasing rate (as relaxation processes are triggered) it does not decrease and, secondly, more dislocations are generated in the reinforced than in the unreinforced case, giving a faster rate of matrix hardening.

4.6 Diffraction studies of plastic deformation in long and short fibre systems

Whilst the rates of development of internal stress and matrix microstructure combine to control the macroscopic stress–strain response of a composite material, it is difficult to identify their separate roles simply in terms of an analysis of overall stress–strain data. Diffraction data, however, provides a direct measure of the average phase stresses via their lattice strains (see §11.3). In this section experimental observations for long and short fibre composites will be discussed in terms of the evolution of internal stress and matrix work-hardening with deformation.

Long fibre composites

Axial stress–strain behaviour is conventionally divided into three stages; *stage I*: elastic matrix/elastic fibres, *stage II*: plastic matrix/elastic fibres, and *stage III*: plastic matrix/plastic or fractured fibres (Fig. 4.22(a)). Because, up until fibre fracture, the total strain in each phase must equal the composite strain, the development of internal straining of long fibre composites is best considered against the total strain as measured by a strain gauge ε_C^{SG} (Fig. 4.22(b)). In stage I, the average elastic phase strains $\bar{\varepsilon}_{M,I}$, as could, for example, be measured by X-ray diffraction $\varepsilon_{M,I}^{XR}$, are both equal to the composite strain

$$\varepsilon_C^{SG} = \varepsilon_M^{XR} = \varepsilon_I^{XR} \tag{4.36}$$

In stage II, the elastic ε_M^{XR} and plastic ε_M^P matrix strains combine to equal the composite strain

$$\varepsilon_C^{SG} = \varepsilon_M^{XR} + \varepsilon_M^P = \varepsilon_I^{XR} \tag{4.37}$$

In this regime, the stress-free misfit, $-\varepsilon^P$, between fibre and matrix is equal and opposite to the matrix plastic strain ε_M^P, but the plastic strain of the composite ε_C^P is considerably less than this because of the constraining effect of the fibres. This means that, in the unloaded state, both phases are internally strained and stressed consistent with stress balance

$$\varepsilon_C^{SG} = \varepsilon_C^P = \varepsilon_M^{XR} + \varepsilon_M^P = \varepsilon_I^{XR} \quad \text{with} \quad \varepsilon_M^{XR} = \langle\varepsilon\rangle_M^P \quad \text{and} \quad \varepsilon_I^{XR} = \langle\varepsilon\rangle_I^P$$

where $\langle\varepsilon\rangle_{M,I}^P$ are the internal mean strains arising from the plastic flow induced misfit, being tensile in the fibre and compressive in the

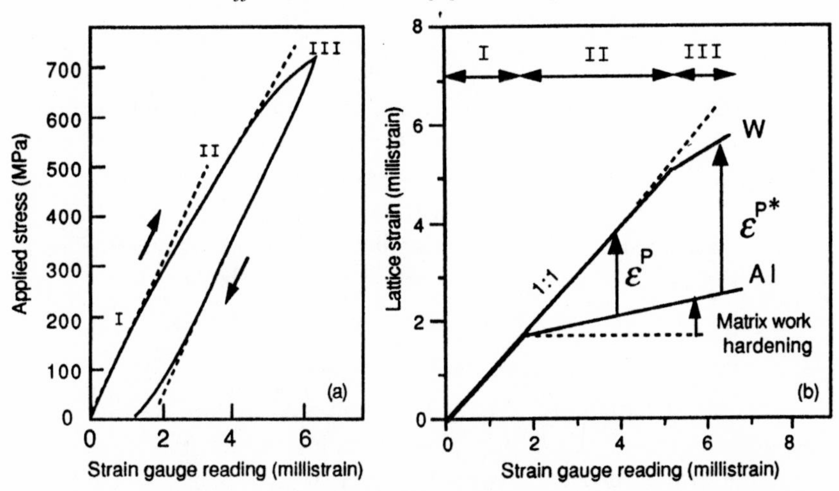

Fig. 4.22 (a) A typical axial stress–strain curve[68] for a long fibre composite (Al–W) and (b) X-ray measurements of axial elastic mean phase strains with composite loading for an Al–W system (deduced from the published stress plots using Young's moduli data quoted in the original paper[68] (Sample 4). In (a) the solid line represents the rule of mixtures composite stress curve calculated from the lattice strains in (b).

matrix. As discussed in §4.4, stress relaxation can reduce the misfit (to $\varepsilon_M^{P*} < \varepsilon_M^P$), thus decreasing the elastic loading of the fibres (i.e. $\varepsilon_I^{XR} < \varepsilon_C^{SG}$).

The Cu–W and Al–W systems have been studied extensively[12,67–69], and in both systems the linearity of stage II is undisputed (Fig. 4.22(a)). Furthermore, as is clear from the diffraction data of Cheskis and Heckel[68] (Fig. 4.22(b)), the elastic matrix strain also exhibits a linear stage II response. For both Cu and Al matrices, the work-hardening rates observed *in situ* are many times those expected on the basis of the work-hardening response of unreinforced matrices. This result has been confirmed by further work comparing the rate of overall work-hardening with the rate of mean stress generation using the Bauschinger test[43]. The explanation for this rather surprising result would seem to lie, not in an increase in matrix substructure strengthening, which in itself could not account for such a large effect, but rather in the increased difficulty experienced by the dislocations in passing the fibres because of the repulsive nature of the local stress field surrounding them (i.e. the source shortening term §4.3.2[43,47]).

With regard to the development of internal stress, the near coincidence of the fibre and composite strain curves (i.e. $\varepsilon_{\mathrm{I}}^{\mathrm{XR}} \approx \varepsilon_{\mathrm{C}}^{\mathrm{SG}}$) during stage II indicates that stress relaxation is negligible in this case (i.e. $\varepsilon^{\mathrm{P}} \approx \varepsilon^{\mathrm{P*}}$), and that internal stress induced work-hardening is linearly related to plastic strain until the beginning of stage III. A compressive analysis[43] of Cu–W systems has shown that, while there is always a linear relationship between ε^{P} and $\varepsilon^{\mathrm{P*}}$, the constant of proportionality varies from specimen to specimen, but generally decreases with increasing fibre diameter. This is indicative of an increasing ease of internal stress relaxation.

Short fibre composites

The analysis of short fibre systems is complicated by the fact that at no stage need the total elastic plus plastic strain in either phase equal the composite strain. Consequently, such systems are best considered in terms of the redistribution of the applied load between the phases through the generation of elastic and plastic misfits (Fig. 4.23(b)), rather than in terms of deviations from the 1:1 strain line, as in Fig. 4.23(a). It is clear from Figs. 4.23(a) and (b) that, in contrast to the long fibre case, neither matrix nor the composite show any appreciable *linear* work-hardening regime. Furthermore, the matrix work-hardening rate is much lower than in Fig. 4.22(b). As to the development of internal stresses, the observed lattice strains in stage I (Fig. 4.23(b)) are in good agreement with the strains predicted by the Eshelby model, and as expected, with the onset of

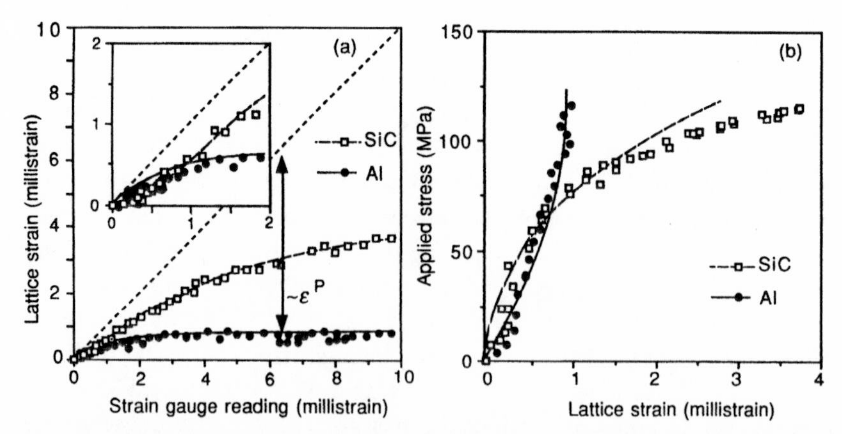

Fig. 4.23 The Al and SiC lattice strain response for Al–5 vol% SiC ($s = 5$) composite with applied loading measured using {111} neutron diffraction peaks and plotted as a function of (a) composite strain and (b) applied load. In (b) the lines are fitted using an Eshelby-type approach.

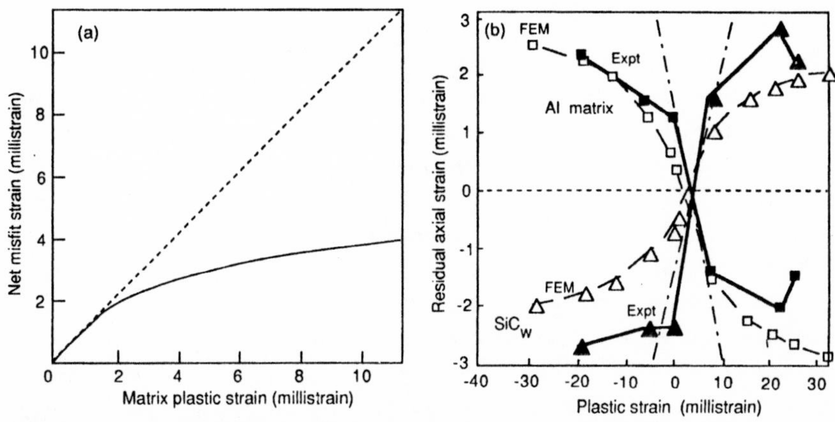

Fig. 4.24 (a) A fit using the data of Fig. 4.23 shows that beyond 0.2% plastic strain the rate of inclusion/matrix misfit generation is less than would be expected for unrelaxed plastic flow (dashed line), i.e. the rate of generation of the inclusion/matrix misfit is no longer equal to the global deformation of the matrix. In (b) similar behaviour is shown for 2009Al–15 vol% SiC_W; here the plastically and thermally induced internal stresses were evaluated in terms of the residual axial elastic strains subsequent to plastic straining (filled triangles represent the whisker strain, filled squares the average matrix strain). The dash–dot lines were calculated using the axial matrix stress per unit plastic strain value of 30 GPa taken directly from Fig. 4.11 (neglecting the small transverse matrix stress). They are in good agreement with the maximum gradients, as are the corresponding finite element predictions. FEM predictions cover the whole response, since primary plastic flow, plastic relaxation and other matrix-related processes are all included within the empirical flow criterion.

plasticity a rapid increase in the rate of load transfer to the reinforcement is observed (§4.3). Eshelby simulations of the diffraction-derived elastic strain curves allow calculation of the retained misfit ε^{P*} in terms of the plastic component of the internal stresses (eqn (4.35)). This is shown in Fig. 4.24(a), and indicates that flow remains essentially unrelaxed only for plastic strains less than 0.2%, beyond which the rate of increase in reinforcement stressing decreases. This means that progressively larger plastic strains are required to generate the same increase in reinforcement stressing (§4.5). Similar observations have been made by Porvik *et al.*[21], who measured the residual average phase strains after plastic deformation. In the current notation, this is equal to $\langle\varepsilon\rangle_M^P + \langle\varepsilon\rangle_M^{\Delta T}$ and, as shown in Fig. 4.24(b), saturates at quite small strains (at a lower strain in tension (RHS) than in compression (LHS) because of the thermal residual stress). Because all the opportunities for flow that are available in the monolithic alloy are implicitly included within the constitutive equations on which finite element models are based, such approaches predict automatically the increasing saturation of internal stress with plastic straining exemplified in Figs. 4.24(a) and (b). The rate of load transfer is thus determined by the dynamic equilibrium between the rate at which plastic deformation transfers load to the whiskers and the rate at which relaxation reduces the stressing of the whiskers (e.g. by diffusion or plastic relaxation).

In summary, diffraction measurements confirm the development of considerable internal stresses in MMCs upon plastic straining. These stresses increase rapidly at first, approximately in accord with the unrelaxed bound on work-hardening capability. Local relaxation processes increasingly act to limit, but not to inhibit completely, the growth of these stresses. This is not to say that matrix strengthening is not important. Depending on the fibre aspect ratio, local strengthening of the matrix may substantially aid the transfer of load (and thereby increase the load borne by the fibres). The important point, noted in §4.3, is that (provided the ceramic content is not very low) the theoretical work-hardening rate from load transfer (W/P in eqn (4.21)) greatly exceeds the intrinsic work-hardening rate of the matrix, even if the latter is significantly higher *in situ* than when unreinforced.

References

1. P. L. Morris and C. Baker (1989) Properties and Applications of Particulate Reinforced Al MMCs, in *Inst. of Metals Conf. – MMCs: Property Optimisation and Applications*, London, pp. 3.1–3.3.
2. D. L. McDanels (1985) Analysis of Stress–Strain, Fracture, and Ductility of Aluminium Matrix Composites Containing Discontinuous Silicon Carbide Reinforcement, *Metall. Trans.*, **16A**, pp. 1105–15.
3. J. B. Borrowdaile, S. Skyervold and W. Ruch (1989) The Secondary Processing of Particulate Reinforced Al Composites, in *Inst. of Metals Conf. – MMCs: Property Optimisation and Applications*, London, pp. 10.1–10.3.
4. R. J. Lederich and S. M. L. Sastry (1982) Deformation Behaviour of SiC_W Reinforced Al Composites, *Mat. Sci. & Eng.*, **55**, pp. 143–6.
5. C. J. Peel, R. Moreton and S. M. Flitcroft (1989) Progress Towards the Optimisation of Particulate Reinforced Composites for Aerospace Applications, in *Inst. of Metals Conf. – MMCs: Property Optimisation and Applications*, London, pp. 4.1–4.3.
6. T. G. Nieh and R. F. Karlak (1983) Hot-Rolled SiC–Al Composites, *J. Mat. Sci. Letts.*, **2**, pp. 119–22.
7. T. Christman, A. Needleman and S. Suresh (1989) An Experimental and Numerical Study of Deformation in MMCs, *Acta Metall.*, **37**, pp. 3029–50.
8. A. Levy and J. M. Papazian (1990) Tensile Properties of Short Fibre-

Reinforced SiC/Al Composites, part II. Finite Element Analysis, *Metall. Trans.*, **21A**, pp. 411–20.
9. T. G. Nieh and D. J. Chellman (1984) Modulus Measurements in Discontinuous Reinforced Al Composites, *Scripta Met.*, **18**, pp. 925–8.
10. P. B. Prangnell, T. J. Warner and W. M. Stobbs (1992) Private communication.
11. L. M. Brown and D. R. Clarke (1975) Work Hardening Due to Internal Stresses in Composite Materials, *Acta Metall.*, **23**, pp. 821–30.
12. O. B. Pedersen (1985) Residual Stresses and the Strength of Metal Matrix Composites, in *Proc. ICCM V*, San Diego, W. C. Harrigan, J. Strife and A. Dhingra (eds.), TMS–AIME, pp. 1–19.
13. O. B. Pedersen (1979) Thermoelasticity and Plasticity of Composite Materials, in *Proc. ICM* – 3, Cambridge, England, K. J. Miller and R. F. Smith (eds.), Pergamon, pp. 263–73.
14. P. J. Withers, W. M. Stobbs and O. B. Pedersen (1989) The Application of the Eshelby Method of Internal Stress Determination for Short Fibre Metal Matrix Composites, *Acta Metall.*, **37**, pp. 3061–84.
15. R. J. Arsenault and M. Taya (1987) Thermal Residual Stresses in MMCs, *Acta Metall.*, **35**, pp. 651–9.
16. A. R. T. DeSilva (1969) Thermal Stresses in Fibre Reinforced Composites, *J. Mech. Phys. Solids*, **17**, pp. 387–403.
17. D. A. Koss and S. M. Copley (1971) Thermally Induced Residual Stress in Eutectic Composites, *Metall.* Trans., **2A**, pp. 1557–60.
18. J. F. Mason (1990) *The Fabrication and Mechanical Properties of Mg–Li Alloys Reinforced with SiC Whiskers*, Ph.D. Thesis, Cambridge.
19. J. F. Mason, C. M. Warwick, P. Smith, J. A. Charles and T. W. Clyne (1989) Magnesium–Lithium Alloys in Metal Matrix Composites – A Preliminary Report, *J. Mat. Sci.*, **24**, pp. 3934–6.
20. P. J. Withers, D. J. Jensen, H. Lilholt and W. M. Stobbs (1987) The Evaluation of Internal Stresses in a Short Fibre MMC, in *Proc. ICCM VI/ECCM 2*, London, F. L. Matthews, N. C. R. Buskell, J. M. Hodgkinson and J. Morton (eds.), Elsevier, pp. 255–64.
21. G. L. Porvik, M. G. Stout, M. Bourke, J. A. Goldstone, A. C. Lawson, M. Lovato, S. R. MacEwen, S. R. Nutt and A. Needleman (1991) Mechanically Induced Residual Stresses in Al–SiC Composites, *Scripta Met.*, **25**, pp. 1883–8.
22. D. B. Zahl and R. M. McMeeking (1991) The Influence of Residual Stress on the Yielding of MMCs, *Acta Met. et Mat.*, **39**, pp. 1117–22.
23. A. K. Vasudevan, O. Richmond, F. Zok and J. D. Embury (1989) The Influence of Hydrostatic Pressure on the Ductility of Al–SiC Composites, *Mat. Sci. & Eng.*, **A107**, pp. 63–9.
24. A. Levy and J. M. Papazian (1991) Thermal Cycling of Discontinuously Reinforced SiC/Al Composites, in *Metal Matrix Composites – Processing, Microstructure and Properties, 12th Risø Int. Symp.*, Roskilde, N. Hansen, D. J. Jensen, T. Leffers, H. Lilholt, T. Lorentzen, A. S. Pedersen, O. B. Pedersen and B. Ralph (eds.), Risø Nat. Lab., Denmark, pp. 475–82.
25. R. J. Arsenault and N. Shi (1986) Dislocation Generation due to Differences between Coefficients of Thermal Expansion, *Mat. Sci. & Eng.*, **81**, pp. 175–87.
26. R. J. Arsenault, L. Wang and C. R. Feng (1991) Strengthening of Composites due to Microstructural Changes in the Matrix, *Acta. Metall.*, **39**, pp. 47–57.

27. W. S. Miller and F. J. Humphreys (1990) Strengthening Mechanisms in Metal Matrix Composites, in *Fundamental Relationships Between Microstructure and Mechanical Properties of Metal–Matrix Composites*, P. K. Liaw and M. N. Gungor (eds.), TMS, Warrendale, Pa., pp. 517–41.
28. N. Hansen (1977) The Effect of Grain Size and Strain on the Tensile Flow Stress of Al at Room Temperature, *Acta Metall.*, **25**, pp. 863–9.
29. E. O. Hall (1951) The Deformation and Ageing of Mild Steel: III. Discussion of Results, *Proc. Phys. Soc.*, **B64**, pp. 747–53.
30. N. J. Petch (1953) The Cleavage Strength of Polycrystals, *J. Iron Steel Inst.*, **174**, pp. 25–8.
31. R. J. McElroy and Z. C. Szkopiak (1972) Dislocation-Substructure-Strengthening and Mechanical-Thermal Treatments of Metals, *Int. Met. Rev.*, **17**, pp. 175–202.
32. F. J. Humphreys (1988) Deformation and Annealing Mechanisms in Discontinuously Reinforced Metal Matrix Composites, in *Mechanical and Physical Behaviour of Metallic and Ceramic Composites, 9th Risø Int. Symp.*, Roskilde, S. I. Andersen, H. Lilholt and O. B. Pedersen (eds.), Risø Nat. Lab., Denmark, pp. 51–74.
33. T. Christman, A. Needleman and S. Suresh (1990) Microstructural Development in an Al-Alloy SiC Whisker Composite, *Acta Metall.*, **37**, pp. 3029–50.
34. L. M. Brown (1985) Internal Stresses in Dispersion Hardened Alloys, in *Proc. IUTAM Eshelby Mem. Symp.*, Sheffield, UK, K. J. Miller, B. Bilby and J. R. Willis (eds.), Cambridge University Press, pp. 357–68.
35. L. M. Brown and W. M. Stobbs (1971) The Work-Hardening of Cu–SiO_2 – I. A Model Based on Internal Stresses, with no Plastic Relaxation, *Phil. Mag.*, **23**, pp. 1185–99.
36. O. B. Pedersen (1983) Thermoelasticity and Plasticity of Composites – I. Mean Field Theory, *Acta Metall.*, **31**, pp. 1795–808.
37. O. B. Pedersen (1978) Transformation Theory for Composites, *Z. ang. Math. Mech.*, **58**, pp. 227–8.
38. K. Tanaka and T. Mori (1970) The Hardening of Crystals by Non-Deforming Particles and Fibres, *Acta Metall.*, **18**, pp. 931–9.
39. L. M. Brown and W. M. Stobbs (1976) The Work-Hardening of Cu–SiO_2 – V. Equilibrium Plastic Relaxation by Secondary Dislocations, *Phil. Mag.*, **34**, pp. 351–72.
40. M. F. Ashby (1966) Work-Hardening of Dispersion-Hardened Crystals, *Phil. Mag.*, **14**, pp. 1157–78.
41. P. J. Withers and T. Lorentzen (1989) A Study on the Relation between the Internal Stresses and the External Loading Response in Al/SiC Composites, in *Proc. ICCM VII*, Guangzou, W. Yunshu, G. Zhenlong and W. Renjie (eds.), Pergamon, New York, pp. 429–34.
42. F. J. Humphreys (1985) Dislocation–Particle Interactions, in *Dislocations and Properties of Real Materials*, Inst. of Metals, London, pp. 175–204.
43. O. B. Pedersen (1991) Thermoelasticity and Plasticity of Composites – II, *Acta Met. et Mat.*, **39**, pp. 1201–19.
44. P. J. Withers, T. Lorentzen and O. B. Pedersen (1991) Effect of Internal Stresses on Deformation at Room Temperature, in *Metal Matrix Composites – Processing, Microstructure and Properties, 12th Risø Int. Symp.*, Roskilde, N. Hansen, D. J. Jensen, T. Leffers, H. Lilholt, T. Lorentzen, A. S. Pedersen, O. B. Pedersen and B. Ralph (eds.), Risø Nat. Lab., Denmark, pp. 189–204.

45. C. Y. Barlow and N. Hansen (1991) Deformation Structures and Flow Stress in Al Containing Short Whiskers, *Acta Met. et Mat.*, **39**, pp. 1971–80.
46. A. Kelly (1966) *Strong Solids*, Oxford University Press.
47. L. M. Brown and D. R. Clarke (1977) The Work Hardening of Fibrous Composites with particular reference to the Cu–W System, *Acta Metall.*, **25**, pp. 563–70.
48. N. Lilholt (1983) Additive Strengthening, in *Deformation of Multi-Phase and Particle Containing Materials*, J. B. Bilde-Sørenson, N. Hansen, A. Horsewell, T. Leffers and H. Lilholt (eds.), Risø Nat. Lab., Roskilde, Denmark, pp. 381–92.
49. R. S. W. Shewfelt and L. M. Brown (1977) High Temperature Strength of Dispersion-Hardened Single Crystals – II. Theory, *Phil. Mag.*, **35**, pp. 945–62.
50. A. J. E. Foreman and M. J. Makin (1967) Dislocation Movement through Random Arrays of Obstacles, *Can. J. Phys.*, **45**, pp. 571–7.
51. J. D. Atkinson, L. M. Brown and W. M. Stobbs (1974) The Work-Hardening of Cu–SiO_2 – IV. The Bauschinger Effect and Plastic Relaxation, *Phil. Mag.*, **26**, pp. 1247–80.
52. F. J. Humphreys (1983) Deformation Mechanisms and Microstructures in Particle-Hardened Alloys, in *Deformation of Multi-Phase and Particle Containing Materials*, J. B. Bilde-Sørensen, N. Hansen, A. Horsewell, T. Leffers and H. Lilholt (eds.), Risø Nat. Lab., Roskilde, Denmark, pp. 41–52.
53. J. D. Eshelby (1961) Elastic Inclusions and Inhomogeneities, in *Prog. Solid Mech.*, I. N. Sneddon and R. Hill (eds.), pp. 89–140.
54. O. B. Pedersen and L. M. Brown (1977) Equivalence of Strain and Energy Calculations of Mean Stress, *Acta Metall.*, **25**, pp. 1303–5.
55. J. F. Mason and T. W. Clyne (1989) Microstructural Development and Mechanical Behaviour of SiC Whisker Reinforced Mg–Li Alloys, in *3rd European Conf. on Comp. Mats. (ECCM3)*, Bordeaux, A. R. Bunsell, P. Lamicq and A. Massiah (eds.), Elsevier, pp. 213–20.
56. P. B. Hirsch (1957) Notes on: 'The Plastic Deformation of Aged Al Alloys', *J. Inst. Met.*, **86**, pp. 13–14.
57. M. Taya and T. Mori (1987) Dislocations Punched-Out Around a Short Fibre in a Short Fibre Metal Matrix Composite Subjected to Uniform Temperature Change, *Acta Metall.*, **35**, pp. 155–62.
58. L. M. Brown and W. M. Stobbs (1971) The Work-Hardening of Cu–SiO_2 – II. The Role of Plastic Relaxation, *Phil. Mag.*, **23**, pp. 1201–33.
59. D. C. Dunand and A. Mortensen (1991) On the Relaxation of a Mismatching Spheroid by Prismatic Loop Punching, *Scripta Met. et Mat.*, **25**, pp. 761–6.
60. D. C. Dunand and A. Mortensen (1991) Dislocation Emission at Fibres – I. Theory of Longitudinal Punching by Thermal Stresses, *Acta Met. et Mat.*, **39**, pp. 1405–16.
61. T. Mori, M. Okabe and T. Mura (1980) Diffusional Relaxation about a Second Phase Particle, *Acta Metall.*, **28**, pp. 319–25.
62. T. J. Warner, P. J. Withers, D. J. Jensen and W. M. Stobbs (1989) Phase Transformations in MMC Reinforcements, in *Fundamental Relationships between Microstructure and Mechanical Properties of MMCs*, Indianapolis, M. N. Gungor and P. K. Liaw (eds.), TMS, pp. 313–24.
63. I. W. Chen and Y. H. Chiao (1983) Martensitic Transformations in ZrO_2 and HfO_2 – An Assessment of Small-Particle Experiments with Metal and Ceramic Matrices, *Advances in Ceramics, Vol. 12: Science and Technology of Zirconia II*, pp. 33–45.

64. Y. L. Liu, N. Hansen and D. J. Jensen (1989) Nucleation and Growth in Cold-Rolled Al Containing SiC Whiskers and Al_2O_3 Particles, in *Proc. ICCM VII*, Guangzou, W. Yunshu, G. Zhenlong and W. Renjie (eds.), Pergamon, New York, pp. 529–34.
65. R. A. Shahani and T. W. Clyne (1991) Recrystallization in Fibrous and Particulate MMCs, *Mat. Sci. & Eng.*, **A135**, pp. 281–6.
66. W. J. Clegg (1988) A Stress Analysis of the Tensile Deformation of Metal Matrix Composites, *Acta Metall.*, **36**, pp. 2141–9.
67. H. P. Cheskis and R. W. Heckel (1968) *In Situ* Measurements of Deformation Behaviour of Individual Phases in Composites by X-ray Diffraction, in *Metal Matrix Composites*, ASTM STP 438, pp. 76–91.
68. H. P. Cheskis and R. W. Heckel (1970) Deformation Behaviour of Continuous-Fibre MMC Materials, *Metall. Trans.*, **1**, pp. 1931–42.
69. H. Lilholt (1977) Hardening in Two-phase Materials – I. Strength Contributions in Fibre-Reinforced Cu–W, *Acta Metall.*, **25**, pp. 571–85.

5

Thermal effects and high temperature behaviour

The behaviour of a metal matrix composite is often sensitive to changes in temperature. This arises for two reasons; firstly, because the response of a metal to an applied load is itself temperature dependent and secondly, because changes in temperature can cause internal stresses to be set up as a result of differential thermal contraction between the phases. In the previous chapter, the thermal stresses were shown to result in yielding asymmetries. Here the implications of these thermal stresses are further explored, both in situations where the misfit strain is elastically accommodated, and when inelastic deformation can occur. This leads to an examination of the creep behaviour of MMCs and allows an understanding of the dramatic effects induced by thermally cycling the material while under load.

5.1 Thermal stresses and strains

5.1.1 Differential thermal contraction stresses

As is displayed in Fig. 5.1, metals generally have larger thermal expansion coefficients (α) than ceramics. Since fabrication of MMCs almost inevitably involves consolidation at a relatively high temperature, it is not surprising that they often contain significant differential thermal contraction stresses at ambient temperatures (e.g. see Fig. 5.17). Assuming the material to be effectively stress-free at some (high) temperature, T_{esf}, the stress state at a lower temperature can be envisaged as arising from the fitting of an oversized inclusion into an undersized hole in the matrix. The misfit strain is then simply $\Delta\alpha\,\Delta T$, where $\Delta T = T_{esf} - T_0$, the ambient temperature. For an isolated spherical particle, the resultant stress field is readily obtained from eqn (3.34), by using the following equation for

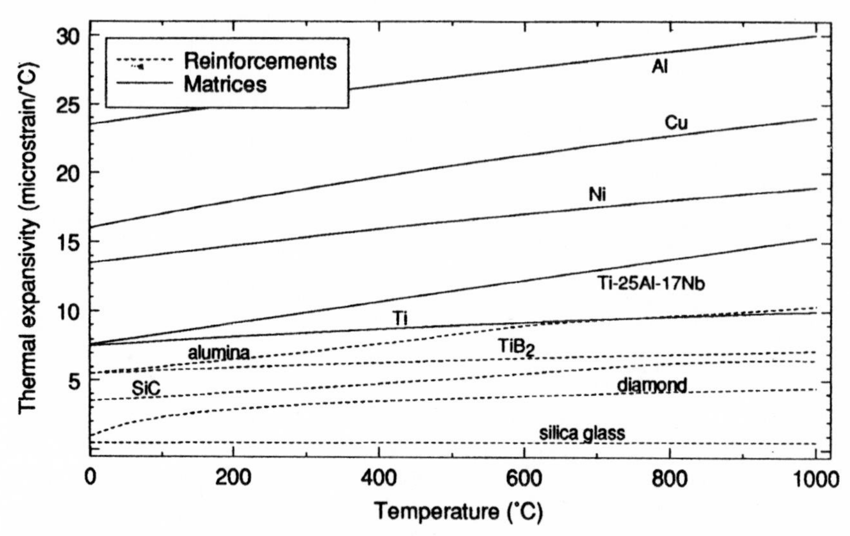

Fig. 5.1 Experimental data for the thermal expansion coefficients of various reinforcements (dashed) and matrices[1]. Note that certain metal/ceramic combinations can be identified for which the expansivity mismatch $\Delta\alpha$ is very small (e.g. Ti–Al_2O_3).

the pressure P within the sphere[2]

$$P = \frac{4G_M \dfrac{(1+\nu_M)}{3(1-\nu_M)}\left(\dfrac{K_I}{K_M}\right)\Delta\alpha\,\Delta T}{\dfrac{(1+\nu_M)}{3(1-\nu_M)}\left(\dfrac{K_I}{K_M}-1\right)+1} \tag{5.1}$$

Naturally, while the stress field is radially symmetric, there are large differences between the hoop and radial matrix stresses in the vicinity of the sphere (Fig. 5.2(a)). These tend to cause local plastic flow of the matrix and equations are available[2] for the prediction of the extent of this plastic zone and the resultant changes in stress distribution (Fig. 5.2(b)).

The stress field around a spherical particle is clearly the easiest to calculate. The stress field in an infinitely long fibre/matrix system (or finite radial extent) can, however, be solved analytically[3,4], as is shown in Fig. 5.3 for a SiC fibre in a Ti matrix. In a short fibre composite, the stress field is more complex still, particularly when the effect of neighbouring fibres is considered. In this case, recourse to numerical methods, with considerable computational effort, is necessary to predict the stress field

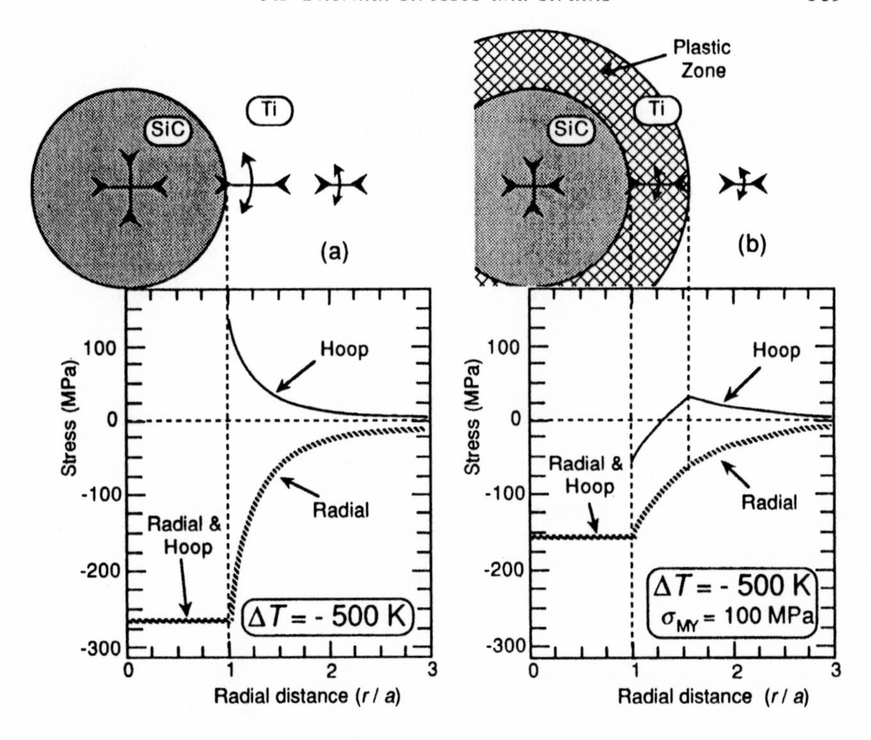

Fig. 5.2 (a) The elastic stress field caused by cooling an infinite matrix of Ti containing a spherical particle of SiC through 500 K. The particle is predicted to be in uniform hydrostatic compression, while in the neighbouring matrix, there is a large deviatoric stress component, which may be sufficient to cause plastic flow. The hydrostatic component of the matrix stress state, however, is constant with distance from the sphere, and in the case of an infinite matrix, has a value of zero. (b) The redistribution of stress brought about by local plastic flow[2], obtained assuming a matrix yield stress of 100 MPa, with no work-hardening. The sharpest change is in the matrix hoop stresses, which are markedly reduced and can become compressive near the surface of the sphere. For the Al–SiC system matrix plasticity can occur even for very small temperature excursions, because of the larger disparity in CTE and the low yield stress of Al alloys.

(see §2.4). However, the general nature of the stress distribution, and some idea of the magnitudes of the stresses, can be inferred from these simpler cases. It is also possible to use the Eshelby method (eqn (4.7) in §4.1.2) to predict the volume-averaged matrix stresses resulting from differential thermal contraction. Before going on to examine how these thermally induced stresses can affect inelastic deformation processes, it is useful to consider how the elastic accommodation of the strain misfit controls the overall coefficient of thermal expansion of the composite.

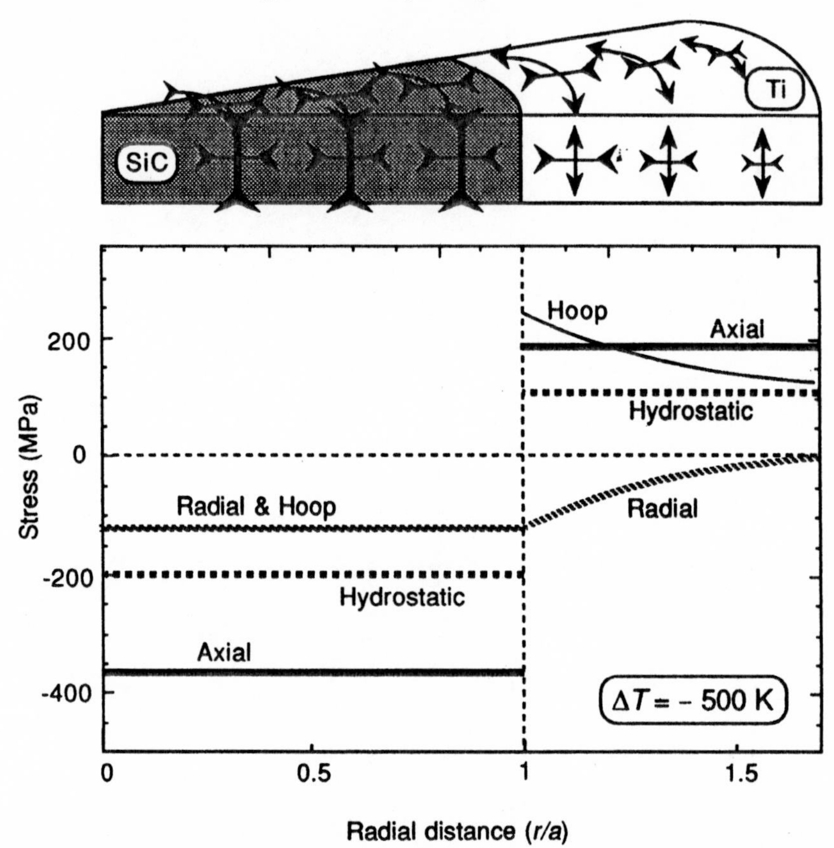

Fig. 5.3 The predicted[4] elastic stress state for a Ti–long fibre SiC composite, represented as two coaxial cylinders, with relative radii corresponding to a fibre content of 35 vol%. While the distribution of stresses is similar to that for the sphere (Fig. 5.2(a)), the matrix stress tends to extend further from the interface and there is now a deviatoric stress in the SiC.

5.1.2 Thermal expansivities

An understanding of the stresses which arise on changing the temperature of a composite allows predictions to be made of its coefficient of thermal expansion (CTE). A temperature change ΔT will produce a misfit strain $\Delta\alpha\,\Delta T$, which, if elastically accommodated, will give rise to a set of internal stresses. These stresses will have associated strains and the net effect of these on the length of the composite in any given direction can be calculated or estimated. This net length change arising from the internal stresses is simply added to the natural thermal expansion of the matrix to give the overall length change and hence the composite expansivity.

This simple view of thermal expansion allows certain points to be identified immediately. For example, a porous material, regarded as a composite of voids in a matrix, will not develop any internal stresses on heating because the 'inclusions' have zero stiffness. Hence, the presence of pores (of whatever shape, size and volume fraction) will not affect the CTE, although they will evidently give rise to sharp reductions in stiffness.

Quantitative prediction of composite CTE requires some sort of modelling of the thermal stresses and strains. Various approaches have been used (see below), but the Eshelby method outlined in Chapter 3 provides a simple and elegant solution to the problem. In §3.2 it was shown that a temperature increase ΔT causes an inclusion misfit strain $\varepsilon^{T*} = (\alpha_I - \alpha_M)\,\Delta T$, which will be isotropic if matrix and inclusion have isotropic expansivities. The appropriate transformation strain of a single 'ghost' inclusion was derived as

$$\varepsilon^{T} = [(C_I - C_M)S + C_M]^{-1} C_I \varepsilon^{T*} \tag{3.7}$$

and it was then shown in §3.5 that, with a finite volume fraction of inclusions, this becomes

$$\varepsilon^{T} = -\{(C_M - C_I)[S - f(S - I)] - C_M\}^{-1} C_I \varepsilon^{T*} \tag{3.21}$$

Now, the composite expansivity may be defined by

$$\alpha_C = \frac{\bar{\varepsilon}_C}{\Delta T} \tag{5.2}$$

where $\bar{\varepsilon}_C$ is the overall average strain of the composite. It is made up of the unconstrained expansion of the matrix and the composite strain $\langle\varepsilon\rangle_C^{\Delta T}$ caused by the thermal misfit stresses

$$\bar{\varepsilon}_C = \alpha_M\,\Delta T + \langle\varepsilon\rangle_C^{\Delta T} \tag{5.3}$$

where $\langle\varepsilon\rangle_C^{\Delta T}$ is obtained by summing the contributions from the two constituents in the 'ghost' composite, as outlined in §3.6,

$$\langle\varepsilon\rangle_C^{\Delta T} = f\varepsilon^{T} \tag{3.29}$$

The final expression is obtained by substituting in eqn (5.2) from eqns (5.3), (3.29) and (3.21)

$$\boxed{\alpha_C = \alpha_M - f\{(C_M - C_I)[S - f(S - I)] - C_M\}^{-1} C_I(\alpha_I - \alpha_M)} \tag{5.4}$$

Various simple features can be demonstrated using this equation – see Fig. 5.4.

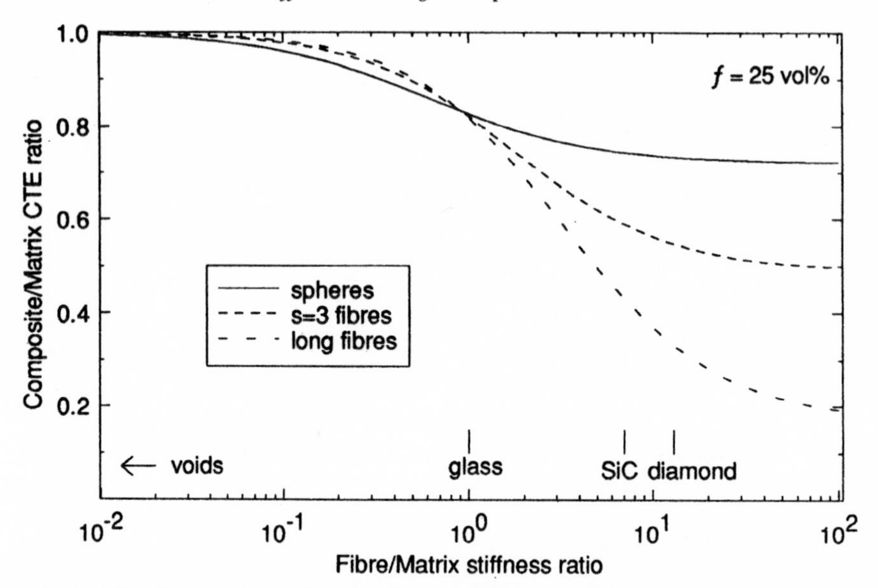

Fig. 5.4 Predictions from the Eshelby model (eqn (5.4)) for the axial thermal expansion coefficient (CTE) of a composite containing 25 vol% of aligned ellipsoidal inclusions. The plots, which were produced using data corresponding to an Al matrix (CTE = 24×10^{-6} K^{-1}) containing a reinforcement with a low CTE (4×10^{-6} K^{-1}), show that the reduction in CTE rises as the stiffness of the reinforcement is raised. Note that only for infinite fibres does the composite CTE tend to that of the reinforcement with increasing stiffness.

Although eqn (5.4) is straightforward to use, it does require evaluation of the Eshelby S-tensor (see Appendix III). As reviewed by Bowles and Tompkins[5], some simpler models have been proposed, at least for spheres and long fibres. For long fibres, one of the more successful is that due to Schapery[6], who used a thermodynamic approach to minimise the difference between upper and lower bound solutions, and derived the following approximations for the axial and transverse expansivities

$$\alpha_{3C} = \frac{E_I \alpha_I f + E_M \alpha_M (1 - f)}{E_I f + E_M (1 - f)} \tag{5.5}$$

$$\alpha_{1C} = (1 + \nu_M)\alpha_M (1 - f) + (1 + \nu_I)\alpha_I f - \alpha_{3C}\nu_{31C} \tag{5.6}$$

A simple rule of mixtures between the Poisson ratios of the two constituents should[7] provide a good approximation for ν_{31C}.

Predictions from the Eshelby and Schapery models are shown in Fig. 5.5 for Al–SiC composites. The increase in transverse CTE on adding a small volume fraction of fibres, predicted by both models, is worthy of

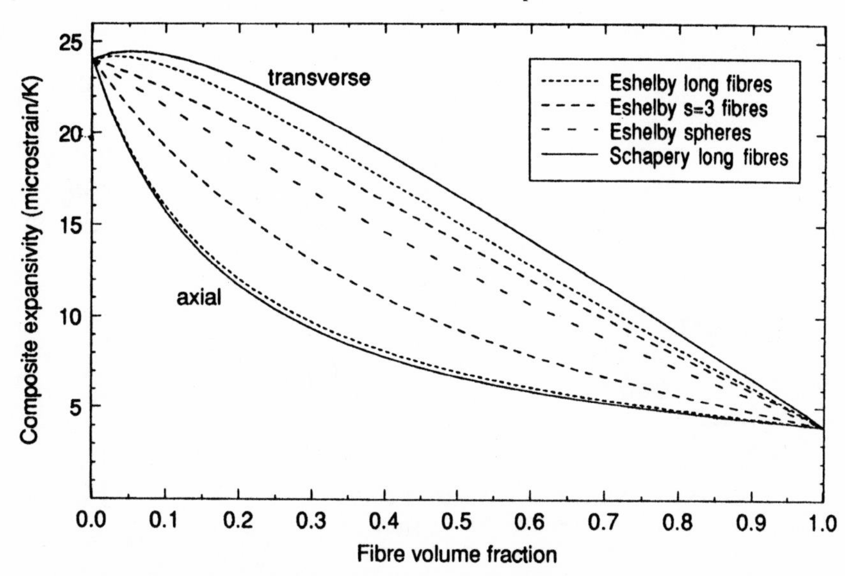

Fig. 5.5 The predicted dependence of CTE on reinforcement content for Al–SiC composites, according to the Eshelby (eqn (5.4)) and Schapery (eqns (5.5) and (5.6)) models. The Eshelby plot for spheres is not very far from a simple rule of mixtures, but this is evidently not appropriate for fibre composites, particularly with respect to the axial values. The Schapery model is seen to give good agreement with regard to the shapes of the plots, although the transverse values are slight overestimates compared with the Eshelby predictions.

note. This arises because, on heating the composite, axial expansion of the matrix is strongly inhibited and the resultant axial compression of the matrix generates a Poisson expansion in the transverse direction, which more than compensates for the reduction in natural transverse thermal expansion induced by the presence of the fibres. Such effects should be borne in mind when designing MMCs for controlled thermal expansion applications (§12.1.6). Experimental data for both axial[8] and transverse[9] CTEs of long fibre MMCs are consistent with the Eshelby and Schapery predictions.

5.2 Isothermal creep

5.2.1 Creep mechanisms and strain rate expressions for metals

Creep is defined as a time-dependent deformation under a constant load. The deformation mechanisms responsible for creep are thermally activated, so that, for most metals, it is expected to become significant

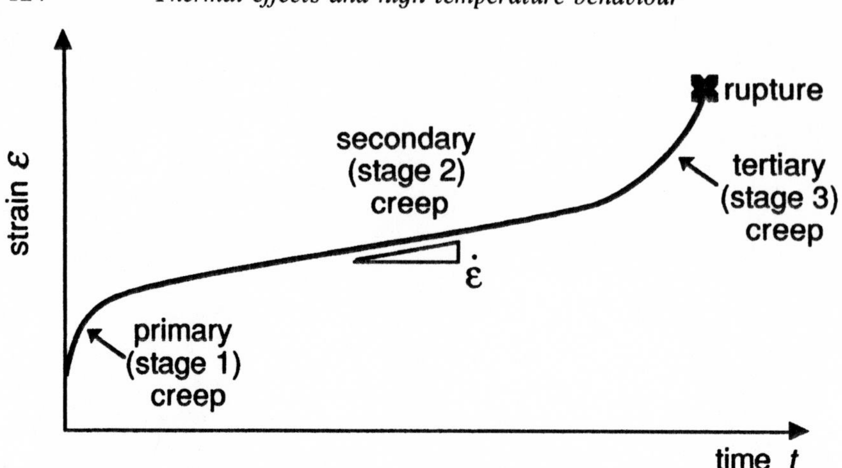

Fig. 5.6 Schematic depiction of a typical strain history during creep deformation under constant load, showing the steady state creep rate $\dot{\varepsilon}$.

only at elevated temperature ($T \geq 0.3$–$0.4T_{mp}$). Typically, the deformation history of a specimen during creep has the form shown in Fig. 5.6. Most materials exhibit creep curves of this form, although the extent of each regime and the mechanisms responsible for the deformation may differ. Broadly, primary creep represents the setting up of some kind of microstructural balance, which is then maintained during the quasi-steady state of secondary creep, before break down begins as the tertiary regime is entered. For example, this balance might be one in which dislocations are surmounting obstacles at a rate dictated by diffusional processes and the breakdown might represent irreparable microstructural damage such as internal void formation. Interest commonly focuses on the strain rate during secondary creep, although properties such as the creep rupture strain may also be of concern.

A distinction can be drawn because cases in which secondary creep is accompanied by some progressive change in the microstructure (e.g. in the dislocation density or the subgrain size) and those in which it remains essentially invariant. Obviously, changes in microstructure tend to affect the creep rate, but it can still remain constant provided the microstructure is evolving at a steady rate. In fact, the microstructure often remains stable during secondary creep under given conditions, but the same material may stabilise to a different microstructure if the stress or temperature is changed. Effects such as these can complicate the issue when making inferences about creep mechanisms from experimental strain rate data.

Several reviews[10] have been published on isothermal creep of metals. Broadly, the main creep mechanisms, and corresponding types of creep rate equations, can be classified in the following way.

Power law creep (dislocation climb)

The rate-determining step in the motion of dislocations is taken to be climb over obstacles of some sort. This is commonly the case and the creep rate is observed to obey an equation of the form

$$\dot{\gamma} = AD_{\text{eff}} \frac{Gb}{kT} \left(\frac{\sigma_{\text{VM}}}{G} \right)^n \tag{5.7}$$

where $\dot{\gamma}$ is the shear strain rate ($=\dot{\varepsilon}\sqrt{3}$, where $\dot{\varepsilon}$ is the elongational strain rate, for uniaxial loading), A is a dimensionless constant, D_{eff} is the effective diffusion coefficient, G is the shear modulus, b is the Burgers vector, k is Boltzmann's constant, T is the absolute temperature, σ_{VM} is the deviatoric (von Mises) stress from the applied load ($=\sigma^{\text{A}}/\sqrt{3}$ for uniaxial loading) and n is the stress exponent ($n \sim 3$–10). This equation actually represents two forms of behaviour, in that D_{eff} is given by[10]

$$D_{\text{eff}} = D_{\text{v}} \left[1 + \frac{10a_{\text{p}}}{b^2} \left(\frac{\sigma_{\text{VM}}}{G} \right)^2 \frac{D_{\text{p}}}{D_{\text{v}}} \right] \tag{5.8}$$

where a_{p} is the cross section of a dislocation core (pipe) and D_{v}, D_{p} are volume and pipe diffusion coefficients. Consequently, it is proportional to D_{v} at high temperature ($\geq 0.6T_{\text{mp}}$) and to D_{p} at low temperature (where D_{v} becomes negligible). It is often possible to differentiate between these two regimes by evaluating the activation energy, Q. Note also that the stress exponent should be ($n + 2$) at low temperature, as against n at high temperature, although this is rarely useful in helping to identify the dominant mechanisms, as there is no sound basis for the prediction of n (see below). A schematic illustration of how creep rate data can be presented to reveal these effects is shown in Fig. 5.7. It may be noted that the temperature dependence of the elastic modulus can cause curvature of $\ln(\dot{\varepsilon})$ *vs* $(1/T)$ plots and can cause the activation energy to appear anomalously high[11].

The theoretical basis for eqn (5.7) is rather weak. A stress exponent of 3 is predicted[12] from simple arguments about how the local stress acts on a single climbing dislocation and a value of 4 has been suggested[13] on the basis of Orowan by-passing effects. In practice, higher values of n than these are common, which may be a consequence of dislocations not acting

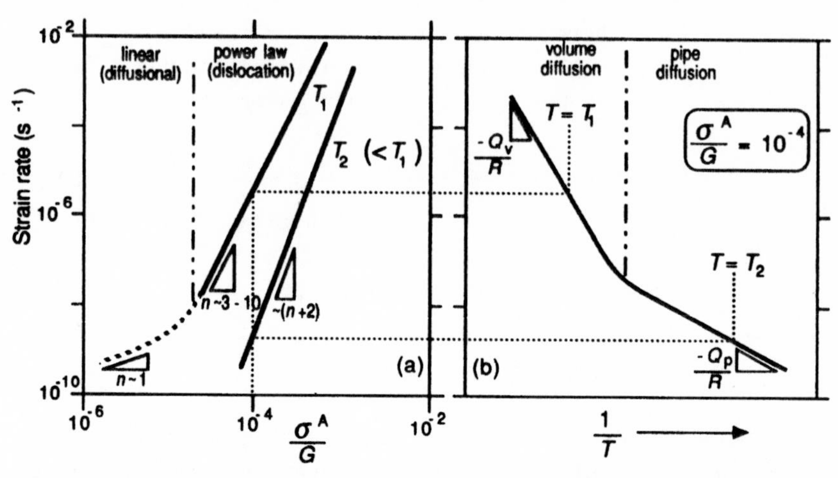

Fig. 5.7 Schematic illustration of typical creep data, plotted (a) as $\ln(\dot{\varepsilon})$ against $\ln(\sigma^{A}/G)$, for a given temperature, revealing the stress exponent n, and (b) as $\ln(\dot{\varepsilon})$ against $1/T$, for a given applied stress, revealing the activation energy for the diffusion mechanism governing the rate of creep. T_1 is a temperature in the range where volume diffusion is dominant, while at T_2 dislocation pipe diffusion is rate-determining.

independently. A further point to note is that changes in microstructure with applied stress level can affect the apparent value of n. For example, Sherby *et al.*[14] have shown that the strain rate is approximately proportional to the cube of the subgrain size in Al; as this stabilises during testing to a value inversely proportional to the applied stress, experimental data apparently giving $n = 5$ actually indicate a microstructure-invariant stress exponent of about 8. Taking such effects into account and expressing in terms of elongational strain rate and applied uniaxial stress, eqn (5.7) may be written

$$\dot{\varepsilon} = A\left(\frac{Gb}{kT}\right)\left(\frac{\lambda}{b}\right)^{p}\left(\frac{\sigma^{A}}{G}\right)^{n} D_0 \exp\left(\frac{-Q}{RT}\right) \tag{5.9}$$

where A is another dimensionless constant and λ is microstructure-related, probably representing a diffusion distance.

While attention is often focused on power law creep, there are several other regimes of creep behaviour, which are summarised below.

Linear viscous (diffusional) creep

Under low applied stresses (or when dislocation mobility is severely impaired), dislocation climb becomes very slow, but creep rates are commonly observed to be higher than predicted by eqn (5.9) – see Fig. 5.7(a). This is because mass transport by the diffusion of vacancies, either along grain boundaries or through the lattice, starts to make a significant contribution to the macroscopic strain. This diffusion takes place under the influence of the gradient of hydrostatic stress set up between different parts of individual grains and occurs so as to change the length of the grain in the direction of applied load. The rate equation in this regime can be written

$$\dot{\varepsilon} = \frac{AV_a}{kTd^2} \sigma^A D_{eff} \tag{5.10}$$

where V_a is the atomic volume and the effective diffusion coefficient is given by

$$D_{eff} = D_v \left[1 + \frac{\pi \delta D_b}{d D_v} \right] \tag{5.11}$$

in which d is the grain size, D_b is the boundary diffusion coefficient and δ is the effective thickness of a boundary. At high temperatures (Nabarro–Herring creep), the creep rate is proportional to D_v/d^2, while at low temperatures (Coble creep) it scales as D_b/d^3. It also appears, at least in some systems, that the variation in creep rate with stress can become linear at low stresses even when dislocation motion is more important than diffusional mass transport[15]. This is often termed Harper–Dorn creep.

Power law breakdown

The creep rate also tends to exceed that given by eqn (5.9) at very high applied stresses ($\geq 10^{-3}\,G$), where the dependence on stress typically becomes exponential. This is thought to be associated with a transition from climb-controlled to glide-controlled dislocation motion[10].

Dynamic recrystallisation

At high temperatures, creep behaviour can be substantially modified by repeated recrystallisation within the material during straining. Recrystallisation dramatically reduces the dislocation density, sweeping away the stabilised subgrain structure and in effect causing a period of primary creep, in which the rate may exceed that during secondary creep by a factor of up to ten. Wide oscillations in creep rate can therefore occur[16].

5.2.2 *Creep of dispersion strengthened metals*

It is well-established that the presence of a dispersion of fine ($\sim$100 nm) insoluble particles in a metal can raise the creep resistance substantially, even when constituting only a small overall volume fraction ($<1\%$). This arises during power law creep primarily through the opposition to dislocation glide offered by the stable obstacles on the slip plane. Orowan bowing[17] is necessary for dislocations to by-pass the particles, requiring an applied stress given approximately by

$$\sigma_{\mathrm{Orow}} \approx \frac{Gb}{L} \tag{5.12}$$

where G is the shear modulus, b is the Burgers vector and L is the particle separation. The magnitude of this Orowan stress is about $10^{-4}\,G$ for particles a micron or so apart, e.g. about 2–3 MPa for aluminium. (For large particles or fibres in an MMC, the associated Orowan stress is small and can usually be neglected – see §4.2.3.)

The requirement of Orowan bowing is associated with the concept of a ***threshold stress*** σ_0, below which there is essentially no creep. Under these circumstances, the strain rate is given by

$$\dot{\varepsilon} = A'\left(\frac{\sigma^{\mathrm{A}} - \sigma_0}{G}\right)^n \exp\left(\frac{-Q}{RT}\right) \tag{5.13}$$

where A' is a constant. The gradient of $\ln(\dot{\varepsilon})$ *vs.* $\ln(\sigma)$ plot, i.e. the apparent stress exponent, n_{app}, will under these circumstances differ from n. It is readily shown[18] that eqns (5.9) and (5.13) can be combined to give

$$n_{\mathrm{app}} = \frac{n}{1 - \left(\dfrac{\sigma_0}{\sigma^{\mathrm{A}}}\right)}\left(1 - \frac{\partial \sigma_0}{\partial \sigma^{\mathrm{A}}}\right) \tag{5.14}$$

Provided the threshold stress is relatively small and does not depend on the applied stress, then n_{app} will be close to the true exponent n. As the applied stress is reduced towards the threshold value, $(\sigma_0/\sigma^{\mathrm{A}})$ will no longer be small and the existence of the threshold stress will be apparent in a conventional plot of $\ln(\dot{\varepsilon})$ against $\ln(\sigma)$ as an increase in the gradient towards infinity as the applied stress approaches the threshold value (i.e. the stress sensitivity becomes infinite if the threshold stress is not taken into account).

In practice, the threshold stress may be less than the Orowan stress, because climb processes can assist by-pass, particularly at higher temperatures[19,20]. Although the threshold stress is expected on this basis to be temperature-dependent, it does not go to zero – even if climb is very easy; this is because extra dislocation line length must be created to allow climb over an obstructing particle.

A further important point to note about dislocation creep in dispersion-strengthened metals is that the presence of the particles tends to stabilise the dislocation substructure by exerting a strong Zener pinning effect[21,22] on grain boundaries and subgrain boundaries – see §10.2. It is clear that this stabilisation reduces dislocation creep by inhibiting subgrain growth[19,23]. Recrystallisation, which can cause a substantial surge in creep rate, is also strongly inhibited.

It has often been reported that dispersion strengthened metals exhibit high stress exponents and activation energies. For example, sintered Al powder (containing fine Al_2O_3 particles) was found[24] to creep with $n_{app} \sim 10$ and an apparent activation energy, Q_{app}, of the order of 1000 kJ mole^{-1} whereas, for pure Al, $n \sim 4$–5 and the activation energy for volume diffusion in Al is[25,26] about 140 kJ mole. While it might be argued that no value of n can be regarded as anomalous (in view of the absence of a reliable model for its prediction), no diffusional process is expected within the material having an activation energy higher than that for volume diffusion. In fact, anomalously high apparent activation energies in dispersion strengthened metals have been largely explained[11,27,28] on the basis of the temperature dependence of the elastic modulus and of the threshold stress.

In a similar manner to the derivation of eqn (5.14), eqns (5.9) and (5.13) may be combined[29] to derive an expression† for the apparent activation energy Q_{app} in terms of the true value Q

$$Q_{app} = Q - \frac{n_{app} R T^2}{(\sigma^A - \sigma_0)} \frac{\partial \sigma_0}{\partial T} - \frac{n_{app} R T^2}{G} \frac{\partial G}{\partial T} \tag{5.15}$$

Because both the threshold stress σ_0 and the shear modulus G tend to decrease with increasing temperature, both the second and third terms in this expression normally make a positive contribution to Q_{app}. Hence a large value of Q_{app} is to be expected, particularly when n_{app} is high. The modulus dependence term usually makes a small but significant contribution to Q_{app}. For example, taking $n_{app} = 15$, $T = 500$ K and

† In the original Nardone and Strife publication[29], the denominator of the second term in their version of this expression (eqn (7) in their paper) has been misprinted as $1 - \sigma/\sigma_R$, i.e. $1 - \sigma^A/\sigma_0$, instead of $1 - \sigma_R/\sigma$.

$\partial G/\partial T = -30\,\mathrm{MPa\,K^{-1}}$ (Al) would give a contribution of about $30\,\mathrm{kJ\,mole^{-1}}$. However, the threshold stress term may be potentially larger. For an applied stress of 20 MPa and $\sigma_0 \sim 5$ MPa, even if the magnitude of $\partial\sigma_0/\partial T$ were as low as $0.02\,\mathrm{MPa\,K^{-1}}$, with the above values of n_{app} and T, a contribution of $\sim 45\,\mathrm{kJ\,mole^{-1}}$ would result. On this basis, most observed values of Q_{app} may well be consistent with true activation energies no larger than that for volume diffusion.

Much of the above discussion is oriented towards dislocation creep. The presence of fine particles can also raise the resistance of the metal to diffusional creep. This is thought to be associated with the effect of the dispersoids on the way in which a grain boundary acts as a source and sink for vacancies. For example, grain boundary dislocations, which can normally act as such a source or sink, can become pinned by fine particles in the boundary and unable to act as efficiently. This pinning can account for both an enhanced creep resistance and the observation of a (temperature-dependent) threshold stress[30–32]. Detailed interpretation of the latter actually appears to require both diffusional and interfacial processes to be invoked[33]. In any event, the effect of fine particles on diffusional creep is less marked than for power law creep, possibly because the Burgers vector of a boundary dislocation is considerably smaller than that of a lattice dislocation[34], so that the pinning force is smaller. One consequence of this is that dispersion strengthened metals often creep by a diffusional mechanism at stresses and temperatures such that the corresponding pure metals would deform (at a considerably greater rate) by power law creep.

5.2.3 Creep of MMCs

Experimental data

Much of the experimental data on the isothermal creep of particulate and short fibre MMCs shows similarities to those for dispersion strengthened metals. Creep rates are in general substantially lower than those for the corresponding unreinforced alloys. This is illustrated by the data in Fig. 5.8, which also demonstrate that whisker-reinforced composites are more creep-resistant than particulate MMCs. However, the high creep resistance of the MMCs shown in this figure is a little misleading, as the composites were made by a powder route and hence also contained oxide dispersoids (not present in the unreinforced material). In contrast to the above results, the creep curves shown in Fig. 5.9 demonstrate that particulate reinforcement can actually impair the creep resistance of an

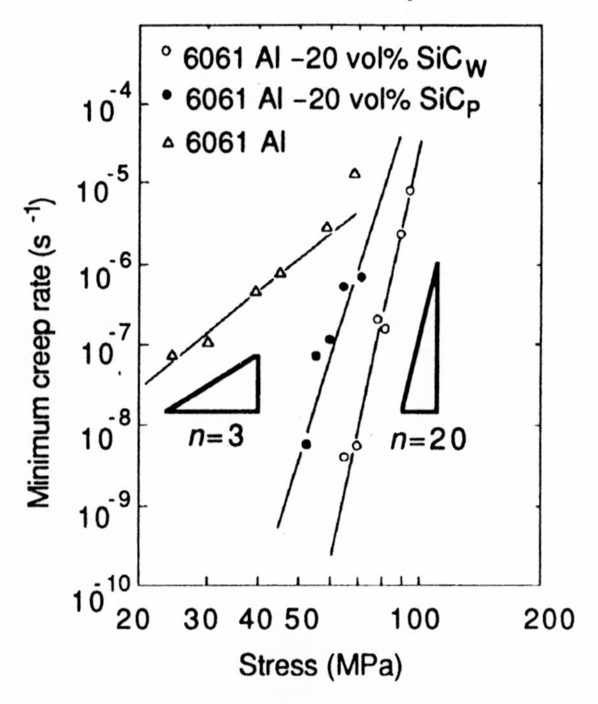

Fig. 5.8 Experimental data[35] recorded at 561 K showing that SiC-reinforced 6061 Al (powder route) has a much higher creep resistance than the unreinforced alloy (ingot route), but with a greater sensitivity to the level of applied stress. Also apparent is the greater efficacy of whiskers compared with particles for enhancing creep resistance.

oxide dispersion strengthened (ODS) matrix, probably by promoting cavitation. It can be seen, however, that the addition of short fibre reinforcement gives a clear improvement in creep strength, presumably because effective load transfer to the fibres can take place.

Clearly, detailed interpretation of experimental MMC data requires a good knowledge of the matrix microstructure, particularly in connection with powder route material. While there are many reports of 'anomalously' high stress exponents and activation energies in MMCs (see below), most papers have neither specified the processing details (such as the prior history of the Al powder) which determine the dispersoid content, nor reported direct microstructural observations of these fine particles. Only a few authors[38] have clearly identified this problem. The compilation of data shown in Table 5.1 illustrates these points. While anomalously high values of n_{app} and Q_{app} have been observed for cast

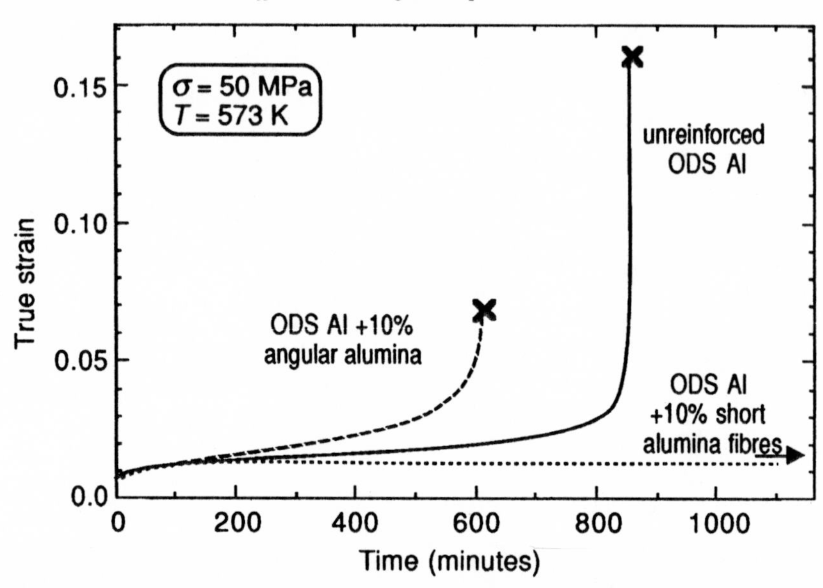

Fig. 5.9 Experimental creep curves[36] for materials produced by blending of Al powder (having a relatively thick surface oxide layer), with and without reinforcing additions, followed by cold isostatic pressing and hot extrusion. While the addition of Saffil® (alumina) short fibres (3 μm diameter, aspect ratio about 5) has led to an extended steady state with a reduced creep rate, the presence of the same volume fraction of alumina particles (diameter about 15 μm) has caused the virtual disappearance of the secondary creep regime and has impaired the overall creep resistance, in agreement with comments by Taya *et al.*[37]

material, the values are generally somewhat lower than for powder route material. Furthermore, it is not clear from the literature whether a threshold stress of some description is to be expected in MMCs in the absence of any fine dispersed phase. Orowan strengthening by such a coarse particle dispersion is not expected to be significant, but it has been proposed[18,39] that the internal stresses from differential thermal contraction (sometimes termed 'resisting stresses') can oppose creep and hence give rise to a threshold stress, σ_0. This would evidently be a function of temperature, so that, from eqns (5.14) and (5.15), high values of n_{app} and Q_{app} would be expected. Certainly, for aligned fibre composites, thermal stresses in the matrix would change the volume-averaged deviatoric stress in the matrix resulting from an applied load.

An estimate can be made of $\partial\sigma_0/\partial T$, if σ_0 is taken as the difference between the average axial and transverse matrix stresses (see §4.1.1). For example,

Table 5.1 *Experimental stress exponent and activation energy data for steady state isothermal creep of Al-based particulate and short fibre MMCs. (Q for lattice diffusion in Al ~ 140 kJ mole^{-1}.)*

Matrix alloy	Reinforcement (by volume)	Fabrication route (and fabricator)	Temperature (K)	n_{app}	Q_{app} (kJ mole^{-1})	Ref.
6061	20% SiC_W	pb,e (ARCO)	500–620	20	390	35
2124 (T4)	20% SiC_W	pb,e (ARCO)	450–550	8 (450 K) 21 (550 K)	280 (450 K) 430 (550 K)	107
2124 (T4)	25% SiC_W	pb,e (ARCO)	650–770	5–8	140–290	39
2124 (T4)	25% SiC_P	pb,e (ARCO)	650–770	4–7	220–300	39
6061	17% SiC_W	pb,e (ARCO)	570–720	10	360	108
2124 (T4)	20% SiC_P	pb,e (DWA)	570–720	20	390	108
6061	10% SiC_P	pb,r (in-house)	623–723	15	250	109
6061	20% SiC_W	sc,e (in-house)	575–775	12	300	110
Al (1100)	5% and 10% Al_2O_3–SiO_2 short fibres	sc,e (Nihon Keikinzoku)	570–770	9–15	160–200	111
6061	10% and 20% SiC_P	Stir-cast (Duralcan)	575–825	6 (825 K) 15 (575 K)	~20–50	112
Al–7% Si	10% SiC_F (Nicalon)	Rheocast (in-house)	625–800	5–10	260	113

pb,e – powder blended and extruded
pb,r – powder blended and rolled
sc,e – squeeze cast and extruded.

Fig. 4.4(b) indicates that $\partial\sigma_0/\partial T$ has a value of about 1.2 MPa K^{-1} (using the Eshelby model for Al–20 vol% SiC_W ($s = 5$)). Substitution of this value into eqn (5.15) would suggest a very large contribution to Q_{app} of several hundred kJ $mole^{-1}$. However, this calculation only gives an upper bound, because of the probable effect of stress relaxation processes in reducing these thermal stresses, particularly at elevated temperature. Furthermore, the loading direction may be relevant. Heating of fibrous MMC specimens which have undergone prolonged stress relaxation at room temperature would normally put the matrix into axial compression (see Figs. 5.15 and 5.16); this would presumably retard creep when tested in tension, but encourage it in compression. The effect will evidently be much weaker than for room temperature testing (see §4.1.2), because the thermal stresses will tend to relax away rapidly before and during the creep test, but the above estimates suggest that detectable effects on Q_{app} might nevertheless be produced. This rationale might account for the tendency (Table 5.1) for values of n_{app} and Q_{app} observed with cast MMCs to be higher than those expected for the unreinforced matrix, but not as high as for powder route MMCs, in which the fine particles give rise to a conventional Orowan threshold stress.

It is interesting to note that data[40–42] for a cast Mg–Li alloy indicate no discrepancy between the n and Q values for matrix and composite (Fig. 5.10). These data also confirm that the presence of the fibres raises the creep resistance substantially (in the absence of any dispersoid phase), and that this resistance is greater for fibres of higher aspect ratio. Similar stress exponent and activation energy results have been obtained by other workers[43] for particulate MMCs based on this matrix. In conjunction with the absence of anomalously high values for n or Q with this system, it should be noted that the Mg–12 wt% Li matrix[44] is one in which diffusion is exceptionally fast (due to the presence of the highly mobile Li atoms) and stress relaxation takes place very rapidly (for example, see Table 4.2). A possible inference here is that significant internal thermal stresses do not persist in this matrix long enough to influence the net matrix stresses under applied load during isothermal creep testing.

Theoretical models

Isothermal creep models can be divided, along similar lines to plasticity models (Chapter 4), into those in which the volume-averaged matrix stress is assumed to determine the deformation behaviour (e.g. Eshelby-type), and those in which the local variations in matrix stress state are taken

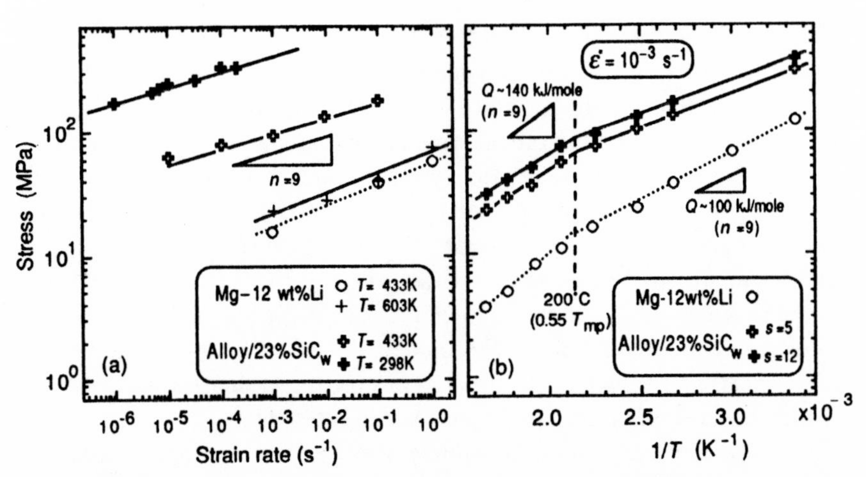

Fig. 5.10 Experimental data[40,41] for a cast-and-extruded Mg–12 wt% Li alloy, with and without 23 vol% of SiC whiskers. These data show (a) stress exponents of about 8–9 and (b) activation energies of about 140 kJ mole^{-1} at high temperature ($\leq 0.55T_{mp}$) and about 100 kJ mole^{-1} at low temperature. The higher temperature value is close to that for volume diffusion of Li in this alloy[42]. Clearly, the fibres are carrying a substantial load (raising the creep resistance) despite the tendency for relaxation. Note that the anomalously high values of n and Q found in Al-based MMCs are not observed here.

into account (e.g. FEM, shear lag). In both cases, it is usually assumed that the fibres or particles remain elastic and that the matrix deforms under the influence of the *in situ* stress in the same way as it would when unreinforced (usually with a power law dependence on the deviatoric stress). A practical concern is therefore with knowing the creep characteristics of unreinforced metal having the same microstructural features as the composite matrix. More important, however, is to recognise the following difference between creep behaviour in a monolithic matrix and in a composite. During dislocation creep[†] in a monolithic matrix, the stress field is taken to be uniform on a macroscale and semi-empirical equations are derived which, for a given microstructure, describe the dynamic equilibrium between processes enabling and obstructing dislocation motion, leading to a steady state creep rate. Such creep rate equations can be applied to the matrix in a composite, either on a volume-averaged basis or on a set of local volume elements. Steady state creep within an MMC

[†] While diffusion creep in a monolithic matrix can be driven by gradients of hydrostatic stress between different parts of each grain, the stress field within the matrix of an MMC is inherently much more inhomogeneous, giving rise to a much greater opportunity for diffusion creep under similar applied stresses.

is, however, more complex than this. It can be thought of as a dynamic equilibrium between the rate at which the matrix undergoes plastic/creep deformation, thus creating a matrix/reinforcement misfit and loading up the reinforcement, and the rate at which (non-damaging) stress relaxation processes† operate to unload the reinforcement. Interfacial sliding and diffusion are important mechanisms in this respect. They are composite-specific, depending on the nature of the interface, the scale of the reinforcement, etc. Such characteristics will not, of course, figure in any model based simply on constitutive equations describing the creep of the unreinforced matrix. Many of the models described below are flawed in this respect, and thus tend to predict an ever-increasing transfer of load to the non-creeping fibres and a decreasing creep rate.

One such model[37] uses the volume-averaged matrix stress, calculated according to the Eshelby method, to calculate the average creep rate of the matrix. Wakashima *et al.*[46] have gone on to include relaxation and have analysed the situation when plastic flow and relaxation processes balance to give steady state creep, in terms of the overall free energy of the system (Fig. 5.11). Although this treatment is helpful in conceptual terms, it does not readily lead to reliable a priori predictions of creep rate. This may be partly because local variations in matrix stress are not considered, but the main problem is simply that the creep response of the matrix to a given *in situ* deviatoric stress, which is in any event a complex function of its microstructure, becomes further complicated in MMCs by factors such as the ease of interfacial sliding and interfacial diffusion, as outlined above. Provided an equation is available for the strain rate of the matrix as a function of the overall *in situ* stress state, then this type of modelling can be useful for exploring factors such as reinforcement content. Some predictions[47] from an Eshelby-based model

† Of course, such a situation may never be set up. In particular, damage processes, such as cavitation and fibre fracture, may initiate before any stable balance is struck, so that tertiary (stage 3) creep would intervene. This is sometimes observed for MMCs, particularly with particulate reinforcement[45].

Fig. 5.11 (*opposite*) Schematic depictions of overall axial stresses and strains during creep loading, showing how a balance may be set up between (time-dependent) matrix plastic deformation, which transfers load to the fibres, and stress relaxation processes, which transfer load back to the matrix. Plastic flow is driven by the deviatoric stress in the matrix and stress relaxation mechanisms by stress gradients in the matrix (strictly through a reduction in the overall system energy). Once the overall stress in the matrix is established from this type of modelling, the composite creep rate can be predicted if the strain rate–stress relationship obeyed by the matrix is known.

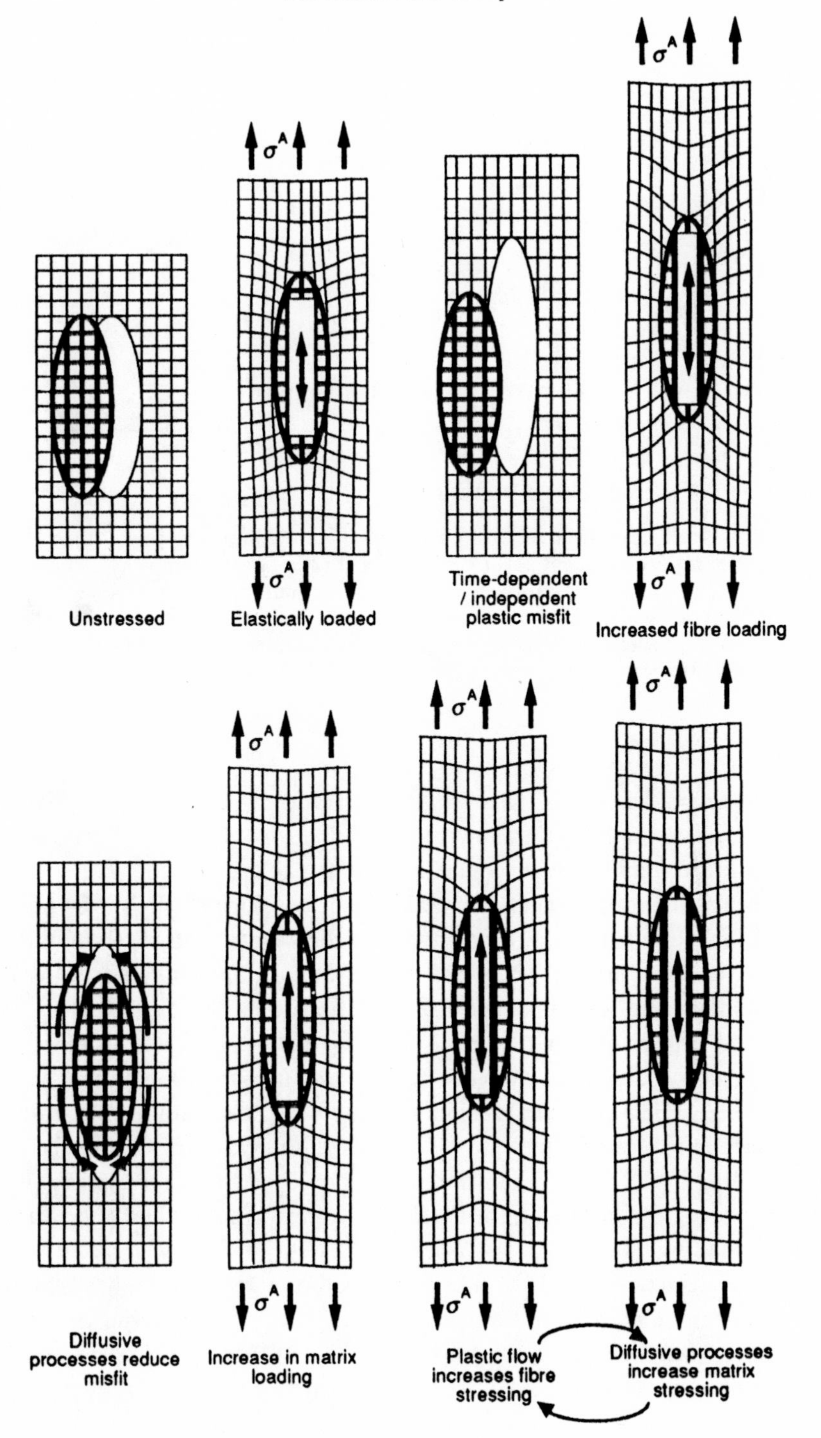
σ^A
Unstressed
Elastically loaded
Time-dependent / independent plastic misfit
Increased fibre loading
Diffusive processes reduce misfit
Increase in matrix loading
Plastic flow increases fibre stressing
Diffusive processes increase matrix stressing

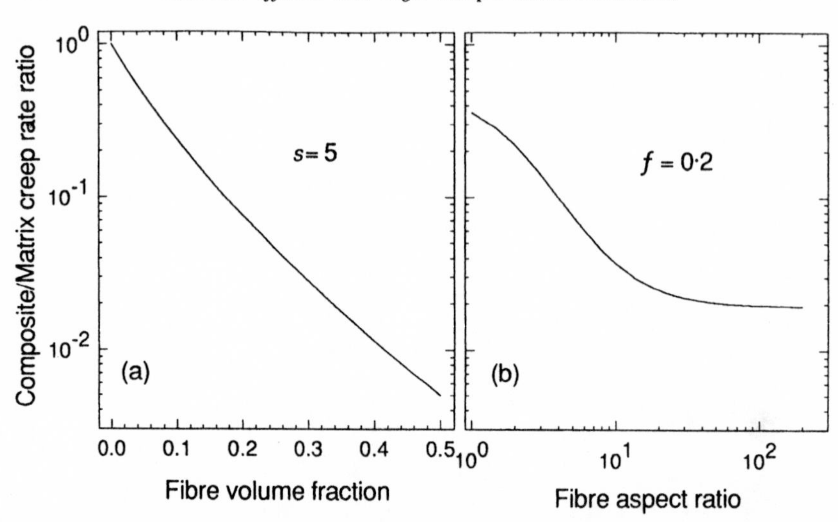

Fig. 5.12 Predictions from an Eshelby model[47] for composite creep rates as a function of (a) fibre content and (b) fibre aspect ratio. These data were calculated for Al–SiC under 20 MPa applied tensile load, assuming a stress exponent, n, of 5. A higher value of n would tend to exaggerate the variations with f and s, which arise from their effect on the *in situ* matrix stress state. No account has been taken of thermal stresses.

are shown in Fig. 5.12. These data illustrate, for example, the rather sharp increase in creep resistance expected on moving from particulate to short fibre reinforcement – cf. Figs. 5.8 and 5.9.

Various simple treatments have also been developed[48–53] on the basis of a shear lag-type assumption concerning the distribution of strain in the matrix (§2.2.1). Most of these assume a radially uniform matrix shear strain, increasing with distance from the fibre mid-plane (normal to its axis) and neglect matrix strains beyond the end of the fibre. Usually, as in the method described above, only the average stress in the matrix is considered. These treatments inherently involve gross simplifications and are unsuited to predicting effects of constraint by neighbouring fibres. These drawbacks of shear lag treatments are in general more serious for modelling of plastic flow and, particularly, creep, than when describing elastic loading.

Finite element methods do not suffer from the geometrical limitations of the shear lag models. In this case the main concern is with the selection of appropriate boundary conditions (see §2.4.3). A detailed FEM study of isothermal creep in MMCs has been carried out by Dragone and

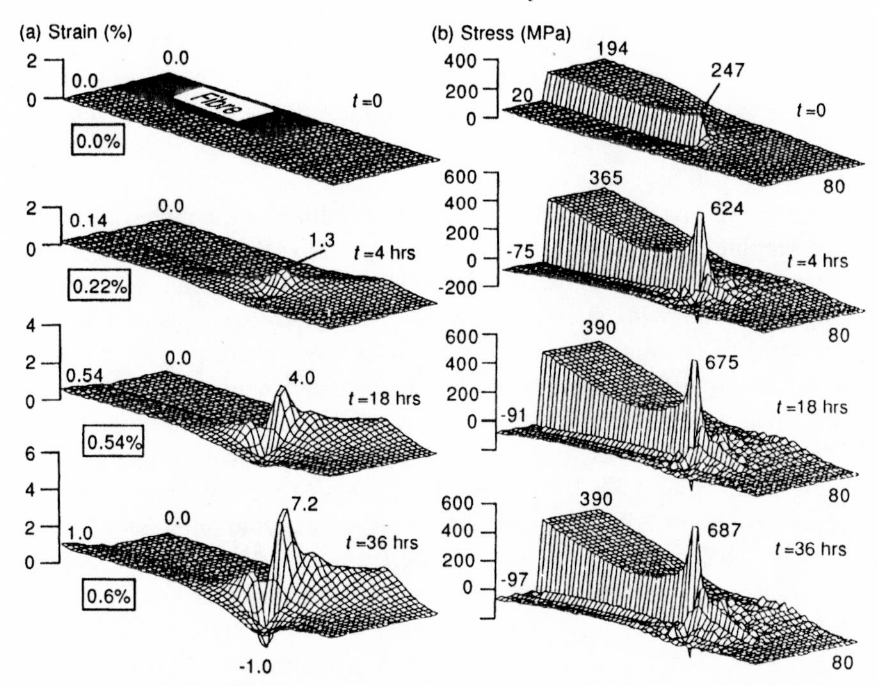

Fig. 5.13 Predicted distributions[54, 55] of (a) equivalent plastic strain and (b) axial stress within an Al–20 vol% SiC ($s = 5$) fibre composite creeping under an 80 MPa applied load. The fibre is cylindrical and the stress distributions are symmetrical radially and about the fibre mid-plane normal to its axis. The numbers shown represent strains in % (stresses in MPa) at particular locations. The absence of stress relaxation processes (cf. Taya *et al.*[37]) means that a steady state is never achieved; instead the creep rate asymptotically approaches zero.

Nix[54,55]. They applied a power law constitutive equation to local regions of the matrix. The approach is equivalent to treating the matrix deformation as if it were taking place by plastic flow, except that the plasticity now becomes time-dependent. The interfacial bond is taken as perfect and extension of the matrix assumed to transfer load progressively to the fibre (Fig. 5.13(b)). As with the other simple models, no steady state is achieved – rather the creep rate asymptotically decreases, as in the long fibre case (below). It crucially ignores the new opportunities for stress relaxation available in a composite. For example, such mechanisms as diffusive mass transport (driven by the very high hydrostatic gradient near the fibre end), and interfacial sliding could allow the matrix to extend without transferring load to the fibre. The effect of incorporating stress relaxation processes would be to set up a steady state with a higher average stress in the matrix, thereby raising the composite creep rate. It should in principle be possible to develop an FEM model incorporating these local stress relaxation processes. (Naturally, a remeshing scheme (see §2.4.2) will be required in which convective motion is allowed, to account for diffusive mass transport.)

An interesting point relating to FEM modelling as attempted hitherto is exemplified by the results of Dragone and Nix[54,55], in that if a single creep stress exponent is used to describe matrix creep for all stress levels (in their case $n = 4$), then the composite creep stress exponent must be the same (although the creep rate will be lower). This is at variance with experimental observations, which indicate an increase in n. This discrepancy may be a result of using a single constitutive equation irrespective of the stress level (at high stresses an exponential law may be more appropriate), or it may be due to activation of stress relieving processes. A contribution may also have come from internal thermal stresses, which were not included in the FEM model. Such FEM studies can, however, give useful insights into various aspects of the creep process. For example, Fig. 5.13(a), giving the distribution of creep strain within the matrix, shows how this tends to become concentrated in the matrix beyond the end of the fibre. In addition, FEM work indicates that, as one might expect given the pronounced tendency for peak stresses to be reduced by creep, the creep response is much less sensitive to fibre clustering[56] than plastic straining is found to be. In general, while FEM modelling is potentially very useful in describing the distribution of internal stresses, more work is needed on characterising the creep response of given materials (especially at the very high stress levels that arise soon after the application of the load, which can lie outside the range and timescale of conventional creep tests) and on incorporating the effects of local matrix microstructure.

Long fibre composites

Continuous fibre composites are expected to exhibit greater creep resistance than short fibre and particulate MMCs and this is reflected in calculations based purely on matrix stress state, such as those in Fig. 5.12(b). However, long fibre composites will tend to exhibit greater advantages than are indicated purely on this basis, because stress relaxation processes will be more difficult, for example, diffusion paths will be very long – see Fig. 5.11. As inhibition of stress relaxation will tend to reduce the *in situ* creep rate of the matrix for a given *in situ* stress, this factor will supplement the effect of the reduced matrix stress in giving a

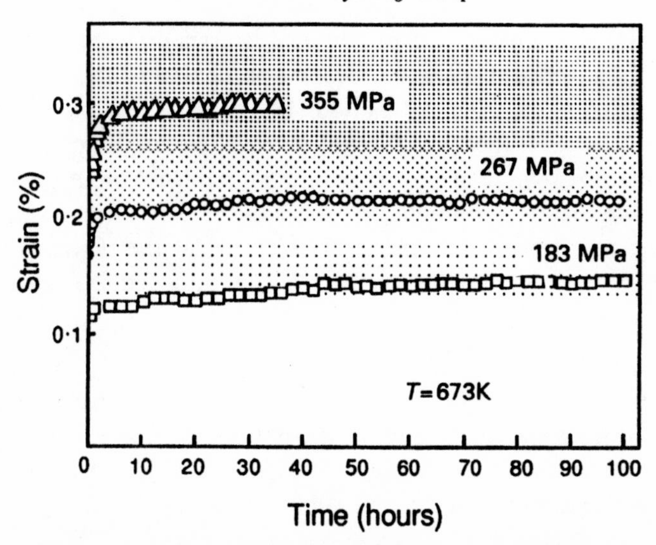

Fig. 5.14 Creep plot taken from the work of Endo *et al.*[57] for Al–49 vol% long fibre SiC, which illustrates the asymptotic approach of the creep strain towards the value corresponding to complete transfer of all the load to the elastically deforming fibres (upper limit of the shaded areas, ε_{max}). The lower limit of the shaded areas represents the predicted instantaneous elastic response, $\varepsilon_{t=0}$.

low creep rate. In principle, assuming the fibre ends to be fixed to the grips, a steady state is expected for continuous fibres in which they bear all the load[53], so that the composite creep rate will depend on the creep characteristics of the fibre. Naturally, in the case of elastically distorting fibres the analysis is especially straightforward. Assuming power law creep, the overall creep strain, ε_C, of the composite rises from the instantaneous value, $\varepsilon_{t=0}$, to asymptotically approach the maximum value, ε_{max}, when the fibres carry all of the load according to

$$\frac{1}{(\varepsilon_{max} - \varepsilon_C)^{n-1}} - \frac{1}{(\varepsilon_{max} - \varepsilon_{t=0})^{n-1}} = \frac{AE_M(1-f)}{E_C}\left(\frac{fE_I}{1-f}\right)^n t \quad (5.16)$$

where the matrix creep rate is given by $A\sigma^n$, $\varepsilon_{t=0} = \sigma^A/E_C$ and $\epsilon_{max} = \sigma^A/(fE_I)$. This kind of asymptotic behaviour is well-illustrated by the data in Fig. 5.14.

5.3 Thermal cycling creep

We have already seen that thermal stresses can affect the mechanical behaviour of MMCs, both in terms of yielding asymmetries (§4.1.2) and through changes in creep performance (§5.2.3). These effects are in general relatively weak, primarily because thermal stresses often become reduced to low levels by stress relaxation processes, particularly if these are allowed to operate over an extended period. However, if the temperature is repeatedly changed, then the thermal stresses are continuously regenerated, so that they can then exert a substantial influence on the mechanical performance. Before examining the mechanical behaviour of an MMC during thermal cycling, it is instructive to consider the effects of thermal cycling-induced internal stress changes on an unloaded specimen.

5.3.1 Thermal cycling

Dilatometry measurements of MMCs usually show evidence of hysteresis – i.e. the heating and cooling values do not coincide. A simple view[58] of the expected behaviour is presented in Fig. 5.15, which shows schematic changes in axial matrix stress and composite strain. Transverse stresses

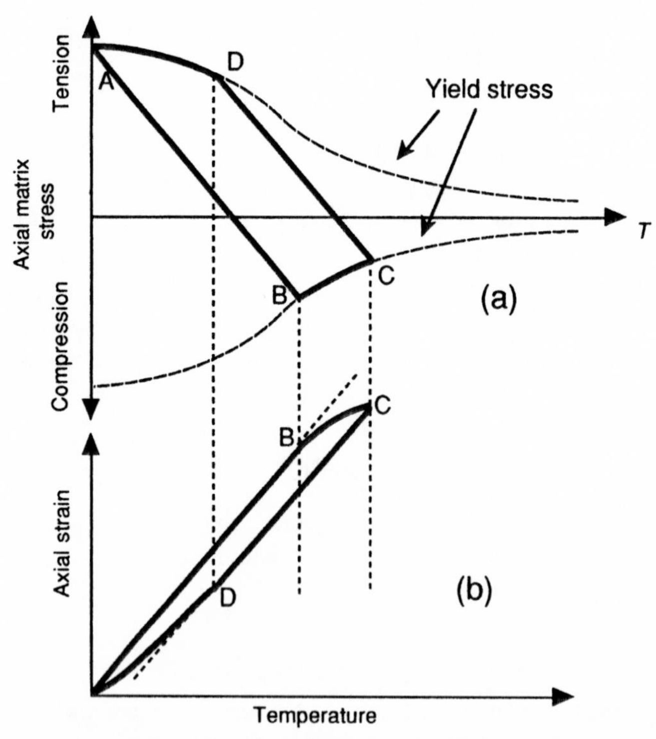

Fig. 5.15 Schematic illustration[58] of the variation in (a) matrix stress and (b) specimen strain during thermal cycling of an aligned fibre composite.

are neglected. The matrix is initially (point A) taken, Fig. 5.15(a), as having a residual tensile stress equal to the yield stress (because of the cooling cycle). On heating, this stress falls, becoming compressive and eventually causing yielding in compression (point B). A period of progressive plastic flow then follows. On cooling (from point C), the matrix yields in tension at point D, before returning to A. The dilatometry traces are predicted to show hysteresis, but no net dimensional change.

Most experimental data for long fibre composites[59–65] are broadly consistent with this view. In fact, a simple analysis[65] of the dilatometer trace can be used to calculate the changing axial matrix stress in a long fibre composite. Assuming no interfacial sliding, fibre yielding, etc., the axial strain of the composite must be equal to that of the fibres, which can be expressed as the sum of their natural thermal expansion and their elastic strain

$$\varepsilon_{3C} = \varepsilon_{3I} = \alpha_I \, \Delta T + \frac{[\sigma_{3I} - \nu_I(\sigma_{1I} + \sigma_{2I})]}{E_I} \tag{5.17}$$

As the radial and hoop strains in the fibre are relatively small compared with the axial strain (e.g. Fig. 5.3) and the Poisson ratio of ceramic fibres fairly low (~0.2), the contribution from the transverse fibre stresses can be neglected for an approximate calculation. The axial force balance, $\bar{\sigma}_{3I} f + \bar{\sigma}_{3M}(1 - f) = 0$, can then be used to find the average axial stress in the matrix

$$\bar{\sigma}_{3M} = \frac{f}{(1 - f)} E_I(\alpha_I \, \Delta T - \varepsilon_{3C}) \tag{5.18}$$

so that $\bar{\sigma}_{3M}$ is simply proportional to the difference, $\Delta\varepsilon$, between the natural thermal strain of the fibre and the measured strain of the composite. The initial thermal stress in the matrix can be deduced by taking the composite up to high temperature ($\sim 0.8 T_{mp}$), where the matrix stress will become very small, and running the fibre thermal expansion line back from this region. This is illustrated in Fig. 5.16, which shows (a) an experimental dilatometer trace and (b) the deduced matrix stress history. The latter is seen to be broadly of the form shown in Fig. 5.15(a). Note also that the elastic portions of the stress history in Fig. 5.16(b) – that at the start of cooling is very short – have gradients close to that predicted by the simple Turner expression[65–67]

$$\frac{\partial \bar{\sigma}_{3M}}{\partial T} = -\frac{(\alpha_M - \alpha_I) f E_I E_M}{(1 - f) E_M + f E_I} \tag{5.19}$$

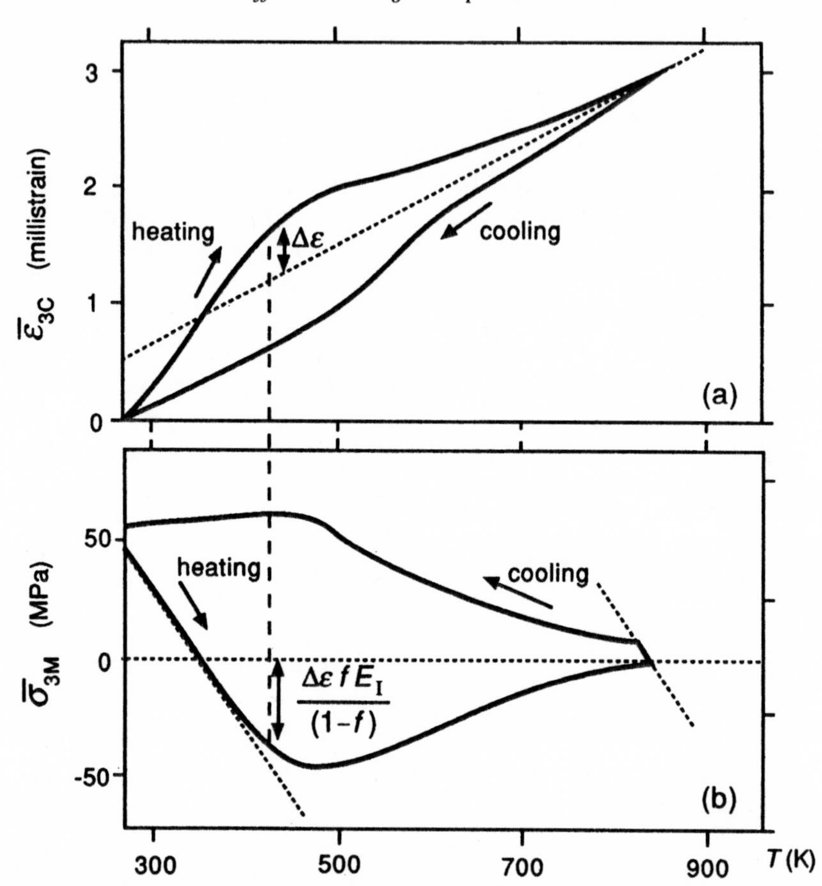

Fig. 5.16 (a) Experimental strain history[65] and (b) deduced variation of matrix axial stress for an Al–3 wt% Mg/30 vol% SiC long fibre composite. The matrix stress is taken, eqn (5.18), as being proportional to the difference between the measured composite strain and the natural thermal expansion of the fibres, which can be obtained by extrapolation of the high temperature data. The initial portions of the stress history curve, at the start of heating and of cooling, have gradients corresponding to elastic behaviour, as predicted by eqn (5.19) and indicated by the dotted lines.

This simple procedure forms a convenient way of studying the initial matrix stress, as well as the high temperature characteristics. For example, Masutti *et al.*[65] showed that a quench in liquid nitrogen, followed by heating to room temperature, generated a compressive residual stress of about 25 MPa in an Al–3 wt% Mg/20 vol% SiC composite, compared with a tensile stress of 50 MPa after heating and cooling back to room temperature.

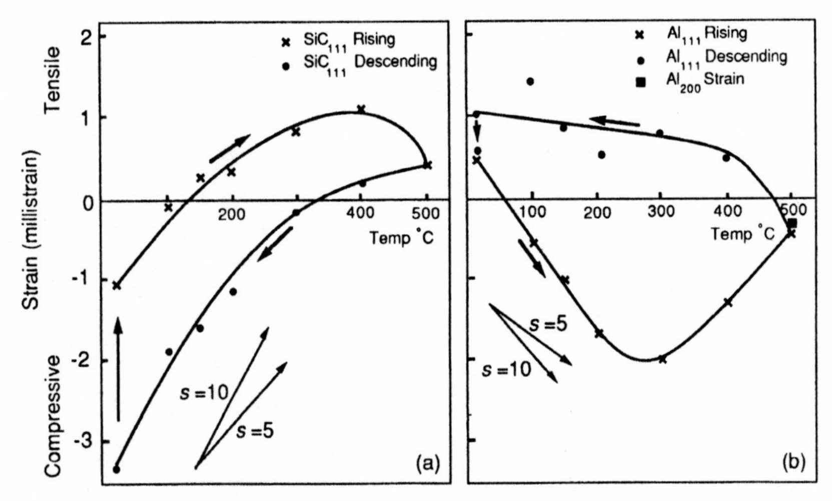

Fig. 5.17 Neutron diffraction measurements[68] of the lattice strain variation over a thermal cycle for 5 vol% aligned SiC_W/Al in (a) the reinforcement and (b) the matrix. The observed strain (stress) relaxation at the end of the cycle took place over a period of 8 hours. The gradients expected for the elastic accommodation of the misfit are shown for aspect ratios 5 and 10.

The behaviour of short fibre composites during thermal cycling is rather similar to that described above, but there is more scope for various stress relaxation processes to operate. This is illustrated by the neutron diffraction data[68] presented in Fig. 5.17, which show the changing axial strains (and hence stresses) in both fibre and matrix. The matrix strain (stress) history is similar to those shown in Figs. 5.15 and 5.16, although the observation of a significant drop in the final residual stress over an 8 hour period at room temperature is suggestive of continuing stress relaxation processes. A number of experimental studies have revealed that thermal cycling of aligned short fibre composites can cause a progressive net change in specimen dimensions on every cycle – an effect often termed ***strain ratchetting***. Although some aspects of this phenomenon remain to be clarified, it seems likely that it is essentially a consequence of asymmetrical stress relaxation during the cycle. The general observation has been for an increase in specimen length along the fibre direction.

Typical dilatometry data for the first few cycles are shown in Fig. 5.18, in which the elastic expansion gradients for the two phases and for the composite are also shown. These traces bear some similarity to those for the long fibre composite (Fig. 5.16(a)). The major differences are that the gradient at high temperature does not fall as low as that of the fibre alone

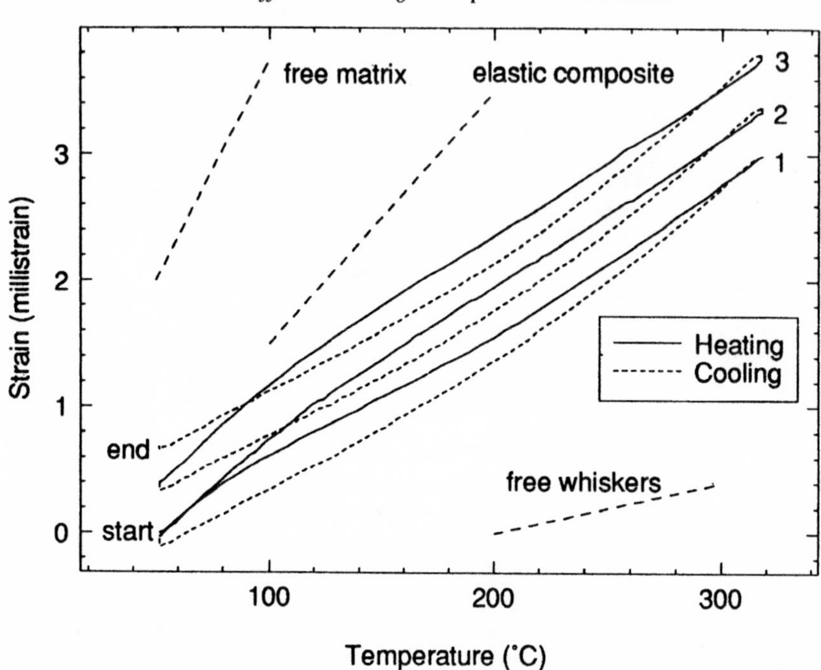

Fig. 5.18 Dilatometer data[71] showing the axial strain as a function of temperature during the first 3 thermal cycles for a Mg–11 wt% Li/20 vol% SiC_W composite. Also shown are the gradients for elastic expansion of matrix, fibre and composite (obtained using eqn (5.4)).

and that a net increase in length per cycle is established (after the first cycle – which tends to differ from the subsequent ones because the initial residual stress may be different). This net strain per cycle has been observed[69,71,72] to remain constant for many cycles, although Toitot *et al.*[72] report that, for a cycle extending from only 50 °C to 150 °C, there is a marked transient response so that a stable strain increment is attained only after about 1000 cycles. They propose a scheme in which plastic flow, rather than creep, is predominant in order to explain this observation and this may well be appropriate when the maximum temperature is relatively low. For cases where the cycle extends to temperatures where the matrix creeps readily, the observed behaviour can be broadly explained in terms of the changing matrix stresses during the cycle, using matrix creep characteristics. Information about the effects of changes to the thermal cycle can be helpful in interpreting the behaviour of particular systems. For example, Fig. 5.19 shows the mean axial strain per cycle observed with Mg–11 wt% Li/20 vol% SiC_W for different thermal history profiles.

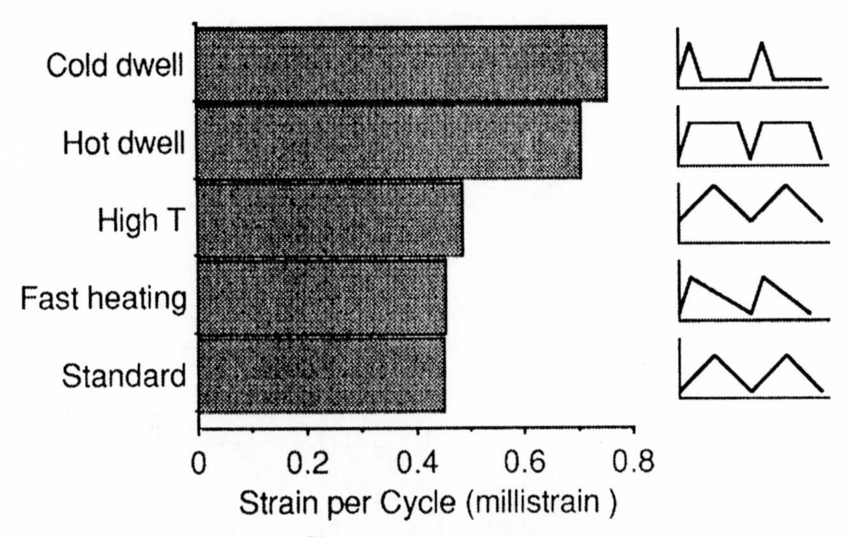

Fig. 5.19 Mean axial strain[71] per thermal cycle (measured after two hundred 100 ⇄ 350 °C cycles) for a Mg–11 wt% Li/20 vol% SiC_W composite, shown for the five different thermal history profiles depicted on the right. In this context it is important to note that even the low temperature (100 °C) is relatively high in terms of the relaxation kinetics of this matrix.

A high strain per cycle was maintained in all cases, but it appeared to be larger when dwells were imposed at either high or low temperature, suggesting that relaxation is an essential component of thermal cycle creep. A schematic illustration of the changing internal stress state during cycling is given in Fig. 5.20(b). While the relative importance of different mechanisms will clearly vary from system to system and with different thermal histories, many aspects of the thermal cycling behaviour of MMCs can be explained on this basis.

5.3.2 Loaded thermal cycling

Many researchers have observed[47,70,74–77] that the creep rates of discontinuous MMCs are substantially enhanced by thermally cycling the specimen while under an applied load. In this context, it is useful to introduce the concept of a 'diffusional mean' temperature, $\bar{T}$, and the corresponding diffusion coefficient, $\bar{D}$, which facilitate comparison with isothermal creep data. Essentially, $\bar{T}$ is the constant temperature for which the isothermal creep rate of unreinforced matrix would be expected to equal that averaged over the prescribed thermal cycle, assuming the creep

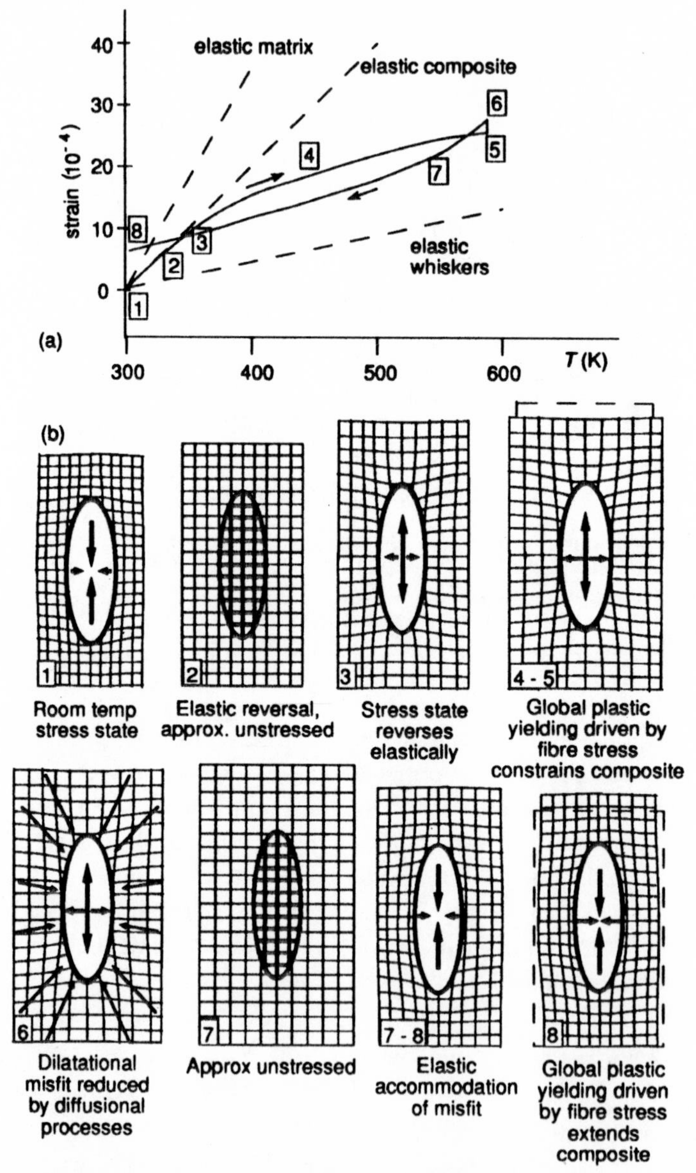

Fig. 5.20 Simplified representation of (a) the strain history observed during a thermal cycle, together with (b) schematic depictions of corresponding changes in stress state. Essentially, plastic flow starts at point [3], followed by increasing contributions from diffusional processes, probably with at least partial relaxation of the hydrostatic stress during the high temperature dwell. On cooling, the elastic regime is more limited, so that inelastic behaviour tending to elongate the composite is more pronounced than the complementary effects during the heating cycle, resulting in a net elongation.

rate to be proportional to D_{eff} – as defined in eqn (5.11). For a thermal cycle extending over a period τ, $\bar{D}$ is defined by the equation

$$\bar{D} = \frac{1}{\tau}\int_0^{\tau} D_{eff}(T(t))\,dt \qquad (5.20)$$

where $T(t)$ is the thermal history function within the cycle. Thus, when examining the average creep rate of a specimen over the thermal cycle, comparison with the rate at $\bar{T}$ will reveal whether creep has been enhanced.

In fact, it is well-established that materials in which anisotropic internal stresses are generated on changing the temperature tend to creep at an enhanced rate when thermally cycled, particularly in the low stress regime. This was first observed[78,79] for pure polycrystalline (non-cubic) metals such as Zn and U, having anisotropic thermal expansivities. A change in temperature leads to internal stresses in these materials in a similar way to the situation in MMCs. Experimental evidence[80] confirms that cycled specimens creep at a higher rate (Fig. 5.23), particularly in the low applied stress regime, where the stress exponent for cycled specimens falls[74] from 5 to about 1 (except at low cycling temperatures[70]), giving rise to superplastic extensions[74,76]. In contrast to low temperature cycling[72], a constant strain increment is observed per cycle with no incubation period[73]. Indeed, no transients are observed even upon changing the applied creep load[76]. Further, it is found[73,76,77] that, beyond a critical threshold, there is an approximately linear relationship between the creep per cycle and ΔT, although there is some evidence that the strain increment eventually reaches a plateau[73]. At low temperatures[70] on the other hand, the strain per cycle varies approximately as ΔT^2.

Experimental evidence suggests that increasing the size and volume fraction of particulate reinforcement increases the creep per cycle, although the effect of volume fraction may reverse at high inclusion contents[81]. Given that strain ratchetting is observed with short fibre MMCs in the absence of an applied load, it is not surprising that creep rates are enhanced in this case. However, creep enhancement is also observed for (equiaxed) particulate MMCs, suggesting that the mechanisms involved need to be considered rather carefully. In contrast, thermal cycling-enhanced creep has not been observed for long fibre composites, except under off-axis loading[82].

Two types of theoretical model have been proposed:

- ***continuously*** or ***repeatedly deforming models***[83,84] (also called enhanced plasticity models[85]) in which the thermal stress combines with the applied load to exceed the material's yield stress, giving rise to a time-independent strain which is biased according to the sense of the applied load

- ***enhanced creep models*** in which the internal stresses are not sufficient to cause matrix yielding, but aid conventional global time-dependent creep

Enhanced plasticity models

As reviewed by Derby[85], a large number of plasticity-based models have been proposed[77,83,84,86–89], but they all have a form similar to that originally developed to describe the effect of the anisotropic expansion of polycrystalline uranium on creep

$$\Delta\varepsilon = \frac{5}{6}\frac{\sigma^{A}}{\sigma_{Y}}\,\Delta\alpha\,\Delta T \tag{5.21}$$

These models assume that the thermal excursion ΔT is sufficient to bring the matrix to the yield point†, such that an applied load biases the plastic strain in each grain, giving rise to a net permanent extension when the strain is averaged throughout the whole matrix over the complete thermal cycle. In its simplest form (eqn (5.21)), it will tend to underestimate the strain increment for aligned systems, because in the original model the grains are randomly oriented and would thus, in the absence of an applied load, flow plastically in random directions. A net extension would therefore only occur under an applied load. This is not the case with aligned systems, for which much of the matrix would satisfy the yield condition parallel to the applied load. In agreement with the experimental results, this model predicts linear relationships between the strain increment and both the applied load and ΔT, and predicts enhanced plasticity even for particulate material (and randomly oriented short fibres). It has been extended to include the variation in matrix yield stress with temperature[89], but the approach can only be valid if the temperature rise is large and rapid enough for local stress relaxation processes to be unable to relax away the internal stresses.

In the variant proposed by Daehn and co-workers[77,87,88] it is assumed that deformation in a short fibre MMC occurs by (plane strain) matrix shear, neglecting regions of the matrix beyond the ends of the fibres (which

† ΔT in eqn (5.21) should really be $\Delta T - \Delta T_Y$, where ΔT_Y is the thermal excursion necessary to provoke yielding[84].

are really taken as plates). This enables them to express the *in situ* shear stress driving the deformation in a simple geometrical form involving fibre volume fraction and aspect ratio.

$$\tau_{31} = \frac{2\sigma^{A}h}{L} \tag{5.22}$$

where L is the length of a whisker (plate) and h is the spacing between them. The internal stresses in the matrix are then related to the temperature change ΔT

$$\sigma_{3M} = -[E_{M}(\Delta\alpha\,\Delta T + \varepsilon_{3M}^{P})] \Bigg/ \left[1 + \frac{(1-f)E_{M}}{fE_{I}}\right] \tag{5.23}$$

where ε_{3M}^{P} is the matrix plastic strain in the axial direction. A simple Mohr circle technique is then employed to establish the magnitude and direction of the maximum shear stress in the matrix. This is then used to calculate the matrix shear strain rate, based on a simple power law creep expression, and hence the composite strain rate. A simple computer program is then used to find the strain rate history during any prescribed thermal cycle, so that the average strain rate can be predicted. There is no simulation of any progressive loading of the fibres as a result of the matrix straining, so that only a steady state is treated. Using this method, surprisingly good agreement has been obtained between theory and experiment for the dependence of strain rate on applied stress – see Fig. 5.21. The low stress exponent (or, equivalently, high strain rate sensitivity exponent) under low applied stress has the effect of inhibiting neck growth and hence favours the development of large (superplastic) strains; Daehn and co-workers have been prominent in proposing how this could be exploited in superplastic forming of MMCs (see §9.3.3).

The approach of calculating the net (deviatoric) stress in the matrix at any point in the cycle and then applying a matrix plastic flow or creep rate expression can also be employed with more realistic geometrical models. For example, finite element modelling[90] has been used to explore the effect of the amplitude of the thermal cycle – see Fig. 5.22. It can be seen in this figure that better agreement is observed with experimental data using this approach than with the simpler analytical model. Note also that the fastest creep rate is produced with an intermediate fibre aspect ratio (Fig. 5.22(b)). This might have been expected in view of the tendency for a higher aspect ratio to raise the (deviatoric) thermal stress in the matrix, but to reduce the proportion of the applied load borne by the matrix. The effect remains to be confirmed experimentally.

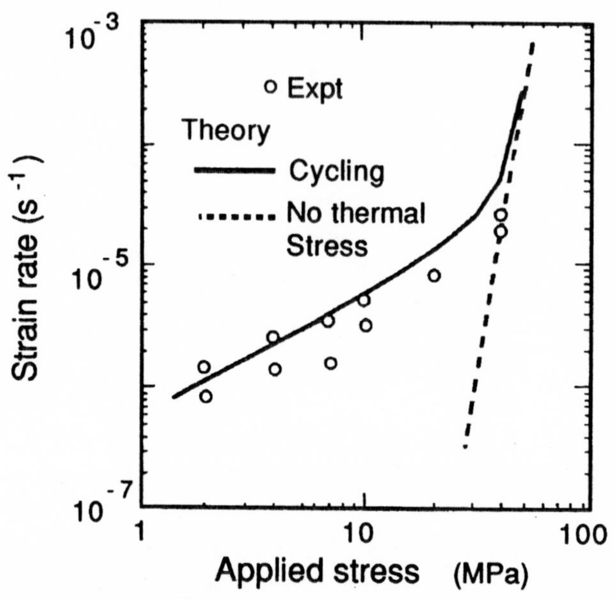

Fig. 5.21 Experimental data[77] for the steady state creep rate of a 6061 Al–20 vol% SiC_W composite, cycled between 373 K to 723 K with a 200 s cycle time, as a function of the applied stress. Also shown are theoretical predictions for this treatment (solid line). The dotted line was obtained by running the model with no thermal expansivity mismatch between fibre and matrix, which is equivalent to an isothermal test at the diffusional mean temperature.

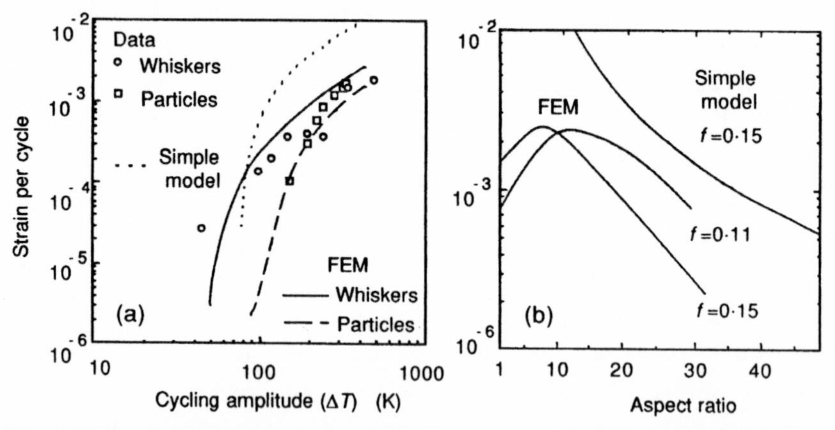

Fig. 5.22 Predictions from finite element modelling[90] and simpler analytical modelling[77] (whiskers only) of the thermal cycling creep of discontinuous MMCs, with (a) experimental data for 6061 Al–11 wt% SiC_W[77] and for 1100 Al–11 vol% SiC_P[89], giving the strain per cycle as a function of the amplitude of the thermal cycle, for an applied stress of 8 MPa. (b) The predicted dependence of the strain per cycle on the fibre aspect ratio, for an applied stress of 8 MPa and a temperature cycle of 350 °C amplitude.

Enhanced creep models

In these models the matrix stress is assumed to be too small for instantaneous plastic flow, but sufficient for time-dependent global matrix creep. For example, in Eshelby-type models (see below), global matrix extension is predicted in terms of the average matrix stress state (applied plus thermal component) and its creep behaviour. Again, as with the isothermal creep models (§5.2.3), the effect of the creep process itself on the level of the internal stresses driving it may or may not be taken into account, depending on the complexity of the model. Clearly, the extent of enhanced creep is much more dependent on the shape and timescale of the cycle than is the case for the enhanced plasticity models. Also, as can be inferred from the creep acceleration ratios calculated in connection with the cycling of uranium[84,91], the enhancement increases sharply as the applied load is reduced and the stress sensitivity of the matrix increases.

Following an approach similar to that used for enhanced plasticity, it is possible to arrive at a simple creep *strain rate* relation for a given internal stress level ($\sigma^A \ll \bar{\sigma}_M < \sigma_Y$)[84,91]

$$\dot{\varepsilon}_C = B\sigma^A\bar{\sigma}_M^n \tag{5.24}$$

where B is a constant. For slow cooling rates, the rate of generation of matrix stress ($\bar{\sigma}_M$) by cooling will be counter-balanced by its rate of reduction via enhanced creep, so that a steady state is possible (just as $\bar{\sigma}_M$ remains constant, and equal to $\bar{\sigma}_Y$, for enhanced plasticity). If, however, the cooling rate is too fast, then the matrix stress will increase until it exceeds the yield stress, giving rise to enhanced plasticity.

Some confusion exists in the literature regarding the strain limit proposed by Young *et al.*[91] and Anderson and Bishop[84] for creep deformation, which is

$$\dot{\varepsilon}_{\mathrm{Cmax}} = B_1 \frac{\sigma^A}{E} \ln\left(\frac{\sigma_Y}{2\sigma^A}\right) \tag{5.25}$$

where B_1 is another constant. This equation refers to the maximum strain that could occur after a thermal cycle has finished, i.e. it is the length change which would occur if an internal deviatoric stress just equal to the yield stress were completely relaxed upon completion of the cycle. This is not the maximum strain per cycle, because, just as with the plasticity model, the internal stress is continually being regenerated by the temperature changes provided the cooling rate is sufficiently slow.

Rather, it is the maximum contribution of enhanced creep to the strain per cycle, in the limit of a square waveform temperature cycle.

Wu *et al.*[74,80] considered the effect of the internal stress $\sigma_0^{\Delta T}$ on dislocation movement, proposing that it assists the motion of one-half of the dislocations and impedes the motion of the rest. Wu *et al.*[80] showed that, for an applied stress which is less than $\sigma_0^{\Delta T}$, internal stress-assisted creep can be described by

$$\dot{\varepsilon} = AD_{\text{eff}}\left(\frac{\sigma_0^{\Delta T}}{E}\right)^{n-1}\frac{\sigma^{A}}{E} \tag{5.26}$$

where n is the stress exponent exhibited when $\sigma^{A} \gg \sigma_0^{\Delta T}$. Wu *et al.*[80] give an equation, based on a hyperbolic sine creep relation, which correctly predicts the creep rates of the thermally cycled material over the complete range of applied stress – see Fig. 5.23. The approach is, however, rather empirical; both the nature of $\sigma_0^{\Delta T}$ and the way in which it is assumed to influence dislocation motion are unclear. Nevertheless, the model does provide an insight into the origin of the $n = 1$ behaviour at low applied loads. Basically, the internal thermal stress dominates the net matrix stress state, drastically reducing the sensitivity to the applied load.

Eshelby modelling has also been used to explore thermal cycling creep[47,72,92,93]. In this case a single average deviatoric stress (e.g. $\bar{\sigma}_{3M} - \bar{\sigma}_{1M}$) can be associated with the matrix and established as a function of temperature (see Fig. 5.24). Combination of the matrix creep strain rate (as a function of the temperature) with the deviatoric stress allows the composite strain rate to be calculated. It can be seen that this could become negative, particularly for high fibre contents and at high temperatures. This plot highlights a particular problem with thermal cycling creep modelling, which is the identification of the initial stress state of the matrix. This is conveniently characterised by specifying the temperature at which the matrix stress would be zero (with no applied load) – the effectively 'stress-free' temperature, T_{esf} – see Fig. 5.24.

The comparisons shown in Fig. 5.25, between prediction and experiment for the variation of instantaneous creep strain rate during the cycle, demonstrate that this type of modelling can give useful insights into the effects responsible for the observed behaviour. Note, however, that under this approach, equiaxed ($s = 1$) particles should not generate any *volume-averaged* deviatoric thermal stress in the matrix, so that thermal cycling is not expected to generate strong effects in particulate MMCs. The observation of significant effects suggests that the presence of *local*

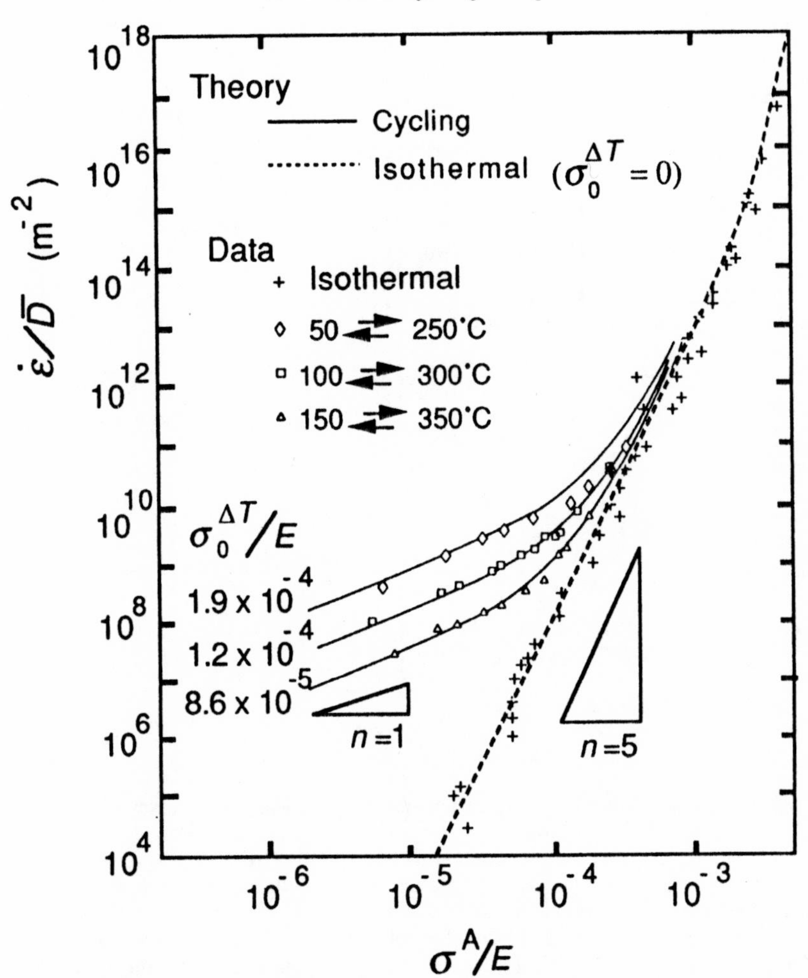

Fig. 5.23 Diffusion-compensated creep strain rates as a function of modulus-compensated applied stress for isothermal and thermally cycled loading of polycrystalline Zn[80]. The dotted lines tend to a gradient $n = 1$ in the low stress regime, in accordance with eqn (5.26).

deviatoric thermal stresses is relevant, although it may be noted that most particulate MMCs actually exhibit particle aspect ratios greater than unity and a degree of alignment.

5.3.3 Effects of prior thermal cycling on properties

There have been some reports of property impairment as a consequence of thermal cycling treatments, particularly for long fibre MMCs. In some

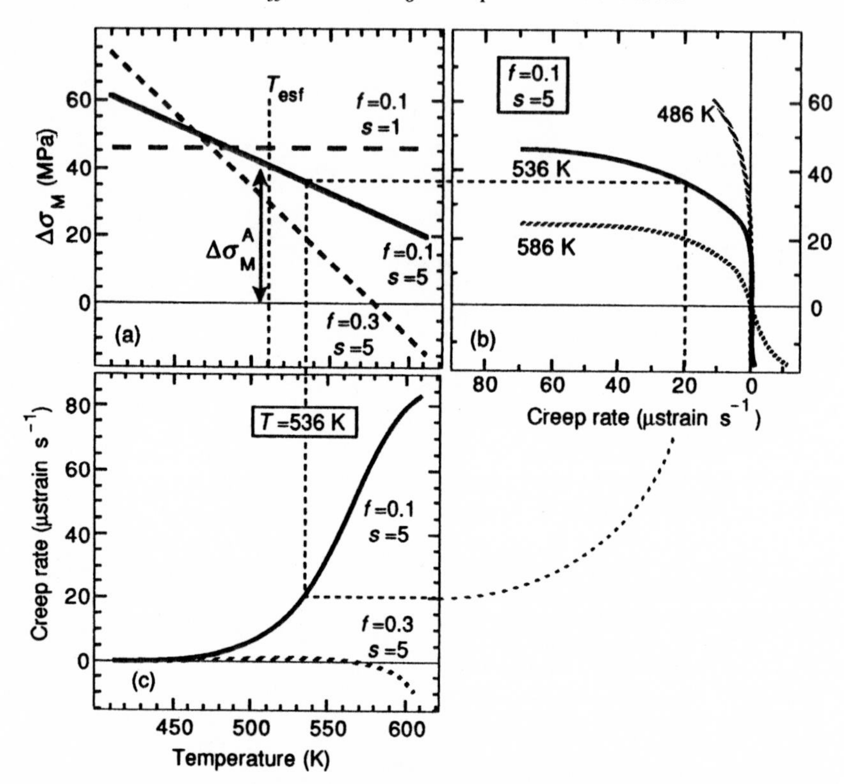

Fig. 5.24 Eshelby-type modelling of volume-averaged matrix stresses and resultant creep rates[47]. In (a) the matrix deviatoric stress $\Delta\sigma_M$ is calculated as a function of T for three Al–Al_2O_3 composites externally loaded to 50 MPa, assuming an effectively stress-free temperature (T_{esf}) of 510 K, while in (b) creep strain rates are plotted for one of the composites as a function of the deviatoric stress at different temperatures. In (c) knowledge of the *in situ* matrix stress taken from (a) is used to predict the creep strain rate as a function of T. Note that above T_{esf} the thermal stresses lower the matrix stress, thus slowing the rate of creep; in this case the peak cycle temperature is only 300 °C (573 K); it would be extremely unlikely for the effective stress-free temperature to remain constant over the cycle at higher temperatures.

cases, this can be attributed to a well-defined change in matrix microstructure or interfacial reaction, caused by the exposure to high temperature, rather than by the cycling per se. For example, in directionally solidified eutectic '*in situ*' composites, there have been reports of property degradation from the precipitation of brittle phases[94] and from changes in fibre morphology[95], but the cycling itself appeared to have little effect. (The generation of thermal gradients, however, can be deleterious – see

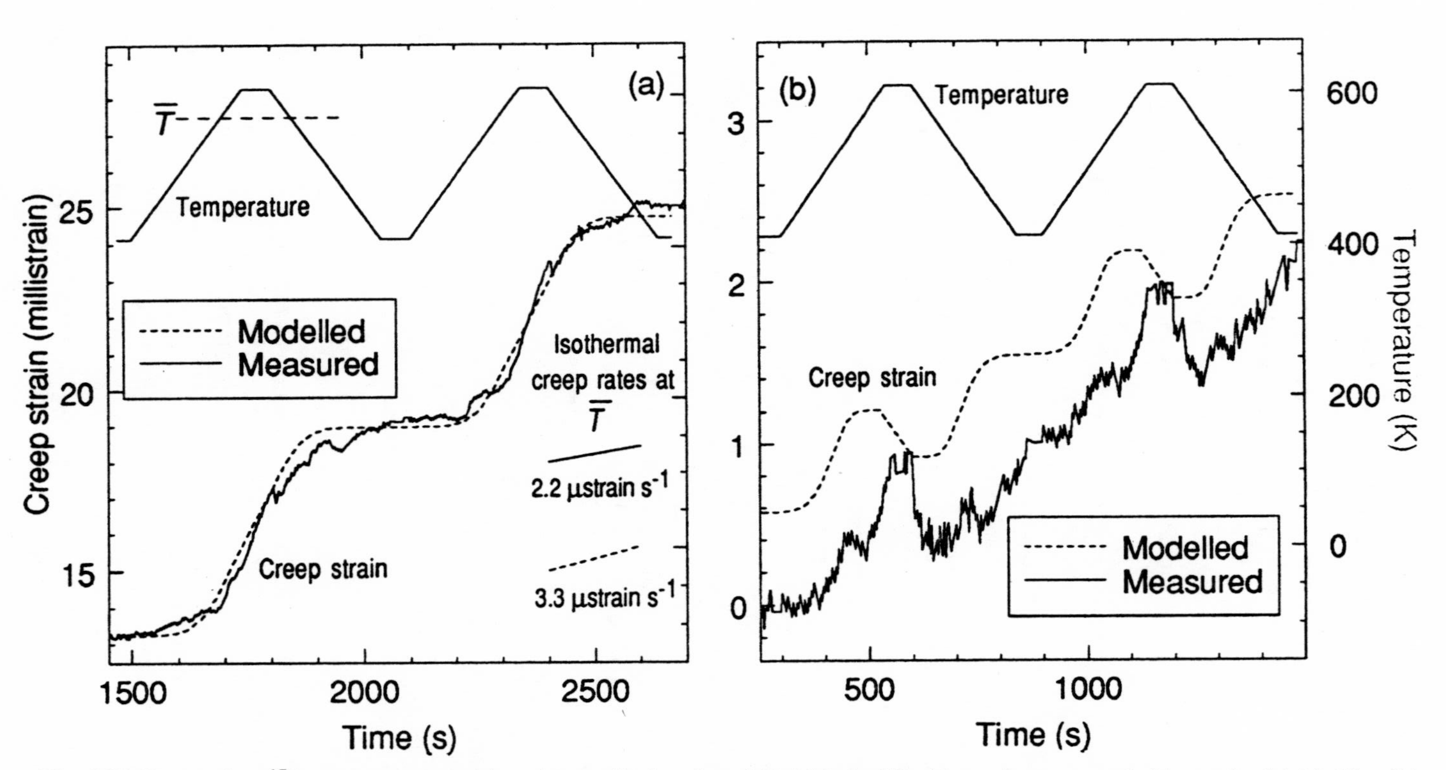

Fig. 5.25 Comparison[47] between measured (continuous line) and modelled (dashed line) in-cycle creep strain history for (a) Al–10 vol% Saffil® (short Al_2O_3 fibre) and (b) Al–20 vol% Saffil®, under 20 MPa load. Comparative data are also shown in (a) for the corresponding isothermal creep rate at the diffusional mean temperature. Note that in (b) periods of reverse creep are both observed and predicted for a part of the cycle.

§9.1.4.) In other long fibre composites, such as B-reinforced Ti, thermal cycling is reported[96] as causing no enhancement of the interfacial reaction. In contrast, a single thermal cycle has been reported[97] to degrade the properties of Ti–SiC laminates, which is largely attributable to interfacial reaction.

Not surprisingly, some long fibre composite systems with a large thermal expansivity mismatch do seem to be susceptible to microstructural damage caused by the internal stresses which arise during thermal cycling. For example, interfacial cavitation and debonding have been observed[98–102] in Al and Ni reinforced with C, B, W and SiC fibres. This has led to impairment of tensile strength and ductility[101], and fractured fibres resulting from thermal stresses have been blamed for fatigue cracks[102]. These reports do, however, relate to systems (such as Al–C) which tend to exhibit low interfacial bond strength. Long fibre Mg–C MMCs have been reported[103] to retain their (desirable) low CTE properties after repeated cycling.

For short fibre and particulate composites, severe microstructural damage as a consequence of cycling seems less common, as expected from the lower levels of thermal stress. Patterson and Taya reported[69] a modest strength reduction due to cycling-induced damage in powder route 2124 Al/SiC_W MMC. Nakanishi *et al.*[104], however, found little or no change in strength or ductility after cycling cast Al–SiC_W, while Nishida *et al.* reported[105] little effect for cast Al–Si_3N_{4W}. These composites would, of course, have had a much softer matrix than the powder route 2124 Al/SiC_W, so that the maximum thermal stresses would have been lower. Warwick[106] has reported a migration of SiC whiskers towards the free surface of Mg–11 wt% Li/SiC_W composites during thermal cycling, which was explained in terms of local thermal stresses.

References

1. S. Krishnamurthy (1989) Interfacial Reactions in Fiber Reinforced Gamma Titanium Aluminide Composites, in *Interfaces in Metal–Ceramic Composites*, R. Y. Lin, R. J. Arsenault, G. P. Martins and S. G. Fishman (eds.), TMS-AIME, Warrendale, Pa, pp. 75–83.
2. J. K. Lee, Y. Y. Earmme, H. I. Aaronson and K. C. Russel (1980) Plastic Relaxation of the Transformation Strain Energy of a Misfitting Spherical Precipitate: Ideal Plastic Behaviour, *Metall. Trans.*, **11A**, pp. 1837–47.
3. Y. Mikata and M. Taya (1985) Stress Field in a Coated Continuous Fibre Composite Subjected to Thermo-Mechanical Loadings, *J. Comp. Mat.*, **19**, pp. 554–79.

4. C. M. Warwick and T. W. Clyne (1991) Development of a Composite Coaxial Cylinder Stress Analysis Model and its Application to SiC Monofilament Systems, *J. Mat. Sci.*, **26**, pp. 3817–27.
5. D. E. Bowles and S. S. Tompkins (1989) Predictions of Coefficients of Thermal Expansion for Unidirectional Composites, *J. Comp. Mat.*, **23**, pp. 370–88.
6. R. A. Schapery (1968) Thermal Expansion Coefficients of Composite Materials based on Energy Principles, *J. Comp. Mat.*, **2**, pp. 380–404.
7. T. W. Clyne (1990) A Compressibility-based Derivation of Simple Expressions for the Transverse Poisson's Ratio and Shear Modulus of an Aligned Long Fibre Composite, *J. Mat. Sci. Letts.*, **9**, pp. 336–9.
8. E. E. Wolf, E. G. Kendall and W. C. Riley (1980) Thermal Expansion Measurements of MMCs, in *3rd Int. Conf. on Comp. Mat.* (*ICCM III*), Paris, A. R. Bunsell and C. Bathais (eds.), Pergamon, pp. 1140–52.
9. K. G. Kreider and V. M. Patrani (1970) Thermal Expansion of Boron Fiber-Aluminium Composites, *Metall. Trans.*, **1**, pp. 3431–5.
10. H. J. Frost and M. F. Ashby (1982) *Deformation Mechanism Maps – The Plasticity and Creep of Metals and Ceramics*, Pergamon, Oxford.
11. R. W. Lund and W. D. Nix (1975) On High Creep Activation Energies for Dispersion Strengthened Metals, *Metall. Trans.*, **6A**, pp. 1329–33.
12. L. M. Brown and M. F. Ashby (1980) On the Power Law Creep Equation, *Scripta Met.*, **14**, pp. 1297–302.
13. G. S. Ansell and J. Weertman (1959) Creep of Dispersion Hardened Al Alloy, *Trans. TMS-AIME*, **215**, pp. 838–43.
14. O. D. Sherby, R. H. Klundt and A. K. Miller (1977) Flow Stress, Subgrain Size and Subgrain Stability at Elevated Temperature, *Metall. Trans.*, **8A**, pp. 843–850.
15. J. G. Harper and J. E. Dorn (1957) Viscous Creep of Al Near its Melting Temperature, *Acta Metall.*, **5**, pp. 654–65.
16. M. J. Luton and C. M. Sellars (1969) Dynamic Recrystallisation in Ni & Ni–Fe Alloys during High Temperature Deformation, *Acta Metall.*, **17**, pp. 1033–43.
17. L. M. Brown (1979) Precipitation and Dispersion Hardening, in *ICSMA 5*, Aachen, P. Hansen, V. Gerold and G. Kostorz (eds.), Pergamon, pp. 1551–71.
18. V. C. Nardone and J. K. Tien (1986) On the Creep Rate Dependence of Particle Strengthened Alloys, *Scripta Met.*, **20**, pp. 797–802.
19. J. H. Hausselt and W. D. Nix (1977) A Model for High Temperature Deformation of Dispersion Strengthened Metals Based on Substructural Observations in Ni–20Cr–2ThO_2, *Acta Metall.*, **25**, pp. 1491–502.
20. R. S. W. Shewfelt and L. M. Brown (1974) High Temperature Strength of Dispersion Hardened Single Crystals II. – Theory, *Phil. Mag.*, **30**, pp. 1135–45.
21. N. Hansen and B. Bay (1972) The Effect of Particle Content, Particle Distribution and Cold Deformation on the Recrystallization of Low Oxide Al–Al_2O_3 Products, *J. Mat. Sci.*, **7**, pp. 1351–62.
22. R. A. Shahani and T. W. Clyne (1991) Recrystallization in Fibrous and Particulate MMCs, *Mat. Sci. & Eng.*, **A135**, pp. 281–5.
23. V. C. Nardone and J. K. Tien (1983) Pinning of Dislocations on the Departure Side of Strengthening Dispersoids, *Scripta Met.*, **17**, pp. 467–70.
24. C. L. Myers, J. C. Shyne and O. D. Sherby (1963) Relation of Properties

to Structure in Sintered Aluminium Powder, *J. Austral. Inst. Metals*, **8**, pp. 171–83.
25. T. A. Trozera, O. D. Sherby and J. E. Dorn (1957) Effect of Strain Rate and Temperature on the Plastic Deformation of High Purity Aluminium, *Trans. ASM*, **49**, pp. 173–85.
26. J. L. Lytton, L. A. Shepard and J. E. Dorn (1958) The Activation Energies for Creep of Single Aluminium Crystals, *Trans TMS-AIME*, **212**, pp. 220–5.
27. R. W. Lund and W. D. Nix (1976) High Temperature Creep of Ni–20Cr–2ThO$_2$ Single Crystals, *Acta Metall.*, **24**, pp. 469–81.
28. S. Purushothaman and J. K. Tien (1978) Role of Back Stresses in the Creep Behaviour of Particle Strengthened Alloys, *Acta Metall.*, **26**, pp. 519–28.
29. V. C. Nardone and J. R. Strife (1987) Analysis of the Creep Behaviour of Silicon Carbide Whisker Reinforced 2124 Al (T4), *Metall. Trans.*, **18A**, pp. 109–14.
30. I. G. Crosland and R. B. Jones (1977) Grain Boundary Diffusion Creep in Mg, *Metal Sci.*, **11**, pp. 504–8.
31. W. J. Clegg and J. W. Martin (1982) Diffusion Creep Threshold in Two-Phase Alloys, *Metal Sci.*, **16**, pp. 65–72.
32. E. Arzt, M. F. Ashby and R. A. Verrall (1983) Interface Controlled Diffusional Creep, *Acta Metall.*, **31**, pp. 1977–89.
33. W. J. Clegg (1984) On the Existence of a Threshold Stress in Diffusional Creep, *Scripta Met.*, **18**, pp. 767–71.
34. T. Schober and R. W. Ballufi (1970) Quantitative Observation of Misfit Dislocation Arrays in Low and High Angle Twist Grain Boundaries, *Phil. Mag.*, **21**, pp. 109–23.
35. T. G. Nieh (1984) Creep Rupture of a Silicon Carbide Reinforced Aluminium Composite, *Metall. Trans.*, **15A**, pp. 139–46.
36. J. A. G. Furness (1991) *Thermal Cycling Creep of Aluminium-based Composites*, PhD thesis, Univ. of Cambridge.
37. M. Taya, M. Dunn and H. Lilholt (1991) Long Term Properties of Metal Matrix Composites, in *Metal Matrix Composites – Processing, Microstructure and Properties, Risø 12th Int. Symp.*, Roskilde, N. Hansen, D. J. Jensen, T. Leffers, H. Lilholt, T. Lorentzen, A. S. Pedersen, O. B. Pedersen and B. Ralph (eds.), Risø Nat. Lab., Denmark, pp. 149–72.
38. K. T. Park, E. J. Lavernia and F. A. Mohamed (1990) High Temperature Creep of Silicon Carbide Particulate Reinforced Aluminium, *Acta Metall.*, **38**, pp. 2149–59.
39. P. Jarry, W. Loué and J. Bouvaist (1987) Rheological Behaviour of SiC/Al Composites, in *Proc. ICCM VI/ECCM 2*, F. L. Matthews, N. C. R. Buskell, J. M. Hodgkinson and J. Morton (eds.), London, Elsevier, pp. 2.350–2.361.
40. J. F. Mason (1990) *The Fabrication and Mechanical Properties of Mg–Li Alloys Reinforced with SiC Whiskers*, PhD thesis, Univ. of Cambridge.
41. J. F. Mason and T. W. Clyne (1990) Diffusive Stress Relaxation During Creep of Aligned Whisker Reinforced Mg–Li Alloy, in *Proc. 7th Int. Conf. on Comp. Mat. (ICCM VII)*, Beijing, China, W. Yunshu, G. Zhenlong and W. Renjie (eds.), Pergamon, Oxford, pp. 45–51.
42. E. N. Protasov, A. A. Gnilomedov and A. A. Lvov (1978) Electrochemical Behaviour of Lithium Magnesium Alloys in LiCl–KCl, LiCl–KCl–CsCl and LiF–LiCl–LiBr Melts, *Elektrokhimiya*, **14**, pp. 1296–300.

43. J. Wolfenstine, G. Gonzalez-Doncel and O. D. Sherby (1990) Elevated Temperature Properties of Mg–14Li–B Particulate Composites, *J. Mater. Res.*, **5**, pp. 1359–62.
44. J. F. Mason, C. M. Warwick, P. Smith, J. A. Charles and T. W. Clyne (1989) Magnesium-Lithium Alloys in Metal Matrix Composites – A Preliminary Report, *J. Mat. Sci.*, **24**, pp. 3934–46.
45. H. Lilholt (1991) Aspects of Deformation of Metal Matrix Composites, *Mat. Sci. & Eng.*, **A135**, pp. 161–71.
46. K. Wakashima, B. H. Choi and T. Mori (1990) Plastic Incompatibility and its Accommodation by Diffusional Flow: Modelling of Steady State Creep of a Metal Matrix Composite, *Mat. Sci. & Eng.*, **A127**, pp. 57–64.
47. J. A. G. Furness and T. W. Clyne (1991) Thermal Cycling Creep of Short Fibre MMCs – Measurement and Modelling of the Strain Cycle, in *Metal Matrix Composites – Processing Microstructure and Properties, 12th Risø Int. Symp.*, Risø, N. Hansen, D. J. Jensen, T. Leffers, H. Lilholt, T. Lorentzen, A. S. Pedersen, O. B. Pedersen and B. Ralph (eds.), Risø Nat. Lab., Denmark, pp. 349–54.
48. A. Kelly and W. R. Tyson (1966) Tensile Properties of Fibre Reinforced Metals II. Creep of Silver–Tungsten, *J. Mech. Phys. Solids*, **14**, pp. 177–86.
49. S. T. Mileiko (1970) Steady State Creep of a Composite Material with Short Fibres, *J. Mat. Sci.*, **5**, pp. 254–61.
50. D. McLean (1972) Viscous Flow of Aligned Composites, *J. Mat. Sci.*, **7**, pp. 98–104.
51. H. Lilholt (1985) Creep of Fibrous Composite Materials, *Comp. Sci. & Tech.*, **22**, pp. 277–94.
52. H. Lilholt and M. Taya (1987) Creep Behaviour of the Metal Matrix Composite Al 2124 with SiC Fibres, in *Proc. ICCM VI/ECCM 2*, F. L. Matthews, N. C. R. Buskell, J. M. Hodgkinson and J. Morton (eds.), London, Elsevier, pp. 2.234–2.244.
53. S. Goto and M. McLean (1991) Role of Interfaces in Creep of Fibre-Reinforced Metal Matrix Composites – I. Continuous Fibres, *Acta Metall.*, **39**, pp. 153–164.
54. T. L. Dragone and W. D. Nix (1990) A Numerical Study of High Temperature Creep Deformation in Metal–Matrix Composites, in *Metal and Ceramic Matrix Composites: Processing, Modelling and Mechanical Behaviour*, Anaheim, California, R. B. Bhagat, A. H. Clauer, P. Kumar and A. M. Ritter (eds.), TMS, pp. 367–80.
55. T. L. Dragone and W. D. Nix (1990) Geometric Factors Affecting the Internal Stress Distribution and High Temperature Creep Rate of Discontinuous Fibre Reinforced Metals, *Acta Metall.*, **38**, pp. 1941–53.
56. N. Sørensen (1991) Effects of Clustering on the Creep Properties of Whisker Reinforced Al, in *Metal Matrix Composites – Processing, Microstructure and Properties, 12th Risø Int. Symp. on Mat. Sci.*, Risø, N. Hansen, D. J. Jensen, T. Leffers, H. Lilholt, T. Lorentzen, A. S. Pedersen, O. B. Pedersen and B. Ralph (eds.), Risø Nat. Lab., Denmark, pp. 667–74.
57. T. Endo, M. Chang, N. Matsuda and K. Matsuura (1991) Creep Behaviour of SiC/Al Composite at Elevated Temperatures, *ibid.*, pp. 323–8.
58. M. Rabinovitch, J. F. Stohr, T. Khan and H. Bibring (1983) Directionally

Solidified Composites for Application at High Temperatures, in *Handbook of Composites, vol. 4 – Fabrication of Composites*, A. Kelly and S. T. Mileiko (eds.), Elsevier, Amsterdam, pp. 295–372.

59. K. Wakashima, M. Otsuka and S. Umekawa (1974) Thermal Expansions of Heterogeneous Solids Containing Aligned Ellipsoidal Inclusions, *J. Comp. Mat.*, **8**, pp. 391–404.
60. S. Noda, N. Kurihara, K. Wakashima and S. Umekawa (1978) Thermal Cycling Induced Deformation of Fibrous Composites with Particular Reference to the Tungsten–Copper System, *Metall. Trans.*, **9A**, pp. 1229–36.
61. M. H. Kural and B. K. Min (1984) The Effects of Matrix Plasticity on the Thermal Deformation of Continuous Fibre Graphite/Metal Composites, *J. Comp. Mat.*, **18**, pp. 519–35.
62. S. Yamada and S. Towata (1985) Thermal Expansion Coefficients of Unidirectional and Angle-Plied Silicon Carbide Fibre Reinforced Aluminium Alloys, *J. Jap. Inst. Metals*, **49**, pp. 376–81.
63. E. G. Wolff, B. K. Min and M. H. Kural (1985) Thermal Cycling of a Unidirectional Graphite–Magnesium Composite, *J. Mat. Sci.*, **20**, pp. 1141–9.
64. X. Dumant, F. Fenot and G. Regazzoni (1988) Non-Linearity in Thermal Expansion of Continuous Fibre Metal Matrix Composites, in *Mechanical and Physical Behaviour of Metallic and Ceramic Composites*, Risø 9th Int. Symp., S. I. Andersen, H. Lilholt and O. B. Pedersen (eds.), Risø Nat. Lab., Denmark, pp. 349–56.
65. D. Masutti, J. P. Lentz and F. Delannay (1990) Measurement of Internal Stresses and of the Temperature Dependence of the Matrix Yield Stress in Metal Matrix Composites from Thermal Expansion Curves, *J. Mat. Sci. Letts.*, **9**, pp. 340–2.
66. A. Kitihara, S. Akiyama, H. Ueno, S. Nagata and K. Imagawa (1983) The Thermal Expansion of an Aluminium/Hollow Glass Microspheres Composite, *J. Jap. Inst. Light Metals*, **33**, pp. 596–601.
67. S. Towata and S. Yamada (1984) Thermal Expansion Behaviour of Silicon Carbide Fiber Reinforced Aluminium Alloys, *J. Jap. Inst. Metals*, **48**, pp. 848–53.
68. P. J. Withers, D. J. Jensen, H. Lilholt and W. M. Stobbs (1987) The Evaluation of Internal Stresses in a Short Fibre MMC, in *Proc. ICCM VI/ECCM2*, London, F. L. Matthews, N. C. R. Buskell, J. M. Hodgkinson and J. Morton (eds.), Elsevier, pp. 2.255–64.
69. W. G. Patterson and M. Taya (1985) Thermal Cycling Damage of SiC Whisker/2124 Aluminium, in *Proc. ICCM V*, San Diego, W. C. Harrigan, J. Strife and A. Dhingra (eds.), TMS-AIME, pp. 53–66.
70. J. C. Le Flour and R. Locicéro (1987) Influence of Internal Stresses Induced by Thermal Cycling on the Plastic Deformation Resistance of an Al/SiC Composite Material, *Scripta Met.*, **21**, pp. 1071–6.
71. C. M. Warwick and T. W. Clyne (1990) The Micromechanisms of Strain Ratchetting During Thermal Cycling of Fibrous MMCs, in *Fundamental Relationships Between Microstructures and Mechanical Properties of Metal–Matrix Composites*, Indianapolis, P. K. Liaw and M. N. Gungor (eds.), TMS, pp. 209–223.
72. D. Toitot, E. Andrieu and P. Jarry (1991) Dimensional Changes and Transient Deformations affecting a Metal Matrix Composite during Thermo-Mechanical Loadings, in *Metal Matrix Composites – Processing, Microstructure and Properties. 12th Risø Int. Symp.*,

N. Hansen, D. J. Jensen, T. Leffers, H. Lilholt, T. Lorentzen, A. S. Pedersen, O. B. Pedersen and B. Ralph (eds.), Risø Nat. Lab., Denmark, pp. 695–700.

73. K. Wakashima, H. Tsukamoto and B. H. Choi (1991) A Mechanism of Thermal Cycling-induced Superplasticity in Discontinuous Fibre Reinforced MMCs, *ibid.*, pp. 725–34.
74. M. Y. Wu and O. D. Sherby (1984) Superplasticity in a Silicon Carbide Whisker Reinforced Aluminium Alloy, *Scripta Met.*, **18**, pp. 773–6.
75. S. H. Hong, O. D. Sherby, A. P. Divecha, S. D. Karmarkar and B. A. MacDonald (1988) Internal Stress Superplasticity in 2024 Al–SiC Whisker Reinforced Composites, *J. Comp. Mat.*, **22**, pp. 103–3.
76. S. M. Pickard and B. Derby (1988) Thermal Cycle Creep of Al/SiC Particulate Composite, in *Mechanical and Physical Behaviour of Metallic and Ceramic Composites, 9th Risø Int. Symp. on Met. and Mat. Sci.*, Røskilde, S. I. Andersen, H. Lilholt and O. B. Pedersen (eds.), Risø Nat. Lab., Denmark, pp. 447–52.
77. G. S. Daehn and G. González-Doncel (1989) Deformation of Whisker-Reinforced Metal Matrix Composites under Changing Temperature Conditions, *Metall. Trans.*, **20A**, pp. 2355–68.
78. R. H. Johnson and E. C. Sykes (1966) Enhancement of Ductility in α-Uranium, *Nature*, **209**, pp. 192–3.
79. R. C. Lobb, E. C. Sykes and R. H. Johnson (1972) The Superplastic Behaviour of Anisotropic Metals Thermally Cycled under Stress, *Metal Science J.*, **6**, pp. 33–39.
80. M. Y. Wu, J. Wadsworth and O. D. Sherby (1987) Internal Stress Superplasticity in Anisotropic Polycrystalline Zinc and Uranium, *Metall. Trans.*, **18A**, pp. 451–62.
81. S. M. Pickard and B. Derby (1990) Superplasticity During Thermal Cycling of Metal Matrix Composites, in *Fundamental Relationships Between Microstructures and Mechanical Properties of Metal Matrix Composites*, Indianapolis, P. K. Liaw and M. N. Gungor (eds.), TMS, pp. 103–13.
82. A. Colclough, B. Dempster, Y. Favry and D. Valentin (1991) Thermomechanical Behaviour of SiC–Al Composites, *Mat. Sci. & Eng.*, **A135**, pp. 203–7.
83. A. C. Roberts and A. H. Cottrell (1956) Creep of α-Uranium during Irradiation with Neutrons, *Phil. Mag.*, **1**, pp. 711–17.
84. R. G. Anderson and J. F. W. Bishop (1962) The Effect of Neutron Irradiation and Thermal Cycling on Permanent Deformations in Uranium under Load, in *Symp. on Uranium and Graphite*, London, Inst. of Metals, pp. 17–23.
85. B. Derby (1991) Thermal Cycling of Metal Matrix Composites, in *Metal Matrix Composites – Processing, Microstructure and Properties, Risø 12th Int. Symp.*, N. Hansen, D. J. Jensen, T. Leffers, H. Lilholt, T. Lorentzen, A. S. Pedersen, O. B. Pedersen and B. Ralph (eds.), Risø Nat. Lab., Denmark, pp. 31–50.
86. G. W. Greenwood and R. H. Johnson (1965) The Deformation of Metals Under Small Stresses during Phase Transformations, *Proc. Roy. Soc. Lond.*, **283A**, pp. 403–22.
87. G. S. Daehn and T. Oyama (1988) The Mechanism of Thermal Cycling Enhanced Deformation in Whisker-Reinforced Composites, *Scripta Met.*, **22**, pp. 1097–102.

88. G. S. Daehn (1989) Plastic Deformation of Continuous-Fibre Reinforced Composites Subjected to Changing Temperature, *Scripta Met.*, **23**, pp. 247–52.
89. S. M. Pickard and B. Derby (1990) The Deformation of Particle Reinforced Metal Matrix Composites During Temperature Cycling, *Acta Metall.*, **38**, pp. 2537–52.
90. H. Zhang, G. S. Daehn and R. H. Wagoner (1991) Simulation of the Plastic Response of Whisker Reinforced Metal Matrix Composites under Thermal Cycling Conditions, *Scipta Met.*, **25**, pp. 2285–90.
91. A. G. Young, K. M. Gardiner and W. R. Rotsey (1960) The Plastic Deformation of α-Uranium, *J. Nucl. Mat.*, **2**, pp. 234–7.
92. M. Taya and T. Mori (1987) Modelling of Thermal Cycling Damage in Metal Matrix Composites, in *Proc. ICCM VI/ECCM2*, London, F. L. Matthews, N. C. R. Buskell, J. M. Hodgkinson and J. Morton (eds.), pp. 2.104–2.112.
93. M. Taya and T. Mori (1987) Modelling of Dimensional Change in a Metal Matrix Composite Subjected to Thermal Cycling, in *Thermomechanical Couplings in Solids*, H. D. Bui and Q. S. Nguyen (eds.), Elsevier, pp. 147–62.
94. H. Yoshizawa, K. Wakashima and S. Umekawa (1982) Microstructural and Dimensional Stabilities of a Potential gamma/gamma′–alpha (Mo) Directionally Solidified Eutectic Superalloy under Cyclic Thermal Exposure to 1000 °C, *J. Mat. Sci.*, **17**, pp. 3484–90.
95. G. Smolka, A. Maciejny and A. Dytkowicz (1988) High Temperature Stability of the Ni(Cr, Al)–TiC in-situ Composite, in *Mechanical and Physical Behaviour of Metallic and Ceramic Composites*, 9th Int. Risø Symp., S. I. Andersen, H. Lilholt and O. B. Pedersen (eds.), Risø Nat. Lab., Denmark, pp. 475–8.
96. K. Nakano, L. Albingre, R. Pailler and J. M. Quenisset (1985) Thermal Cycling of Titanium-based Composites Reinforced by B (B_4C) Filaments, *J. Mat. Sci. Letts.*, **4**, pp. 1046–50.
97. R. A. Naik, W. D. Pollock and W. S. Johnson (1991) Effect of a High Temperature Cycle on the Mechanical Properties of Silicon Carbide/Titanium Metal Matrix Composites, *J. Mat. Sci.*, **26**, pp. 2913–20.
98. C. A. Hoffman (1973) Effect of Thermal Loading on Fibre Reinforced Composites with Constituents of Different Thermal Expansivities, *J. Eng. Mat. Tech.*, **95**, pp. 55–62.
99. M. A. Wright (1975) The Effect of Thermal Cycling on the Mechanical Properties of Various Aluminium Alloys Reinforced with Unidirectional Boron Fibres, *Metall. Trans.*, **6A**, pp. 129–34.
100. M. Sakai and K. Watanabe (1982) Effect of Thermal Cycle on the Phase Boundary of SiC/Ni Monofilament Composite, *J. Jap. Inst. Metals*, pp. 993–9.
101. T. Kyono, I. W. Hall, M. Taya and A. Kitamwa (1986) Thermal Cycling Damage in Carbon Fiber–Aluminium Composites, in *Composites 86: Recent Advances in Japan and the United States*, Tokyo, Jap. Soc. Comp. Mats., pp. 553–61.
102. G. S. Zhong, L. Rabenberg and H. L. Marcus (1990) *In Situ* Thermal Fatigue of Al/Graphite Metal Matrix Composites, in *Fundamental Relationships Between Microstructures and Mechanical Properties of*

Metal–Matrix Composites, Indianapolis, P. K. Liaw and M. N. Gungor (eds.), TMS, pp. 289–300.

103. S. S. Tompkins, K. E. Ard and G. R. Sharp (1986) Thermal Expansion Behaviour of Graphite/Glass and Graphite/Magnesium, in *Materials for Space – the Gathering Momentum*, Seattle, Washington, USA, SAMPE, Covina, CA, USA, pp. 623–37.
104. M. Nakanishi, Y. Nishida, H. Matsubara, M. Yamada and Y. Tozawa (1990) Effect of Thermal Cycling on the Properties of SiC Whisker-Reinforced Aluminium Alloys, *J. Mat. Sci. Letts.*, **9**, pp. 470–2.
105. Y. Nishida, M. H. Masaru and M. Y. Nakanishi (1989) The Influence of Thermal Cycling on the Properties of Si_3N_4 Whisker Reinforced Aluminium Alloy Composites, in *Proc. 3rd European Conf. on Comp. Mats.* (*ECCM3*), Bordeaux, A. R. Bunsell, P. Lamicq and A. Massiah (eds.), Elsevier, pp. 145–50.
106. C. M. Warwick (1990) *Metal Matrix Composites Based on Magnesium–Lithium*, PhD thesis, Univ. of Cambridge, UK.
107. V. C. Nardone and J. R. Strife (1987) Analysis of the Creep Behaviour of SiC Whisker-Reinforced 2124 (T4), *Metall. Trans.*, **18A**, pp. 109–14.
108. K. Xia, T. G. Nieh, J. Wadsworth and T. G. Langdon (1990) The Creep Properties of Aluminium Composites Reinforced with SiC, in *Fundamental Relationships Between Microstructures and Mechanical Properties of Metal–Matrix Composites*, Indianapolis, P. K. Liaw and M. N. Gungor (eds.), TMS, pp. 543–56.
109. A. B. Pandey, R. S. Mishra and Y. R. Mahajan (1990) Creep Behaviour of an Aluminium SiC Particulate Composite, *Scripta Met. et Mat.*, **24**, pp. 1565–70.
110. Z. Xiong, L. Geng and C. K. Yao (1990) Investigation of High-Temperature Deformation Behaviour of a SiC Whisker Reinforced 6061 Aluminium Composite, *Comp. Sci. & Tech.*, **39**, pp. 117–25.
111. K. Matsuura and N. Matsuda (1988) Creep Behaviour of Short Fibre Reinforced Al–Al_2O_3 Alloys, in *Proc. 8th Int. Conf. on Strength of Metals and Alloys* (*ICSMA 8*), Tampere, Finland, P. O. Kettunuen, T. K. Lepisto and M. E. Lehtonen (eds.), Pergamon, pp. 1409–14.
112. F. R. Tuler, J. T. Beals, C. Demetry, D. Zhao and D. J. Lloyd (1988) Deformation Mechanism Mapping of SiC/Al Metal Matrix Composite Materials, in *Cast Reinforced Metal Composites*, Chicago, S. G. Fishman and A. K. Dhingra (eds.), ASM, pp. 321–6.
113. B. Coutard, F. Girot, Y. Lepetitcorps and J. M. Quenisset (1989) Hot Working Behaviour of Discontinuous SiC/Al Composites Obtained by Rheocasting, in *Proc. 3rd European Conf. on Comp. Mat.* (*ECCM3*), Bordeaux, A. R. Bunsell, P. Lamicq and A. Massiah (eds.), pp. 233–42.

6
The interfacial region

In composites the role of the interface is crucial. Stiffening and strengthening rely on load transfer across the interface, toughness is influenced by crack deflection/fibre pull-out, and ductility is affected by relaxation of peak stresses near the interface. Unfortunately, however, a great deal of confusion surrounds the question of how best to characterise, and then optimise, the mechanical response of the interface to stresses arising from an applied load. In this chapter, a brief outline is given of the meaning and significance of interfacial bond strength, followed by a summary of the methods used to measure interfacial mechanical properties, with particular reference to fibrous MMCs. Some attention is then devoted to interfacial chemical reactions. Finally, the production and characteristics of fibre coatings are briefly examined.

6.1 The significance of interfacial bond strength

In the previous two chapters it has become clear that many important phenomena can take place at the matrix/reinforcement interface. For polymer-based composites, although the chemistry involved may be complex, the objectives in terms of interfacial properties are often the rather straightforward ones of a high bond strength (to transfer load efficiently to the fibres) and a good resistance to environmental attack. In designing ceramic composites, on the other hand, the aim is usually to make the interface very weak, as the prime concern is in promoting energy dissipation at the interface so as to raise the toughness. For MMCs, a strong bond is usually desirable, but there may be instances where inelastic processes at the interface can be beneficial. Furthermore, the avoidance of environmental attack is often less important than the need for control over chemical reactions between the constituents themselves, occurring during fabrication or under service conditions at high temperature.

Broadly speaking, the problem of how to characterise the mechanical performance of the interface has been approached in two different ways. The most common approach has been to establish critical stress levels at which inelastic processes initiate there. This is complicated by the fact that various combinations of normal and shear stresses might trigger the same process. Furthermore, a critical interfacial stress value does not always lead directly to useful predictions about the mechanical performance of the composite. The alternative approach, which has received considerable attention recently, involves measuring an interfacial fracture toughness. This may require careful experimental procedures, but it offers promise for establishing correlations with composite performance. Before examining this in detail, it is instructive to consider the nature of the interfacial stresses and the interfacial processes they bring about.

6.1.1 Interfacial stresses and inelastic processes

Interfacial stresses can arise from differential thermal contraction and from prior plastic flow of the matrix, as well as by the application of an external load. For a particular case, analytical or numerical methods can be used to explore the interfacial stress state (see, for example, Figs. 2.10–2.15). These stresses can activate a variety of inelastic processes, as is illustrated schematically in Fig. 6.1. Unfortunately, the interplay between microstructure, stress state and the nature of the inelastic processes can be highly complex. It is further complicated in many MMCs by the scope for progressive interfacial chemical reaction, which can substantially alter the response of the interface (see §6.3.3).

The first problem is to identify the various parameters which may be significant. For the most part, the values of these are peculiar to the specimen being examined, rather than being characteristic of the matrix/reinforcement combination concerned. In general, the interfacial properties are dependent on processing route and thermo-mechanical history. Among the relevant properties are the critical stress levels to cause interfacial debonding/cavitation (at or in the immediate neighbourhood of the interface). Related to the debonding stresses, but more general and potentially more useful, is the critical value of the strain energy release rate necessary to cause a crack to propagate along the interface, $\mathcal{G}_{ic}$. In the same way that the normal stress to cause debonding will differ from that required in shear, the value of $\mathcal{G}_{ic}$ will vary with the proportion of

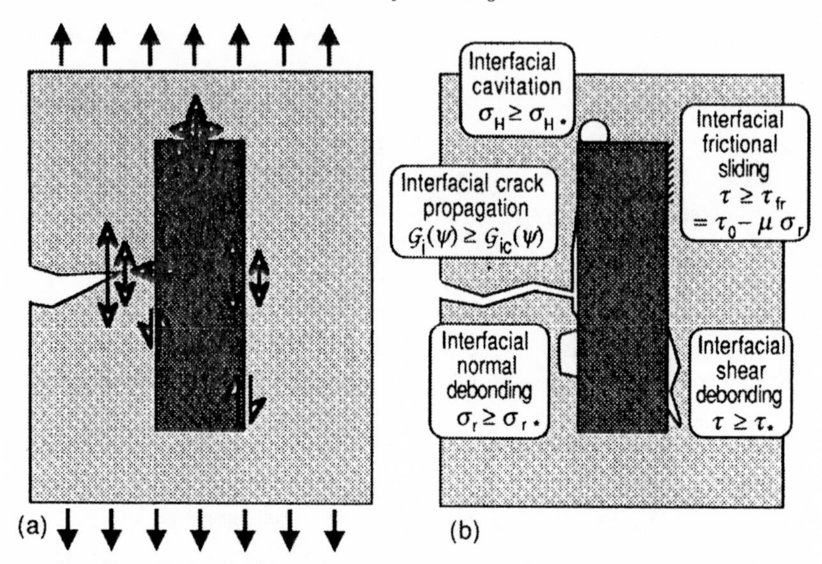

Fig. 6.1 Schematic illustration for a short fibre composite of (a) stresses near the interface, arising from an applied load and from differential thermal contraction, and (b) the various inelastic interfacial processes they initiate. These and other processes (see §4.4.2) occurring at or near the interface can act to relax the stress state by unloading the fibre. In addition, energy is absorbed when these processes are activated, and this may make a significant contribution to the toughness of the composite (see §6.1.4 and §7.3.1).

crack opening and shearing stress intensities, characterised by the phase angle ψ (see §6.1.3 and Fig. 6.5). Also of interest are the conditions under which interfacial sliding can occur after debonding. This is expected to require predominantly shear loading and it is of particular importance in view of the scope for substantial absorption of energy by frictional dissipation.

6.1.2 Critical stress levels

It is common to define a critical shear stress for shear debonding, τ_*. This is usually taken to be independent of the normal (radial) stress σ_r. The shear stress necessary for subsequent interfacial frictional sliding, τ_{fr}, on the other hand, is commonly assumed to rise as the radial stress σ_r becomes more compressive. Even in the absence of residual radial compression, however, τ_{fr} is expected to have a non-zero value, as a consequence of surface roughness effects. This can be represented[1,2] as a contribution to σ_r at the onset of sliding, but it is simpler to express τ_{fr}

in the form

$$\tau_{fr} = \tau_0 - \mu\sigma_r \tag{6.1}$$

where μ is the coefficient of sliding friction and τ_0 is some base level shear stress. Under an applied load parallel to the fibre axis (z-direction), differential Poisson contraction effects may cause the value of σ_r to vary along the length (z) of a (discontinuous) fibre,

$$\sigma_r(z) = \sigma_r^{\Delta T} + \sigma_r^{\Delta \nu}(z) \tag{6.2}$$

where $\sigma_r^{\Delta T}$, the radial stress from differential thermal contraction, is normally compressive in MMCs at room temperature. With an applied axial tensile load, the sign of $\sigma_r^{\Delta \nu}$ will also be negative (compressive) whereas it will be tensile if it is only the fibre that is being loaded. This is the case, for example, with a fibre bridging a crack (or in the single fibre pull-out test – see §6.2.1). However, Poisson effect transverse stresses are usually rather small compared with those from differential thermal contraction (e.g. see Fig. 4.4), so that σ_r usually remains compressive in MMCs for fibres aligned with the stress axis.

While the ease of shear debonding and frictional sliding are important, there is also interest in the conditions under which normal decohesion may occur, causing interfacial cavitation and/or the opening of a crack along the interface. It might be expected that it would be possible to identify a minimum (tensile) value of the radial stress, σ_{r*}, required to cause normal debonding. For cavitation, a critical hydrostatic component of the stress state, σ_{H*}, is expected. Certainly, the ends of fibres aligned parallel to the stress axis, where peak hydrostatic tension is generated, are the preferred sites for cavitation (§7.2.2).

6.1.3 Interfacial fracture toughness

During interfacial frictional sliding, the energy dissipated can be estimated from τ_{fr}, the contact area and the sliding distance (§6.1.4). Much more complex, and of greater general significance, is the analysis of the energy absorbed during the initial debonding process, i.e. the propagation of a crack along the interface. For an interfacial crack to advance, the strain energy thus released, $\mathcal{G}_i$, must exceed a critical value $\mathcal{G}_{ic}$, characteristic of the interface. There has been significant progress made in recent years concerning interfacial fracture mechanics, particularly for bimaterial interfaces[3–8]. In a homogeneous material, any crack will tend to follow a path for which the stress intensity at the crack tip is purely mode I (crack opening). For interfaces which are significantly less tough than the neighbouring bulk materials, a crack can continue to follow an interface, even though the stress intensity is not purely mode I and may have a substantial mode II (shearing) component[9–11]. The ***mixity*** of crack tip loading mode is characterised by the so-called '***phase angle***', ψ, which can vary between $\psi = 0°$ (pure mode I) and $\psi = 90°$ (pure mode II), depending on the loading geometry and elastic properties of the two materials[†]. The value of ψ is defined by

$$\psi = \tan^{-1}\left(\frac{K_{II}}{K_I}\right) \tag{6.3}$$

where K_I and K_{II} are the mode I and mode II stress intensities. This is important because $\mathcal{G}_{ic}$ can vary quite markedly with ψ (see Fig. 6.5). A large shear component tends to result in the crack tip being shielded (particularly if the interface is geometrically rough), with more frictional work being done in the wake of the crack and hence a larger $\mathcal{G}_{ic}$ value[7,9,12]. In practice, strongly mixed mode loading ($\psi \sim 30$–$60°$) is often generated at both planar interfaces and those in fibre/matrix or particle/matrix systems.

Use of ψ to characterise the mode mixity has evolved from the rather complex mathematics of interfacial fracture mechanics. While it is an elegant representation, it does cause certain minor conceptual difficulties, particularly in terms of reconciling the fracture mechanics and critical stress level viewpoints. A negative value of ψ could in principle arise from a change in the direction of shearing, and hence in the sign of K_{II}, but this has no significance as the initial choice of sign for K_{II} is arbitrary. Mathematically, a negative ψ would also arise if K_I became negative (compressive normal stress), but in fact all such cases must be taken as $\psi = 90°$, because there is then zero crack opening stress intensity. However, a compressive normal stress might be expected to hamper crack propagation and raise the energy absorbed; certainly this is assumed to be the case during frictional sliding (eqn (6.1)). This would suggest that a value of ψ greater than 90°, and a correspondingly higher $\mathcal{G}_{ic}$ value than that for $\psi = 90°$, would be appropriate. Clearly this is impossible with the above definition of ψ.

In practice, much of the work on interfacial toughness measurement

[†] Mode III loading (antiplane shearing) is usually omitted from this rationale. In fact, for a fibre/matrix interface this would only arise from rotation of the fibre about its axis, which would be extremely unusual.

has been oriented towards planar interfaces free from significant residual normal stresses (e.g. thin coatings), with applied loadings such that no compressive normal stress is generated. Treatment of fibre/matrix systems, in which residual compressive stress across the interface is likely, has often been aimed at ceramic matrix composites (CMCs), for which the low thermal expansion mismatch and absence of matrix plasticity means that such stresses are usually small. These theoretical considerations do not, of course, prevent measurement of $\mathcal{G}_{ic}$ in MMCs, but they should be noted in making any comparisons with data from other systems. In fact, experimental $\mathcal{G}_{ic}$ data are, as yet, in very short supply generally (see §6.2.2 and §6.2.3), although this is likely to change in the near future.

6.1.4 Composite performance

Relatively few quantitative correlations have been established for MMCs between interfacial characteristics and composite performance indicators. However, poor interfacial bonding has been shown to result in a reduced elastic modulus[13] and work-hardening rate[14] and this is readily explained in terms of reduced load transfer. Examples of the progressive impairment of load transfer, and associated reductions in elastic modulus, have also been detected as a consequence of prior plastic strain[15] and of progressive interfacial reaction (see §6.3.3).

A technique giving information about inelastic processes, which can be particularly useful for long fibre MMCs, is to monitor the changing Poisson's ratios during loading, via strain gauge measurements. The three Poisson's ratios exhibited by a uniaxial composite can be predicted analytically for elastic loading (see §2.1.4 and Fig. 2.4). Consider the data[16] shown in Fig. 6.2, referring to Ti–SiC laminae. For the as-fabricated composite under axial loading, (a), ν_{32} is close to the elastic value ($\sim$0.30) up to a strain of about 0.5%, above which it increases. This corresponds to the onset of plastic deformation in the matrix, which will raise the apparent Poisson's ratio. Under transverse loading, (b), a value of ν_{23} close to that for elastic deformation ($\sim$0.23) is again observed initially, but inelastic behaviour sets in at a low strain ($\sim$0.1%) and ν_{23} falls sharply beyond this point. This is due to the onset of interfacial damage, since the debonding and opening up of interfaces generates longitudinal extension without the corresponding lateral contraction.

Substantial heat treatment (giving $\sim$5–6 μm interfacial reaction layer) changes the behaviour. The axially loaded plot, Fig. 6.2(c), is similar to that of Fig. 6.2(a), but with the expected reduction in ductility and very

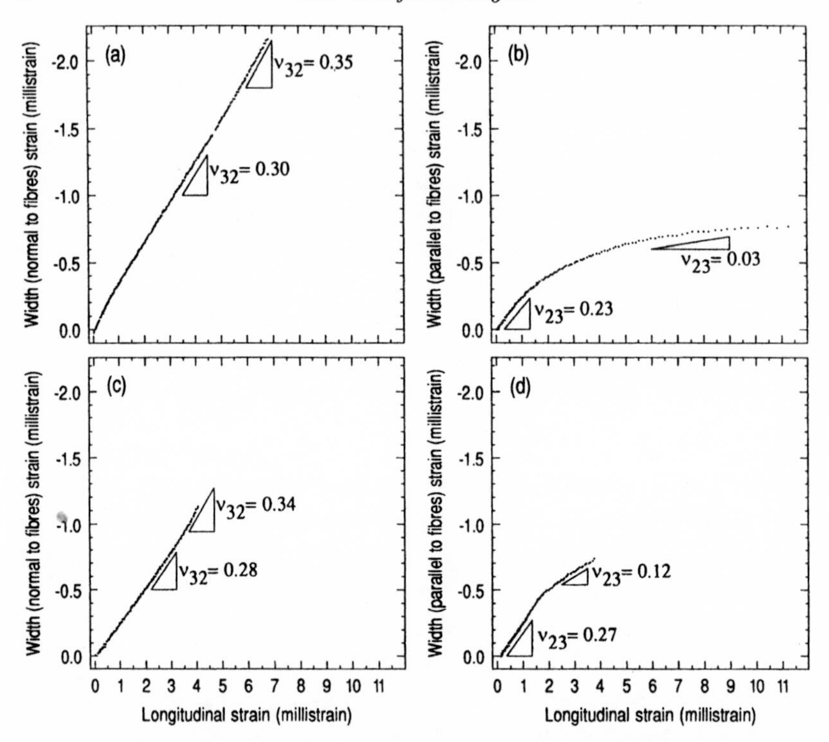

Fig. 6.2 Poisson plots for Ti–6Al–4V/32 vol% SiC (C/TiB_2) monofilament composites (a) as-fabricated under axial loading, (b) as-fabricated under transverse loading, (c) heat-treated (26 hours @ 865°C) under axial loading, (d) heat-treated (26 hours @ 865°C), transverse loading. Monitoring of the changes in Poisson's ratios as deformation becomes inelastic can give information about processes occurring at the interface. For example, increases in ν_{32} during axial tensile loading suggest plasticity throughout the matrix, while decreases in ν_{23} during transverse tension can arise as interfacial debonding and disengagement occur.

limited plastic flow in the matrix. The transversely loaded plot, (d), however, differs significantly from that for the as-fabricated specimen, having substantially higher values of ν_{23} in both regimes. In the elastic portion, this may also reflect the elastic properties of the reaction zone, but for the inelastic regime the change is presumably due to the reduced capacity of the interface to open up without causing failure of the specimen.

Clear correlations between interfacial parameters and composite strength or toughness properties are more difficult to establish. However, some simple analyses have been developed. One approach is to assume that an interfacial region, being composed of brittle constituents, will contain a defect equal in size to the thickness of the region. This has been used to explain the deterioration in axial tensile strength of long fibre composites usually observed on increasing the thickness of an interfacial zone. Early work in this area by Metcalfe[17,18] has since been extended by further studies[19–22] aimed at developing a simple fracture mechanics basis for the effect. The model is based on treating a fibre with reaction layer of thickness δ as it if had a circumferential notch of this depth. Neglecting a geometrical factor with a value normally close to unity[19], the fracture toughness (critical stress intensity factor – see §7.3.1) of the fibre[†] can be related to this flaw size, δ, and the consequent fibre failure stress σ_{I*}

$$K_{Ic} = \sigma_{I*}(\pi\delta)^{1/2} \tag{6.4}$$

The effective fibre strength *in situ* will therefore fall as the reaction zone thickness rises, provided this is greater than the inherent maximum flaw size of the fibre. If it is assumed that the fracture toughness of the fibre can also be expressed in terms of its critical strain energy release rate and stiffness

$$K_{Ic} = (\mathcal{G}_{Ic} E_I)^{1/2} \tag{6.5}$$

then the fibre strength can be written as

$$\sigma_{I*} = \left(\frac{\mathcal{G}_{Ic} E_I}{\pi\delta}\right)^{1/2} \tag{6.6}$$

This equation can be rearranged to give the critical reaction layer thickness, above which the fibre strength will be impaired, in terms of the strength of the as-received fibre, σ_{I0*}

$$\delta_0 = \frac{\mathcal{G}_{Ic} E_I}{\pi(\sigma_{I0*})^2} \tag{6.7}$$

The *in situ* fibre strength will therefore be given by eqn (6.6), using the actual value of δ but subject to a minimum of δ_0. Typically, predicted values of δ_0 for various fibres are of the order of 0.5–1 μm.

Some kind of constitutive law is needed in order to predict the composite strength from the *in situ* fibre strength. The simplest, which is probably appropriate in view of the crude nature of the model, is to use a rule of mixtures expression

$$\sigma_{C*} = f\sigma_{I*} + (1 - f)\sigma_{M*} \tag{6.8}$$

where σ_{M*} is the failure strength of the matrix. Combination of eqns (6.8)

[†] Note that the subscript I refers here to the inclusion (fibre) and not to mode I loading. In fact, it is understood that the quoted fracture toughness of a homogeneous material refers to mode I loading, but the subscript confirming this is omitted in the present treatment.

and (6.6) allows the axial tensile strength of a long fibre composite to be predicted as a function of the interfacial reaction layer thickness. The model is usually extended to consider whether the strength contribution from the reinforcement should go to zero at large values of δ. This is implemented by specifying a minimum composite strength, corresponding to the stress level needed to cause the reaction layer to fracture

$$\sigma_{C*\infty} = \varepsilon_{\delta*} E_C \tag{6.9}$$

where $\varepsilon_{\delta*}$ is the failure strain of the reaction layer (taken as a constant) and E_C is the composite stiffness. There is often some uncertainty about values of $\varepsilon_{\delta*}$, but this limit is only likely to affect the predicted curve for high fibre contents, when it might exceed the matrix contribution in eqn (6.8) and hence constitute a lower bound for thick reaction layers.

Available experimental data are in surprisingly good agreement with this very simple model. For example, Fig. 6.3 shows results[23] for Ti

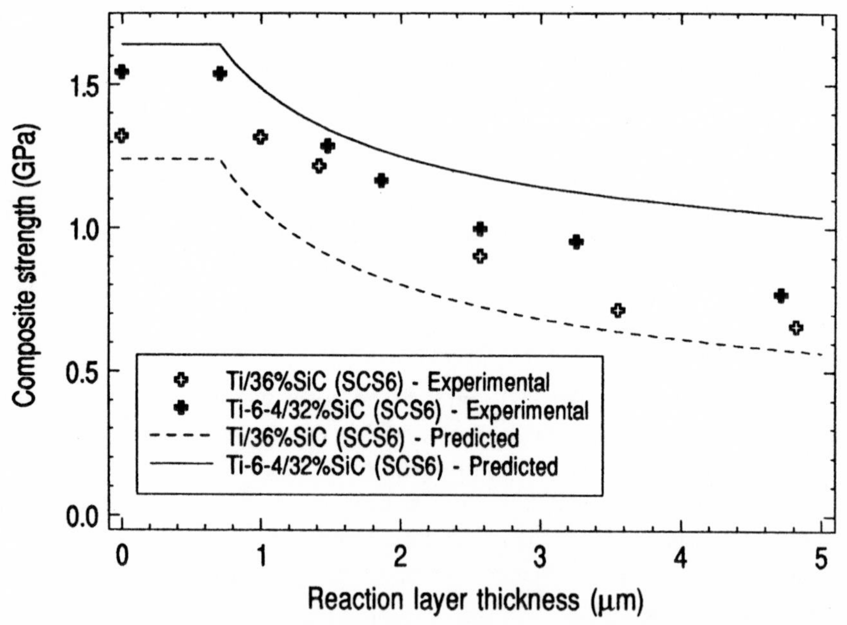

Fig. 6.3 Experimental[23] and predicted variation of composite strength with interfacial reaction layer thickness for axial tensile loading of two long fibre composites – CP Ti and Ti–6Al–4V matrices reinforced with the SCS-6 monofilament. The predicted curves were obtained with eqns (6.6)–(6.8), using the following approximate data for the fibre: $\mathcal{G}_{\mathrm{Ic}} \sim 50\ \mathrm{J\,m^{-2}}$, $E_{\mathrm{I}} \sim 400$ GPa, $\sigma_{\mathrm{I0}*} \sim 3$ GPa. The matrix strength, $\sigma_{\mathrm{M}*}$, was taken as 0.25 GPa for CP Ti and 1.0 GPa for Ti–6Al–4V.

reinforced with SiC monofilaments. Some of the data needed for the predictions, such as $\mathcal{G}_c$ for the fibre, are not readily available, but use of expected approximate values leads to fairly good agreement for both the form of the dependence and the absolute magnitude of the strengths. It is sometimes stated that this type of behaviour is expected when the interfacial bonding (shear strength) is good, while the composite strength is expected to be less sensitive to reaction layer thickness if the bonding is poor. Evidence for this is less convincing, although this may be partly because poor bonding is often associated with lack of wetting and difficulties in promoting progressive reaction. Furthermore, it is often difficult to promote interfacial reaction while retaining a low interfacial shear strength.

The above model only incorporates fracture mechanics in an attempt to establish critical macroscopic stress levels. In order to understand and optimise effects such as the significant increase in the toughness of zinc alloys reported[24,25] on introducing long carbon fibres, the energy absorbed at the interface, and its contribution to the toughness of the composite, must be studied in detail. The geometry of interfacial cracking and subsequent frictional sliding is illustrated in Fig. 6.4 for long fibre composites under axial and transverse loading. Depending on the $\mathcal{G}_c$ values of interface and fibre, the approach of a matrix crack may trigger an interfacial crack before the fibre fractures. While there have been several proposals[7,26,27] concerning the critical conditions to ensure crack deflection, there is still a degree of uncertainty in this area. It seems likely, however, that a critical value of the ratio of the toughness of the interface to that of the deflecting bulk material (fibre) must not be exceeded if deflection is to occur. Estimates[27,28] of this critical value are expected to be less than unity and are typically around 15%. While $\mathcal{G}_{ic}$ values are in short supply, particularly for MMCs, they are often expected to fall substantially below $\mathcal{G}_c$ values for ceramic (fibres), which are typically $\sim$10–100 J m^{-2}, so that such interfacial crack triggering should be quite common.

Calculation of the energy absorbed is then complicated by the variation in ψ around the fibre and with distance from the crack plane. However, if the effect of this on $\mathcal{G}_{ic}$ is neglected, then the contribution to energy absorption at a single fibre may be written

$$\Delta U = 2\pi d z_* \mathcal{G}_{ic} \tag{6.10}$$

where d is the fibre diameter and z_* is the debonded length. The number of fibres intersecting unit area is given by $(4f/\pi d^2)$, so that the total

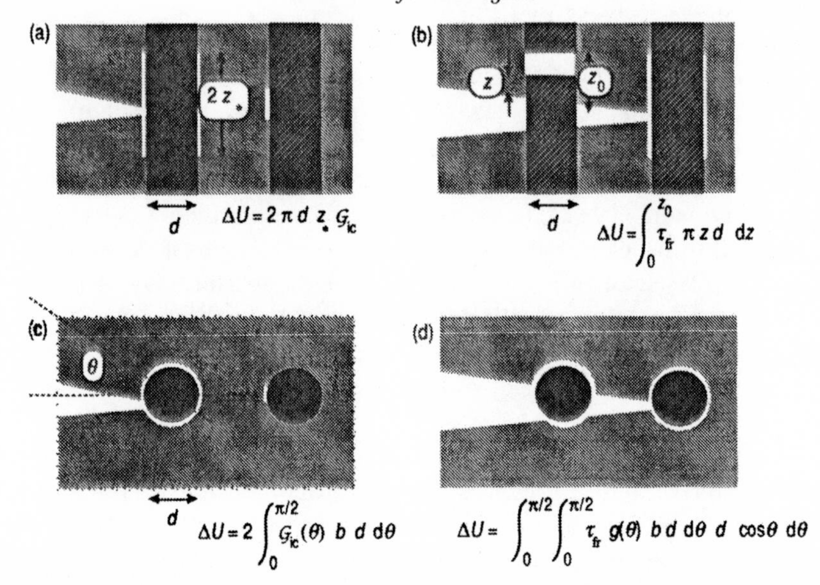

Fig. 6.4 Energy absorption from interfacial processes near the crack tip during tensile loading of fibre composites. During axial loading, interfacial debonding, (a), is triggered under mixed mode conditions which vary with position around the fibre and with distance from the crack plane. This is followed by frictional pull-out under predominantly shear loading, (b), in the wake of the crack. During transverse loading, on the other hand, both initial debonding, (c), and subsequent interfacial disengagement, (d), occur under mixed mode conditions which vary widely with the angle θ. The function $g(\theta)$ in (d) falls quickly to zero with rising θ, so that frictional sliding is very limited.

contribution to $\mathcal{G}_c$ for the composite is given by

$$\Delta \mathcal{G}_c = \frac{8z_* f \mathcal{G}_{ic}}{d} \tag{6.11}$$

The debonded aspect ratio, (z_*/d), might in practice vary up to, say, 20–30, so that in a high fibre content composite the total contribution could reach about $100 \times \mathcal{G}_{ic}$. However, the above condition for initial crack deflection means that the maximum value of $\mathcal{G}_{ic}$ if debonding is to occur might typically be ~ 10 J m^{-2}, depending on the toughness of the fibre. Consequently, the maximum contribution from this mechanism is around 1 kJ m^{-2}, with a typical value substantially less than this. Such values are unlikely to make a significant contribution in MMCs.

There is, however, more scope for energy absorption during subsequent fibre pull-out (Fig. 6.4(b)). The contribution from a single fibre may be written

$$\Delta U = \int_0^{z_0} \tau_{fr} \pi d z \, dz \tag{6.12}$$

where z_0 is the pull-out length and τ_{fr} is the shear stress for frictional sliding (see §6.1.2). Carrying out this integration and multiplying by $(4f/\pi d^2)$ as before leads to

$$\Delta \mathcal{G}_c = 2f\tau_{fr}s_0^2 d \tag{6.13}$$

where s_0 is the pull-out aspect ratio $(=z_0/d)$. Prediction of the average value of s_0 is rather difficult; it will depend on the debonded length and the flaw distribution along the fibre (i.e. on its Weibull modulus) as well as on τ_{fr} and, for short fibres, on the fibre aspect ratio distribution. A crude indication of the expected value of s_0 can be obtained from the shear lag treatment (eqn (2.25)) by assuming the fibre stress peaks in the crack plane and falls off with distance at a rate proportional to τ_{fr}. This suggests that s_0 should range up to about $[\Delta\sigma/(4\tau_{fr})]$, where $\Delta\sigma$ is the variation in fibre strength along its length. Given that $\Delta\sigma$ might be up to several hundred MPa, while τ_{fr} can range from tens to hundreds of MPa, substitution of suitable numbers indicates that, particularly for large diameter fibres (monofilaments, $d \sim$ 100–150 μm), potentially significant contributions of at least several kJ m^{-2} are possible. (The contribution from fibre fracture, $f\mathcal{G}_{lc}$, would normally be negligible for fibres used in MMCs.)

For transverse loading of long fibre composites (Fig. 6.4(c) and (d)), the analysis of interfacial energy absorption is more complex than the axial case, but the broad conclusion is straightforward. For initial debonding, the single fibre contribution is given by integrating the contributions around the interface, which exhibits a wide range of $\mathcal{G}_{ic}$ values

$$\Delta U = 2\int_0^{\pi/2} \mathcal{G}_{ic}(\theta) b d \, d\theta \tag{6.14}$$

where b is the length of the specimen in the fibre direction and the angle θ defines the position around the fibre circumference (Fig. 6.4(c)). Summing this over the composite requires an assumption about the geometry of the fibre array. For a square array with the crack passing through a plane of fibres, the overall contribution to $\mathcal{G}_c$ becomes

$$\Delta \mathcal{G}_c = \frac{4\overline{\mathcal{G}_{ic}} L d(\pi/2)}{L d\sqrt{(\pi/d)}} = 2\overline{\mathcal{G}_{ic}}\sqrt{(\pi f)} \tag{6.15}$$

The appropriate mean value $\overline{\mathcal{G}_{ic}}$ is difficult to ascertain, but it will not be very high, as much of the interface is under predominantly mode I loading. It is therefore clear that the contribution will be of the order of a few tens of J m^{-2} at most and hence entirely negligible.

A similar conclusion can be drawn for subsequent frictional sliding. This will only occur at regions of the fibre circumference close to the crack plane (small θ) and over short sliding distances, depending on the interfacial roughness. The energy contribution could be estimated using a function $g(\theta)$ having a value of unity at $\theta = 0°$ but falling very sharply to zero with increasing θ.

$$\Delta U = \int_0^{\pi/2} \left[\int_0^{\pi/2} \tau_{\mathrm{fr}} g(\theta) b d \, \mathrm{d}\theta \right] d \cos\theta \, \mathrm{d}\theta \tag{6.16}$$

This leads to an expression for the overall contribution having the form

$$\Delta \mathcal{G}_{\mathrm{c}} = 2\tau_{\mathrm{fr}} d \sqrt{(f/\pi)} F \tag{6.17}$$

where F, a dimensionless factor resulting from the integration of $g(\theta)$, has a value which is dependent on interfacial roughness but must always be $\ll 1$. As $(\tau_{\mathrm{fr}} d)$ is expected to be a few kJ m^{-2} at most, the overall contribution will always be negligible in MMCs.

In summary, only during axial loading can interfacial processes absorb significant amounts of energy, with pull-out likely to be more important than debonding. Under transverse loading (and, by implication, also for particulate MMCs), interfacial energy absorption will be negligible. Toughness optimisation for such cases should therefore be aimed at providing a strong bond, but more particularly at avoiding interfacial defects, stress raisers, etc., and at ensuring a uniform distribution of reinforcement, so as to eliminate regions of high triaxial constraint where voids and cracks may nucleate (see §7.2.2).

6.2 The characterisation of bond strength

A variety of tests have been developed to characterise the mechanical response of the interface in composites. These may be divided into tests aimed at establishing critical stress values for debonding or frictional sliding and those designed to measure the critical strain energy release rate for interfacial cracking, $\mathcal{G}_{ic}$. Some of the tests devised to measure $\mathcal{G}_{ic}$ are illustrated in Fig. 6.5. Also shown is a schematic illustration of how the value of $\mathcal{G}_{ic}$ is expected to vary with the mode mixity (as defined by the phase angle ψ) and an indication of the possible range of ψ values

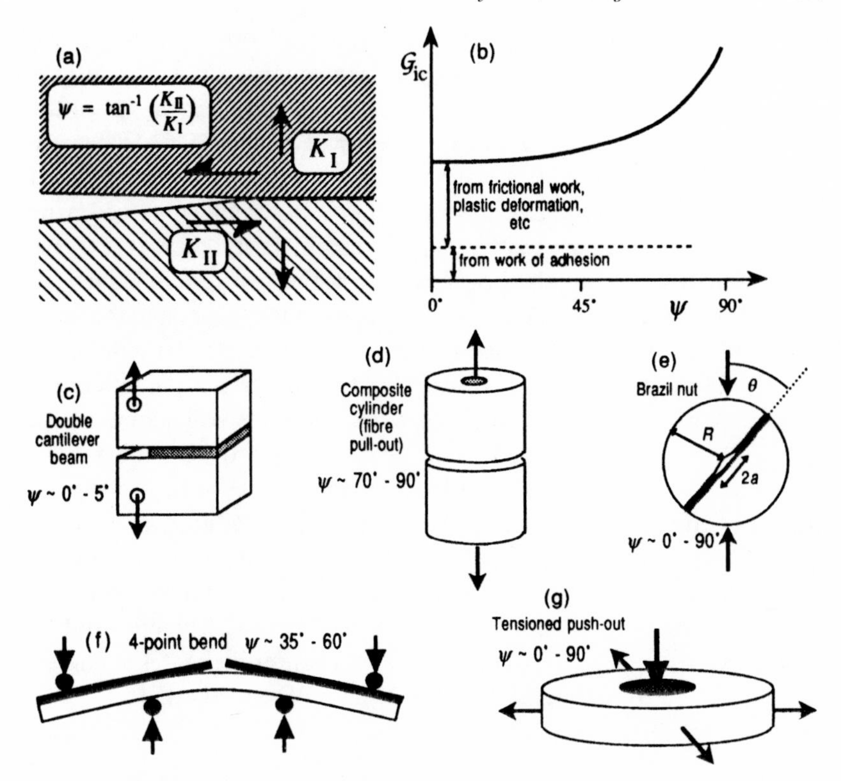

Fig. 6.5 (a) The phase angle ψ, used to characterise the mixity of crack tip loading mode, is defined in terms of the stress intensity factors, K_I and K_{II}. The value of ψ can vary from 0° (pure crack opening mode) to 90° (pure shearing mode). (b) A schematic illustration of the dependence of the critical strain energy release rate of the interface, $\mathcal{G}_{ic}$, on the phase angle ψ, which characterises the mixity of crack tip loading mode. Higher values of ψ, representing greater proportions of shear loading at the interface, give rise to higher interfacial toughnesses. This is because more frictional and plastic work is done at the tip and in the wake area when the loading is predominantly in shear. (c)–(g) Various interfacial test geometries, with corresponding ranges of ψ.

operating at the crack tip during each test. Of the tests shown, the cracked Brazilian disc[29,30] (or 'Brazil nut'), the 4-point bend test[6,31–34] and the well-established double cantilever beam all require planar interfaces and are not suitable for fibre composites.

A number of tests have been developed involving pull-out or push-out of single fibres from the surrounding matrix. These have been most commonly used to establish critical stress levels for debonding or sliding (see §6.2.1), but they can also be employed[2] to measure $\mathcal{G}_{ic}$ values –

particularly if continuous load–displacement monitoring is carried out. Other single fibre tests, such as the full fragmentation test[35], cannot be used to measure $\mathcal{G}_{ic}$. A disadvantage of all single fibre tests has been that they involve predominantly shear loadings ($\psi \sim 90°$). This is because, while the Poisson contraction of the fibre does generate a radial tensile stress, and hence a mode I component, this is normally small compared with the shear stress, so that ψ values remain fairly close to 90° (see §6.2.1). (Furthermore, residual compressive radial stresses are likely to be present[36]). The recently suggested tensioned push-out test[37], however, should allow both critical stress and $\mathcal{G}_{ic}$ values to be established over a wide range of mode mixities. This test involves the simultaneous application of transverse tensile and axial compressive loadings to the fibre, so that any value of ψ should be possible (provided the residual stresses are known). A further point worthy of note is that many test geometries are such that the specimen must be produced in a special operation which differs from the normal composite manufacturing route. Such specimens may thus contain different interfacial microstructures and residual stresses from those in the composite of interest. The push-out test is one of the few applicable to normal fibrous composite material and this has contributed to the high levels of interest in the method.

6.2.1 Single fibre loading tests

Fibre pull-out

This test has been extensively applied to polymer composites. It comprises the extraction of a single fibre, half embedded within a matrix, under an axial tensile stress. Until recently, the interpretation of the resulting load-displacement data has generally been carried out according to a minor adaptation of the shear lag theory[38,39]. A schematic illustration is given in Fig. 6.6 of the axial distributions of normal stress in the fibre and shear stress at the interface. These distributions correspond to three stages in the process; elastic distortion up until debonding, propagation of the debonded portion and subsequent pure frictional sliding. Basic assumptions of the shear lag model (Chapter 2), such as no shear strain in the fibre and no transfer of normal stress across the fibre end, are retained in simple treatments of this problem[38,39].

Analysis is usually divided into two distinct parts; the first corresponds to the point of debonding and the second to the subsequent frictional sliding process. It is conventional to assume that the peak in the load–displacement plot corresponds to the debonding event, occurring at

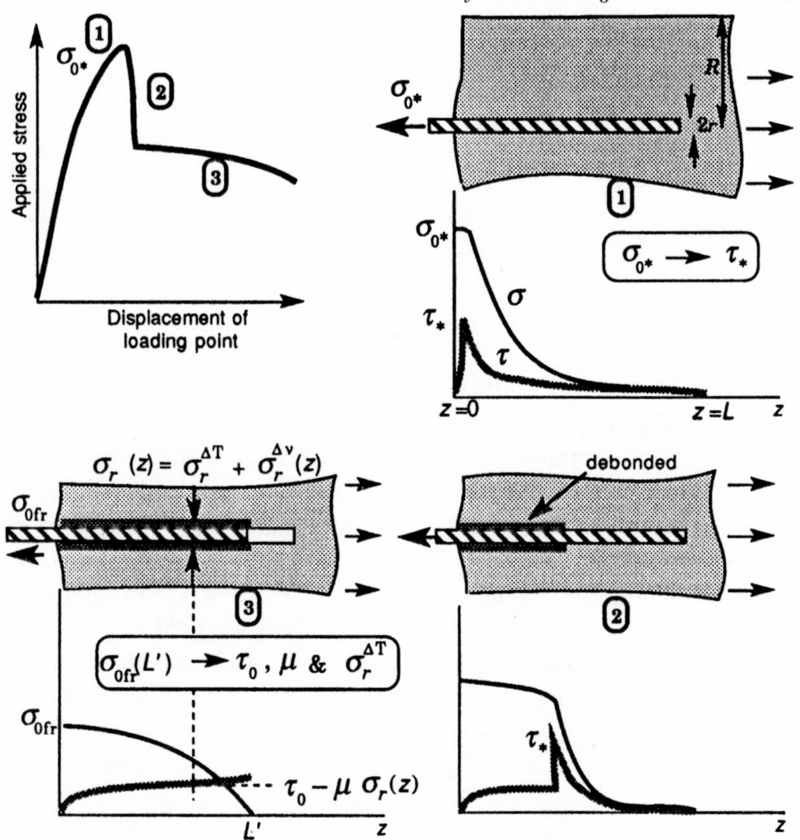

Fig. 6.6 A schematic depiction of the stress distributions and load–displacement plot during the single fibre pull-out test. The applied load generates an interfacial shear stress, which has a peak near to the front surface. At some critical applied load, σ_{0*}, this shear stress causes the interface to debond. Debonding then spreads along the interface and subsequent interfacial motion is by frictional sliding. The residual radial compressive stress (from differential thermal contraction) is partially offset by a Poisson effect, as the axial stress in the fibre causes it to contract more than the surrounding, lightly stressed, matrix. This may cause the shear stress to vary somewhat along the length of the fibre.

an applied stress σ_*^A. Solution of the basic second order linear differential equation governing the variation of σ_I along the length of the fibre (eqn (2.27)) under the boundary conditions $\sigma_I(0) = \sigma^A$ and $\sigma_I(L) = 0$, and assuming the matrix remote from the fibre to be unstrained, leads to

$$\sigma_I = \sigma^A \left\{ \frac{\sinh[n(L - z)/r_0]}{\sinh(nL/r_0)} \right\} \qquad \text{with} \qquad n^2 = \frac{E_M}{E_I(1 + \nu_M)\ln(R/r_0)}$$

The interfacial shear stress, according to the basic equation of the shear lag model (eqn (2.25)), then becomes

$$\tau = \frac{r_0}{2}\frac{d\sigma}{dz} = \frac{-n\sigma^{A}}{2}\cosh\left[\frac{n(L-z)}{r_0}\right]\operatorname{cosech}\left(\frac{nL}{r_0}\right) \tag{6.18}$$

Applying this at $z = 0$, the debonding shear stress is readily deduced from the peak fibre stress (taken as σ_{0*})

$$\tau_* = \frac{n\sigma_{0*}\coth(nL/r_0)}{2} \tag{6.19}$$

For the relatively high values of the fibre aspect ratio, $s\ (=L/r_0 = 2L/d)$, typical of this kind of test, the stresses are very low along most of the fibre length. Some variants of the basic model have been proposed. For example, Hsueh[40] incorporated the possibility of stress transfer across the fibre end, ensuring that the load carried by the free fibre is balanced by that in the composite. This model leads to more complex equations, but the general form of the curves are similar. In particular, the ratio of τ_* to σ_{0*} is usually very close to that for the basic shear lag treatment. Neither of these models take account of the fact that the shear stress in the matrix should fall to zero at the free surface from which the fibre emerges.

The frictional sliding behaviour has also been analysed, taking account of the effect of the Poisson contraction of the fibre in reducing the thermal radial compressive stress. On the basis of a crude assumption that only the fibre carries an axial normal stress (with the resulting radial strain at the interface producing a reduction in radial stress proportional to the matrix stiffness, $\Delta\sigma_r = E_M\nu_I\sigma_I/E_I$), the following relationship is obtained between pull-out stress and remaining embedded length, L'

$$\sigma_{0fr} = \frac{-\sigma_r^{\Delta T}E_I[1-\exp(-BL')]}{\nu_I E_M} \tag{6.20}$$

in which the constant B is given by

$$B = \frac{2\mu\nu_I E_M}{r_0 E_I} \tag{6.21}$$

where μ is the coefficient of friction. It is therefore possible for μ and $\sigma_r^{\Delta T}$ to be evaluated from a single pull-out load–displacement curve, simply by finding the combination of these two parameters which gives the best fit. In practice, it may be difficult to obtain data of sufficient accuracy for this to be carried out. (In any event, this analysis can lead to inconsistencies with boundary conditions.)

The pull-out test can also be used[7,41] to obtain $\mathcal{G}_{ic}$ values for the interface. A numerical analysis has been presented[41] allowing evaluation of the phase angle, ψ, from fibre and matrix properties. If residual stresses are neglected, the value of ψ will typically be around 70–80°. (In practice, a lower mode I component even than represented by these values is expected in MMCs, as a result of the residual radial compressive stress at the fibre/matrix interface.) By considering an energy balance during advance of the interfacial crack as debonding occurs, it can be shown[32,41] that $\mathcal{G}_{ic}$ is given by

$$\mathcal{G}_{ic} = \frac{(E_M/E_I)d(\sigma_{0*})^2}{8E_I[(E_M/E_I) + f(1 - f)]} \tag{6.22}$$

where f is the fibre volume fraction, d is the fibre diameter and σ_{0*} is the applied stress on the fibre needed to propagate the interfacial crack. The expression is only valid when the debonded length is greater than the fibre diameter. Stable crack propagation is therefore needed for the method to be used and this does not always occur[2].

Although single fibre pull-out testing (to obtain critical stress values) has been applied to MMCs[42], there are practical difficulties in specimen preparation and handling when the matrix is relatively stiff, with premature fibre fracture a common problem. The push-out test is more convenient in practical terms.

Fibre push-out

Fibre push-out (and push-down) testing has received considerable attention recently, in terms of both experimental[43,44] and theoretical[40,45–50] work. The stress distributions during the push-out operation are illustrated schematically in Fig. 6.7. Debonding is stimulated near the free surface under essentially pure shear loading. In fact, the applied load does change the normal stress on the interface, differential Poisson expansion tending to increase the radial compressive stress. Once debonding has started, it will propagate along the length of the fibre and subsequent motion will occur purely by frictional sliding at the interface. While debonding is taking place, the load may rise or fall, depending on the relative magnitude of the debonding and frictional sliding stresses. This can be observed with continuous load–displacement monitoring. An example of some experimental data for a ceramic composite is shown in Fig. 6.8, illustrating that stable crack growth can occur in some cases. In fact, such progressive crack growth is very unlikely with the low fibre

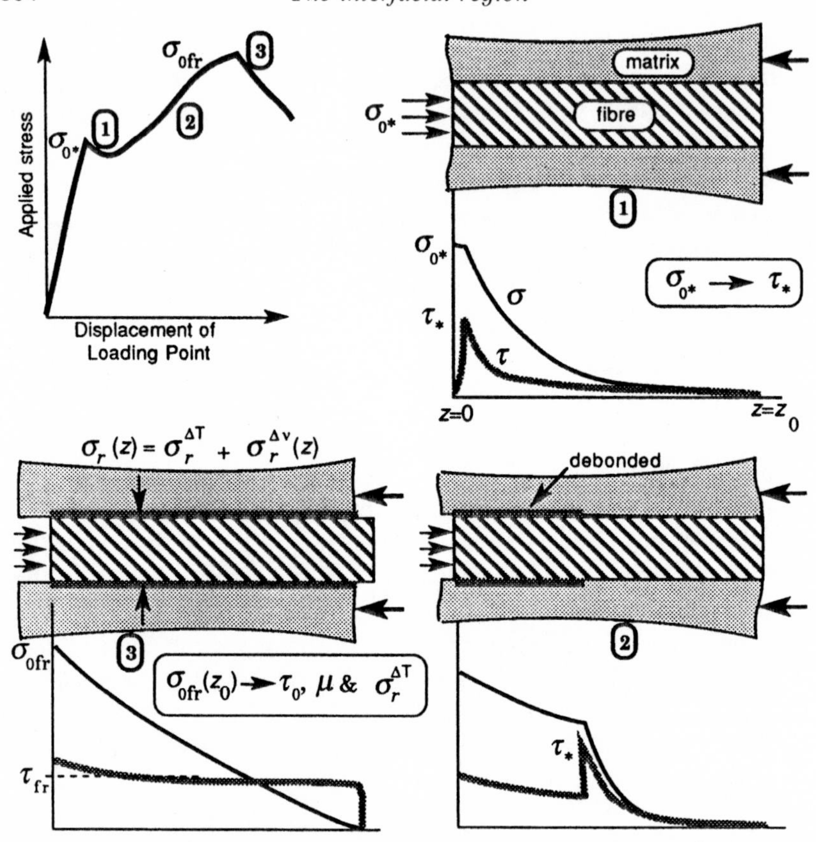

Fig. 6.7 A schematic depiction of the stress distributions and load–displacement plot during the single fibre push-out test. One difference from the pull-out test is that the Poisson effect causes the fibre to expand (rather than contract), which raises (rather than offsets) the radial compressive stress across the interface due to differential thermal contraction. In its standard form, therefore, there is no possibility of generating any mode I (crack opening) load on the interface. However, this can be done by simultaneously imposing in-plane biaxial tension and in this variant of the test[37] there is scope for examining the interfacial toughness over a range of ψ values.

aspect ratios normally found to be necessary in order to test MMCs in view of their relatively high bond strengths.

Shear lag based models have been developed for push-out testing[48,49], which are very similar to those for pull-out. However, comparisons with predictions from more rigorous finite element modelling[51] and with photoelastic measurements[52] (Fig. 6.9) indicate that the actual distribution of shear stress along the fibre is more uniform than the shear

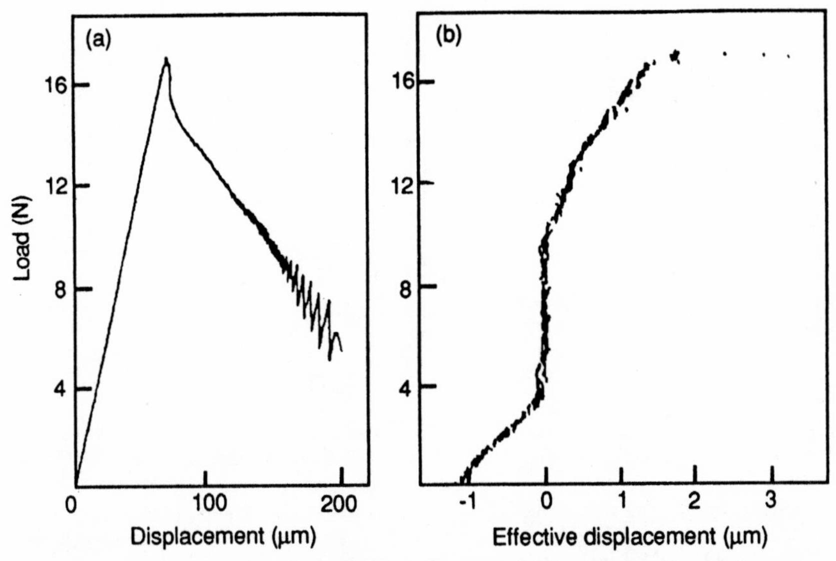

Fig. 6.8 Experimental load–displacement data for a push-out test applied to a SiC monofilament of aspect ratio about 12 in a glass matrix[53], showing the data (a) as-collected and (b) after subtraction of the machine compliance, revealing the displacement of the fibre relative to the matrix. It is clear that in this case initial debonding has been followed by a period of progressive crack growth along the interface. A period of such stable crack growth may or may not occur, depending on fibre aspect ratio, interfacial toughness, etc.

lag treatments suggest, particularly for the low aspect ratios commonly used for push-out with MMCs. If the shear stress is taken as constant along the fibre length during frictional sliding, then a simple force balance can be used to obtain τ_{fr} from the stress applied to the fibre σ_{0fr}

$$\sigma_{0fr}\pi r_0^2 = \tau_{fr} L 2\pi r_0$$
$$\therefore \; \tau_{fr} = \frac{\sigma_{0fr}}{4s} \tag{6.23}$$

It is, however, often difficult to evaluate separately the parameters which determine the value of τ_{fr}

$$\tau_{fr} = \tau_0 - \mu\sigma_r \tag{6.1}$$

Procedures have been suggested[49] for evaluation of μ and σ_r, assuming τ_0 to be negligible. In fact, such an assumption may well be unjustified. In any event, since σ_r is relatively unaffected in MMCs by external loading, as it is dominated by thermal stress, it might be argued that τ_{fr} can be taken as the material property of interest.

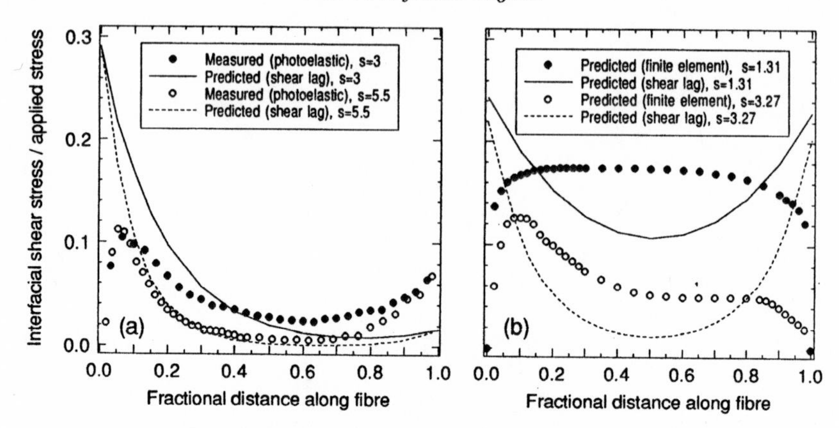

Fig. 6.9 Simple shear lag models for the elastic stress state during push-out apparently exaggerate the variation in interfacial shear stress along the fibre length. Comparisons are shown here for the Hsueh shear lag model[49] with (a) photoelastic measurements[52], for two transparent resins ($s = 3$ and 5.5, $E_I/E_M = 1.47$, $\nu_I/\nu_M = 1$, $R/r_0 = 5.5$, where R is the matrix radius) and with (b) FEM calculations[51], for sapphire fibres in niobium ($s = 1.31$ and 3.27, $E_I/E_M = 4.0$, $\nu_I/\nu_M = 0.76$, $R/r_0 = 2.3$, where R is taken as the horizontal distance from the axis to the edge of the specimen support surface). The finite element calculations[51] indicate that the size of the gap in the support surface is of some significance, particularly with regard to the region near the lower regions (fractional distance along fibre close to 1).

In practice, indentation equipment allowing load–displacement curves to be measured can be expensive. However, it is possible to obtain useful information with a conventional microhardness indenter, provided the testing can then be carried out for a range of fibre aspect ratios. This is most conveniently done using a wedge-shaped specimen, as shown in Fig. 6.10. The dependence of push-out load on aspect ratio can be used to infer whether debonding or frictional sliding requires the larger load. For example, a linear variation of push-out load with aspect ratio indicates that the peak load corresponds to frictional sliding, with little variation in τ_{fr} along the fibre length[52].

The push-out test can also be used to obtain $\mathcal{G}_{ic}$ data. Various analyses have been presented[2,46,54] of the energy balance during push-out. An equation can be obtained relating fibre displacement u with respect to the matrix to interfacial properties

$$u = \frac{2W^2}{\pi^2 \tau_{fr} d^3 E_I} - \frac{\mathcal{G}_{ic}}{\tau_{fr}} \tag{6.24}$$

where W is the applied load (N) while the crack propagates. Again, this analysis is not always suitable because stable crack growth may not occur.

6.2.2 Other tests

Apart from the single fibre loading tests, the procedures shown in Fig. 6.5 are applicable only to planar interfaces and hence cannot be used for most MMCs in their normal form. However, several other techniques which can be applied to MMC systems have been suggested which give mechanical information about the interface.

Full fragmentation

This procedure for deducing an interfacial shear strength involves embedding a single fibre in a matrix, heavily straining the matrix in tension and then measuring the mean aspect ratio of the resulting fibre segments. Analysis is based on a constant interfacial shear stress τ, with the Weibull modulus of the fibre taken into account[35]. One of the criticisms of the method is that it is unclear precisely what interfacial characteristic is being measured, although it is presumably related to τ_*.

Debonding observations

Estimates based on thermodynamic arguments and debonding observations[55,56] have suggested a high normal debonding stress σ_{r*} (several GPa) for Al–SiC: it certainly appears that this bond is normally stronger than that, for example, in Ti–SiC, although the details are probably sensitive to interfacial contamination or reaction. More experimental data are needed in order to confirm the validity of such deductions.

Fibre protrusion

Fibre protrusion/intrusion during thermal cycling has been used[57] to investigate interfacial characteristics. Observations made on the relative displacement of fibre and matrix free surfaces as thermal stresses are changed can allow deductions to be made about the interfacial shear strength and the nature of the initial residual stress state.

Macroscopic interfacial shear strength testing

Apparatus has been developed[58,59] for bringing two components into contact under a selected normal load and measuring the shear force needed to cause sliding. The operation can be carried out in a controlled

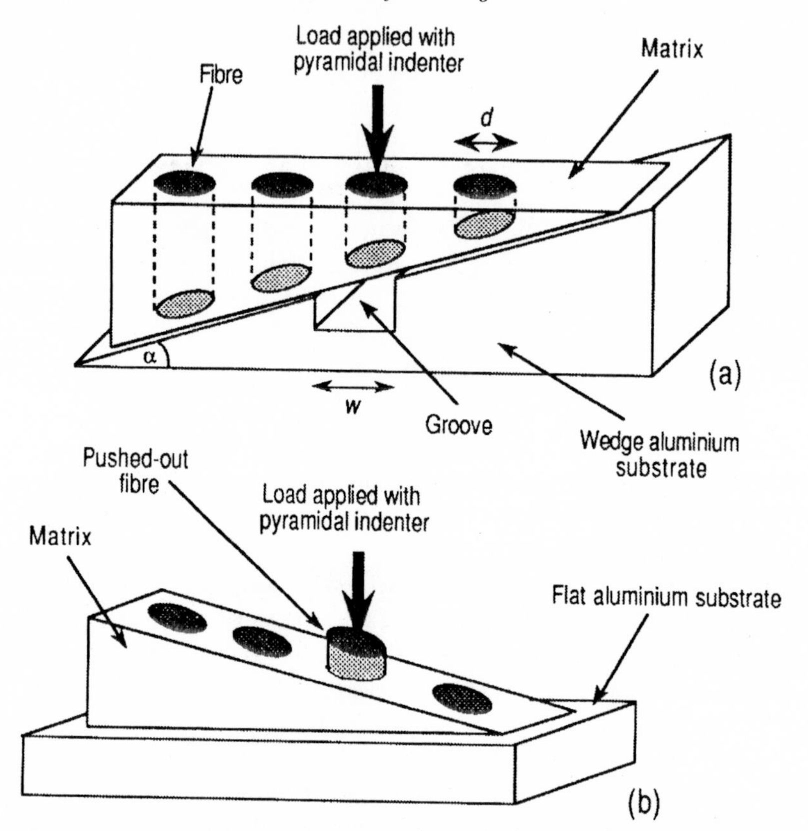

Fig. 6.10 The single fibre push-out test[52] using a wedge-shaped specimen, showing (a) push-out and (b) push-back operations. A convenient way to measure the push-out load for a range of fibre aspect ratios is to produce a composite wedge with a small, well-defined angle, α ($\sim 3°$), and to use the distance from the edge to determine the aspect ratio of each fibre tested. (*cont.*)

atmosphere and the interface subjected to heat treatments *in situ*[59]. There are difficulties in relating the measured shear force to data from other tests, notably because of uncertainties about the true contact area. Nevertheless, clear trends can be revealed, such as a much reduced shear strength in Al–SiC when a graphitic layer is present at the interface[59].

Table 6.1 shows a compilation of experimental interfacial bond strength data obtained for various Ti–SiC long fibre composites, using a range of the tests outlined above. It is difficult to make any clear deductions from these data about the validity of the test procedures or the implications

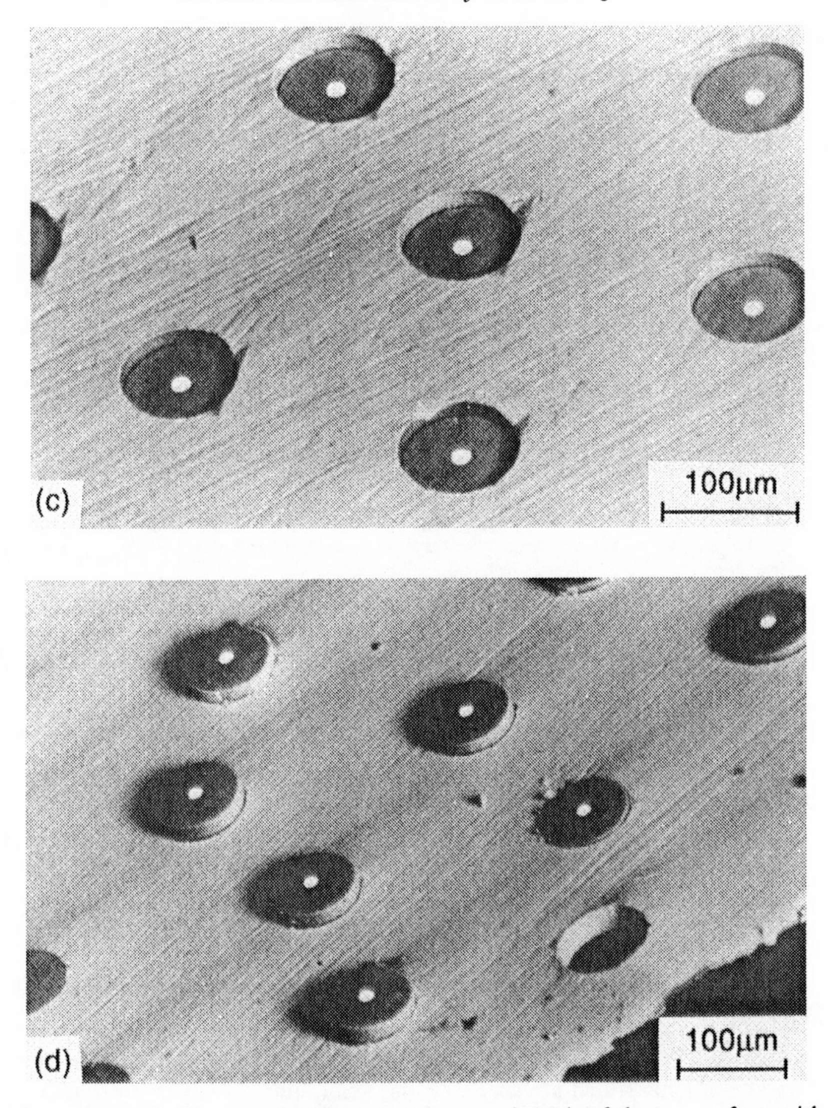

Fig. 6.10 (*cont.*) The scanning electron micrograph (c) is of the top surface, with some pushed-down fibres visible, while (d) shows the underside near the thin end of a Ti–6Al–4V/35 vol% SiC monofilament composite wedge after testing. One very low aspect ratio fibre has been pushed out completely.

of the values obtained, since they are from composites in which various fibres, matrices and fabrication conditions were used and several different interfacial parameters were measured. More comprehensive data are expected to become available in the near future.

Table 6.1 *Interfacial shear strength data*§ *for Ti/SiC monofilament composites, obtained using various tests.*

Matrix	SiC fibre	Processing	Test method	τ_* (MPa)	τ_{fr} (MPa)	Ref.
Ti	Sigma-1†	spray deposit	pull-out	50	12	60
Ti–6Al–4V	Sigma-1†	diffusion bond	fragmentation	345	—	35
Ti–6Al–4V	Sigma-2‡	diffusion bond	push-out	—	80	52
Ti–6Al–4V	Sigma-2‡	diffusion bond, 59 h @ 815 °C	push-out	—	211	52
Ti–25Al–10Nb–3V	SCS-6#	powder hot press	push-out	110	57	44
Ti–25Al–10Nb–3V	SCS-6#	powder hot press, 100 h @ 800 °C	push-out	99	59	44
Ti–6Al–4V	SCS-6#	diffusion bond	push-out	150	90	44
Ti–15V–3Al–3Cr-3Sn	SCS-6#	diffusion bond	push-out	120	80	44
Ti–25Al–10Nb–3V	SCS-6#	diffusion bond	push-out	120	50	44
Ti	SCS-6#	spray deposit	pull-out	5	1	60
Ti–6Al–4V	SCS-6#	diffusion bond	fragmentation	180	—	35
Ti–6Al–4V	BorSiC$	diffusion bond	fragmentation	240	—	35

§ Values for the spray-deposited matrix are low, but this material has little or no reaction layer and also exhibits porosity (~5–10%), leading to a reduced radial compressive stress. The other fabrication methods all involve a heat treatment sufficient to form reaction layers. For the Sigma-2 fibre a substantial heat treatment leads to a rise in τ_{fr} as the C interlayer became consumed. The fragmentation test appears to give rather high values relative to other tests on comparable composites.
† $d \sim 100$ μm, no coat, W core.
‡ $d \sim 100$ μm, 1 μm C/1 μm TiB_2 coat, W core.
# $d \sim 140$ μm, 4 μm C coat, C core.
$ $d \sim 100$ μm, boron fibre, SiC coat, W core.

6.3 Interfacial chemistry

6.3.1 Thermodynamics and kinetics of interfacial reactions

Interfacial reaction may occur both during composite fabrication and under service conditions. Depending on the fabrication process and reaction kinetics, it may be possible to make good practical composites from constituents which are thermodynamically unstable. Thermodynamic and kinetic data are shown in Table 6.2 for various chemical reactions which might take place at the interface for different MMC systems. A large negative value for ΔG^0 (at T_{mp}) is indicative of a reaction with a large thermodynamic driving force at the specified temperature†. If ΔG^0 is positive, then the reaction will not take place at all, but a large negative value does not necessarily mean that it will occur quickly.

† ΔG^0 will be different at other temperatures, but the figure given can be used as a comparative guide.

Table 6.2 *Summary of thermodynamic and kinetic data*[15,61–71] *for various reactions*[§] *that could occur in the interfacial regions of MMCs.*

Reaction				
Matrix + ceramic	Reaction products	ΔG^0 (kJ mole^{-1})	Rate constant (nm s$^{-1/2}$)	Q (kJ mole^{-1})
$\frac{8}{3}Ti + SiC$	$\frac{1}{3}Ti_5Si_3 + TiC$	−900 (1200 K)	22 (1200 K)	200
$Ti + C$	TiC			
$Ti + TiB_2$	$2TiB$	−30 (1200 K)	8 (1200 K)	220
$Ti + B_4C$	$4TiB + TiC$		12 (1200 K)	220
$\frac{3}{2}Ti + Al_2O_3$	$\frac{3}{2}TiO_2 + 2Al$			
$2Ni + SiC$	$Ni_2Si + C$	−75 (1600 K)	12 (1200 K)	184
$2Co + SiC$	$Co_2Si + C$		2.2 (1200 K)	204
$\frac{4}{3}Al + SiC$	$\frac{1}{3}Al_4C_3 + Si$	−88.5 (900 K)		
$\frac{4}{3}Al + C$	$\frac{1}{3}Al_4C_3$	−24 (900 K)		
$\frac{4}{3}Al + SiO_2$	$\frac{2}{3}Al_2O_3 + Si$	−210 (900 K)		
$\frac{3}{4}Mg + Al_2O_3$	$\frac{3}{4}MgAl_2O_4 + \frac{1}{2}Al$	−13 (900 K)	†	103
$2Mg + SiC$	$Mg_2Si + C$	−7 (900 K)		
$2Mg + Li + SiC$	$Mg_2Si + \frac{1}{2}Li_2C_2$	−18 (900 K)		
$MgO + Al_2O_3$	$MgAl_2O_4$	−28 (900 K)		
$4Mg + SiO_2$	$2MgO + Mg_2Si$	−131 (900 K)		
$Al + \frac{1}{2}Mg + SiO_2$	$\frac{1}{2}MgAl_2O_4 + Si$	−219 (900 K)	†	

§ Standard free energy changes for the reactions are quoted per mole of ceramic reactant, with all constituents at unit activity. In practice, dissolution effects (e.g. O in Ti and Si and Al) may be important, as they could allow appreciable attack to occur before the quoted reaction products are formed. The rate constant values given are necessarily approximate, as they have been distilled from published data which are incomplete and partially inconsistent. Some published activation energy data have been in error due to omission of the factor of 2 in eqn (6.26).

† Spinel formation has been observed[66] to follow a linear growth law, with an incubation period.

The reaction layer thickness, x, after a time t, can be deduced from the rate constant, k, given in Table 6.2 for each reaction at a specified temperature.

$$x = k\sqrt{t} \tag{6.25}$$

Confusion occasionally arises with regard to rate constants and activation energies for reactions. The activation energy, Q, is conventionally that for the diffusional process governing the reaction. Since the diffusion distance is expected to exhibit a $\sqrt{(Dt)}$ dependence, then the expected form for the dependence of k, defined as above, on temperature is[63]

$$k = k_0 \exp\left(\frac{-Q}{2RT}\right) \tag{6.26}$$

where R is the molar gas constant and the temperature T is in Kelvin. The rate constants are given in units of nm/$\sqrt{}$s. From the value of Q, the rate constant can be evaluated for any temperature, since a new rate constant, k_2, is related to the given one, k_1, by

$$k_2 = k_1 \exp\left(\frac{Q(T_2 - T_1)}{2RT_2T_1}\right) \tag{6.27}$$

Titanium and other high temperature matrices

Interfacial chemical reactions are of particular concern for titanium composites, because titanium and its alloys tend to react with most candidate reinforcements and the extent of reaction during fabrication is frequently quite substantial†. Furthermore, there is interest in using titanium MMCs at elevated service temperatures (see Chapter 12). The diffusion bonding process commonly used in fabricating Ti-based composites (§9.2.2) can give rise to significant interfacial reaction when the reinforcement is SiC monofilament[62,63,70,72–78], typically generating a reaction layer around a micron or so in thickness. Such a layer is expected to have a significant deleterious effect on properties, particularly under transverse loading. Processes such as spray deposition (§9.1.2) and vapour deposition (§9.2.3) on the other hand, involve less severe or prolonged high temperature metal/ceramic contact. Subsequent hot processing is necessary to complete the consolidation, but since metal flow distances are shorter than in the diffusion bonding of foils, there may be scope for a reduction in final reaction layer thickness.

It is important to note that there are significant differences between the reaction characteristics of different Ti–ceramic systems. Firstly, while kinetic data are far from complete, it is clear that some ceramics react with Ti faster than others. For example, although TiB_2 undergoes a well-defined reaction[79] with Ti which is thermodynamically favoured (albeit rather marginally) over the complete temperature range, the rate of reaction is significantly slower[15,62,65] than for SiC. The rate of reaction of B_4C with Ti appears[62] to be intermediate between those for TiB_2 and SiC. Furthermore, it is now clear that all reaction layers of the same thickness do not have the same effect on the mechanical behaviour. Presumably, factors such as the grain size and grain boundary structures

† This is partly due to the fact that the surface oxide film tends to dissolve in the matrix at temperatures above about 600 °C, leading to loss of the protection which is so important for other reactive metals such as Al.

are of some relevance. It also seems likely that the volume change accompanying the reaction, which will influence the interfacial stress state, can be of some importance (see §6.3.3).

The present situation for Ti is that no ideal ceramic reinforcement has yet emerged. Silicon carbide is apparently far from ideal, as the reaction layer both forms relatively quickly and has a markedly deleterious effect on composite properties. Unfortunately, however, SiC is the most readily available of the candidate long fibre reinforcements (needed for the good creep resistance which is often one of the primary aims). This has given rise to a marked interest in producing coated SiC fibres, with coatings which prevent or delay deleterious interfacial reaction (see §6.4).

Of the other high temperature metals, there has been interest in nickel, which has a lower thermodynamic driving force for reaction with SiC. However, interfacial degradation in this system has been confirmed as being pronounced (often aggravated by the presence of carbide-forming alloying elements such as Cr[80]). Earlier interest[81,82] in reinforcing Ni with tungsten wires revealed a strong tendency for recrystallisation, and hence substantial weakening, to be stimulated in the W as a result of interfacial reaction. Similar reaction problems with W were experienced for cobalt[83] and steel[84], although Co apparently reacts more slowly with SiC than does Ni. Substantial reaction was also observed[85] between SiC and a FeCrAlY alloy. A W coating applied by CVD to the SiC fibres gave some protection, but the coating also reacted with the FeCrAlY matrix[85].

Aluminium and other low temperature matrices

Aluminium is another highly reactive metal, which will reduce most oxides and carbides and is therefore expected to react with the vast majority of reinforcements. However, in the case of Al the reaction rate is often relatively slow, usually because of the presence of some protective layer, commonly alumina, which either persists through processing or is formed *in situ*. In many cases, the fibre–matrix reaction is limited by the resultant slow kinetics, relative to the thermal exposure involved in fabrication and service. Most Al–SiC composites fall into this category, although a fabrication route involving prolonged exposure of SiC to the Al melt can cause extensive reaction[67]. The extent of reaction can be reduced by raising the level of Si in the Al alloy, which favours the reverse reaction by raising the activity of silicon in the melt (see §9.1.3). Consequently, cast Al MMCs are commonly made using Si-containing alloys, or alternatively are reinforced with Al_2O_3. In most cases, only very thin reaction layers (~few nm) are detectable in Al-based composites, although such reactions may nevertheless affect the mechanical behaviour (see §6.3.2).

Of the other low temperature matrices, the thermodynamic stability of many magnesium composites often appears to be rather marginal. This is primarily because, although Mg is a reactive metal with a high affinity for oxygen, it does not have a stable carbide. It therefore does not react with carbon and the reaction with SiC is very marginal thermodynamically (see Table 6.2). The same is true when lithium is present and experimental observations[86] indicate that SiC whiskers remain unattacked in Mg–Li alloys at high temperature. Magnesium does, however, react strongly with oxides, producing MgO or spinels such as $MgAl_2O_4$. Its presence in Al alloys therefore often substantially raises the interfacial reactivity, particularly if silica is available (see Table 6.2 and §6.3.2). In practice silica is often present as an additive, impurity or surface layer in the reinforcement, or in a binder added to a preform (§9.1.1). Magnesium has certainly tended to figure prominently in attempts to promote spontaneous infiltration or melt entry under reduced applied pressure (§9.1.1). Other metals have received only limited attention. Copper-based composites appear to have good potential for applications requiring high conductivity. Copper-based systems containing tungsten and carbon fibres, Al_2O_3, SiC and B_4C have all successfully been produced[87,88], although preliminary evidence suggests that at least for the Al_2O_3, SiC and B_4C systems the interface is weak. Whether this is due to interfacial reaction products has yet to be confirmed.

6.3.2 *Effect of reaction on mechanical behaviour*

There is considerable evidence that chemical reaction, provided it remains limited to a thin layer, can help to ensure intimate interfacial contact and hence raise the bond strength. In some cases the extent of the chemical reaction is limited and well-defined because one of the reactants becomes consumed. An example is provided by 'Saffil®' alumina fibres, which contain a few % of SiO_2, concentrated at the grain boundaries and on the free surface. This silica is readily attacked by a strong reducing agent, such as Mg, present during fabrication. The surface analysis data[89] shown in Fig. 6.11(a) give concentrations in the top few nm of the fibres after various treatments. These data confirm that, during infiltration, Mg from the melt penetrates the fibres to a depth of a few nm, corresponding to the silica-enriched layer. This localised attack of the fibre surface does not appear to occur with Mg-free melts, and this has been correlated with a significantly lower interfacial bond strength exhibited by such composites[91], although this inference is made solely on the basis of fracture surface observations (Fig. 6.11(b) and (c)).

It also appears likely that the rather limited interfacial reactions typical of other Al-based composites contribute to the high interfacial bond strength commonly observed. For example, it has been shown[92] that, in heat treated Al–Mg/SiC composites, even a very thin layer of reaction product Al_4C_3 particles (see Fig. 6.12(b)) appears to inhibit interfacial sliding and apparently increases the elastic modulus. However, in systems prone to progressive interfacial reaction, such as Ti/SiC[15,63,93,94], there is evidence that the interface becomes a preferred site for cracking, both in-plane and through-thickness relative to the layer. Details of the changes in interfacial response characteristics need to be carefully examined. Fibre push-out data generally indicate that the interfacial shear strength (and frictional sliding stress) continue to be raised as reaction proceeds[43,44] and the interface becomes progressively more rough. This is seen, for

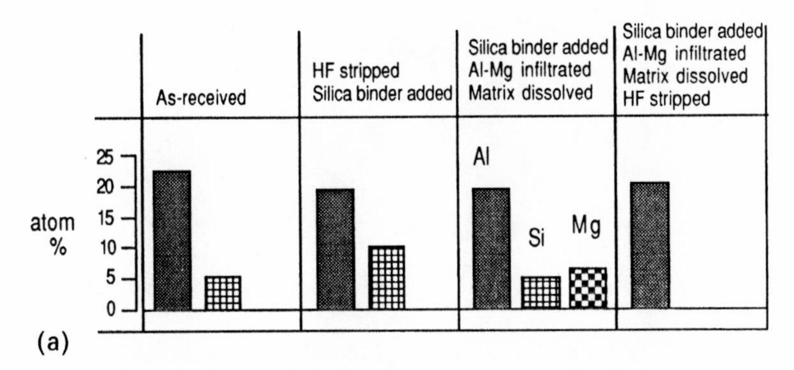

Fig. 6.11 (a) X-ray photoelectron spectroscopy (XPS) analyses[89] of Al, Si and Mg levels in the surface layers of Saffil® fibres after various treatments. The presence of a silica-based binder, commonly used with such fibres, complicates the issue[90]. A brief exposure to HF acid is used to remove a thin surface layer on the fibre, while prolonged immersion of a composite in sodium hydroxide solution dissolves away all the matrix. These data confirm that the thin silicon-rich surface of the fibre becomes impregnated with Mg during manufacture of the composite. That this local attack appears to raise the bond strength appreciably can be seen from the two fracture surfaces[91], which are from Saffil® composites based on (b) an Al alloy containing 2.5% Mg and (c) high purity aluminium. (*continued overleaf.*)

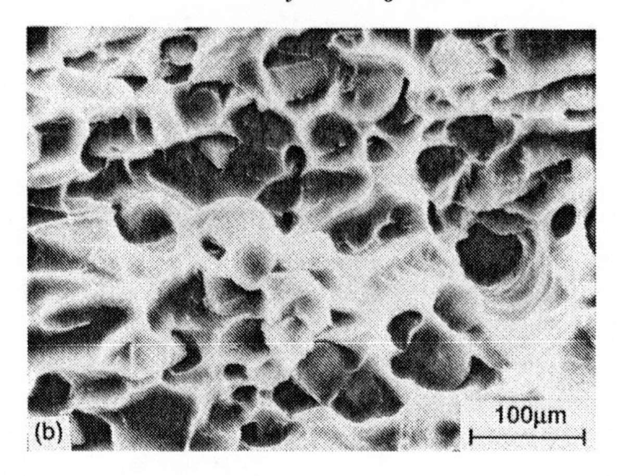

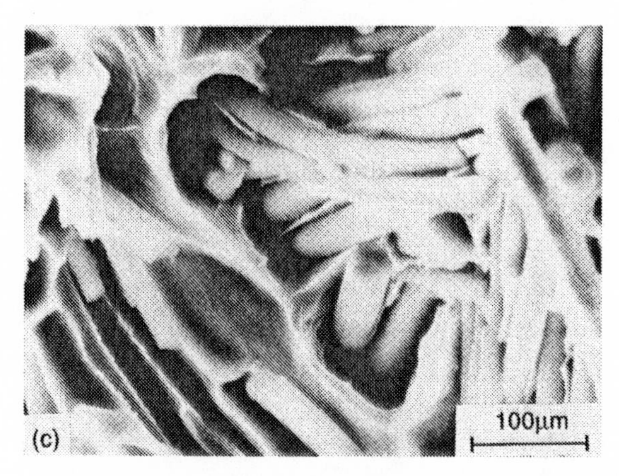

Fig. 6.11 (*cont.*).

example, in the data[52,95] shown in Fig. 6.13. In this case, the fibre had a 1 μm C/1 μm TiB_2 coating. The rather sharp rise in frictional shear strength τ_{fr} after about 3–4 μm of reaction product had formed is thought to correspond to the final disappearance of the C layer, within which interfacial sliding tends to occur preferentially. The reaction was accompanied by a fall in axial composite tensile strength as expected (e.g. Fig. 6.3), but the work of fracture was little affected. Fracture surface examination[16] showed that fibre pull-out was significant initially, but was eliminated when the reaction layer had become thick (Fig. 6.14).

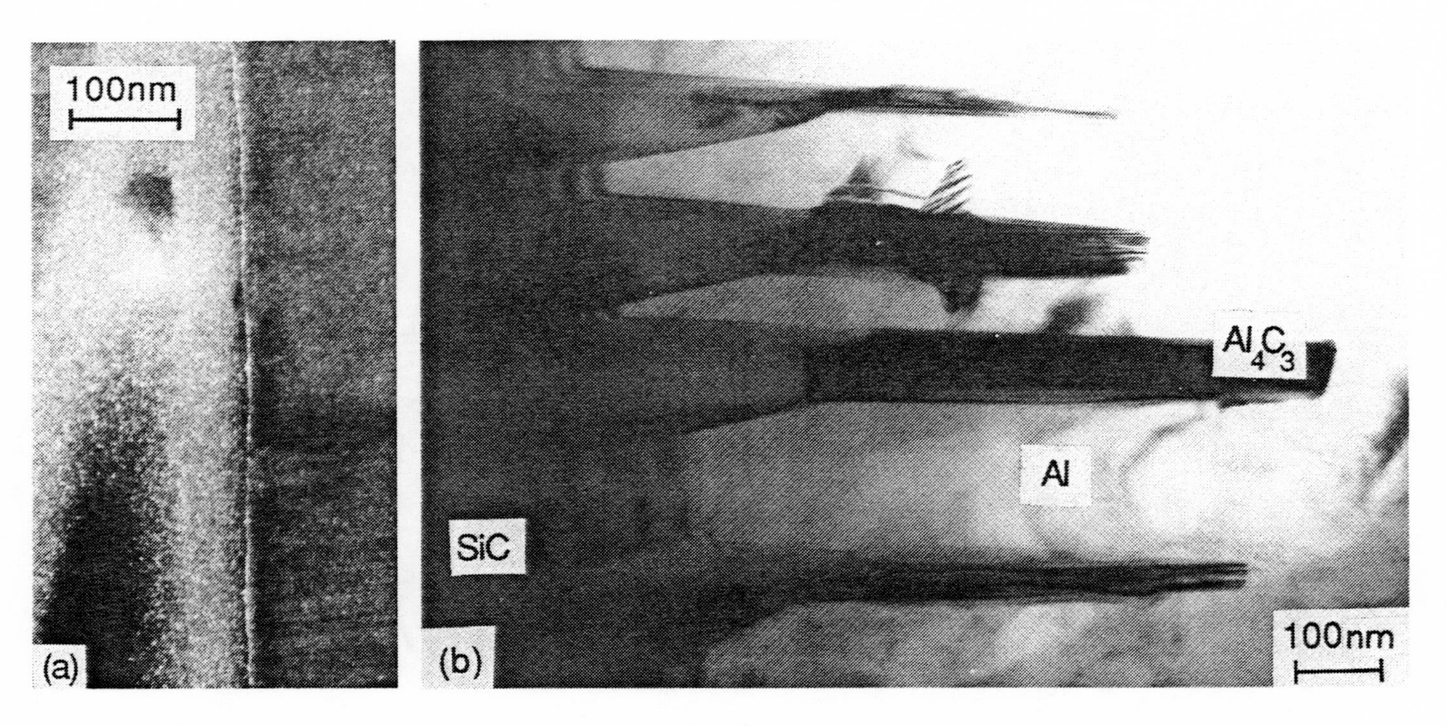

Fig. 6.12 (a) Dark field TEM image of an Al/SiC interface taken with the objective aperture positioned at a scattering angle so as to include electrons from an amorphous halo in the diffraction pattern. The layer thickness is ~ 1.5 nm. (b) The 'keying-in' of Al_4C_3 particles appears to inhibit interfacial sliding[92] in 2124 Al–14 vol% SiC_P composite after a heat treatment of 20 hours at 625 °C.

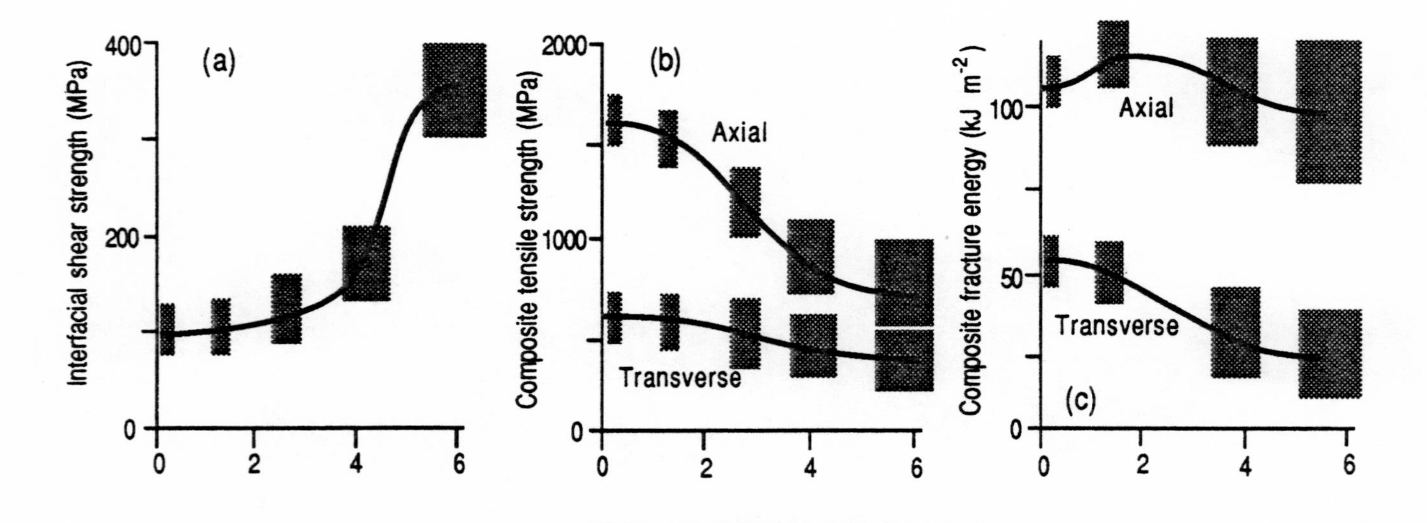

Fig. 6.13 (a) Interfacial shear strength, (b) composite tensile strength and (c) composite work of fracture, for Ti–6Al–4V/35 vol% SiC monofilaments (with a duplex C/TiB_2 coating), as a function of the reaction layer thickness[95,96]. The τ_{fr} values were measured using the single fibre push-out test and the work of fracture data were obtained using a Charpy-like impact test on un-notched rectangular bars.

Substitution of appropriate values ($f \sim 35\%$, $s_0 \sim 1$–1.5, $\tau_{fr} \sim 100$ MPa, $r_0 \sim 50$ μm) in eqn (6.13) suggests that the contribution from fibre pull-out should be of the order of 10–15 kJ m^{-2}. This represents a small, but not insignificant, contribution in the context of the relatively high works of fracture observed here, with matrix plasticity apparently making a large contribution. It may be that the apparent small increase in work of fracture after a certain amount of reaction reflects a slightly larger contribution from continuing pull-out, with a larger τ_{fr}, until pull-out becomes negligible with a heavily reacted interface.

For the transverse loading case, interfacial processes cannot contribute to energy absorption (see §6.1.4) and the effect of the reaction layer on the work of fracture is one of promoting relatively brittle failure of the matrix. It has been shown[16,96] by careful monitoring of apparent Poisson's ratios during tensile loading that interfacial damage occurs early during transverse loading of Ti–SiC long fibre composites, whereas matrix plasticity is significant during axial loading (see Fig. 6.2). The fall in both tensile strength and work of fracture as reaction proceeds can therefore be attributed to the brittle reaction layer promoting brittle failure of the matrix. That axial cracks readily form in the reaction layer can readily be seen in Fig. 6.14(c), and this may be at least partly due to the volume reduction accompanying reaction in this system (see §6.3.3).

6.3.3 Reaction transformation strain effects

Among the effects expected to result from interfacial reaction is a change in stress state arising from any transformation strain. These strains are readily estimated from density data. The case of Ti–SiC may be taken as an example. The relatively high volume change ($\sim -4.6\%$) for Ti–SiC may be partly responsible for the decreases in strength and toughness, and changes in crack path, observed[64,65,94] to result from interfacial reaction in this system, in both particulate and fibrous composites. It can be shown[64] using an Eshelby calculation for the appropriate misfit strains that, for a typical particle size of some 20 μm, a reaction layer about 2 μm in thickness would generate a radial tensile stress sufficient to cancel out the compression from differential thermal contraction ($\sim$200 MPa), allowing for some stress relaxation during cooling.

Further evidence[15] of the ease of interfacial damage in Ti–SiC is shown in Fig. 6.15. These stiffness data show that a fall is observed when reaction

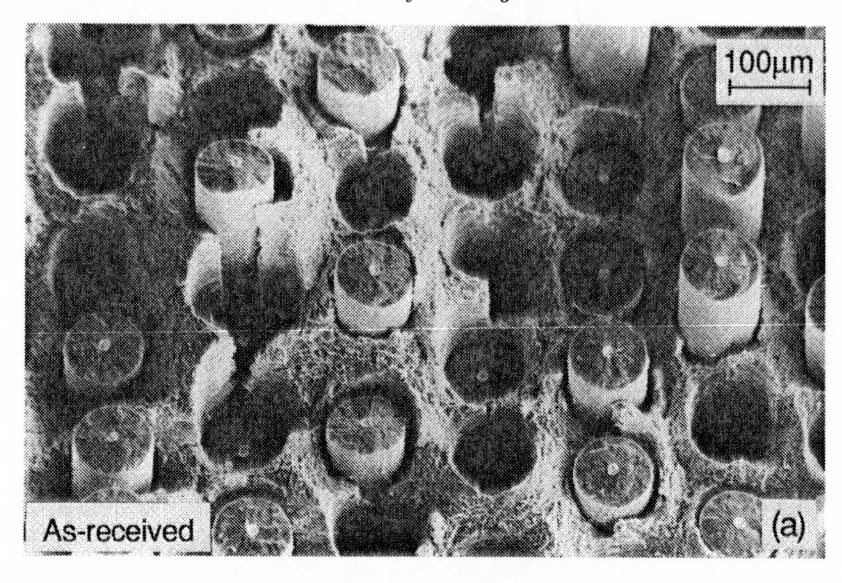

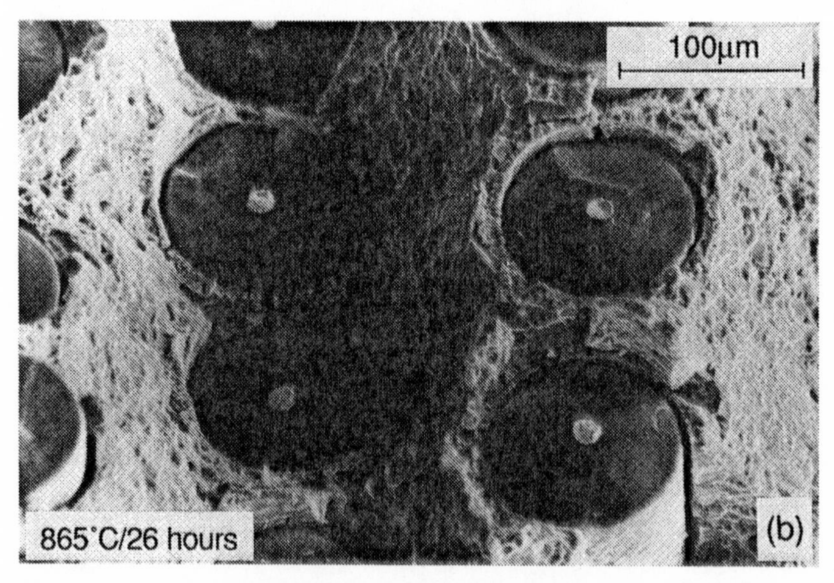

Fig. 6.14 Fracture surfaces from specimens[95,96] used to obtain the work of fracture data (under axial loading) in Fig. 6.13(c), showing significant fibre pull-out in the as-fabricated composite (a), but virtually none in the heat treated (865°C for 26 hours) specimen (b), which had a reaction layer about 5 μm thick.

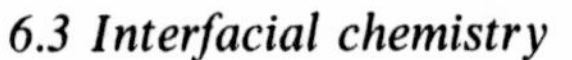

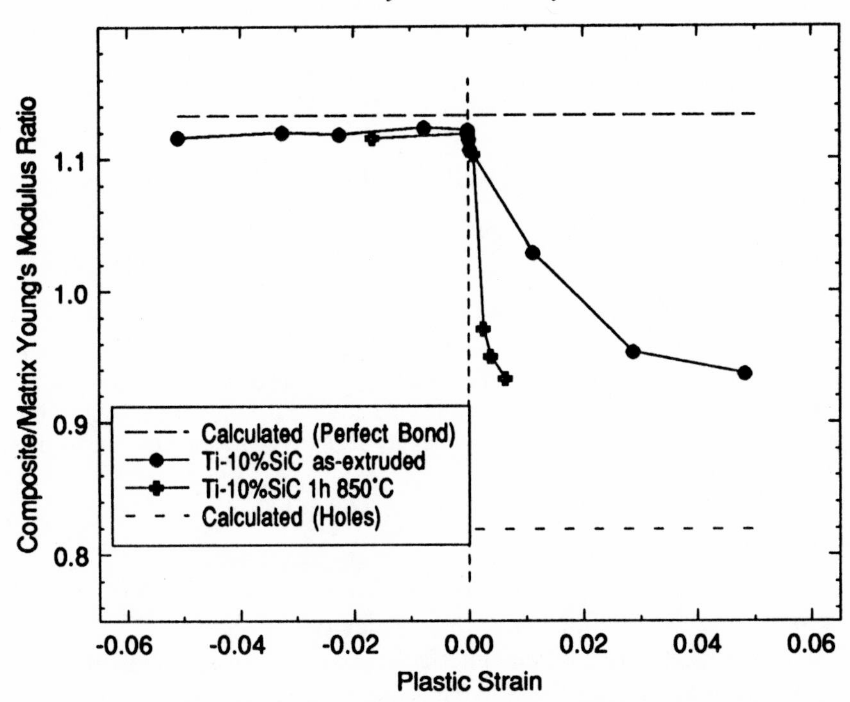

Fig. 6.15 Composite stiffness, measured using fatigue loading methods (§11.1), as a function of prior plastic strain, for particulate-reinforced Ti, with and without a prior heat treatment[15]. The predicted stiffness levels were obtained using the Eshelby method for spherical inclusions. The interface in the Ti–SiC composites appears to be prone to damage during tensile loading, particularly when substantial interfacial reaction has occurred. Straining of the reacted composite leads to a reduction in stiffness to near the level expected if the particles were holes. This must correspond to virtually complete interfacial debonding, as even debonded particles would lead to a higher stiffness than holes, due to the inhibition of Poisson contraction.

is promoted. Furthermore, the stiffness of both as-extruded ($\sim$0.2–0.4 μm reaction layer) and heat treated Ti–SiC composites falls as plastic tensile strain is imposed on the composite. Such a fall is not observed during compressive loading. This is consistent with the interface opening up under tensile loading. The micrographs shown in Fig. 6.16 illustrate the change in crack path as a thick reaction zone is produced. These may be compared with schematic stress distributions around the particle with and without an effective volume contraction of the particle (Fig. 6.17).

Something of a contrast to the Ti–SiC case appears to be provided by

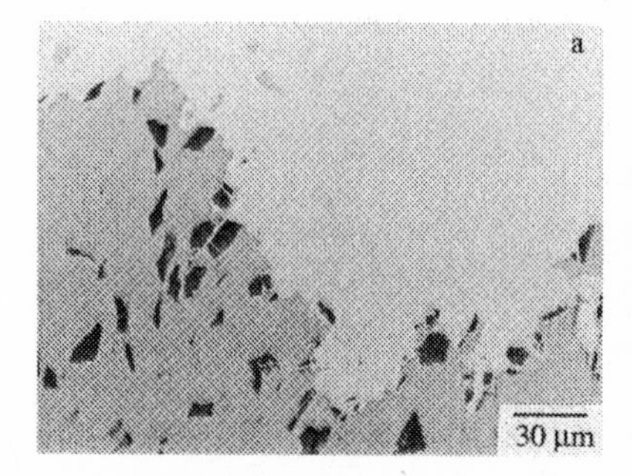

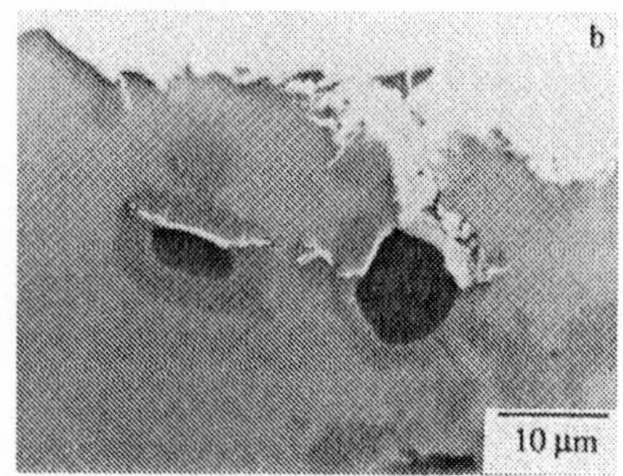

Fig. 6.16 Scanning electron micrographs[64] of Ti–SiC particulate composites after impact testing, sectioned normal to the fracture surface, showing specimens with reaction layer thicknesses of about (a) 0.2 μm and (b) 5 μm. Interfacial debonding between the reaction layer and the remnant SiC core becomes pronounced as the reaction proceeds.

Ti–TiB_2. The reaction in this system, which involves a volume reduction of only 1.4%, is less damaging to mechanical properties[15]. A further indication of better retention of interfacial integrity is provided by thermal conductivity values[97] of Ti-based particulate composites, which are higher than for the unreinforced matrix with TiB_2 reinforcement, whereas they are lower with SiC. (Both ceramics have higher conductivities than Ti.)

6.4 Fibre coatings

6.4.1 Coating techniques

Several techniques are available for deposition of thin coatings on long fibres and, to a much lesser extent, on short fibre and particulate reinforcement. These may be divided into the following groups.

Chemical vapour deposition

The CVD process usually involves a hot fibre being traversed through a reaction zone. The process consists of a vaporised species either decomposing thermally or reacting with another vapour so as to form a deposit on a substrate. The technique is used to manufacture SiC (and B) monofilaments, so that the prospect of incorporating the coating operation into the fibre manufacturing process is commercially attractive.

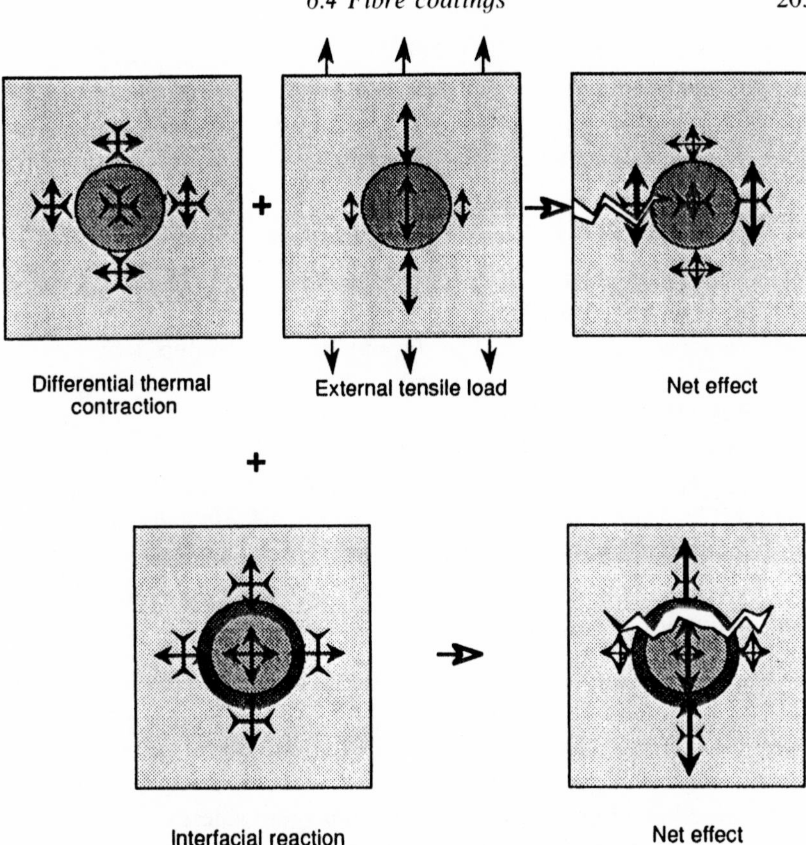

Fig. 6.17 Schematic illustration[64] of the stress distributions around a SiC particle in a Ti matrix under external load, with and without the effect of a transformation strain arising from the interfacial reaction. A volume contraction during the reaction leads to a misfit strain similar to that which would arise on raising the temperature. This causes tensile stresses to be set up normal to the interface, offsetting the residual compressive stresses from differential thermal contraction. This will tend to promote interfacial cracking.

Problems may arise in finding a suitable carrier gas and establishing a deposition procedure for particular coating materials. In practice there are often problems of safety (as many of the gases involved are aggressive and toxic) and of inducing sufficiently high deposition rates, although this is obviously less critical for a thin coating than for the fibre itself. In addition, control over factors such as deposition temperature, and the microstructure and stress state of the deposit, will in general be rather limited. Reviews have been published[98–100] of the factors involved in the use of CVD to produce coatings on fibres, as well as descriptions[101–103] of the equipment employed and technical details. In certain cases the deposition process is promoted or modified by generating an electric discharge plasma, leading to plasma-assisted CVD.

Physical vapour deposition

PVD constitutes a similar group of processes to CVD, the primary difference being that the vapour is formed by a technique which does not involve a chemical reaction. The vapour can be produced by simple heating (evaporation) or by bombarding a target with high energy ions (sputtering)[104]. It is also possible to generate a discharge which causes the deposit to be bombarded as it is formed (ion plating)[105]. Using this technique to clean the fibre before deposition of the coating leads to better adhesion[105]. These processes can also be carried out with a reactive atmosphere. For example, Umekawa *et al.*[82] deposited various coatings on W wires by reactive evaporation and Kieschke and Clyne[106] used reactive sputtering to form yttria on SiC.

Other methods

Other procedures which have been used to coat fibres include plating and spraying techniques. Most of these have been aimed at promoting wetting on fabricating the composite, so that there is less concern about the integrity and structure of the coating than when a protective layer is being formed. Several methods have been suggested in which fibres are passed through a plating bath, notably for depositing Al[107] or Pb[108] on C, and Al on SiC[109] fibres. These processes all involve promotion of wetting by sodium metal or a sodium salt. The plating process can also be encouraged by producing an ion in a suitable electrolyte and generating an electric current (electroplating). This is particularly well-suited to metals like copper and nickel; carbon fibres coated with thin layers ($\sim$0.5–3 μm) of these are available commercially[100]. Some of the problems encountered during electrodeposition have been reviewed[110]. Techniques are also available involving chemical deposition from solution, such as by sol-gel precipitation. An example of this is provided by the method described by Teng and Boyd[111] for sol-gel coating of SiC particles with thin ($\sim$50 nm) layers of Al_2O_3 or MgO, to act as reaction barriers. Thermal spraying has been used to form deposits on fibres, but this is normally used only as a method of composite manufacture (§9.2.1), rather than to form thin coatings.

Table 6.3 *Thermodynamic and diffusion data relevant for selection of barrier layer materials*[61,112–114].

	Property for oxide of X				
X	Cation charge	Cation radius (pm)	$\Delta G_{1000\,K}$ (kJ mol^{-1})	Diffusivity of X (at 1000 K) (m^2 s^{-1})	Diffusivity of O (at 1000 K) (m^2 s^{-1})
Ti	+4	68	−710	—	—
Al	+3	51	−893	—	—
Mg	+2	66	−996	—	—
Y	+3	89	−1080	1.3×10^{-21}	1.0×10^{-16}
Zr	+4	79	−840	1.8×10^{-26}	4.1×10^{-14}
Hf	+4	78	−892	Not available	Not available

6.4.2 Diffusion barrier coatings

A wide range of coatings and deposition methods have been proposed in order to protect fibres from chemical attack. The first problem is to identify a material which will not itself react with matrix or reinforcement. From a thermodynamic point of view, highly stable oxides emerge as strong contenders for barrier coatings. As an example, consider the Gibbs free energy of formation of several candidate oxides (Y_2O_3, HfO_2 and ZrO_2 in Table 6.3) for use in Ti matrix composites. (The free energy of formation of TiO_2 is also shown.) It can be seen that Y_2O_3 is the most stable of the candidate oxides. Further data concerning entropy changes during oxygen dissolution in Ti are needed for detailed predictions, but it should be noted that there is some evidence[115] that both ZrO_2 and HfO_2 can undergo appreciable reaction with solid Ti, while Y_2O_3 is essentially stable. These thermodynamic data give no information on the possible formation of other compounds such as mixed oxides or the possibility of reaction with the fibre (e.g. carbide formation). Information on possible fibre/coating reactions are incomplete (especially for yttrium compounds), although in most cases the high stability of the oxide will prohibit any such reaction.

Yttria has been explored[80,116,117] as a barrier coating for fibres and has been found effective. However, the range of deposition processes applicable is rather limited. In practice, many barrier layers have been explored which slow down interfacial reaction to a suitable degree, rather than being thermodynamically stable. An example is provided by the development of CVD deposition of TiB_2 coatings[118], including the duplex

C/TiB_2 coating developed commercially by British Petroleum (BP)[65] for SiC monofilaments to be used in titanium MMCs. A similar principle applies to various coatings explored for W wires to be used in Ni and steel[82], and SiC on B (BORSIC). The TiB_2 deposited on B fibres for use in Al MMCs[119], on the other hand, should be thermodynamically stable and the same applies to B_4C layers on B[120].

In addition to thermodynamic stability, or at least slow reaction kinetics, the barrier must impair transport of reactants through it. This aspect has been considered by Kieschke *et al.*[116] In general, migration through grain boundaries (and other defects, such as porosity) in the barrier layer is expected to predominate, so that a dense deposit with a large grain size is preferable. Even if this can be produced, a very thin layer ($< \sim 0.1$ μm) is unlikely to inhibit migration of reactants sufficiently[116]. This is consistent with the observed inefficiency of a variety of very thin barriers in Ti/SiC[74].

Other coatings have been designed purely to protect the fibre prior to incorporation in the matrix. An example is provided by the CVD formation of a 50 nm SiC layer on C fibres, to protect them from oxidation[121].

6.4.3 Wetting promotion coatings

Among the possible roles of fibre coatings is the encouragement of wetting[122]. Promotion of wetting is usually brought about through the stimulation or modification of some local reaction[98,123]. Fluxing action can be promoted by the presence of a suitable reactive salt; for example, K_2ZrF_6 coatings have been used for C and SiC fibres[124]. The critical stage is probably dissolution of the surface Al_2O_3 layer by a fluoride complex[125,126]. Various other wetting agents have been explored, particularly for carbon fibres in Al[101,103] and Mg[123].

6.4.4 Mechanics of coatings

In practice, for protective coatings the avoidance of mechanical damage is often the most important objective. For example, large differential thermal contraction stresses can be generated during fabrication. The stress field calculations shown in Fig. 6.18(a) indicate that large tensile hoop and axial stresses will tend to arise in a yttria coating on a SiC filament during cooling. Cracking caused by such stresses is apparent in Fig. 6.18(b). Matrix plastic deformation close to the interface during

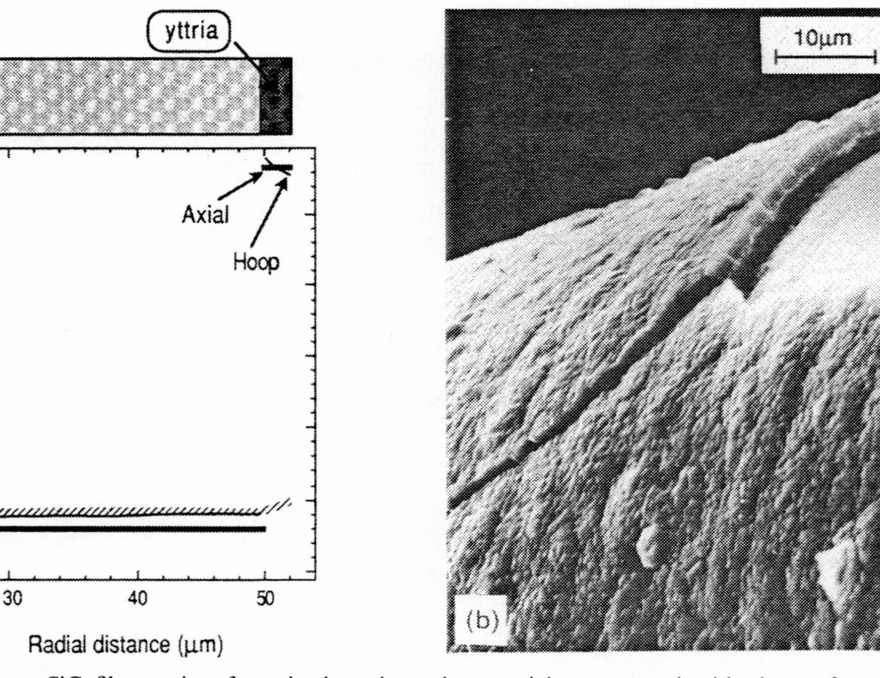

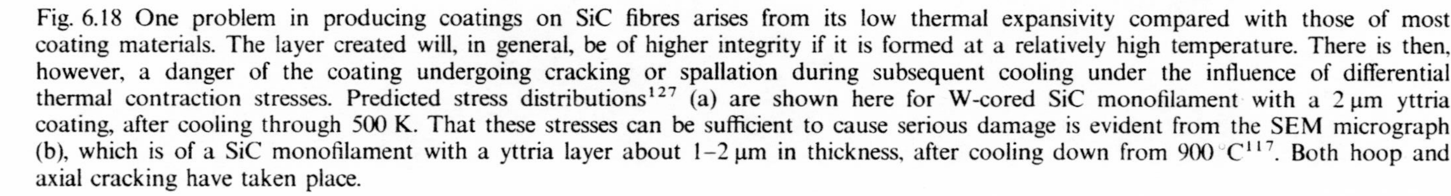

Fig. 6.18 One problem in producing coatings on SiC fibres arises from its low thermal expansivity compared with those of most coating materials. The layer created will, in general, be of higher integrity if it is formed at a relatively high temperature. There is then, however, a danger of the coating undergoing cracking or spallation during subsequent cooling under the influence of differential thermal contraction stresses. Predicted stress distributions[127] (a) are shown here for W-cored SiC monofilament with a 2 μm yttria coating, after cooling through 500 K. That these stresses can be sufficient to cause serious damage is evident from the SEM micrograph (b), which is of a SiC monofilament with a yttria layer about 1–2 μm in thickness, after cooling down from 900 °C[117]. Both hoop and axial cracking have taken place.

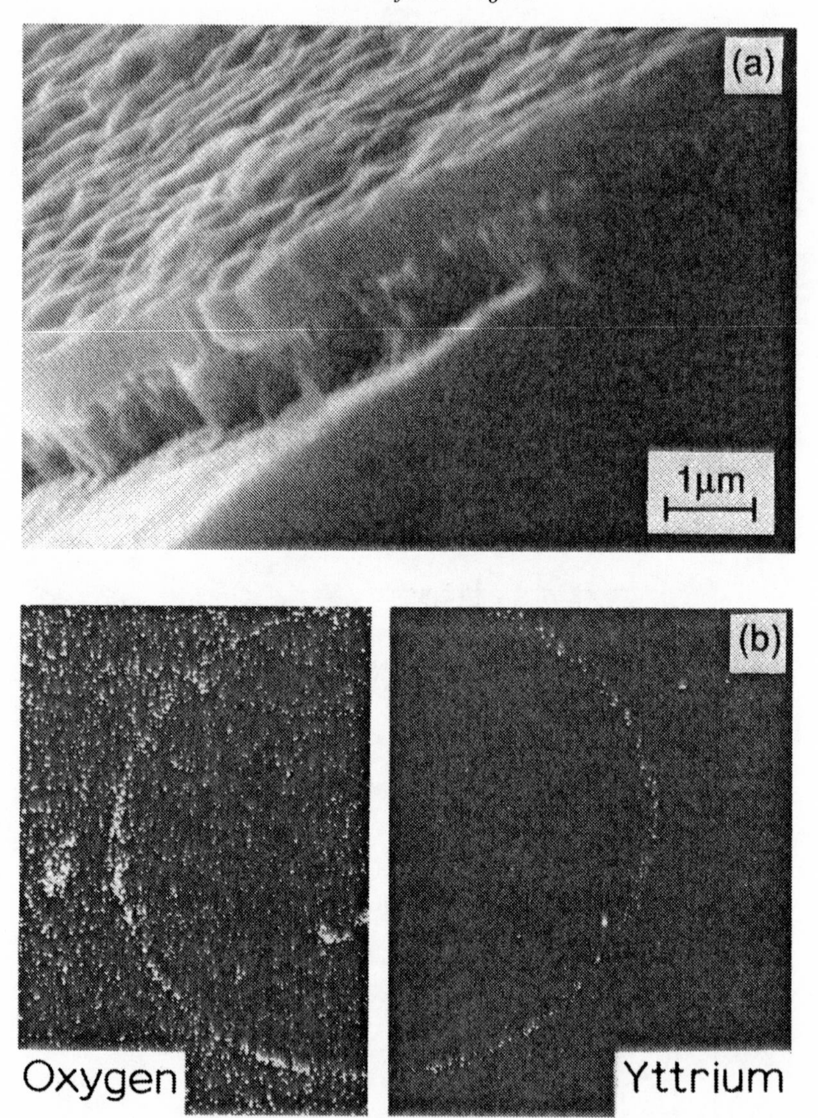

Fig. 6.19 There are potential advantages in having a metallic underlayer and an oxide barrier overlayer as a fibre coating. This is particularly so for Ti matrices, which inevitably contain dissolved oxygen; if a Y_2O_3 barrier becomes mechanically damaged, then exposure of this matrix to an underlayer of metallic yttrium will stimulate a gettering action which will result in reformation of a protective yttria layer. Shown here are (a) an SEM micrograph of a duplex Y/Y_2O_3 coating on a SiC monofilament, produced by heating a Y-coated fibre in a limited oxygen supply and (b) X-ray maps of a Y-coated SiC fibre (100 μm diameter) in Ti after 2 hours at 950 °C, showing the distributions of yttrium and oxygen.

thermal cycling may also cause damage to a brittle coating. Mehan *et al.*[80] demonstrated the value of a Y_2O_3 layer in preventing attack of SiC in the presence of a Ni–Cr alloy, but also showed that (on a planar substrate) this layer underwent spallation after only one thermal cycle. There is scope during PVD processes such as sputtering and ion plating for these stresses to be offset by the so-called 'atomic peening' effect[128], in which bombardment of the deposit can generate a large compressive residual stress. It has been confirmed[129] that a net compressive stress can be induced in a sputtered coating, giving good mechanical stability.

There is interest in the creation of coatings designed, not to eliminate undesirable chemical reactions or encourage wetting, but to promote desirable mechanical behaviour at the interface. In particular, there have been attempts to encourage interfacial sliding (e.g. see Evans[10]) or to provide a compliant or soft layer in contact with the fibre. Unfortunately, the obvious layers for such a purpose, such as graphitic carbon and (hexagonal) boron nitride, tend to react fairly readily with the constituents likely to be present. This has given rise to interest in multi-layer coatings, in which the layers serve different purposes concerned with mechanical and/or chemical behaviour. Examples are provided by the duplex C/TiB_2 coating[65] developed by BP and the duplex Y/Y_2O_3 coating suggested by Kieschke *et al.*[116], in which the underlying metal layer would allow repair of damage to the protective yttria by the gettering of oxygen from the Ti matrix. Two methods of generating a Y/Y_2O_3 structure are illustrated in Fig. 6.19. The area of tailored multi-layer fibre coatings is likely to receive further study in the future.

References

1. H. J. Oel and V. D. Frechette (1986) Stress Distribution in Multi-Phase Systems: II, Composite Disks with Cylindrical Interfaces, *J. Amer. Ceram. Soc.*, **69**, pp. 342–6.
2. R. J. Kerans and T. A. Parathasarathy (1991) Theoretical Analysis of the Fiber Pullout and Pushout Tests, *J. Amer. Ceram. Soc.*, **74**, pp. 1585–96.
3. J. W. Hutchinson, M. E. Mear and J. R. Rice (1987) Crack Paralleling an Interface Between Dissimilar Materials, *J. App. Mech.* (*Trans. ASME*), **54**, pp. 828–32.
4. J. R. Rice (1988) Elastic Fracture Mechanics Concepts for Interfacial Cracks, *J. App. Mech.* (*Trans. ASME*), **55**, pp. 98–103.
5. M. S. Hu, M. D. Thouless and A. G. Evans (1988) The Decohesion of Thin Films from Brittle Substrates, *Acta Metall.*, **36**, pp. 1301–7.
6. M. Hu (1989) The Cracking and Decohesion of Thin Brittle Films, *Mat. Res. Soc. Symp.*, **130**, pp. 213–18.

7. A. G. Evans, M. Rühle, B. J. Dalgleish and P. G. Charalambides (1990) The Fracture Energy of Bimaterial Interfaces, *Mater. Sci. & Eng.*, **A126**, pp. 53–64.
8. Z. Suo and J. W. Hutchinson (1990) Interface Crack Between Two Elastic Layers, *Int. J. Fract.*, **43**, pp. 1–18.
9. A. G. Evans and J. W. Hutchinson (1989) Effects of Non-planarity on the Mixed Mode Fracture Resistance of Bimaterial Interfaces, *Acta Metall.*, **37**, pp 909–16.
10. A. G. Evans (1988) The Mechanical Performance of Fiber Reinforced Ceramic Matrix Composites, in *Mech. & Physical Behav. of Met. & Cer. Composites, 9th Risø Int. Symp. on Met. & Mat. Sci.*, S. I. Anderson, H. Lilholt and O. B. Pedersen (eds.), Risø Nat. Lab., Denmark, pp. 497–502.
11. M.-Y. He and J. W. Hutchinson (1989) Kinking of a Crack Out of an Interface, *J. Appl. Mech.*, **56**, pp. 270–8.
12. C. F. Shih and R. J. Asaro (1988) Elastic–Plastic Analysis of Cracks on Bimaterial Interfaces: Part I – Small Scale Yielding, *J. Appl. Mech.*, **55**, pp. 299–316.
13. H. Takehashi and T. W. Chou (1988) Transverse Elastic Moduli of Unidirectional Fibre Composites with Interfacial Debonding, *Metall. Trans.*, **19A**, pp. 129–35.
14. J. Aboudi (1988) Constitutive Equations for Elastoplastic Composites with Imperfect Bonding, *Int. J. Plasticity*, **4**, pp. 103–25.
15. A. J. Reeves, W. M. Stobbs and T. W. Clyne (1991) The Effect of Interfacial Reaction on the Mechanical Behaviour of Ti Reinforced with SiC and TiB_2 Particles, in *Metal Matrix Composites – Processing, Microstructure and Properties, 12th Risø Int. Symp.*, Roskilde, N. Hansen, D. J. Jensen, T. Leffers, H. Lilholt, T. Lorentzen, A. S. Pedersen, O. B. Pedersen and B. Ralph (eds.), Risø Nat. Lab., Denmark, pp. 631–6.
16. M. C. Watson and T. W. Clyne (1992) Reaction-induced Changes in Interfacial and Macroscopic Mechanical Properties of SiC Monofilament Reinforced Titanium, *to be published* in *Composites.*
17. A. G. Metcalfe (1967) Interaction and Fracture of Titanium–Boron Composites, *J. Comp. Mat.*, **1**, pp. 356–65.
18. A. G. Metcalfe and M. J. Klein (1974) Effect of the Interface on Longitudinal Tensile Properties, in *Interfaces in Metal Matrix Composites*, K. G. Krieder (ed.), Academic Press, New York, pp. 310–30.
19. S. Ochiai and Y. Murakami (1979) Tensile Strengths of Composites with Brittle Reaction Zones at Interfaces, *J. Mat. Sci.*, **14**, pp. 831–40.
20. S. Ochiai, S. Urakawa, K. Ameyama and Y. Murakami (1980) Experiments on Fracture Behaviour of Single Fibre-Brittle Zone Model Composites, *Metall. Trans.*, **11A**, pp. 525–30.
21. M. K. Shorshorov, L. M. Ustinov, A. M. Zirlin, V. I. Olefirenko and L. V. Vinogradov (1979) Brittle Interfacial Layers and the Tensile Strength of Metal Matrix Fibre Composites, *J. Mat. Sci.*, **14**, pp. 1850–61.
22. Y. Le Petitcorps, M. Lahaye, R. Pailler and R. Naslain (1988) Modern Boron and SiC CVD Filaments: a Comparative Study, *Comp. Sci. and Tech.*, **32**, pp. 31–55.
23. T. Onzawa, A. Suzumura and J. H. Kim (1991) Influence of Reaction Zone Thickness on Tensile Strength for Titanium Matrix Composites Reinforced with SiC Fiber, in *Composites: Design, Manufacture and Application*, Hawaii, USA, S. W. Tsai and G. S. Springer (eds.), SAMPE, pp. 19J/1–19J/10.

24. F. Vescera, J. P. Keustermans, M. A. Dellis, B. Lips and F. Delannay (1991) Processing and Properties of Zn–Al Alloy Matrix Composites Reinforced by Bidirectional Carbon Tissues, in *Metal Matrix Composites – Processing, Microstructure and Properties, 12th Risø Int. Symp.*, Roskilde, N. Hansen, D. J. Jensen, T. Leffers, H. Lilholt, T. Lorentzen, A. S. Pedersen, O. B. Pedersen and B. Ralph (eds.), Risø Nat. Lab., Denmark, pp. 719–24.
25. W. Chengfu, Y. Meifang and W. Zhangbao (1987) Analysis of Fracture Features of Carbon Fibre Reinforced Zinc Base Alloy Composite, in *Proc. ICCM VI/ECCM2*, London, F. L. Matthews, N. C. R. Buskell, J. M. Hodgkinson and J. Morton (eds.), Elsevier, pp. 2.476–2.480.
26. J. Cook and J. E. Gordon (1964) A Mechanism for the Control of Crack Propagation in All-Brittle Systems, *Proc. Roy. Soc.*, **A282**, pp. 508–20.
27. M.-Y. He and J. W. Hutchinson (1989) Crack Deflection at an Interface between Dissimilar Elastic Materials, *Int. J. Solids Struct.*, **25**, pp. 1053–67.
28. K. Kendall (1975) Transition between Cohesive and Interfacial Failure in a Laminate, *Proc. Roy. Soc.*, **A344**, pp. 287–302.
29. C. Atkinson, R. E. Smelser and J. Sanchez (1982) Combined Mode Fracture via the Cracked Brazilian Disk Test, *Int. J. Fracture*, **18**, pp. 279–91.
30. J.-S. Wang and Z. Suo (1990) Experimental Determination of Interfacial Fracture Toughness Curves Using Brazil-Nut-Sandwiches, *Acta Met. et Mat.*, **38**, pp. 1279–90.
31. P. G. Charalambides, J. Lund, A. G. Evans and R. M. McMeeking (1989) A Test Specimen for Determining the Fracture Resistance of Bimaterial Interfaces, *J. App. Mech.* (*Trans. ASME*), **56**, pp. 77–82.
32. H. C. Cao and A. G. Evans (1989) An Experimental Study of the Fracture Resistance of Bimaterial Interfaces, *Mech. Mater.*, **7**, pp. 295–304.
33. O. Sbaizero, P. G. Charalambides and A. G. Evans (1990) Delamination Cracking in a Laminated Ceramic Matrix Composite, *J. Amer. Ceram. Soc.*, **73**, pp. 1936–40.
34. P. G. Charalambides, H. C. Cao, J. Lund and A. G. Evans (1990) Development of a Test for Measuring the Mixed Mode Fracture Resistance of Bimaterial Interfaces, *Mech. Mater.*, **8**, pp. 269–83.
35. Y. Le Petitcorps, R. Pailler and R. Naslain (1989) The Fibre/Matrix Interfacial Shear Strength in Titanium Alloy Matrix Composites Reinforced by SiC or B CVD Filaments, *Comp. Sci. & Tech.*, **35**, pp. 207–14.
36. L. S. Sigl and A. G. Evans (1989) Effects of Residual Stress and Frictional Sliding on Cracking and Pullout in Brittle Matrix Composites, *Mech. Mater.*, **8**, pp. 1–12.
37. M. C. Watson and T. W. Clyne (1992) The Tensioned Push-Out Test for Measurement of Fibre/Matrix Interfacial Toughness under Mixed Mode Loading, to be published in *Mat. Sci. & Eng.*
38. P. Lawrence (1972) Some Theoretical Considerations of Fibre Pullout from an Elastic Matrix, *J. Mat. Sci.*, **7**, pp. 1–6.
39. P. S. Chua and M. R. Piggott (1985) The Glass Fibre–Polymer Interface: I. Theoretical Considerations for Single Fibre Pullout Tests, *Comp. Sci. & Tech.*, **22**, pp. 33–42.
40. C. H. Hsueh (1988) Elastic Load Transfer from Partially Embedded Axially Loaded Fibres to Matrix, *J. Mat. Sci. Letts.*, **7**, pp. 497–500.
41. P. G. Charalambides and A. G. Evans (1989) Debonding Properties of Residually Stressed Brittle Matrix Composites, *J. Amer. Ceram. Soc.*, **72**, pp. 746–53.

42. R. R. Kieschke and T. W. Clyne (1989) Pull-Out Testing of SiC Monofilaments in a Spray-Deposited Ti Matrix, in *Interfacial Phenomena in Composite Materials*, F. R. Jones (ed.), Butterworth, pp. 282–93.
43. J. I. Eldridge and P. K. Brindley (1989) Investigation of Interfacial Shear Strength in a SiC Fibre/Ti–24Al–11Nb Composite by a Fibre Push-out Technique, *J. Mat. Sci. Letts.*, **8**, pp. 1451–4.
44. C. J. Yang, S. M. Jeng and J. M. Yang (1990) Interfacial Properties Measurement for SiC Fiber-Reinforced Titanium Alloy Composites, *Scripta Met. et Mat.*, **24**, pp. 469–74.
45. D. B. Marshall (1984) An Indentation Method for Measuring Matrix–Fibre Frictional Stresses in Ceramic Composites, *J. Amer. Ceram. Soc.*, **67**, pp. C259–60.
46. D. B. Marshall and W. C. Oliver (1987) Measurement of Interfacial Mechanical Properties in Fiber-Reinforced Ceramic Composites, *J. Amer. Ceram. Soc.*, **70**, pp. 542–8.
47. T. P. Weihs and W. D. Nix (1988) In Situ Measurements of the Mechanical Properties of Fibres, Matrices and Interfaces in Metal Matrix and Ceramic Matrix Composites, in *Metallic and Ceramic Composites*, 9th Risø Int. Symp., S. I. Andersen, H. Lilholt and O. B. Pedersen (eds.), Risø Nat. Lab., Denmark, pp. 497–502.
48. D. K. Shetty (1988) Shear Lag Analysis of Fibre Push-out (Indentation) Tests for Estimating Interfacial Friction Stress in Ceramic–Matrix Composites, *J. Amer. Ceram. Soc.*, **71**, pp. C107–9.
49. C. H. Hsueh (1990) Evaluation of Interfacial Shear Strength, Residual Clamping Stress and Coefficient of Friction for Fibre-Reinforced Ceramic Composites, *Acta Met. et Mat.*, **38**, pp. 403–9.
50. R. N. Singh and M. Sutcu (1991) Determination of Fibre–Matrix Interfacial Properties in Ceramic Matrix Composites by a Fibre Push-out Technique, *J. Mat. Sci.*, **26**, pp. 2547–56.
51. M. N. Kallas, D. A. Koss, H. T. Hahn and J. R. Hellman (1992) Interfacial Stress State Present in a 'Thin Slice' Fiber Push-out Test, *J. Mat. Sci.*, **27**, pp. 3821–6.
52. M. C. Watson and T. W. Clyne (1992) The Use of Single Fibre Pushout Testing to Explore Interfacial Mechanics in SiC Monofilament-Reinforced Ti. Part I: A Photoelastic Study of the Test, *Acta Met. et Mat.*, **40**, pp. 135–40.
53. T. A. Parthasarathy, P. D. Jero and R. J. Kerans (1991) Extraction of Interface Properties from a Fiber Push-out Test, *Scripta Met.*, **25**, pp. 2457–62.
54. R. J. Kerans, R. S. Hay and N. J. Pagano (1989) The Role of the Fiber Matrix Interface in Ceramic Composites, *Ceram. Bull.*, **68**, pp. 429–42.
55. Y. Flom and R. J. Arsenault (1986) Interfacial Bond Strength in an Al Alloy 6061–SiC composite, *Mat. Sci. & Eng.*, **A77**, pp. 191–7.
56. S. Li, R. J. Arsenault and P. Jena (1988) A Quantum Chemical Study of Adhesion in Al–SiC, *J. Appl. Phys.*, **64**, pp. 6246–53.
57. B. N. Cox (1990) Interfacial Sliding Near a Free Surface in a Fibrous or Layered Composite During Thermal Cycling, *Acta Met. et Mat.*, **38**, pp. 2411–24.
58. S. V. Pepper (1976) Shear Strength of Metal-Sapphire Contacts, *J. Appl. Phys.*, **47**, pp. 801–8.
59. R. S. Haaland (1990) Mechanical Behaviour of 7091 Aluminium/SiC

Interfaces, in *Fundamental Relationships Between Microstructure and Mechanical Properties of Metal Matrix Composites*, Indianapolis, M. N. Gungor and P. K. Liaw (eds.), TMS, pp. 779–91.
60. R. R. Kieschke and T. W. Clyne (1990) Control over Interfacial Bond Strength in Ti/SiC Fibrous Composites, in *Fundamental Relationships Between Microstructures and Mechanical Properties of Metal–Matrix Composites*, Indianapolis, P. K. Liaw and M. N. Gungor (eds.), TMS, pp. 325–40.
61. R. Stull (1971) *JANAF Thermochemical Tables*, US Department of Commerce, NSRDS-NBS, publication 37.
62. E. Jong and H. Flower, private communication.
63. P. Martineau, M. Lahaye, R. Pailler, R. Naslain, M. Couzi and F. Cruege (1984) SiC Filament/Titanium Matrix Composites Regarded as Model Composites. Part 2. Fibre/Matrix Chemical Interactions at High Temperatures, *J. Mat. Sci.*, **19**, pp. 2749–70.
64. A. J. Reeves, H. Dunlop and T. W. Clyne (1991) The Effect of Interfacial Reaction Layer Thickness on Fracture of Ti–SiC_p Composites, *Metall. Trans.*, **23**, pp. 970–81.
65. C. M. Warwick and J. E. Smith (1991) Interfacial Reactions in Ti Alloys Reinforced with C/TiB_2 – Coated SiC Monofilament, in *Metal Matrix Composites – Processing, Microstructure and Properties*, 12th Risø Int. Symp., Risø, N. Hansen, D. J. Jensen, T. Leffers, H. Lilholt, T. Lorentzen, A. S. Pedersen, O. B. Pedersen and B. Ralph (eds.), Risø Nat. Lab., Denmark, pp. 735–40.
66. A. D. McLeod (1990) Kinetics of the Growth of Spinel, $MgAl_2O_4$, in Alumina-Containing Aluminium–Magnesium Alloy Matrix Composites, in *Fabrication of Particulates Reinforced Metal Composites*, J. Masounave and F. G. Hamel (eds.), ASM, pp. 17–22.
67. D. J. Lloyd (1989) The Solidification Microstructures of Particulate Reinforced Al/SiC Composites, *Comp. Sci. & Tech.*, **35**, pp. 159–80.
68. R. Y. Lin and K. Kannikeswaran (1990) Interfacial Reaction Kinetics of Al/SiC Composite during Casting, in *Interfaces in Metal–Ceramics Composites*, Anaheim, California, R. Y. Lin, R. J. Arsenault, G. P. Martins and S. G. Fishman (eds.), TMS, Warrendale, pp. 153–64.
69. O. Kubaschewski and C. B. Alcock (1979) *Metallurgical Thermochemistry*, Pergamon, Oxford.
70. Z. M. Zhang and K. T. Wei (1990) Interfacial Stability of SiC Fiber Reinforced Nickel and Cobalt Composites, in *Interfaces in Metal–Ceramics Composites*, Anaheim, California, R. Y. Lin, R. J. Arsenault, G. P. Martins and S. G. Fishman (eds.), TMS, Warrendale, pp. 259–69.
71. J. F. Mason and T. W. Clyne (1989) Microstructural Development and Mechanical Behaviour of SiC Whisker Reinforced Mg–Li Alloys, in *Proc. 3rd European Conf. on Comp. Mat. (ECCM3)*, Bordeaux, A. R. Bunsell, P. Lamicq and A. Massiah (eds.), Elsevier, pp. 213–20.
72. C. G. Rhodes, A. K. Gosh and R. A. Spurling (1987) Ti–6Al–4V–2Ni as a Matrix for a SiC Reinforced Composite, *Metall. Trans.*, **18A**, pp. 2151–6.
73. C. G. Rhodes and R. A. Spurling (1985) Fibre–Matrix Reaction Zone Growth Kinetics in SiC Reinforced Ti–6Al–4V as Studied by TEM, in *Recent Advances in Composites in the United States and Japan*, J. R. Vinson and M. Taya (eds.), ASTM-STP 684, Philadelphia, pp. 585–99.
74. M. Nathan and J. S. Ahearn (1990) Interfacial Reactions in Ti/SiC Layered Films with and without Thin Diffusion Barriers, *Mat. Sci & Eng.*, **A126**, pp. 225–30.

75. W. J. Whatley and F. E. Wawner (1985) Kinetics of the Reaction Between SiC (SCS-6) Filaments and Ti(6Al–4V) Matrix, *J. Mat. Sci. Letts.*, **4**, pp. 173–5.
76. E. P. Zironi and H. Poppa (1981) Micro-Area Auger Analysis of a SiC/Ti Fibre Composite, *J. Mat. Sci.*, **16**, pp. 3115–21.
77. C. Jones, C. J. Kiely and S. S. Wang (1989) The Characterization of an SCS-6/Ti–6Al–4V MMC Interphase, *J. Mater. Res.*, **4**, pp. 327–35.
78. C. Jones, C. J. Kiely and S. S. Wang (1990) The Effect of Temperature on the Chemistry and Morphology of the Interphase in an SCS-6/Ti–6Al–4V Metal Matrix Composite, *J. Mater. Res.*, **5**, pp. 1435–42.
79. P. B. Prangnell, A. J. Reeves, T. W. Clyne and W. M. Stobbs (1992) A Comparison of Interfacial Reaction Mechanisms in the Al–SiC and Ti–TiB_2 MMC Systems, in *Proc. 2nd European Conf. on Adv. Mats. and Processes, Euromat '91*, Cambridge, UK, T. W. Clyne and P. J. Withers (eds.), Inst. of Metals, pp. 215–29.
80. R. L. Mehan, M. R. Jackson and M. D. McConnel (1983) The Use of Y_2O_3 Coatings in Preventing Solid State Si-Based Ceramic/Metal Reaction, *J. Mat. Sci.*, **18**, pp. 3195–205.
81. I. Kvernes and P. Kofstad (1973) High Temperature Stability of Ni/W Fibre Composites in Oxygen Atmosphere, *Scand. J. Metall.*, **2**, pp. 291–7.
82. S. Umekawa, C. H. Lee, J. Yamatomo and K. Wakashima (1985) Effect of Coatings on Interfacial Reaction in Tungsten/Nickel and Tungsten/316L Composites, in *Recent Advances in Composites in the United States and Japan*, J. R. Vinson and M. Taya (eds.), ASTM STP 864, Philadelphia, pp. 619–31.
83. I. Ahmad and J. Barranco (1980) W–2% ThO_2 Filament Reinforced Cobalt Base Alloy Composites for High Temperature Application, in *Advanced Fibers and Composites for Elevated Temperatures*, I. Ahmad and B. R. Noton (eds.), TMS-AIME, Warrendale, pp. 183–204.
84. R. Warren, L. O. K. Larsson and T. Garvare (1979) Fibre–Matrix Interactions in a W-Wire Reinforced Stainless Steel Composite, *Composites*, **10**, pp. 121–5.
85. I. Ahmad, D. N. Hill, J. Barranco, R. Warenchak and W. Hefferman (1980) Reinforcement of FeCrAlY with Silicon Carbide (Carbon Core) Filament, in *Advanced Fibers and Composites for Elevated Temperatures*, I. Ahmad and B. R. Noton (eds.), TMS-AIME, Warrendale, pp. 156–74.
86. J. F. Mason, C. M. Warwick, P. Smith, J. A. Charles and T. W. Clyne (1989) Magnesium–Lithium Alloys in Metal Matrix Composites – A Preliminary Report, *J. Mat. Sci.*, **24**, pp. 3934–6.
87. D. J. Matthewman, P. J. Withers and J. Mason (1991) Particulate Reinforced Copper, in *Metal Matrix Composites – Exploiting the Investment*, London, Inst. of Metals, p. 25.
88. A. F. Whitehouse, C. M. Warwick and T. W. Clyne (1991) The Electrical Resistivity of Copper Reinforced with Short Carbon Fibres, *J. Mat. Sci.*, **26**, pp. 6176–82.
89. G. R. Cappelman, J. F. Watts and T. W. Clyne (1985) The Interface Region in Squeeze-Infiltrated Composites Containing δ-Alumina Fibre in an Aluminium Matrix, *J. Mat. Sci.*, **20**, pp. 2159–68.
90. C. H. Li, L. Nyborg, S. Bengtsson, R. Warren and I. Olefjord (1989) Reactions between SiO_2 Binder and Matrix in δ-Al–Mg Composites, in *Interfacial Phenomena in Composite Materials 1989*, Sheffield, F. R. Jones (ed.), Butterworth, pp. 253–7.

91. W. J. Clegg, I. Horsefall, J. F. Mason and L. F. Edwards (1988) The Tensile Deformation and Fracture of Al–Saffil–MMCs, *Acta Metall.*, **36**, pp. 2151–9.
92. T. J. Warner and W. M. Stobbs (1989) Clean or Dirty Interfaces? The Influence of Interfacial Morphology on the Mechanical Properties of MMCs, in *Proc. 7th Int. Conf. on Comp. Mat.* (*ICCM7*), Guangzou, W. Yunshu, G. Zhenlong and W. Renjie (eds.), Pergamon, New York, pp. 503–8.
93. W. J. Wheatley and F. W. Wawner (1985) Kinetics of the Reaction Between SiC (SCS-6) Filaments and Ti–6Al–4V Matrix, *J. Mat. Sci. Letts.*, **4**, pp. 173–5.
94. M. H. Loretto and D. G. Konitzer (1990) The Effect of Matrix–Reinforcement Reaction on Fracture in Ti–6Al–4V–Base Composites, *Metall. Trans.*, **21A**, pp. 1579–87.
95. T. W. Clyne, A. J. Phillipps, S. J. Howard and M. C. Watson (1992) Mechanical Characterisation of Metal/Ceramic Interfaces in Composites and Laminates, in *Proc. 2nd European Colloquium on Designing Ceramic Interfaces*, Petten, Netherlands, S. D. Peteves (ed.), to be published.
96. P. K. Wright, R. Nimmer, G. Smith, M. Sensmeier and M. Brun (1990) The Influence of the Interface on Mechanical Behaviour of Ti–6Al–4V/SCS6 Composites, in *Interfaces in Metal–Ceramic Composites*, Anaheim, California, R. Y. Lin, R. J. Arsenault, G. P. Martins and S. G. Fishman (eds.), TMS, pp. 559–81.
97. S. Turner, R. Taylor and T. W. Clyne (1992) Thermal Conductivities of Ti–SiC and Ti–TiB_2 Particulate Composites, *submitted* to *J. Mat. Sci.*
98. F. E. Wawner and S. R. Nutt (1980) Investigation of Diffusion Barrier Materials on SiC Filaments, *Ceram. Eng. Sci. Proc.*, **1**, pp. 709–19.
99. M. K. Alam and S. C. Jain (1990) The CVD Coating of Fibers for Composite Materials, *J. Metals*, **42**, pp. 56–8.
100. R. K. Everett (1991) Deposition Technologies for MMC Fabrication, in *Metal Matrix Composites: Processing and Interfaces*, R. K. Everett and R. J. Arsenault (eds.), Academic Press, Boston, pp. 103–19.
101. M. F. Amateau (1976) Progress in the Development of Graphite–Al Composites using Liquid Infiltration Technology, *J. Comp. Mat.*, **10**, pp. 289–96.
102. A. A. Baker (1972) *Fibre Sci. Techn.*, **5**, pp. 213–18.
103. L. Aggour, E. Fitzer, M. Heym and E. Ignatowitz (1977) Thin Coatings on C Fibres as Diffusion Barriers and Wetting Agents in Al Composites, *Thin Solid Films*, **40**, pp. 97–105.
104. S. M. Rossnagel (1989) Film Modification by Low-Energy Ion-Bombardment during Deposition, *Thin Solid Films*, **171**, pp. 125–42.
105. T. Ohsaki, M. Yoshida, Y. Fukube and K. Nakamura (1977) The Properties of C Fibre Reinforced Al Composites Formed by Ion-Plating Process and Hot Pressing, *Thin Solid Films*, **45**, pp. 563–68.
106. R. R. Kieschke and T. W. Clyne (1991) Development of a Diffusion Barrier for SiC Monofilaments in Titanium, *Mat. Sci. & Eng.*, **135A**, pp. 145–9.
107. D. M. Goddard (1978) Interface Reactions during Preparation of Al-Matrix Composites by Sodium Process, *J. Mat. Sci.*, **13**, pp. 1841–8.
108. D. M. Goddard and E. G. Kendall (1977) Fibreglass-Reinforced Pb Composites, *Composites*, **8**, pp. 103–9.

109. S. Kohara and N. Muto (1985) Fabrication of Silicon Carbide Fiber-Reinforced Aluminium Composites, in *Recent Advances in Composites in the United States and Japan*, J. R. Vinson and M. Taya (eds.), ASTM STP 864, Philadelphia, pp. 456–64.
110. A. G. Kulkarni, B. C. Pai and N. Balasubramanian (1979) The Cementation Technique for Coating Carbon Fibres, *J. Mat. Sci.*, **14**, pp. 592–8.
111. Y. H. Teng and J. D. Boyd (1990) Development of Particulate Coatings to Control Interface Properties, in *Fabrication of Particulates Reinforced Metal Composites*, J. Masounave and F. G. Hamel (eds.), ASM, pp. 125–34.
112. M. F. Berard and D. R. Wilder (1963) Self Diffusion in Polycrystalline Yttrium Oxide, *J. Appl. Phys.*, **34**, pp. 2318–27.
113. M. F. Berard and D. R. Wilder (1969) Cation Self Diffusion in Polycrystalline Y_2O_3 and Er_2O_3, *J. Amer. Ceram. Soc.*, **52**, pp. 85–91.
114. P. Kofstadt (1972) *Nonstoichiometry, Diffusion and Electrical Conductivity in Binary Metal Oxides*, Wiley Interscience, New York.
115. R. E. Tressler (1974) Interfaces in Oxide Reinforced Metals, in *Interfaces in Metallic Matrix Composites*, K. G. Krieder (ed.), Academic Press, New York, pp. 286–301.
116. R. R. Kieschke, R. E. Somehk and T. W. Clyne (1991) Sputter Deposited Barrier Coatings on SiC Monofilaments for Use in Reactive Metallic Matrices – Part I. Optimization of Barrier Structure, *Acta Met. et Mat.*, **39**, pp. 427–36.
117. R. R. Kieschke, C. M. Warwick and T. W. Clyne (1991) Sputter Deposited Barrier Coatings on SiC Monofilaments for Use in Reactive Metallic Matrices – Part III. Microstructural Stability in Composites Based on Mg and Ti, *Acta Met. et Mat.*, **39**, pp. 445–52.
118. H. O. Pierson and E. Randich (1978) Titanium Diboride Coatings and their Interaction with the Substrates, *Thin Solid Films*, **54**, pp. 119–28.
119. J. Bouix (1986) TiB_2-Coated B Fibre for Al Matrix Composites, *J. Less Common Metals*, **117**, pp. 83–9.
120. D. Morin (1976) Boron Carbide Coated Boron Filament Reinforcement in Aluminium Alloy Matrices, *J. Less Common Metals*, **47**, pp. 207–13.
121. H. Vincent, C. Vincent, J. L. Ponthenier, H. Mourichoux and J. Bouix (1989) Elaboration en Continu d'un Depot Mince de Carbure Refractaire en Surface des Fibres de Carbone: Caracterisation de la Fibre C/SiC, in *Proc. 3rd European Conf. on Comp. Mats. (ECCM3)*, Bordeaux, A. R. Bunsell, P. Lamicq and A. Massiah (eds.), Elsevier, pp. 257–64.
122. F. Delannay, L. Froyen and A. Deruyterre (1988) Wetting of Solids by Liquid Metals in Relation to Squeeze Casting of MMCs, in *Cast Reinforced Metal Composites*, Chicago, S. G. Fishman and A. K. Dhingra (eds.), ASM, pp. 81–4.
123. H. A. Katzman (1987) Fibre Coatings for the Fabrication of Graphite Reinforced Mg Composites, *J. Mat. Sci.*, **22**, pp. 144–8.
124. J. P. Rocher, J. M. Quenisset and R. Naslain (1989) Wetting Improvement of Carbon or SiC by Al-Alloys Based on a K_2ZrF_6 Surface Treatment – Application to Composite-Material Casting, *J. Mat. Sci.*, **24**, pp. 2697–703.
125. J. P. Rocher, J. M. Quenisset and R. Naslain (1985) A New Casting Process for Carbon (or SiC-Based) Fibre-Aluminium Low Cost Composite Materials, *J. Mat. Sci. Letts.*, **4**, pp. 1527–9.
126. S. Schamm, J. P. Rocher and R. Naslain (1989) Physicochemical Aspects

of the K_2ZrF_6 Process Allowing the Spontaneous Infiltration of SiC (or C) Preforms by Liquid Aluminium, in *Proc. 3rd European Conf. on Comp. Mats. (ECCM3)*, Bordeaux, A. R. Bunsell, P. Lamicq and A. Massiah (eds.), Elsevier, pp. 157–61.

127. C. M. Warwick and T. W. Clyne (1991) Development of a Composite Coaxial Cylinder Stress Analysis Model and its Application to SiC Monofilament Systems, *J. Mat. Sci.*, **26**, pp. 3817–27.
128. D. W. Hoffmann (1982) Film Stress Diagnostic in the Sputter Deposition of Metals, in *Proc. 7th ICVM*, Iron & Steel Inst. Japan, pp. 145–57.
129. C. M. Warwick, R. R. Kieschke and T. W. Clyne (1989) Measurement and Control of the Stress State in Sputtered Diffusion Barrier Coatings on Monofilaments, in *Interfacial Phenomena in Composite Materials*, Sheffield, F. R. Jones (ed.), Butterworth, pp. 267–75.

7
Fracture processes and failure mechanisms

In earlier chapters, the mechanics of elastic and plastic deformation has been treated through an evaluation of the average stresses in each phase. Unfortunately, this is not an adequate basis for the treatment of fracture and failure, which depend on the local processes controlling the initiation and propagation of a crack. Great sensitivity is thus expected to local parameters such as the reinforcement distribution, morphology and size, as well as to interfacial strength. These factors can vary considerably, often in a poorly controlled manner, and this has led to apparently contradictory experimental data. In this chapter, long fibre composites are considered first. These materials have low ductilities, because of the constraint imposed by the fibres, and simple treatments developed for polymer-based composites can be used to predict the failure stress. The situation is less simple for discontinuous reinforcement, for which the stress state is complex and there is considerable scope for plastic flow. In order to gain an insight into the influence of composite microstructure on failure, aspects of damage initiation and propagation are first examined through a consideration of simple models. While some useful predictions can be made, theoretical treatments in this area cannot yet be regarded as mature.

7.1 Failure processes in long fibre MMCs

7.1.1 Failure of laminae

Under an arbitrary stress state, a lamina (a sheet containing unidirectionally aligned long fibres) can fail in one of three ways, as depicted in Fig. 7.1. Very large tensile stresses parallel to the fibres, σ_3, may result in their fracture, but the composite is often more vulnerable to either tensile or shear failure on surfaces containing the fibre direction, under low stress levels of transverse tension (σ_1) or shear (τ_{31}). In these cases, failure may

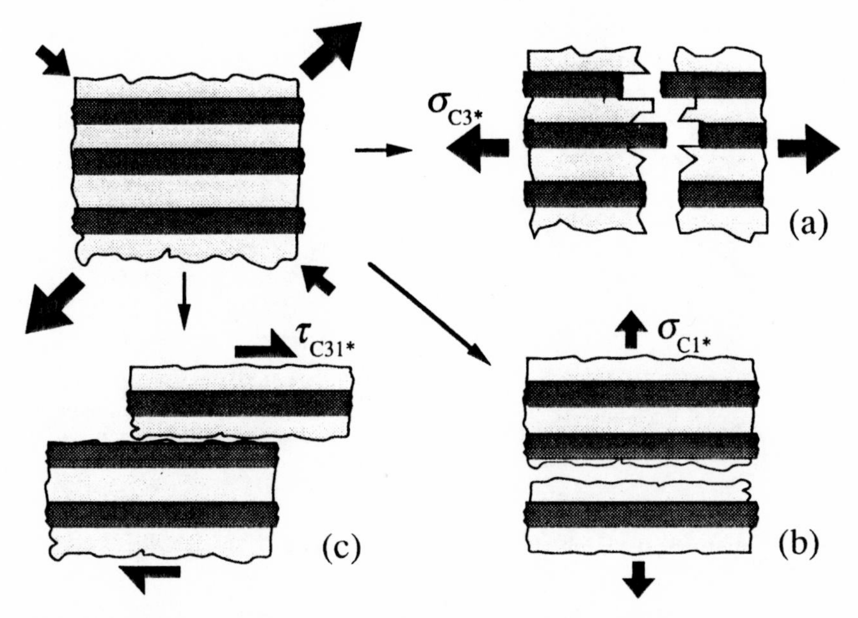

Fig. 7.1 Schematic illustration of three modes of lamina failure. These modes occur when the stress state is such that critical values of (a) an axial tensile stress σ_{C3*}, (b) a transverse tensile stress σ_{C1*} or (c) a shear stress τ_{C31*} are exceeded.

occur within the matrix, at the fibre/matrix interface, within the fibre (relatively uncommon), or across some combination of these. In order to predict how a lamina or laminate will fail, critical stress values (σ_{C3*}, σ_{C1*} and τ_{C31*}) must be identified for the composite. In practice, these values can vary over a wide range, depending on the fibre/matrix combination and, in many cases, on how the composite has been manufactured.

Axial tensile failure

For a brittle matrix/brittle fibre system under an applied axial stress, the analysis of tensile failure is straightforward; both constituents experience the same axial strain and hence sustain stresses in proportion to their Young's moduli. When the matrix exhibits some ductility, then the situation is more complex. A simple method of predicting the stress at which the composite will fail, σ_{C*}, for this case is illustrated in Fig. 7.2. The strain at which the matrix starts to undergo plastic deformation, ε_M^P, will in general be less than the failure strain of the fibres, ε_{I*}, but the matrix failure strain, ε_{M*}, will be greater than that of the fibres. The

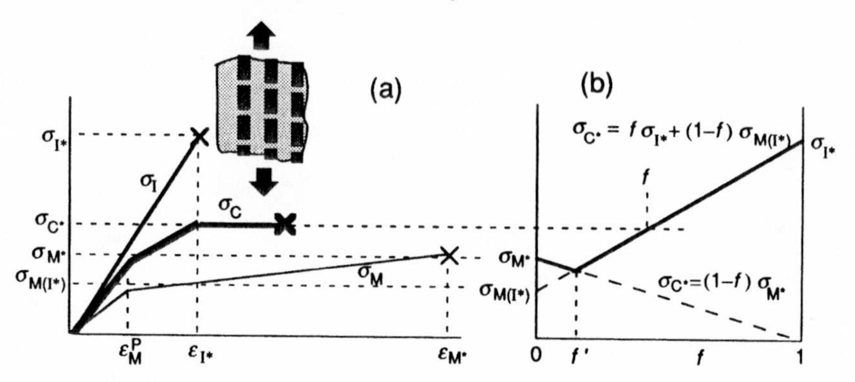

Fig. 7.2 Failure of an MMC lamina loaded parallel to the fibre axis. (a) Idealised stress–strain plots for fibre, matrix and composite and (b) variation of composite tensile strength with fibre volume fraction.

composite stress at which the fibres start to break is given by

$$\sigma_{C(I*)} = f\sigma_{I*} + (1 - f)\sigma_{M(I*)} \tag{7.1}$$

where $\sigma_{C(I*)}$ and $\sigma_{M(I*)}$ are the average stresses in the composite and the matrix at a strain of ε_{I*}. In calculating the failure stress, it is assumed that the composite stress does not rise after initial fibre fracture, so that $\sigma_{C*} = \sigma_{C(I*)}$. During this period, the fibres will break into progressively shorter lengths. This assumption is obviously crude, although the rising matrix stress in this regime will tend to be offset by the reduction in mean load carried by the fibres as their aspect ratio decreases (see §2.2). At very low fibre volume fractions, the predicted failure stress is less than that corresponding to a matrix containing holes instead of fibres, $(1 - f)\sigma_{M*}$; this is rather unrealistic, so the latter expression is often used in this regime (see Fig. 7.2(b)). The fibre volume fraction, f', above which eqn (7.1) applies is readily shown to be

$$f' = \frac{\sigma_{M*} - \sigma_{M(I*)}}{\sigma_{I*} - \sigma_{M(I*)} + \sigma_{M*}} \tag{7.2}$$

Although this model is very crude, some experimental data, shown in Fig. 7.3(a) and (b), are in surprisingly good agreement with it.

Transverse tensile failure

It is not possible to make a simple estimate of the transverse failure stress (σ_{C1*}), comparable with the usage of eqn (7.1) for the axial case (σ_{C3*}). The transverse strength is influenced by many factors, such as the nature

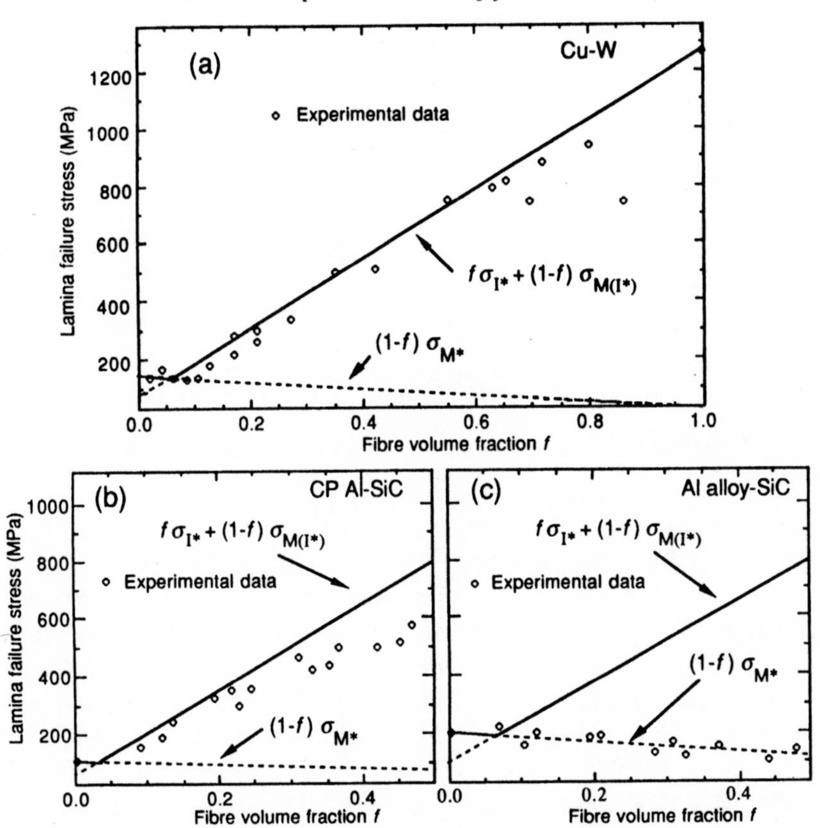

Fig. 7.3 Comparisons between the predicted dependence of failure stress on fibre volume fraction and experimental data for (a) tungsten wires in copper[1], and SiC monofilaments in (b) CP Al and in (c) A384 (an Al casting alloy)[2]. For the Cu–W system (in which the fibres behave in an essentially brittle manner, despite being metallic) these data are in excellent agreement with simple theory. However, while the CP Al-based data are also consistent with the model, the failure stress of the A384-based MMCs appears to fall linearly with increasing f, suggesting that the fibres do not contribute to the strength. This appears to have resulted from the presence of interfacial reaction layers, which were relatively thick in these composites as a consequence of differences in processing conditions and matrix composition[2]. This caused the fibres to fracture readily, reducing their load-bearing capacity.

of the interfacial bonding, the fibre distribution, the presence of voids, etc. In general, however, the strength will be less than that of the unreinforced matrix, often very significantly so, and the strain to failure can be even more dramatically reduced. This is illustrated by the experimental data shown in Fig. 7.4, which compares transverse stress–strain plots for

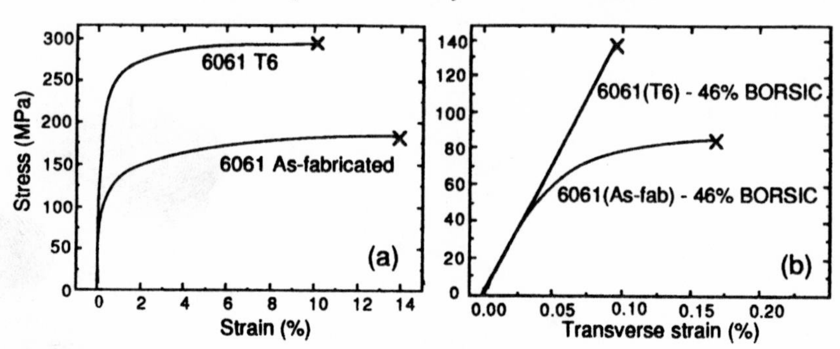

Fig. 7.4 Stress–strain curves for (a) an unreinforced aluminium alloy in the as-fabricated state and after a T6 treatment, and (b) two laminae containing 46 vol% BORSIC fibres, with matrices in the same state as for (a), tested in transverse tension[3]. Note the different scales used for the reinforced and unreinforced systems.

6061 Al–BORSIC laminae with the behaviour of the corresponding unreinforced alloy. These data show that when a full aging (T6) heat treatment is applied to the BORSIC (boron coated with a SiC layer) fibre containing composite, an interfacial chemical reaction leads to increased bond strength, resulting in higher transverse failure strengths than prior to treatment, but even lower ductility; failure strains can be as low as 0.1%. Similar trends have been observed[4] with B fibres in an unalloyed Al matrix, with a transition from interfacial debonding to fibre splitting as the heat treatment time is increased: the failure strains observed, however, were higher at between 0.4% and 0.8%.

Although the interplay of effects determining the transverse fracture behaviour of long fibre composites is very complex (often with a marked sensitivity to flaws arising from inadequate fabrication procedures), there is an inherent tendency for high local stresses and strains to develop in the matrix. If the interfacial bonding is weak, then cracks tend to form at the interface and link up through highly stressed sections of matrix. If, on the other hand, there is strong resistance to interfacial decohesion, cracks can form in the fibres. In either case, the fibres tend to make little or no contribution to the strength. A crude indication of the likely effect of the presence of the fibres on the transverse strength (but not the stiffness) can therefore be obtained by treating them as a set of cylindrical holes and considering the reduction in load-bearing cross section thus introduced. This is not a realistic model, as the presence of even completely debonded fibres would lead to a different stress distribution in the matrix as it deformed, but the approach is useful as a guide. For a simple square

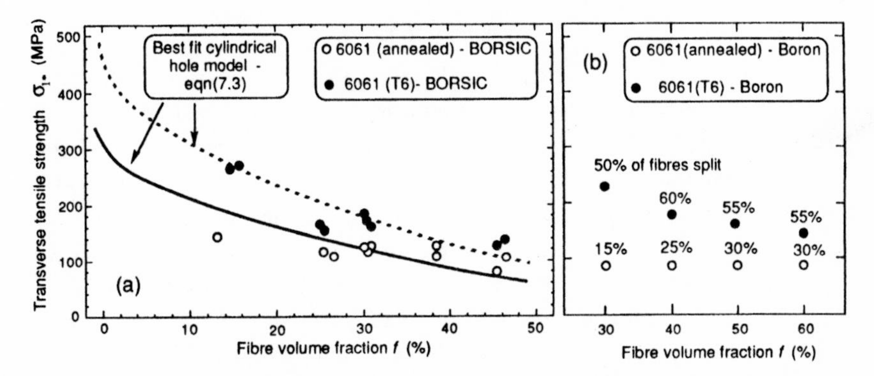

Fig. 7.5 Experimental data for the transverse tensile strength of 6061 Al-based laminae containing (a) BORSIC fibres[3], with predictions from eqn (7.3); good agreement is observed, although it was obtained using values for σ_{M*} somewhat higher than those corresponding to the tensile strengths of the unreinforced matrix[†], (b) uncoated boron fibres, with an indication of the fraction of fibres observed in the failed specimen to have split axially[5] (UTS $\approx$ 200 MPa (B fibre (trans)).

array of holes, consideration of the maximum reduction in matrix cross-sectional area leads to the following expression for the transverse strength of a lamina having a volume fraction f of fibres

$$\sigma_{C1*} = \sigma_{M*}\left[1 - 2\left(\frac{f}{\pi}\right)^{1/2}\right] \tag{7.3}$$

A comparison is shown in Fig. 7.5(a) between predictions from this equation and experimental data[3] for 6061 Al–BORSIC composites. Good agreement is observed, although it was obtained using values for σ_{M*} somewhat higher than those corresponding to the tensile strengths of the unreinforced matrix. Data[5] in Fig. 7.5(b) indicate similar behaviour for boron fibres, although the reduction in composite strength on increasing the fibre content is less obvious. This may be the result of a stronger interface. In Fig. 7.5(b), the higher fraction of fibres noted as having split in the age-hardened matrix indicates that more load is transmitted to the fibres in a stronger matrix. This suggests that, with fibres of high transverse strength and a high interfacial bond strength, the σ_{C1*} value of MMC laminae can be considerably improved upon that given by eqn (7.3).

[†] Normally 6061 Al alloy has tensile strengths of around 100 MPa and 300 MPa in the annealed and T6 conditions, although it is unclear what properties should be ascribed to the matrix *in situ*, see Chapter 10.

Shear failure

As with tensile fracture, shear failure tends to occur on planes dictated by the fibre direction. Possible orientations are illustrated in Fig. 7.6. Normally, there is considerable resistance to the shear fracture of the fibres, so that the modes denoted τ_{13} (and τ_{23}) are unlikely to occur. Of the other two modes, involving shear parallel to the fibres either axially (τ_{31}) or laterally (τ_{21}), it is not clear whether one is inherently more likely than the other. However, when considering a thin lamina in the '3–1' plane, stresses of the τ_{21} type will not be significant and the concern is with the magnitude of τ_{C31*}. Broadly speaking, this will be affected by the same factors that dictated the transverse tensile strength, because shear stresses and strains become concentrated in the matrix between the fibres in a manner similar to that outlined above for tensile stresses and strains.

No simple analytical expression is available to predict the effect of fibre content on τ_{C31*}. Adams and Doner[6] have used finite difference methods to deduce how the shear stress concentration factor should vary with fibre volume fraction (Fig. 7.7). It is evident that, unless the fibre volume fraction is very high (when constraint on matrix deformation becomes severe), this factor is quite close to unity and τ_{C31*} is expected to have a value little reduced below τ_{M*} for the matrix. It might thus be expected that, whilst the axial strength (σ_{C3*}) (and the transverse strength (σ_{C1*}) if the fibres or interfacial bond are weak) of an MMC lamina would be strongly influenced by the presence of the fibres and would be comparable with the values for similarly reinforced polymeric laminae, the shear strength (τ_{C31*}) (and the transverse strength if the fibres and interfacial bond are strong) would be comparable with that of the matrix. There is a shortage of systematic data on MMC laminae, but those shown in Table 7.1 for Al–silica fibre composites suggest that both transverse and shear strengths are appreciably higher than those of a comparable

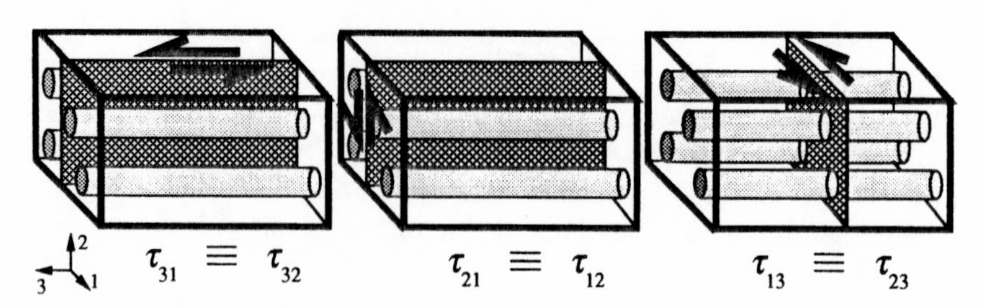

Fig. 7.6 Nomenclature and orientation of shear stresses acting within an aligned fibre composite.

Table 7.1 *Experimental failure data for two laminae*

Failure stress or strain	Polyester–50% glass (typical)	Al–50% silica[7]
σ_{C3*} (MPa)	700	800
σ_{C1*} (MPa)	20	60
τ_{C31*} (MPa)	50	75
ε_{C3*} (%)	2.0	0.7
ε_{C1*} (%)	2.0	0.03

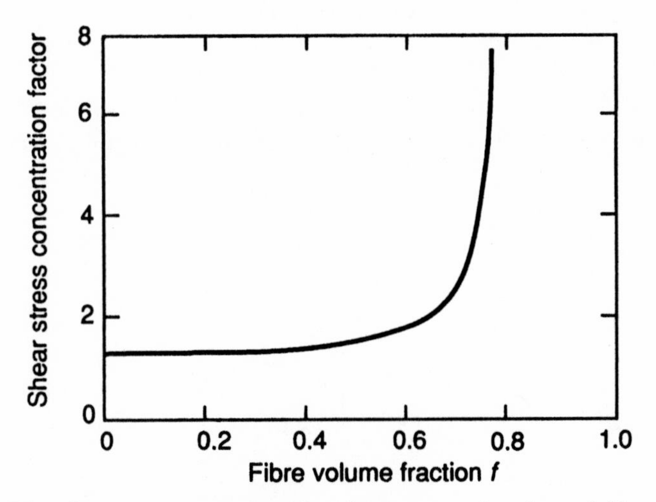

Fig. 7.7 The shear stress concentration factor as a function of fibre fraction predicted[6] for a square array of fibres.

polymer-based composite. Other results[8] indicate that this is also the case for Al–boron laminae ($\tau_{C31*} \sim 100$ MPa), but not for Al–graphite ($\tau_{C31*} \sim 30$ MPa). Graphite fibres normally have a low transverse strength.

7.1.2 Off-axis loading

Extending the ideas of the previous section to the failure of a lamina subjected to an arbitrary in-plane stress state, the main concern is with the operation of the three failure mechanisms (with defined values of σ_{C3*}, σ_{C1*} and τ_{C31*}) shown in Fig. 7.1. In order to determine when and how the lamina will fail, it only remains to define a *criterion* for the failure mechanisms to come into operation. The principal issue is whether the critical stress to trigger one mechanism is entirely unaffected by the stresses tending to cause the others – i.e. whether there is any *interaction* between the modes of failure. Two criteria are commonly employed, the first being based on the supposition that there is no such interaction and the second incorporating such an effect.

Maximum stress theory

The simplest criterion may be stated as failure being expected when a stress in a principal direction reaches the appropriate critical value, that is, when any one of the following equations is satisfied

$$\sigma_{C3} \geq \sigma_{C3*} \qquad \sigma_{C1} \geq \sigma_{C1*} \qquad \tau_{C31} \geq \tau_{C31*} \tag{7.4}$$

For a general in-plane system (σ_x^A, σ_y^A and τ_{xy}^A) being applied to the lamina, evaluation of these stresses can be carried out using this simple transform

$$\begin{vmatrix} \sigma_3 \\ \sigma_1 \\ \tau_{31} \end{vmatrix} = |T| \begin{vmatrix} \sigma_x^A \\ \sigma_y^A \\ \tau_{xy}^A \end{vmatrix} \tag{7.5}$$

in which $|T|$ is given by

$$|T| = \begin{vmatrix} c^2 & s^2 & 2cs \\ s^2 & c^2 & -2cs \\ -cs & cs & c^2 - s^2 \end{vmatrix} \qquad (c = \cos\phi, s = \sin\phi) \tag{7.6}$$

and ϕ is the angle between the x (loading) direction and the 3 (fibre) axis.

Monitoring of the values of the σ_{C3}, σ_{C1} and τ_{C31} stresses as the applied stress is increased then allows the onset of failure to be identified as the point when one of the inequalities in eqn (7.4) first becomes satisfied. Noting the form of $|T|$, and considering a simple applied stress state such as uniaxial tension, the magnitude of σ_x^A necessary to cause failure can be plotted as a function of misorientation ϕ between the stress axis and the fibre axis, for each of the three failure modes

$$\sigma_{x*}^A = \frac{\sigma_{C3*}}{\cos^2\phi} \tag{7.7}$$

$$\sigma_{x*}^A = \frac{\sigma_{C1*}}{\sin^2\phi} \tag{7.8}$$

$$\sigma_{x*}^A = \frac{\tau_{C31*}}{\sin\phi\cos\phi} \tag{7.9}$$

This theory has been found to be broadly appropriate for many polymer composites, but it tends to be rather inaccurate in the transition regime ($\phi \sim 20°$–$45°$) between shear and transverse failure, where a '***mixed mode***' fracture is frequently observed. This involves cracks parallel to the fibres simultaneously opening up and acting as shearing planes, under an applied stress lower than those predicted for either type of failure. The alternative treatment outlined below is often used to account for this.

Tsai–Hill (maximum work) theory

Failure criteria developed on the basis of energy arguments (in a simple extension of the von Mises strain energy criterion for yielding of metals) have proved effective in taking account of mixed mode effects. The most widely used expression arose from the suggestion of Hill[9] that the yield condition for anisotropic metals could be expressed as a quadratic function of the stress components. The application of this concept to long fibre composite laminae under plane stress was outlined by Tsai[10], who derived the critical condition for failure as

$$\frac{\sigma_3^2}{\sigma_{C3*}^2} - \frac{\sigma_3\sigma_1}{\sigma_{C3*}^2} + \frac{\sigma_1^2}{\sigma_{C1*}^2} + \frac{\tau_{31}^2}{\tau_{31*}^2} = 1 \qquad (7.10)$$

This equation, which is known as the Tsai–Hill criterion, defines an envelope in stress space: if the stress state (σ_3, σ_1 and τ_{31}) lies outside of this envelope, i.e. if the sum of the terms on the left-hand side is equal to or greater than unity, then failure is predicted. (Note that the failure mechanism is not specifically identified, although inspection of the relative magnitudes of the terms in eqn (7.10) will give an indication of the likely contribution of the three modes.) The values of σ_3, σ_1 and τ_{31} are obtained, for a given misorientation ϕ between fibre axis and applied stress axis, from eqns (7.5) and (7.6). For a single applied tensile stress σ_ϕ^A ($=\sigma_x^A$ at an angle ϕ to the fibre axis) eqn (7.10) can be written as

$$\sigma_\phi^A = \left[\frac{\cos^2\phi(\cos^2\phi - \sin^2\phi)}{\sigma_{C3*}^2} + \frac{\sin^4\phi}{\sigma_{C1*}^2} + \frac{\cos^2\phi\sin^2\phi}{\tau_{C31*}^2}\right]^{-1/2} \qquad (7.11)$$

and this expression will give the applied stress at which failure is predicted, as a function of the angle ϕ between the fibre axis and the loading direction.

Few off-axis strength data are available for MMC laminae. In Fig. 7.8 a set of off-axis tension results are shown for aluminium laminae[7] containing 50% silica fibres, together with maximum stress and Tsai–Hill theoretical curves. These data do suggest that the Tsai–Hill criterion gives

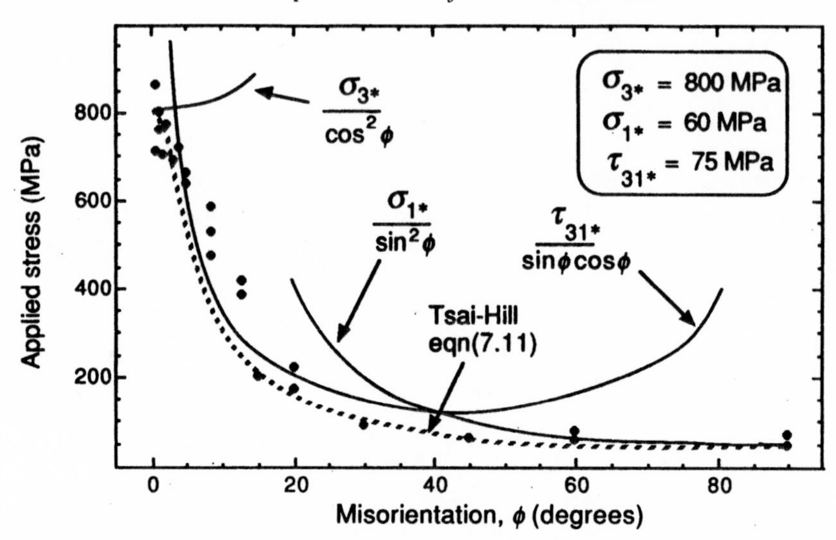

Fig. 7.8 Experimental data[7] on the lamina failure stress as a function of angle between the tensile stress axis and the fibre axis for testing of Al–50 vol% silica fibre composites, together with theoretical curves from the maximum stress and the Tsai–Hill models.

the more reliable guide to off-axis strength, but more information is needed in this area.

7.1.3 Failure of laminates

In a laminate (a number of laminae, or plies, bonded together), the first step is to calculate the stresses, referred to the fibre axis, in each individual ply. There is a straightforward procedure for this (outlined in standard composite textbooks such as Jones[11]) which involves evaluating the stiffness of the laminate, using this to obtain its strain under an applied load, and then working back from this strain to deduce the stresses in any particular ply. A failure criterion can then be applied to see if and how that particular ply will fail. However, a complication often arises in the sense that 'failure' of an individual ply within a laminate does not necessarily mean that the component is no longer usable, as other plies may be capable of withstanding considerably greater loads without catastrophic failure. Analysis of the behaviour beyond the initial, fully elastic stage is, however, complicated by uncertainties as to the degree to which the damaged plies can continue to bear some load.

The simple example of a (0°/90°) crossply laminate being loaded in

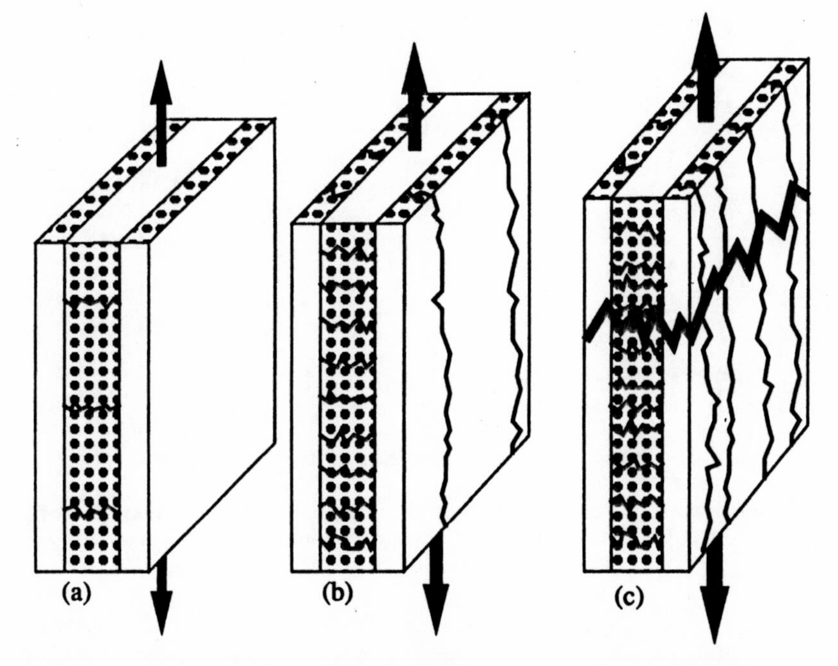

Fig. 7.9 Schematic illustration of the changes taking place during loading of a cross-ply laminate parallel to one of the fibre directions. (a) Cracking of transverse plies as $\sigma_{1(90°)}$ reaches σ_{1*}, (b) onset of cracking parallel to the fibres in the axial plies as $\sigma_{1(0°)}$ (from inhibition of Poisson contraction) reaches σ_{1*} and (c) final failure as $\sigma_{3(0°)}$ in the axial plies, now taking virtually the entire load, reaches σ_{3*}. This sequence is generally expected for polymer based composites, but process (b) may not occur for MMCs.

tension along one of the fibre directions serves to illustrate some basic effects. A sequence of damage evolution commonly observed in resin-based laminates is shown in Fig. 7.9. The transverse plies start to crack first, often followed by cracking parallel to the fibres in the axial plies, before they crack transversely – causing the laminate to fail. This behaviour can be understood by considering Fig. 7.10, which shows the stresses in one ply of an Al–50% boron fibre laminate for different angles between the fibre axis and the applied stress. While the stress in the axial ply (125% of σ^A at 0°) is nearly double that in the transverse ply (75% of σ^A at 90°), the transverse failure stress is likely to be lower than the axial value by a larger factor than this (e.g. see Table 7.1). Parallel cracking of the axial plies (Fig. 7.9(b)) can then arise from the resistance to Poisson contraction offered by the transverse plies; this generates (Fig. 7.10) a tensile stress of about 5% σ^A normal to the fibres, which is a factor of 25

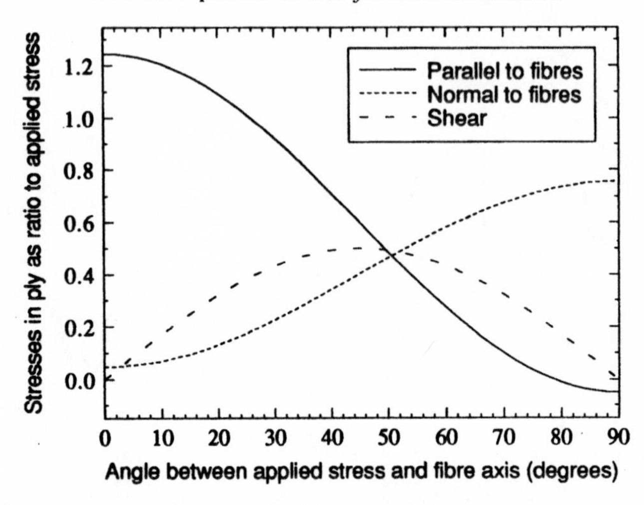

Fig. 7.10 Calculated stresses in one ply of a crossply (0°/90°) aluminium–50 vol% B fibre laminate under uniaxial tensile loading, as a function of the angle between the fibre direction in the ply concerned and the direction of the applied stress.

less than that parallel to the fibres. This suggests that parallel cracking before final failure is unlikely in MMCs. This is in contrast to polymer-based laminates, which have a larger matrix Poisson contraction and a greater difference between axial and transverse failure stresses. These estimates are very approximate and there is as yet little experimental evidence to confirm expected laminate failure characteristics for MMCs.

7.2 Failure processes in discontinuous MMCs

With short fibre or particulate reinforcement, failure processes are more complex than in long fibre MMCs. Fracture is thought to occur largely by the formation and link-up of voids within the matrix. Classical treatments of fracture in metals containing isolated, poorly bonded inclusions (aimed at systems like sulphide particles in steel) allow predictions about the nucleation and growth of cavities, but this may be of limited relevance to MMCs, which contain closely spaced, well-bonded fibres or particles.

7.2.1 Micro-damage processes

Before looking at ways in which existing models of damage accumulation and ductility can be adapted, or new models proposed, to describe discontinuous MMCs, it is important to examine the mechanisms by which damage can initiate at a microscopic level.

Reinforcement cracking

The simplest way to examine the likelihood of the fracture of the reinforcement is to consider the inclusion stresses that occur in MMCs. Traditionally, for reasonably long fibres, a condition for fibre fracture has been derived by assuming that load is transferred to the fibre via interfacial shear stresses (see §2.2 and Fig. 2.5 describing the shear lag approach). Since the interfacial shear stress is limited by the matrix shear stress, the shortest fibre which can be stressed to its failure strength (σ_{I*}) has an aspect ratio s_* given approximately by

$$s_* \approx \sigma_{I*}/2\tau_{MY} \approx \sigma_{I*}/\sigma_{MY} \tag{7.12}$$

For Al alloy matrices, even taking a high yield stress ($\sim$300 MPa) and a low whisker strength ($\sim$3 GPa), this criterion leads to an aspect ratio ($s_* \sim 10$) greater than those typically retained in real whisker composites. In addition, other relaxation processes may operate to reduce the fibre stress at a given plastic strain and thus pre-empt fibre fracture.

Experimental observations also suggest that short fibres rarely fracture[12]. Fracture of SiC particles in Al, however, does occur (see Fig. 7.11), although this is predominantly when the particles are large[14,15]. Since the inclusion stress is approximately independent of size, this suggests that fracture occurs when the particle contains a sufficiently large flaw, which is more likely for large particles†. For Al–SiC_P systems, Lloyd[14] has observed that particle cracking is an important failure mechanism only for composites containing particles greater than around 20 μm in size.

Matrix cavitation and interfacial debonding

The nucleation of cavities will normally take place in the immediate vicinity of the reinforcement (Fig. 7.12), promoted by the high levels of triaxial constraint and the increased degree of matrix work-hardening, resulting from both localisation of the applied plastic strain and from differential thermal contraction effects (see §5.1.1). The stress state needed to promote voiding will depend on the precise nature of the void nucleation event. If the void originates solely within the matrix, which is

† Some of the larger particles may have become cracked before loading, possibly during fabrication, but this does not appear to be entirely responsible for this observation.

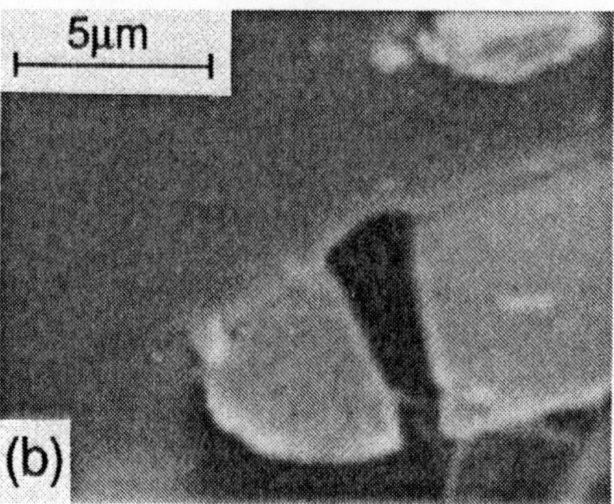

Fig. 7.11 In Al–SiC_W composites, whisker failure is a rare occurrence. The fractured whisker seen in (a) was observed by TEM just below the fracture surface of a tensile tested specimen. On the other hand, for SiC particle containing composites (b) particle failure is often observed[13] at or near the fracture surface. Note how the void formed at the cracked particle is expanding into the matrix.

presumably more likely for strongly bonded interfaces, one might expect that a critical hydrostatic stress, σ_{H*}, characteristic of the matrix material, will need to be exceeded (Fig. 7.12(a) and (b)). With a more weakly bonded interface, it is probable that a critical normal (i.e. radial (particle) or axial (fibre end)) stress, σ_{r*}, should be exceeded for debonding (Fig. 7.12(c)). Partly because of the difficulty of interfacial testing (§6.2), there is a shortage of experimental data in this area and reliable values for σ_{H*} or σ_{r*} are not available for any composite system. However, useful estimates

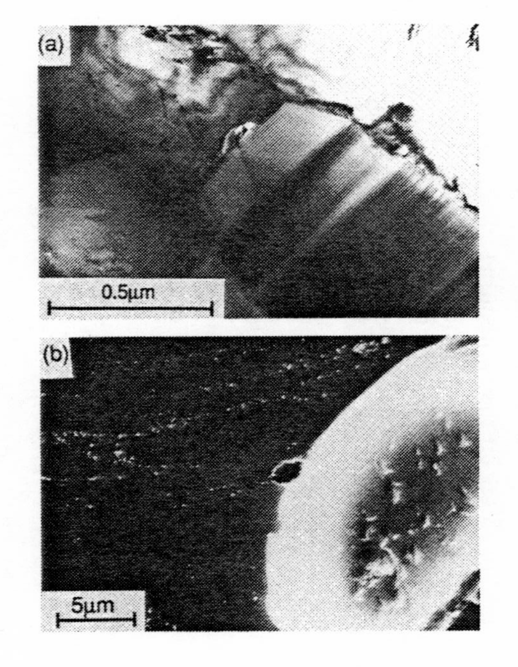

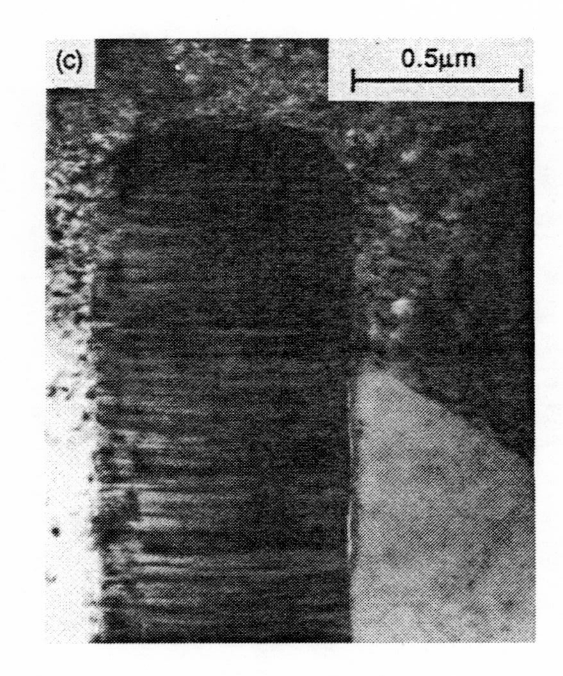

Fig. 7.12 Voids formed near well-bonded interfaces usually form by a cavitation process giving rise to nearly spherical voids. This is well illustrated here for (a) a SiC whisker in a tensile fractured specimen and (b) an alumina particle in a composite subjected to cold drawing. In both cases the matrix is Al. Note that in (b) the fine scale oxide inclusions have also affected cavity nucleation. The flatter shape of voids initiated by debonding of a weak Al–SiC interface is exemplified by the TEM micrograph of an Al–SiC_W MMC shown in (c).

of the rates of void growth as a function of local stress state for a given overall strain rate, can be made – see §7.2.2. In any event, the factors which tend to favour the formation of voids have been listed[16] as:

- large reinforcement size
- high matrix flow stress
- large imposed plastic strain
- reinforcements located on grain boundaries
- particle clustering
- low work of adhesion
- large particle surface normal to load
- small grain size
- high particle aspect ratio along stress state

In addition, the interface region is also of importance, with interfacial layers often having a strong influence on the behaviour (§6.1.4 and §6.3). In general, formation of a thick layer of (brittle) reaction product will tend to favour interfacial debonding and cracking, whilst also raising the shear stress necessary for frictional sliding.

7.2.2 Modelling of damage evolution and composite ductility

Whatever the mechanism of micro-damage initiation, composite ductility is expected to be sensitive both to the strain at which damage nucleates and to the rate at which it accumulates and causes failure. A reliable model for the prediction of ductility in MMCs remains to be developed. In this section, simple existing treatments are outlined, despite their poor predictive capability, in order to understand their deficiencies and to point the way towards the derivation of better models.

The macroscopic strain required to nucleate a void

Models for void nucleation usually focus on the state of strain or stress necessary to nucleate a void in the vicinity of an isolated inclusion (i.e. at low volume fractions). Furthermore, a uniform distribution of equally sized spherical particles is commonly assumed, so that all the particles nucleate a void at the same time. In reality things are often very different with MMCs. Local variations in the particle population and shape tend to promote early voiding in certain regions[17]. Furthermore, the strain required is in practice sensitive to the interfacial bond strength. Historically, most studies have concentrated on systems containing weakly bonded inclusions (e.g. sulphides in steel), which initiate voids at

Table 7.2 *Void nucleation strain criteria based on debonding at isolated inclusions*

Nucleation strain equation	Basis of model	Critical nucleation condition	Ref.
$\varepsilon_n = \dfrac{3\gamma}{G_I b}$	Energy-based	Interfacial energy $\gamma <$ strain energy	20
$\varepsilon_n = \dfrac{3\gamma}{G_I b} \dfrac{1-\cos\theta}{4\sin^2\theta}$	Energy-based	Debonded caps subtending angle 2θ at particle centre	21
$\varepsilon_n = \dfrac{1}{60}\left(\dfrac{7\sigma_{r*}}{G_I}\right)^2 \dfrac{d}{b}$	Stress-based	Critical normal interface stress σ_{r*} or particle fracture stress	20

very low strains. MMCs, on the other hand, often contain inclusions which are well-bonded to the matrix, so that much larger strains may be required to produce local stresses sufficient to cause debonding or matrix cavitation.

There are various criteria for the nucleation of a void. One approach is to take an energy-based viewpoint: clearly the decrease in strain and/or potential energy arising from the nucleation of a void must be greater than the increase in surface energy. Whilst this criterion must necessarily be satisfied, it can be shown[18,19] that it is a sufficient condition only for inclusions smaller than 10 nm. In most models the energy criterion is supplemented by an interfacial stress criterion of some kind, the nature of which depends on the precise mechanism (e.g. matrix cavitation, particle cracking, debonding, etc. – see §7.2.1). A number of nucleation criteria are given in Table 7.2. However, none of these models involve any dependence on the volume fraction of inclusions, so that, intuitively, one might not expect them to be applicable to the high reinforcement contents typical of MMCs.

The Eshelby method has also been applied[19], using the simplification that the stress state at the interface is dominated by the plastically induced misfit, rather than the applied stress. The following equation is obtained

$$\varepsilon_n = \frac{(1-f)(7-5\nu_M)(1+\nu_I)E_M + f(7-5\nu_M)(1+\nu_M)E_I + (8-10\nu_I)(1+\nu_M)E_I}{(7-5\nu_M)E_I}\sigma_I \quad (7.13)$$

where σ_I is the stress in the inclusion. The nucleation strain is predicted to vary with inclusion content, f, but there is no dependence on particle size.

Acoustic emission[22,23] (see §11.12) and metallographic evidence[13] (see Fig. 7.17) suggest that damage nucleates continuously throughout plastic straining, rather than at a unique strain. These observations and other work suggest that, alone, nucleation-based models should be unable to predict the ductilities of MMCs, although they may be helpful in exploring the onset of damage.

The macroscopic strain required to propagate a void

The most promising approach for the prediction of the ductility of discontinuous MMCs lies in considering the void coalescence process which is thought to cause failure – see Fig. 7.13. Cavities form in locations of high hydrostatic stress and, when a certain condition is established, these link up by a ductile tearing mechanism. Several models have been developed to characterise the attainment of the critical state and the composite strain (ductility) at this point.

Evensen and Verk[24] considered crack propagation to arise from particle fracture. Matrix ligaments between two cracked particles are treated as notched bars, the fracture of which is predicted to occur when the length of the two plastic zones is equal to the interparticle spacing. Taking the plastic zone to be the region surrounding the crack strained more than ~0.3 (see Fig. 7.19), the crack opening displacement (see §7.3.1), δ, has a

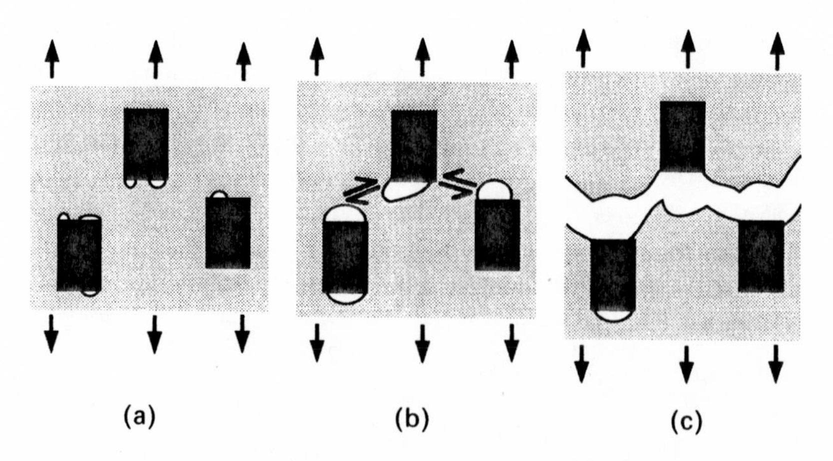

Fig. 7.13 Schematic illustration of the sequence of events during tensile failure of most discontinuous MMCs, showing how voids form in regions of high hydrostatic tension and then coalesce by a ductile tearing mechanism.

value ~1.2 times the dimension of the plastic zone. This leads to

$$2.4\delta = r_0\left(\sqrt{\frac{2\pi}{3f}} - \sqrt{\frac{8}{3}}\right)$$

where the particle separation is written in terms of the particle radius and volume fraction. Assuming δ to increase directly in proportion to the overall composite strain, the strain ε_g needed for the void to grow sufficiently for fracture to occur is given by

$$\varepsilon_g = 0.10\left(\sqrt{\frac{2\pi}{3f}} - \sqrt{\frac{8}{3}}\right) \tag{7.14}$$

A slightly different approach was taken by Brown and Embury[25], who proposed that voids formed by debonding can grow by plastic extension only until their length becomes equal to their spacing (Fig. 7.14). At this point the ligaments are no longer subject to lateral plastic constraint (i.e. 45° shear lines can be drawn) and the voids spontaneously coalesce by local necking to give failure. The strain increment during coalescence is assumed to be negligible. The model allows prediction of the strain needed for (catastrophic) void growth ε_g, as a function of particle content

$$1 + \varepsilon_g = \sqrt{\frac{\pi}{6f}} - \sqrt{\frac{2}{3}} \tag{7.15}$$

Thomason[26] proposed a modification to the above model, based on the internal necking of cuboidal voids (aspect ratio z_v/x_v). This model incorporates the effect of decreased void spacing due to reduced composite cross-section, as well as the geometrical extension of the voids referred to above. Using the same criterion for internal necking as before, the strain

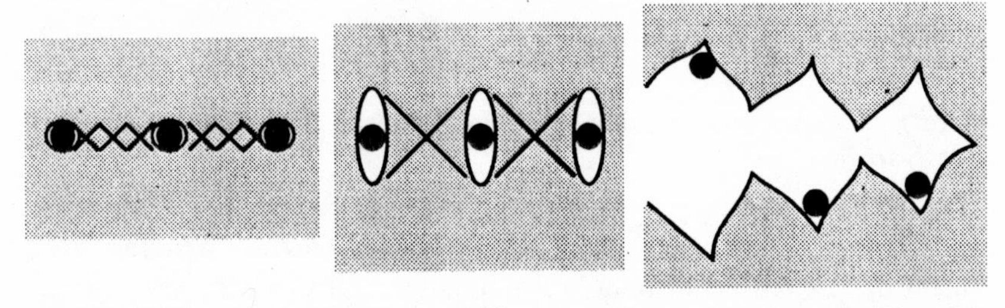

Fig. 7.14 Brown and Embury's model[25] of void growth by internal necking. The initial shape of the voids is based on complete interfacial debonding. Growth then occurs by plastic extension of the voids in direct proportion to the composite strain, until their length is equal to their spacing. Internal necking is then predicted to occur spontaneously with little further increase in strain.

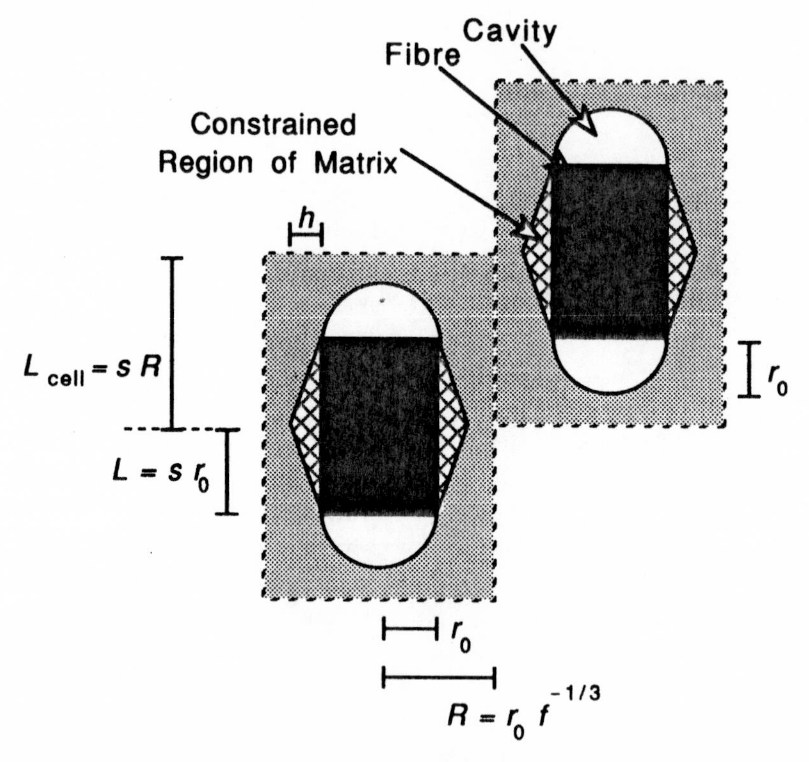

Fig. 7.15 Schematic illustration of the geometry assumed in the Whitehouse and Clyne model[27,28] for the composite strain at the point where a critical fraction (perhaps all) of the cavitation sites have been activated, which is taken as occurring immediately prior to failure by void coalescence.

to failure is given by

$$\varepsilon_g = \frac{1}{2} \ln\left(\frac{z_v}{x_v} \frac{1 - \sqrt{f_v}}{\sqrt{f_v}}\right) \tag{7.16}$$

where f_v is the volume fraction of voids (taken as the volume fraction of reinforcement).

An alternative model has recently been proposed by Whitehouse and Clyne[27,28], following their measurements of the changing void content during tensile straining of particulate and short fibre reinforced aluminium[13,27,28]. They proposed that voids grow to a size dictated by the adjacent reinforcing particle and that void coalescence will occur when a certain fraction F of the available cavitation sites have been activated. The composite strain at this point is then expressed as the product of terms accounting for the volume occupied by the reinforcement, the contribution from the cavitation itself and the region of the matrix constrained from deforming plastically by the presence of the reinforcement. The situation is illustrated in Fig. 7.15. By making some arbitrary geometrical assumptions[27,28], the following expression can be derived for the failure strain

$$\frac{\varepsilon_{C*}}{\varepsilon_{M*}} = (1 - f)\left(1 + \frac{Ff^{4/3}}{s}\right)\left[1 - \frac{2s}{5(f^{-1} - 1)}\right] \tag{7.17}$$

Typical values of F were found experimentally to be quite close to unity for powder route MMCs based on commercial purity Al, although lower values were obtained with cast material (having a high interfacial bond strength).

None of the above models incorporate any effect of reinforcement size.

Void growth rate models

For many systems, reported estimates of interfacial strength are high (e.g. ~1.7 GPa[29], ~7.2 GPa[30] for Al–SiC), so that, for well-consolidated material, interfacial decohesion is unlikely. Thus, in the absence of thick reaction layers (Ti–SiC systems often have such a layer) or other reasons for poor bonding such as persistent prior oxide layers, cracks propagate predominantly by progressive void formation within the matrix at regions of high stress (e.g. near the 'polar' regions of spherical particles – Fig. 7.28(b)). Models predicting the rate of growth of small cavities nucleated in the matrix *near* reinforcing particles have not yet been proposed. However, Rice and Tracey[31] have considered the rate of growth of a pre-existing isolated spherical cavity under a net remote tensile extension field ($\sigma_3^A > (\sigma_1^A = \sigma_2^A)$). For highly triaxial fields ($\sigma_1^A, \sigma_2^A \neq 0$) they predict almost spherical dilatation, and, even for low degrees of triaxiality, while the cavity does change in shape, its rate of *volume* growth is given to a good approximation by the expression appropriate for spherical dilatation

$$\frac{\dot{d}_0}{d_0} = 0.283\dot{\varepsilon}\exp\left(\frac{\sigma_1^A + \sigma_2^A + \sigma_3^A}{2\sigma_{YM}}\right) \tag{7.18}$$

where $\dot{\varepsilon}$ is the far field strain rate.

This model can be used to compare the relative rate of growth of voids under various loading conditions. For example, in an Al-based MMC ($\sigma_{YM} \sim 300$ MPa), under an applied stress of 450 MPa, we have a rate of volume growth

$$3\frac{\dot{d_0}}{d_0} = 0.85\dot{\varepsilon}\exp\left(\frac{450}{2 \times 300}\right)$$

$$= 1.8\dot{\varepsilon} \qquad (7.19a)$$

Were the void to form near the tip of a sharp crack, then the local stress would be[32] $\sigma_3 \approx (2 + \pi)\sigma_{YM}/2$, σ_1 (in the crack propagation direction) $\approx (1 + \pi)\sigma_{YM}/2$ and $\sigma_2 \approx \pi\sigma_{YM}/2$. Using eqn (7.18), the rate of growth of the void would be

$$3\frac{\dot{d_0}}{d_0} \approx 19\dot{\varepsilon} \qquad (7.19b)$$

For a void forming at a whisker end, one would expect it to grow most rapidly just inside the 'corner', where the hydrostatic component of the local stresses is predicted to be the greatest. For 2124 Al ($\sigma_{YM} = 300$ MPa) containing 13% SiC_W (aspect ratio 5), Christman *et al.*[33] used finite element modelling and predicted the composite plastic strain to be ~0.4% at an applied stress of 450 MPa and the hydrostatic stress in the matrix near the fibre corner to be ~600 MPa. This local stress would give a rate of void growth of

$$3\frac{\dot{d_0}}{d_0} \approx 17\dot{\varepsilon} \qquad (7.19c)$$

While this approach necessarily incorporates a number of simplifications and does not yet allow any predictions concerning ductility, it does show that the stresses at a fibre end, even at low overall strains, are large and of a magnitude approaching those at the tip of a sharp crack in unreinforced material!

7.2.3 Experimental data on damage evolution and ductility

Predictions from three void growth models and the Eshelby-based void nucleation model are compared in Fig. 7.16(a) with a selection of experimental fracture strain data for Al–SiC_P MMCs. It is immediately clear that the nucleation-based model is not appropriate for modelling of ductility, as it predicts an increase with reinforcement content. This arises because individual particles are indeed less highly stressed when they comprise more of the volume (see Fig. 3.12), but the model takes no account of the increased plastic constraint, and hence greater tendency for voids to link up, when the reinforcement content is greater. The void

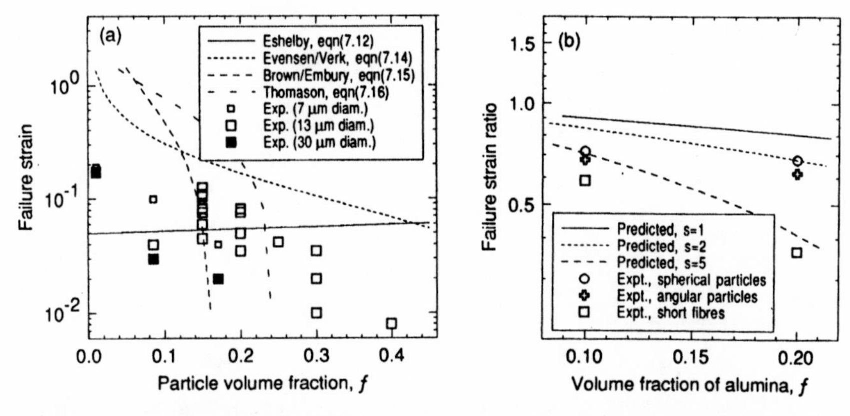

Fig. 7.16 Comparisons between experimental and predicted ductilities. (a) Measured[34–36] strains to failure for Al–SiC_P composites are shown with predictions from one void nucleation and three void growth models. The nucleation model is clearly incorrect when regarded as dictating the ductility, although it is broadly consistent with data[13] for the onset of void formation. (b) Failure strain data[13,27,28] for extruded powder route commercially pure Al matrix composites, expressed as a ratio to the value for unreinforced Al made in the same way (=24.5%), covering a range of reinforcement aspect ratios and two volume fractions. The predicted curves correspond to eqn (7.17), with the value of F set to unity.

growth models are more consistent with experiment in that they all predict a decrease in ductility with increasing reinforcement content. However, none of them gives good quantitative agreement.

This is perhaps not surprising, since these models are all based purely on geometrical concepts, with no account taken of the inherent ductility of the matrix. The Whitehouse and Clyne model, on the other hand, is concerned with how the matrix ductility is modified by the presence of the reinforcement. In order to explore the validity of this model, reliable experimental ductility data are required covering a wide range of ceramic contents (including zero) for a given matrix, with consistently homogeneous reinforcement distributions and interfacial bond strengths. Such data are in short supply, but the limited comparison between theory and experiment shown in Fig. 7.16(b) is encouraging.

There have also been experimental studies aimed at monitoring the development of internal voids[13] and particle fracture[14] during progressive plastic straining (Fig. 7.17). Not surprisingly, the forms of the two graphs are similar. In fact, several of the voids observed by Whitehouse *et al.*[13] were at cracked particles (Fig. 7.11(b)). Le Roy *et al.*[17] have modelled void development in terms of two populations; the first population nucleates

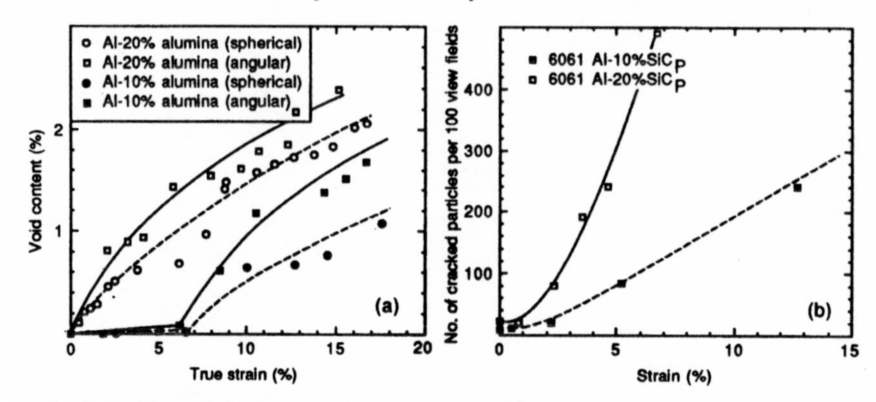

Fig. 7.17 Plots of (a) increase in void content[13] (from density measurements) of Al–Al_2O_{3P} MMCs, with angular and spherical particles, and (b) increase in incidence of cracked particles[14] in Al–SiC_P MMCs, as straining continues.

immediately at elongated particles, while the second population forms at a certain critical nucleation strain, ε_n, around equiaxed particles. Both populations then grow according to a Brown Embury-type model. Particle cracking, on the other hand, has been modelled from a Weibull point of view[37], taking into account the reduction in particle loading as a result of plastic relaxation. The strengthening contribution of the particles can then be related to the fraction of uncracked particles. This model predicts an increase in the proportion of cracked particles with increasing particle size, particle aspect ratio and composite straining, in accordance with experimental observations.

7.3 Fracture toughness and fatigue crack growth in MMCs

While an extensive body of experimental data is now available, many of the results are dependent on microstructural details and hence on fabrication procedures. For example, void initiation is affected by clustering, porosity and poor bonding. Though processing improvements have reduced these vagaries, it is still difficult to differentiate between universal characteristics and observations of more limited value. In this section, after an outline of some simple fracture mechanics, suggested approaches to the interpretation of MMC data are briefly discussed.

7.3.1 An introduction to fracture toughness

Griffith[38] was the first to relate the difference between fracture strengths predicted from interatomic bond strengths and those actually attained in practice to the presence of micro-defects. Furthermore, through a

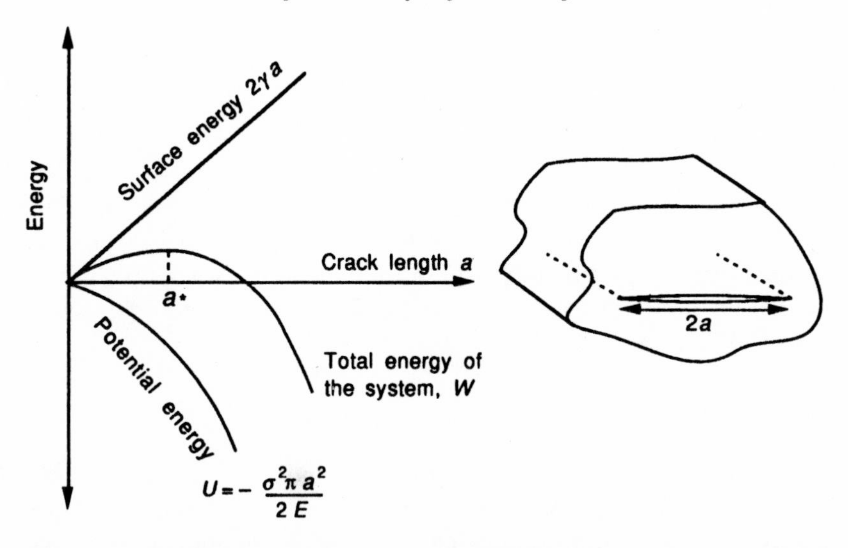

Fig. 7.18 Schematic plot of the energy of a through-thickness crack in an ideally brittle material. If a defect sized $2a_*$ or larger already exists, then the material can reduce energy by spontaneous fracture under the applied stress.

consideration of the energy changes which occur during the propagation of a crack, he was able to turn attention from the difficult problem of the precise stress field at the crack tip towards the energy change of the whole body. He proposed that the crack would only propagate if the total energy of the system was thereby decreased†.

Griffith considered a crack-like defect of length $2a$ within an infinite body. For ideally brittle materials, the energy dissipated upon extending the crack is simply the energy required to produce the new surface area. It can be shown[32] that under fixed grip (constant displacement) and fixed load conditions, the decrease in stored elastic strain and potential energy is the same, and is given by

$$\Delta U = -\frac{\sigma^{A^2}\pi a}{E}\Delta a \tag{7.20}$$

The ***strain energy release rate*** $\Delta U/\Delta a$ is usually denoted by the symbol $\mathcal{G}$. It is clear from Fig. 7.18 that, at a given applied stress, the maximum energy corresponds to a crack half-length a_* ($\Delta W/\Delta a = 0$) beyond which

† In fact, this is a necessary, but not sufficient, condition for fracture. For example, insufficient crack tip stresses might prohibit the extension of a blunt crack, even though from an energy point of view cracking was favourable. Nevertheless, it simplifies fracture analysis considerably, because it is not concerned with the exact details of the crack tip.

spontaneous fracture should occur. Conversely, for a given crack length, there is a critical stress for spontaneous fracture

$$\sigma_*^A = \sqrt{\frac{E2\gamma}{\pi a}} \tag{7.21}$$

Naturally, the approach must be modified for more ductile materials. One way of looking at this problem is to note that, for materials exhibiting plasticity at the crack tip, the plastic work done at the crack tip (γ_P) will be much greater than the surface energy (2γ), so that one might expect the critical fracture stress to be

$$\sigma_*^A = \sqrt{\frac{E\gamma_P}{\pi a}} \tag{7.22}$$

Irwin[39] took a somewhat different approach and used the critical energy release rate at which propagation occurred, $\mathcal{G}_c$, as a measure of the capacity for energy absorption by all the energy dissipating mechanisms operating during crack extension (heat, crack branching, etc.). The value of $\mathcal{G}_c$ is thus a measure of the resistance of the material to crack extension, i.e. its ability to soak up energy as the crack extends.

Provided the crack tip region is much smaller than the body, but larger than the scale of the microstructure, it is possible to consider the stress and energy of the extending crack using linear elasticity theory. The variation in stress normal to the plane of the crack ($\sigma_3(r)$) with distance in front of the crack tip is given by[32]

$$\sigma_3(r) = \sigma^A \sqrt{\frac{a}{2r}} = \frac{K}{\sqrt{2\pi r}} \qquad \text{(when } r \ll a\text{)} \tag{7.23}$$

where K is defined as the stress intensity factor ($\approx \sigma^A \sqrt{\pi a}$). This is a useful parameter because σ_3 is proportional to $1/\sqrt{r}$ whatever the loading geometry. Different types of loading are denoted by K_I, K_{II} and K_{III} for the modes of crack opening, and shearing parallel and normal to the crack advance direction. The energy released during crack extension can then be related to K using eqn (7.20), so that

$$\mathcal{G} = \frac{K^2}{E} \tag{7.24}$$

from which it is possible to define a critical stress intensity factor (e.g. $K_{Ic} = \sqrt{(\mathcal{G}_c E)}$)† beyond which the material is unable to resist crack

† In the following the I symbol will be dropped so that K will be taken as the stress intensity for mode I.

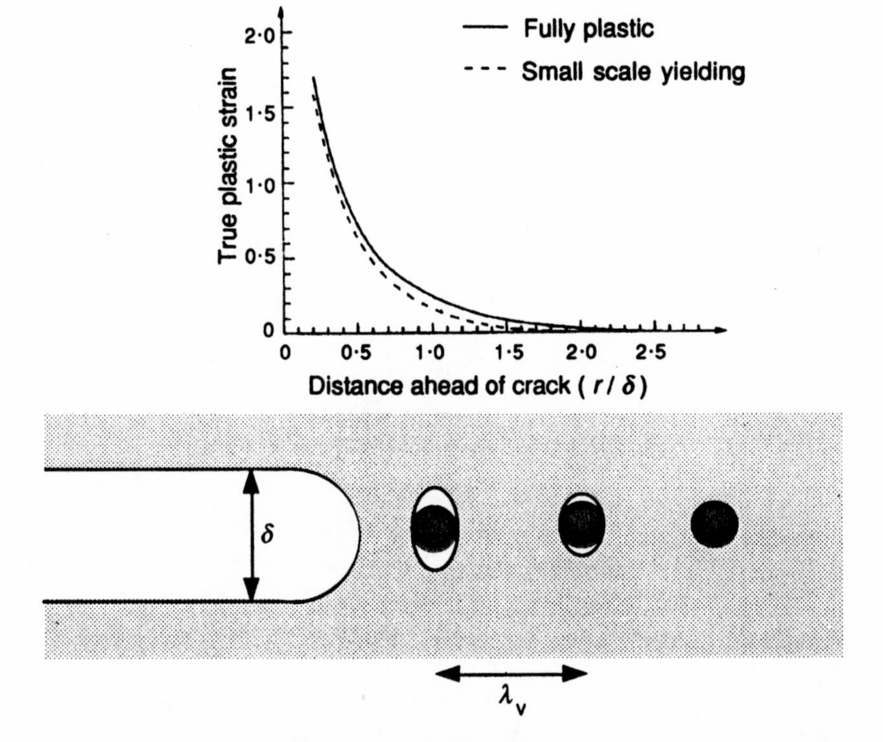

Fig. 7.19 Schematic showing the geometry of the crack tip, the crack opening displacement, the fall off in true strain with distance ahead of the crack[40], and its relation to the microstructure of MMCs. Rice and Johnson[40] predicted that failure would occur spontaneously when the applied stress was such that the critically plastically strained region enveloped microstructural dimensions characteristic of the failure process, i.e. when it became equal to the micro-crack spacing λ_v. When calculating the toughness dependence on volume fraction it is usually assumed that a micro-crack is associated with each particle.

extension. This parameter is commonly called the ***fracture toughness***, and is, for specified loading and crack geometries, a material constant.

For brittle (linear elastic) materials (in which the ***plastic zone*** (r_Y) at the crack tip is small compared with the length of the crack), K, the stress intensity factor, uniquely defines the stress and strain field local to the crack. It can be shown that[32]

$$r_Y = \frac{1}{2\pi}\left(\frac{K}{\sigma_{YM}}\right)^2 \tag{7.25}$$

Another important parameter is the ***crack opening displacement*** ($COD - \delta$) – see Fig. 7.19. This is often used for cases where valid K_c

measurements would require unrealistically large specimens. The critical crack opening displacement at fracture characterises the crack tip region. It can be shown[32] that δ and K are related by the equation

$$\delta = B\frac{K^2}{\sigma_Y E} \tag{7.26}$$

where B is a dimensionless constant, which can be estimated for different cases using finite element methods[39,41]; its value is of the order of unity (0.4–1.0).

7.3.2 *Modelling of the toughness of MMCs*

Crack path models

A possible approach to the prediction of fracture toughness is to make assumptions about conditions ahead of the crack tip. Rice and Johnson[40] considered the true strain ahead of a blunt crack (Fig. 7.19) and assumed that metals would fracture at strains between 0.2 and 1. According to their model, fracture occurs when the high strain region attains a size comparable with a microstructural dimension characteristic of the fracture process. From Fig. 7.19, for microcracks a distance λ_v apart (at every inclusion), this occurs when δ is about 37% to 100% of λ_v (depending on the fracture strain criterion). Taking $\delta = 0.5\lambda_v$ and assuming that a microcrack is associated with every reinforcing particle†, Hahn and Rosenfield[42] calculated the fracture toughness in terms of the volume fraction from eqn (7.26)

$$K_c = (0.5\lambda_v \sigma_Y E)^{1/2} \tag{7.27a}$$

$$\therefore K_c = \left[2\sigma_Y E\left(\frac{\pi}{6}\right)^{1/3} d\right]^{1/2} f^{-1/6} \tag{7.27b}$$

where d is the diameter of the (equiaxed) reinforcement. There is some confusion as to whether the composite or matrix yield stress should be used in this equation. This is really a question concerning the stress field sampled at the crack tip. When the plastic zone is large, the size of the zone (eqn (7.25)) is affected by the properties of the composite and σ_{YC} should be used. When the zone is small, few (if any) reinforcing

† This is not necessarily the case for MMCs containing strongly bonded inclusions.

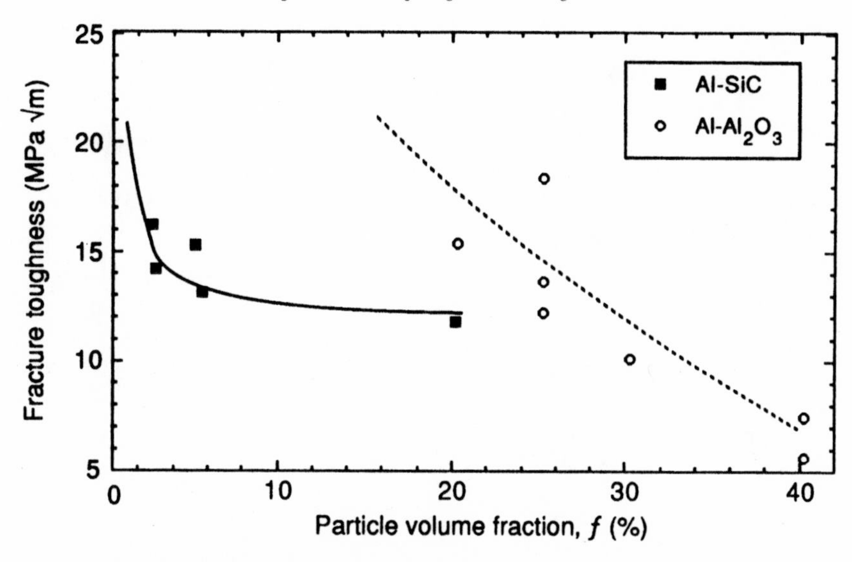

Fig. 7.20 Experimental fracture toughness data for Al–SiC_P[43] and Al–Al_2O_{3P}[44] as a function of the particle volume fraction.

particles are sampled and the yield stress of the matrix is probably more appropriate†.

Substitution into eqn (7.27b) of data for Al with 20 vol% of 10 μm diameter particles gives a value of about 25–30 $MPa\sqrt{m}$. This figure is larger than most experimental values (see Fig. 7.20), and the model obviously breaks down as f becomes very small. Furthermore, the sense of the dependence on σ_Y is contrary to that generally observed. While this approach does lead to predicted values of K_c which are correct in their order of magnitude, the model is evidently far from satisfactory.

Fractography-based models

To avoid the rather implausible prediction that fracture toughness rises with increasing matrix yield strength, it has been proposed that, while the local plastic strain must exceed a critical value, this value should be determined by the local stress state. This critical strain has been estimated in terms of the local strain derived from the geometry of the initial (d_v)

† Even if the plastic zone contains no reinforcing particles, increases in dislocation density may result in a 'matrix' yield stress which is somewhat higher than that of the unreinforced alloy – see Chapter 10.

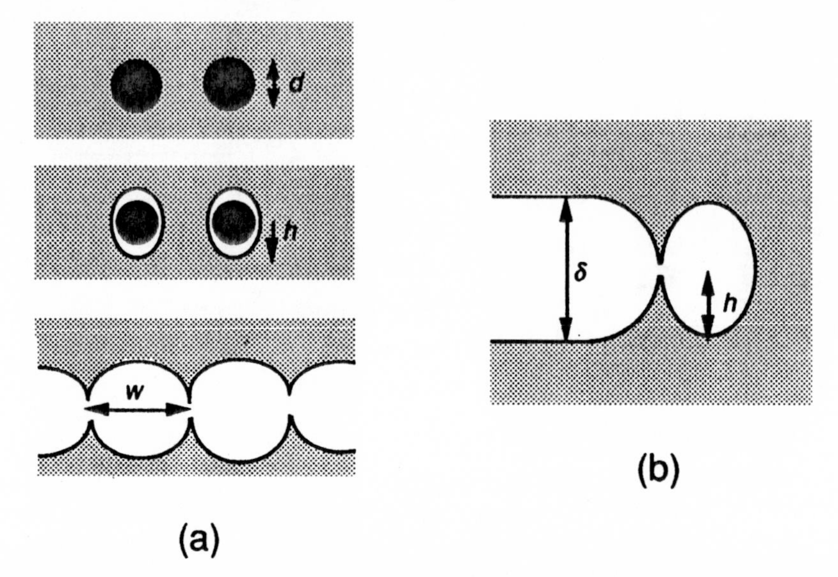

Fig. 7.21 Schematic depiction of crack geometry, showing the parameters pertinent to (a) the strain localisation model[40] and (b) the simple dimple height model[42]. In (a), the change in shape of the void is related to the local strain by $\varepsilon_{T,local} \approx \ln(h/d)$ if the void grows at the same rate as the imposed strain. Thus, as the ductility decreases, the dimples become flatter and the surface more microstructurally flat. By considering the number of voids within the fracture region, h/d can be recast as $\sqrt[3]{(h/w)^2/3f}$, which can be measured directly from the fracture surface.

and the coalesced void[45] (Fig. 7.21(a)), leading to

$$K_c = \left[\tfrac{1}{3}\sigma_{YC}E_C d\,\frac{\pi^{1/3}}{6}\ln\left(\frac{h}{d_v}\right)\right]^{1/2} f^{-1/6} \tag{7.28a}$$

or equivalently in terms of the height/width ratio (h/w) of dimples observed on the fracture surface[46]

$$K_c = \left[\tfrac{1}{3}\sigma_{YC}E_C d\,\frac{\pi^{1/3}}{6}\ln\left(\frac{(h/w)^2}{3f_v}\right)\right]^{1/2} f_v^{-1/6} \tag{7.28b}$$

where f_v and d_v are the volume fraction and diameter of the voids. Using the experimental parameters in Table 7.3 and assuming that $f_v = f$ (and $d_v = d$), the local strain, $\approx 0.33\ln[(h/w)^2/3f_v]$ is negative! It is thus clear that in this case complete debonding could not have taken place around every particle.

An alternative, somewhat simpler approach[47] is to assume that the crack extends by coalescence of the nearest void with the crack tip, to

Table 7.3 *Comparison between measured*[47] *and fractography-based estimates of fracture toughness for some* 6061 *Al-based composites*

Reinforcement	σ_{YC} (MPa)	E_C (GPa)	d (μm)	h (μm)	w (μm)	K_c (MPa$\sqrt{m}$) from eqn (7.28c)	K_c (MPa$\sqrt{m}$) measured
25% SiC_P	350	99	3.0	1.3	3.8	12.2	15.1
20% SiC_W Axial	330	108	2.5	1.6	1.8	13.8	22.4
20% SiC_W Trans	350 (est)	80 (est)	0.8	0.8	1.45	8.7	15.4

give the geometry shown in Fig. 7.21(b). In this case, the critical crack opening displacement, $\delta_* = 2h$, so that from eqn (7.26)

$$K_c = (3.33h\sigma_{YC}E_C)^{1/2} \tag{7.28c}$$

Predicted values obtained using the above relations in combination with observed dimple dimensions, are compared in Table 7.3 with directly measured K_c values.

In summary, these expressions have the advantage that they are related to the microscopic ductility as well as the strength. Furthermore, since they depend on the details of the fracture surface, they also reflect changes in fracture mechanism. However, because these models rely on post-mortem examination of the fracture surface, they are interpretive rather than truly predictive.

Energy-based models

It was shown in §6.1.4 that the energy absorbed during debonding of axial fibres (eqn (6.11)) could contribute up to 1 kJ m^{-2} or so to the fracture energy, while that absorbed during fibre pull-out might range up to a few kJ m^{-2} for monofilaments, but would be less for finer fibres. A low interfacial bond strength τ_{fr} (and a low fibre Weibull modulus) will encourage a large pull-out aspect ratio, s_0, and hence raise the energy absorption. This might make a significant contribution to the fracture energy in some MMCs, particularly those based on metals with low toughness. For example, the presence of carbon fibres in zinc leads to extensive fibre pull-out and enhanced toughness[48]. In this respect, carbon fibres usually exhibit poor bonding and more pronounced pull-out than other reinforcements – see Fig. 7.22. For most MMCs, however,

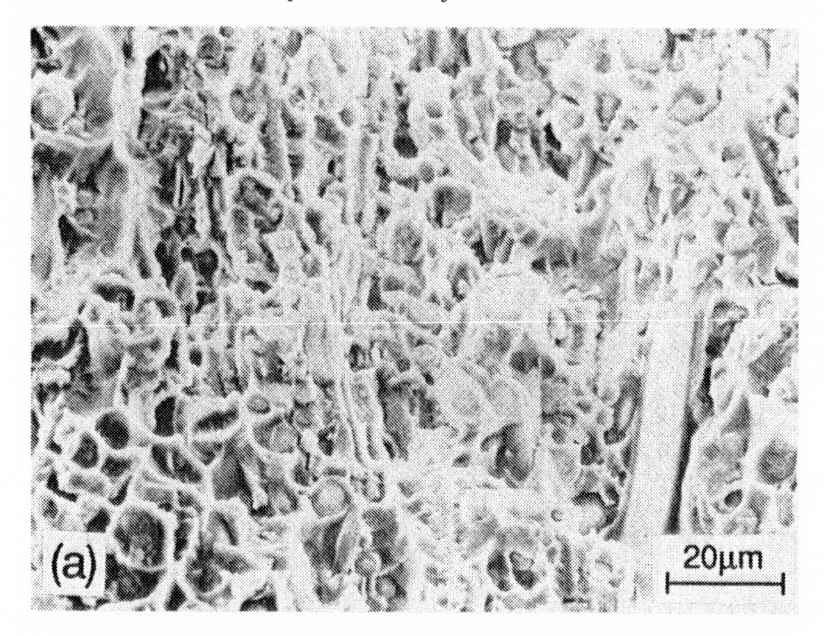

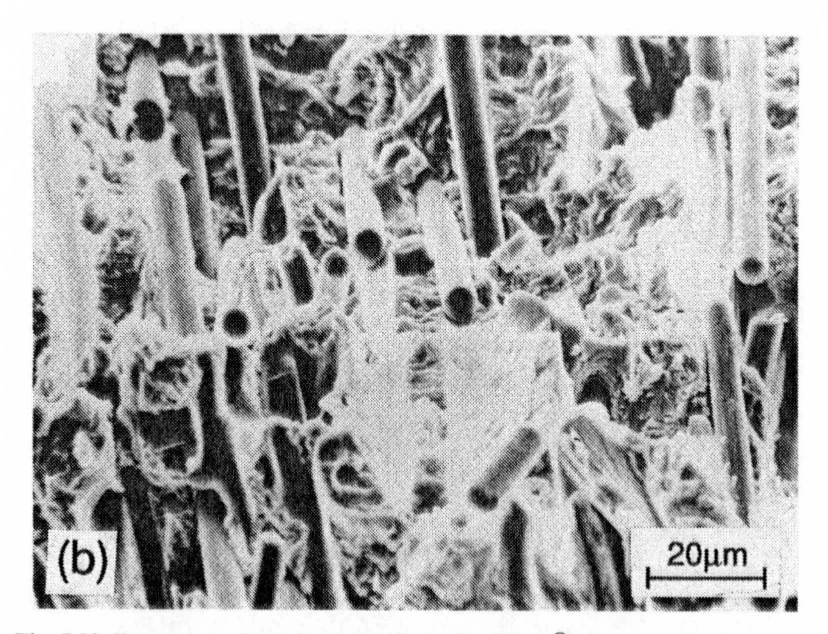

Fig. 7.22 Fracture surfaces from (a) Al–Al_2O_3 (Saffil®) fibre and (b) Al–carbon fibre composites. The greater degree of fibre pullout in (b) can be clearly seen – a consequence of the weaker interfacial bond strength in this case.

contributions from these processes are unlikely to be significant. Furthermore, energy absorption from debonding and pull-out during loading transverse to the fibres (and for particulate MMCs) can usually be neglected altogether (§6.1.4).

Davidson[49] has taken a different energy-based approach, aimed initially at particulate composites. He divided the energy $\mathcal{G}_c$ $(=K_c^2/E)$ into three distinct terms; the work done in the plastic zone of the crack, the mechanical work expended in creating the voided surface and the surface energy itself. For an Al–4% Mg–15% SiC_P system, the plastic work term was calculated as contributing 7.95 MPa$\sqrt{m}$ (calculated from the stress–strain curve) to the fracture toughness, the void creation term contributed 0.5 MPa$\sqrt{m}$ (calculated from a post-mortem examination of the void dimensions) and the surface energy represented 0.01 MPa$\sqrt{m}$. The sum of these compares favourably with an experimental value of around 8.8 MPa$\sqrt{m}$, which is much reduced from the unreinforced value (29 MPa$\sqrt{m}$). This is as a consequence of the reduced plastic straining at the crack tip as a result of the constraining effect of the particles. The dominance of plastic work over void formation in this particular case† is corroborated by the lack of correlation between surface roughness and fracture toughness[50].

In summary, fracture toughness models can, for certain cases, produce answers in fair agreement with experimental results. However, no current model is really satisfactory. The crack path model does not show the correct variation with parameters such as volume fraction and strength, fractography-based models are not predictive, pull-out evidently plays little part in determining the toughness for many systems of interest and attempts to predict the other contributions to the fracture energy are still in embryonic stages.

7.3.3 Fatigue and sub-critical crack growth in MMCs

Fatigue crack propagation

When the stress intensity at a crack is below K_c, immediate failure will not occur. This is not to say that crack growth cannot take place, but simply that it will not happen catastrophically. The most common cause of sub-critical crack growth is fatigue loading. When considering fatigue, the difference in stress intensity between the maximum and minimum

† Davidson reports unusually fine dimples – 0.4 μm, perhaps because of the presence of sub-micron MgO inclusions.

loading (ΔK) is an important parameter. This is because, whilst at sufficiently high K_{max} fast fracture will take place, when K_{max} is smaller the cyclic dissipation of energy is dependent on ΔK. The resistance of the material to crack extension is then given in terms of the crack growth rate per loading cycle (da/dN). At intermediate ΔK, the crack growth rate can be described by the Paris–Erdogan[51] relation

$$\frac{da}{dN} = \beta(\Delta K)^n \tag{7.29}$$

where β is a constant. Hence, a plot of crack growth rate (m/cycle) against ΔK, with log scales, should give a straight line in the Paris regime, with a gradient equal to n. At low stress intensities, there will be a threshold, ΔK_{th}, below which no crack growth occurs, while the crack growth rate usually accelerates as the level for fast fracture, K_c, is approached.

Fatigue data are sometimes presented in the form of S/N curves, showing the number of cycles to failure (N) as a function of the stress amplitude (S). Many materials show a roughly sigmoidal curve, with very rapid crack growth when the stress amplitude is high, a central portion corresponding to the Paris regime and a ***fatigue limit***, which is a stress amplitude below which failure does not occur even after large numbers of cycles. This corresponds to a stress intensity below ΔK_{th}.

Closure is the term used to describe phenomena which prevent attainment at the crack tip of the K_{min}, and thus reduce the crack propagation rate through a failure to attain the full ΔK range†. Three types of closure are commonly encountered: (i) roughness-induced closure, (ii) oxide-induced closure, and (iii) plasticity-induced closure. All three limit the actual K_{min} attained by holding open the crack tip. For MMCs, the effect of closure has been found to be a reduction of about 1.5 MPa$\sqrt{\text{m}}$ in ΔK for K_{max} in the range 4–8 MPa$\sqrt{\text{m}}$. This has been attributed to asperity-induced closure[33], though oxide debris was also evident. Closure is generally reduced at greater R ratios,

$$R = \frac{\text{min. load}}{\text{max. load}} = \frac{K_{min}}{K_{max}} \tag{7.30}$$

because the crack is then held more widely open throughout the cycle.

† The mixity of loading mode also affects closure. A large mode II component shields the tip if the surfaces are rough.

Fatigue in long fibre MMCs

The fatigue resistance of long fibre MMCs loaded in tension along the fibre axis has repeatedly been shown[52–55] to be superior to that of the unreinforced metal. The fatigue limit, for example, is typically increased by a factor of about two. This is broadly as expected on the basis of load transfer to the fibres from the matrix, through which the fatigue cracks must propagate. An elementary model to account for this effect was put forward by Dvorak and Tar[56], who proposed that the fatigue limit $\Delta\sigma_{fat*}$, could be related to the matrix yield stress and the composite/matrix stiffness ratio

$$\Delta\sigma_{fat*} \approx 2\sigma_{YM} \frac{E_C}{E_M} \tag{7.31}$$

It was shown[53] that this 'shakedown stress' corresponded to the level at which microstructural damage started to become significant and the model was later extended to include laminates[55]. Improved fatigue resistance in Mg–5Zn/55 vol% alumina fibre (with strong interfacial bonding) over a wide range of stress amplitude was also exhibited[55] for loading at various angles to the fibre axis. These improvements in fatigue limit are equivalent to increases in ΔK_{th} which are greater than those observed for discontinuous MMCs (see below).

In reporting enhanced fatigue resistance as a result of the presence of long fibres, it should be noted that there is evidence for a strong dependence on the presence of defects. Several workers[54,58] have observed pronounced scatter in their results, traced directly to varying levels of defects such as regions of fibre agglomeration and/or matrix porosity. The behaviour also tends to be sensitive to interfacial bond strength. It is generally observed that classical *S/N* curves and Paris law behaviour are exhibited when cracks propagate normal to the stress axis, but deviations occur when there is extensive interfacial cracking. This is dictated by the ratio of the toughness of the interface to that of the fibre, which must be less than about 15–20% if a matrix crack is to be deflected in this way (see §6.1.4). When this occurs, fibre bridging[59,60] can take place in the wake of the crack, acting to shield the crack tip very efficiently. Ibbotson *et al.*[60] have even observed a decrease in crack growth rate in Ti–6Al–4V/35 vol% SiC (SCS-6) with increased stress intensity factor (increased crack length), attributed to enhanced fibre bridging and a reduced effective ΔK at the crack tip.

Sensmeier and Wright[59] have developed a numerical model for the extent of crack bridging, based essentially on the matrix stress treatment of Dvorak and Tar[56], but using an analysis of interfacial sliding proposed by Marshall, Cox and Evans[61]. This model has been used to predict crack growth rates as a function of stress amplitude, assuming a uniform fibre strength. In practice, the Weibull modulus of the fibre is likely to be important, as a low value will favour earlier fibre fracture outside the crack plane and subsequent fibre pull-out[62] (§6.1.4). A further point worthy of note is that the ease of interfacial debonding will depend on factors such as the degree of chemical reaction (§6.3.2) and the residual stresses (§6.1.1). For example, a sharply increased degree of interfacial debonding at elevated temperature has been attributed[60] to a reduction in the normal compressive stress across the interface from differential thermal contraction.

Fatigue in discontinuous MMCs

The fatigue response of discontinuous MMCs is best considered by relating the magnitude of ΔK to the plastic zone size (r_Y), the scale of the microstructure and the failure mechanism. Below ΔK_{th}, cracks are unable to grow at all (Fig. 7.23). For Al–SiC_P MMCs, ΔK_{th} is typically around 2–4 $MPa\sqrt{m}$, which is approximately double that for unreinforced alloys (1–2 $MPa\sqrt{m}$). A number of explanations for this have

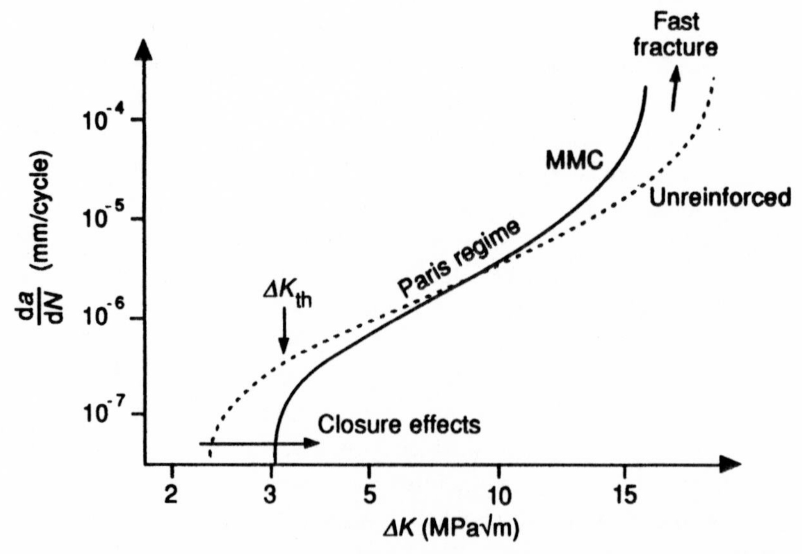

Fig. 7.23 Schematic depiction of fatigue crack growth rate as a function of applied stress intensity factor for a typical discontinuous MMC and the corresponding unreinforced alloy, illustrating the effect of reinforcement on the fatigue response.

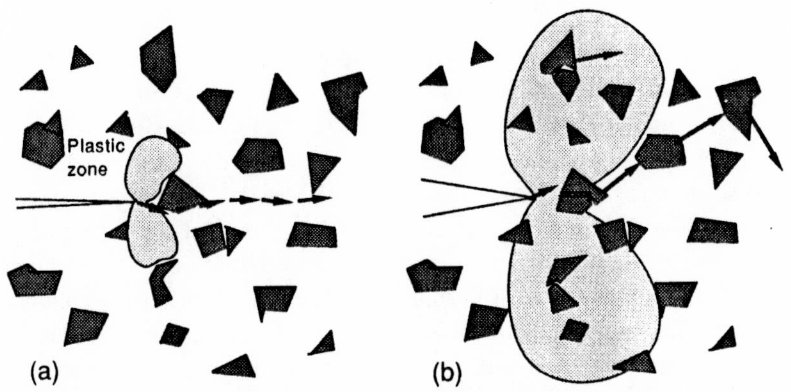

Fig. 7.24 The crack path in discontinuous MMCs depends upon ΔK. (a) For low ΔK the plastic zone is small. This, combined with the fact that the constraining effect of the stiff particles acts to deflect the crack somewhat[63], causes the crack to meander through the matrix (few cracked particles are observed). (b) At high ΔK the plastic zone envelops many particles and particle cracking may become more important[64]. Particle cracking ahead of the crack will accelerate the growth rate. The increase in zone size upon increasing K_{max} or ΔK means that many more particles lie within the highly stressed region, which increases the distance ahead of the crack in which cracking occurs.

been proposed. These include crack deflection by the reinforcement, a reduction in slip band formation† due to the particles[65], and a decrease in $\delta_{max} - \delta_{min}$ (at a given stress intensity)[66] (see eqn (7.26)). Under certain measuring conditions, threshold values can be affected by closure, giving rise to a dependence of ΔK_{th} upon R ratio. Unless a correction is made for closure, artificially high ΔK_{th} values may be recorded.

Because of the relatively low fracture toughness of MMCs, the Paris regime is usually short. The exponent n is often around 5–6, which is higher than those typical of unreinforced systems (~ 4)[67]. In this regime ($\Delta K \sim 8$ MPa$\sqrt{}$m), a number of particles are encompassed by the plastic zone ($r_Y \sim 50$ μm), so that the crack tip 'sees' a continuum with a higher yield stress. The fatigue resistance is thus superior to that of the corresponding unreinforced matrix (Fig. 7.23), although not dramatically so.

At high ΔK (>12 MPa$\sqrt{}$m, $r_Y > 100$ μm), many particles are within the plastic zone (Fig. 7.24). In this regime, fast fracture failure modes become operative and the fatigue resistance of MMCs becomes appreciably worse than that of unreinforced alloys, reflecting their generally inferior fracture toughness (Fig. 7.20).

† Slip bands accelerate growth by providing a favoured crack path.

Stress corrosion cracking

Few studies have been undertaken of the influence of reinforcement on stress corrosion-aided cracking. For Al-based systems, it has been found that, whilst moist or salt-laden moist air cause an acceleration of fatigue crack growth of between 5–10 times[68,69], this degradation in response is no worse than for unreinforced Al alloys. This is true for SiC_P and SiC_W reinforced aluminium[69–71] as well as for long aramid fibre reinforced composites[68,72]. In this case it has been found that in the alignment direction the fibres give rise to good retention of fatigue resistance, whereas in the transverse direction the fatigue resistance is affected by the corrosive environment to a degree directly related to the proportion of matrix occupied by the composite.

In general, much of the stress corrosion cracking observed in high strength Al alloys is dominated by hydrogen-related processes[73,74]. The most critical controlling feature is the effect of second phase on the nature of slip and the associated ease of transport of hydrogen within dislocations. In MMCs, a further factor is introduced in the form of very strong traps for hydrogen at the matrix/reinforcement interface. Using straining electrode tests, Bernstein and Dollar[75] found that, while the strength levels of 2124 Al/20% SiC_P composites were unaffected by testing in a hydrogen environment, the ductility was sharply reduced, irrespective of matrix aging. This was attributed to strong trapping of hydrogen at the interface, where voids were observed (absent from air-tested specimens). The unreinforced alloy was slightly embrittled when underaged (hydrogen concentration in slip-bands), but suffered no embrittlement when peak- or over-aged. The MMC embrittlement is largely reversible if the hydrogen can be driven off[75].

7.4 Effects of microstructural variables

The toughness and fatigue behaviour of MMCs can be quite sensitive to the microstructural details, particularly with discontinuous reinforcement. This sensitivity covers the reinforcement, the interface and the matrix.

7.4.1 General crack path observations

Crack path *initiation* studies[65], using smooth-surfaced MMC specimens under 4-point bend fatigue loading, have identified various preferred failure initiation sites. These include cracked and touching particles and

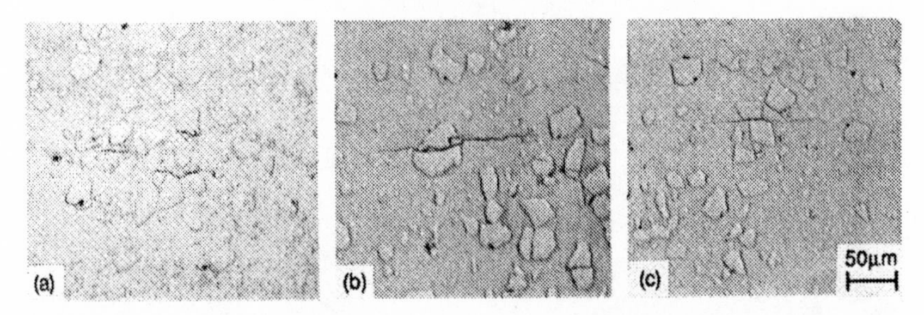

Fig. 7.25 Examples of damage initiation in Al–SiC MMCs; (a) angular stress concentrators[65], (b) particles cracked prior to loading[76] and (c) touching particles[76].

the ends of fibres and whiskers. Examples of such sites are shown in Fig. 7.25. Clearly, stress concentration plays an important role in damage initiation. Resistance to damage might thus be increased by improved processing and the incorporation of more rounded, defect-free (smaller) inclusions.

Cracks can *propagate* by a variety of micro-mechanisms, depending on factors such as interface strength, matrix work-hardening and reinforcement integrity. For example, Fig. 7.26, under fatigue loading the propagation of short cracks has been shown to occur via interface failure in Al–Al_2O_3 (fibre) systems[77,78] when the interface is relatively weak, but predominantly by matrix voiding near particle poles for Al–20 μm SiC_P. In addition, particle cracking ahead of the crack, followed by linkage with the main crack, has also been observed[79,80] (see Fig. 7.26(b)).

The effect of particle clusters on the crack path is exemplified by the map[49] (contours of plastic shear strain) shown in Fig. 7.27[49]. The high plastic strains associated with the cluster in the bottom right are partly due to the presence of pre-existing pores. It has been suggested from such observations that the plastic zone size in MMCs is actually smaller (by a factor of about four) than is expected from conventional fracture mechanics (eqn (7.25)), with strain falling off in front of the crack more quickly than conventionally expected ($r^{-1.1}$ as against $r^{-0.5}$)[32]. Together, these effects sharply reduce the plastic work done, by confining and reducing the plastic strain.

7.4.2 Reinforcement shape

There have been some difficulties in separating the effects of reinforcement shape and size. One reason is that, for the most common reinforcement

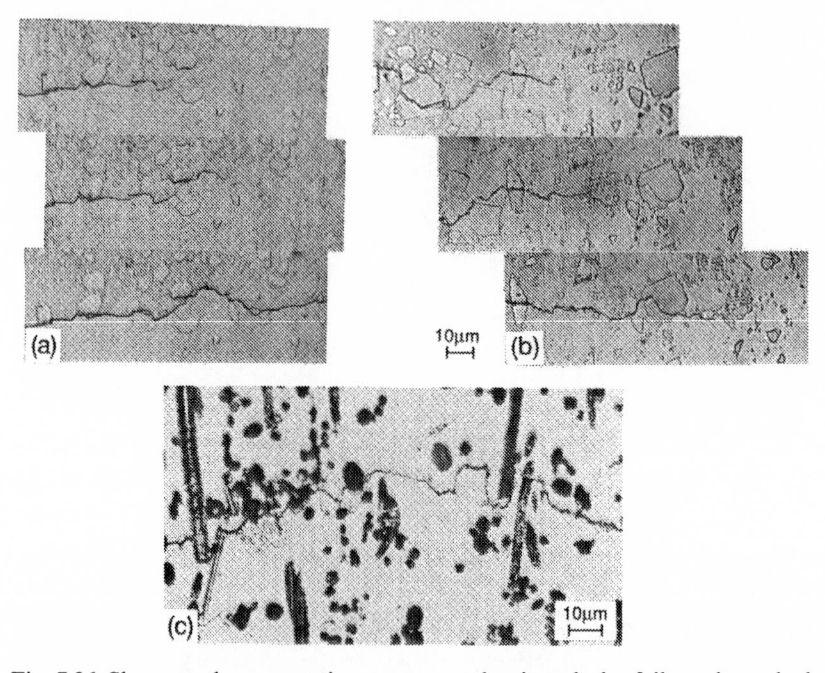

Fig. 7.26 Short crack propagation occurs predominantly by failure through the matrix. As illustrated by this replica series[76] (a) there is a tendency for the crack to follow the poles of the larger particles. Particle cracking ahead of the crack tip also occurs[79] (b), but this mechanism of crack propagation is in general less common. In short fibre MMCs, (c), the crack path is often influenced by stress fields around fibre ends.

(SiC), readily available equiaxed particles are at least an order of magnitude bigger than the diameter of most available SiC whiskers. From the viewpoint of stress concentration, one wouid expect short cracks to initiate preferentially at long whiskers and elongated particles. Statistically rigorous evidence is lacking, but (as was shown in Fig. 7.25) the nucleation of a crack does seem to be encouraged by particles elongated parallel to the stress direction and by flat surfaces normal to this direction. As to whether spherical, as opposed to angular, particles are beneficial, recent evidence[13,28] suggests that rates of void initiation and growth are clearly lower for spheres, although not dramatically so (Fig. 7.17). This is probably because, whilst angularities give rise to high local stresses, these can be relieved by microplastic flow, whereas the high hydrostatic stresses developed at flat interfaces normal to an applied tensile load are much more likely to induce voiding[28] (Fig. 7.28). In this sense, a (large) sphere is almost as efficient as a flat-ended fragment.

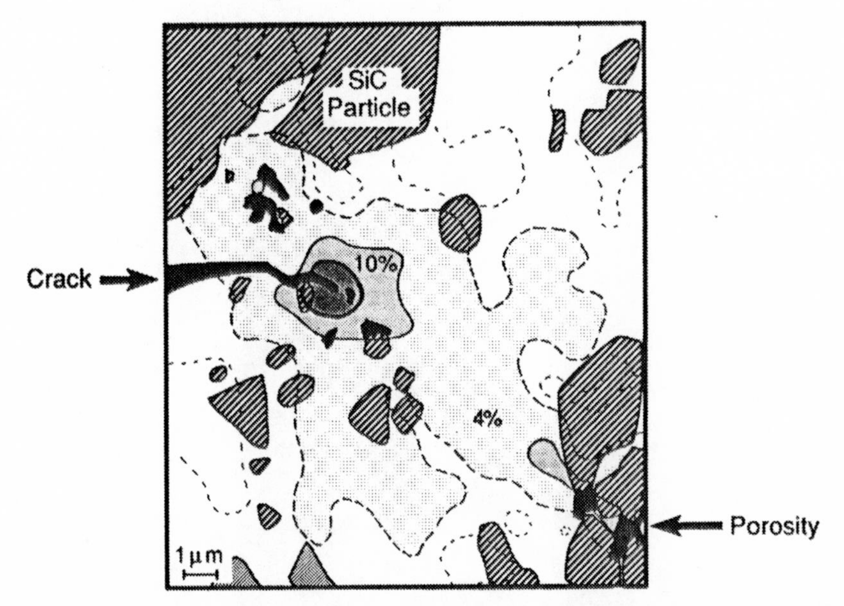

Fig. 7.27 A schematic of the crack tip region in an Al–15% SiC_P composite[49], upon which contours of maximum shear strain (measured under fatigue loading by photographic comparison of the surface at maximum and minimum load) are superimposed. Subsequent crack growth was towards the particle cluster (bottom right), illustrating the effect of poorly consolidated clustered regions on crack growth.

During fast fracture, aligned whisker material (6061 Al–20% SiC_W) has been shown to offer a greater resistance to crack propagation when the crack is grown transverse to the whisker direction, with K_c values of 22 (transverse) and 14 MPa$\sqrt{m}$ (axial)[45]. This result has been explained in terms of differences in the projected area of the reinforcement (leading to greater λ_v, see eqn (7.27a)), as well as in terms of the ease of propagating cracks along whisker/matrix interfaces. These results are mirrored in fatigue[68,69,81], with fatigue crack growth rates for particulate material lying generally between the axial and transverse (most resistant) extremes with whiskers. Fatigue thresholds have been found to be highest in particulate material, but upon correcting for closure, $\Delta K_{th,eff}$ was lower than for an axial crack in whisker material.

7.4.3 Reinforcement size

Experimentally, the effect of particle size on particle cracking is well documented for the Al–SiC system[14,15,63]. In populations with a large

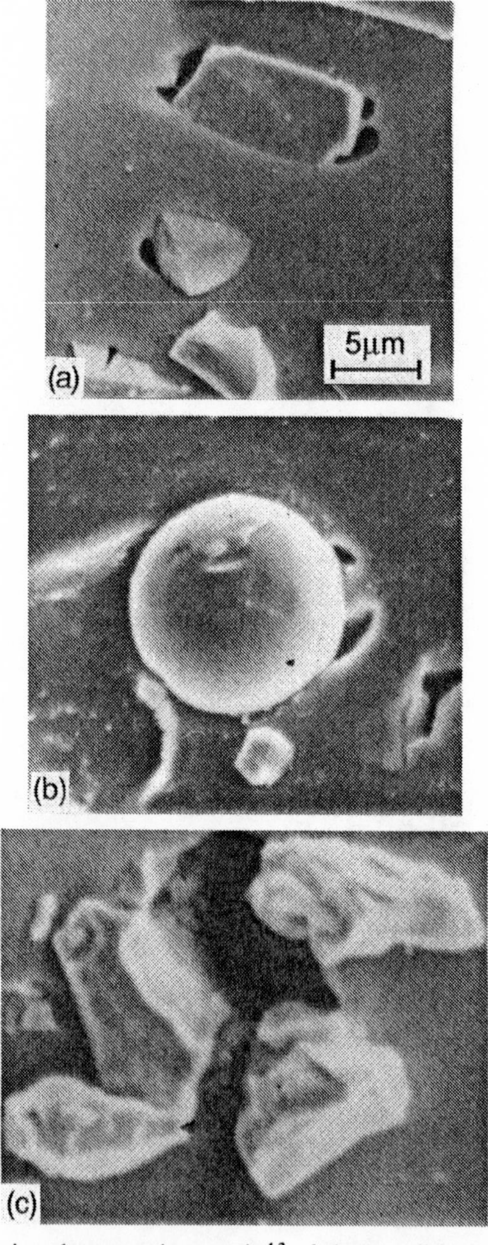

Fig. 7.28. Scanning electron micrographs[13] of electropolished surfaces from Al–10 vol% Al_2O_{3P} specimens subjected to about 17% plastic strain, reinforced with (a) angular and (b) spherical particles. Voids are shown forming in the matrix adjacent to the particles. (c) Voids often form within particle clusters, although some such porosity can arise during fabrication.

Table 7.4 *The variation of fracture toughness with particle size*[83]

Particle size	5 μm	7 μm	16 μm	21 μm
Fracture toughness (MPa$\sqrt{m}$)	14.5	16.5	16.8	18.7

scatter in size, it is the largest particles which tend to fracture[35]. Consequently, in material containing coarse particles, particle fracture is commonplace, whereas for fine particulate, cracking is rare. However, in spite of this difference, the fracture toughness appears to be nearly the same whatever the degree of particle cracking, which might suggest that there is little advantage to be gained in using higher performance particles. This is substantiated by the similar elongations observed with coarse and fine particle containing systems.

Considerable effort has been expended in investigating the size dependency of fracture toughness for the Al–SiC_p system. Flom and Arsenault[82] found fracture toughness was largely independent of particle size for all but the very largest particles (250 μm), whereas Hunt[83] found the progressive increase with increasing size shown in Table 7.4. Similar results have been explained in terms of a change in failure mechanism from being dominated by particle cracking at large sizes to debonding at particle poles for smaller particles[84].

In fatigue, threshold values of ΔK_{th} have been shown to increase with particle size in Al–SiC_p composites[85]. This has been attributed to surface roughness-induced closure effects being nearly three times greater for coarse particle reinforced systems compared with fine[34], although crack trapping as opposed to deflection has also been suggested[85]. In view of the stiffness-induced crack deflecting tendency of the particles, the latter explanation seems improbable for particulate material, although it may be an important mechanism for cracks grown perpendicular to whisker reinforcement.

7.4.4 Bond strength

Given that the majority of failure studies on discontinuous MMCs have centred on the Al–SiC system, for which matrix/inclusion bonding is good, evidence for the effect of bond strength on failure is sparse. The most systematic study to date has looked at the bond strength/composite ductility relationship for the Fe–Al_2O_{3P} system[86]. Using alloying additions to increase the strength of the interfacial bond, a direct correlation between bond strength and improved ductility (up to a factor of four) was observed. Similar findings have been made for the Al–B_4C_P system[87], while a progressive reduction in toughness has been observed for Ti–SiC as the interfacial reaction proceeds[88] (§6.3). A low strength interface can be achieved using surface oxidised SiC_P in Al, but in contrast to the above cases, this apparently leads to *increased* ductility[89] (see §6.1.1). This may be because the layer aids interfacial sliding (as against debonding) and so promotes stress relaxation†, thus reducing the UTS and delaying the onset of catastrophic failure. As was demonstrated in §6.1.4, there is little opportunity for significant energy absorption by interfacial processes themselves for particulate MMCs and only limited potential with fibre reinforcement.

7.4.5 Reinforcement distribution

Recent advances in the quantification of the effects of local reinforcement arrangements have been stimulated by improved quantification of spatial distributions via Dirichlet Tesselations[87,90] using image analysis (§11.7.4). As predicted[16–18], experimental evidence points to the preferential nucleation of cracks in regions of locally high volume fraction. During crack growth, clustered regions can behave in one of two distinct ways. If the particles move independently of one another under the stresses at the crack tip, then the constraint gives rise to very large plastic strains (for example, the cluster in the right-hand corner of Fig. 7.27) and the crack displays a preference for passing through clustered regions[47,90]. This is encouraged by incompletely infiltrated regions and voids are often present within such clusters – an example can be seen in Fig. 7.28(c). However, if the cluster moves collectively, then it behaves as a single large particle, the region between the particles is less strained and the cluster may deflect the crack. This is exemplified by the cluster just below the crack tip in Fig. 7.27.

7.4.6 Matrix aging

The magnitude of the stress that can be supported within the matrix is of great importance when considering failure by void nucleation, as well as its ability to flow under highly constrained conditions. The former is

† The oxide layer may also soak up Mg from the matrix and thus reduce the matrix strength slightly.

dependent upon a combination of yield stress and work-hardening rate. Many studies have been undertaken on the effect of aging condition and a number of common features are beginning to emerge. With respect to ductility, Lloyd[14] has shown that, for Al–SiC_P in low strength tempers, the gauge length plastically strains approximately uniformly, whereas, for high strength tempers, the elongation is localised to the failure region. Consequently, in naturally aged (T4) material, particle cracking is widespread, whereas in T6 material cracking is confined to the fractured region. The tensile elongation is similarly related to matrix condition, decreasing with increasing strength and strain localisation.

Lewandowski *et al.*[35] have reported that changes in aging condition can bring about significant differences in fracture micro-mechanism. They studied over-aged (OA) and under-aged (UA) Al–Zn–Mg–Cu–SiC_P composites (7000 alloy–20% SiC (4 μm particle size)) of comparable microhardness and flow stress. In the UA composite, the area fraction of particles at the fracture surface was approximately equal to that expected if fracture were to take place by particle cracking, whereas in the OA composite it was significantly less, indicative of fracture predominantly in the matrix. This has been attributed to weakening of the interfacial bonding, or of the matrix near the reinforcement, due to a precipitate-free zone in the OA material. These changes in micro-mechanism were not reflected in changes in fracture toughness, though notched bend tests were significantly affected.

Fracture toughness values show a strong dependence on aging condition for Al_2O_{3P}[91], SiC_P[47] and SiC_W[92] reinforced Al (see Fig. 7.29). Whilst the decrease in fracture toughness with increasing yield stress is expected, the failure of the toughness response to improve upon further aging† is not yet fully understood. Possible explanations centre on the role of precipitates at the metal/ceramic interface and in other cases the formation of precipitate-free zones (e.g. in Al–Li alloys), but these require confirmation. In Al–Al_2O_{3P}, failure to recover fracture toughness upon over-aging was ascribed to strain localisation and the reduction of deformation of ligaments formed at cracked particles by large second phase precipitates. This causes a decrease in energy absorption with increasing precipitate size, i.e. upon aging[91].

† The data of Crowe *et al.*[47] show a slight increase in K_c with increasing yield strength (6061 Al–SiC_W).

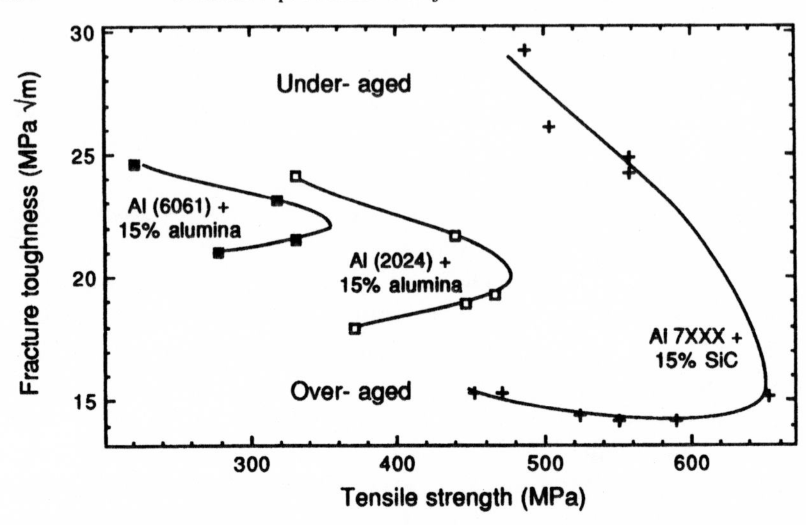

Fig. 7.29 Experimentally observed variations in fracture toughness plotted against corresponding yield strength for specimens subjected to various heat treatments in Al–15 vol% SiC_P[35] and Al–15 vol% Al_2O_{3P}[91] MMCs.

7.5 Effects of testing variables

7.5.1 Strain rate

In most cases the ductility has been found to increase with increasing strain rate (e.g. 2124 Al/SiC_W[93], Al/Al_2O_3 (50 nm particles[94])). This increase has been accompanied by an increase in the number of voids observed, as well as by an increase in the distance from the fracture surface at which they are initiated[93,95]. These factors, when combined with the lower matrix dislocation densities observed at higher strain rates[93], suggest that stabilisation of void growth at high strain rates is responsible for the increases in fracture strain, rather than plastic deformation. Whisker-reinforced MMCs tend to undergo homogeneous plastic flow during quasi-static loading, but at high strain rates ($1000\ s^{-1}$), strong slip markings at 45° are observed on the tensile specimen surface, with cracks at their intersection. In this case final failure occurs without necking, the specimen often breaking into more than two pieces[93]. As to the mechanism of failure, under static loading, voiding or decohesion is responsible, whilst under impact loading, fracture of whiskers lying on lines of localised flow occurs.

Other effects observed on increasing the strain rate include increases in modulus (Al/SiC_W[93]), ultimate tensile stress (2124 Al/SiC_W[93], Al/Al_2O_3 (50 nm particles)[94]), and fracture toughness (Al/SiC_W). The rate-dependent changes in UTS, which occur in the stress–strain curve and, for Al/SiC, only at very high strain rates, are difficult to explain in view of the increased microcracking. (A similar problem is encountered in the interpretation of pressure dependence – see §7.5.3. The explanation probably lies in the delayed initiation or activation of voids.) Further, bearing in mind the increased recovery under quasi-static loading for the Al/Al_2O_3 (50 nm particles)[94] system, one would expect the use of higher strain rates to limit recovery and hence raise initial work-hardening.

7.5.2 Temperature

Changes in tensile strength and ductility with temperature are shown in Fig. 7.30 for a selection of discontinuous MMCs. Composite UTS data reflect changes in the unreinforced alloy, but with an increase of around 100 °C in the temperature at which a sharp drop in strength is observed[36]. Decreases in tensile strength at around 350 °C are accompanied by an increased tendency to neck down prior to failure[36]. Furthermore, as one might expect at high temperatures, the tensile strength becomes less sensitive to the inclusion content, but more sensitive to its shape, increasing with aspect ratio[94,96]. Sharp increases in ductility are observed at around 250 °C for both matrix and composite (Fig. 7.30(b)).

7.5.3 Hydrostatic pressure

Fracture strain

In materials which fail by void nucleation and growth, the superimposition of hydrostatic compression inhibits void dilatation, thus extending ductility[99]. Consequently, high pressure experiments provide an opportunity to study the void growth process and its effect on flow behaviour. Experimental observations can be divided into those on long fibre and those on discontinuously reinforced systems (see Fig. 7.31). For the Al–Ni (eutectic) long fibre system[100], the fracture strain increases linearly with pressure until a plateau is reached at around $\sigma_H \sim 350$ MPa ($\varepsilon_{C3*} \approx 30\%$). This is indicative of a change in mechanism from dilatation-dependent microvoiding and growth (Fig. 7.32(a)), to the operation of a shearing-related mechanism (Fig. 7.32(c)). For discontinuously reinforced Al-based systems, no fall-off in the pressure dependency of fracture strain has been observed, at least up to a pressure of 800 MPa[100–102] (Fig. 7.31(a)). Sectioning and fractography suggest that brittle fracture of the particles initiates voiding, but that catastrophic

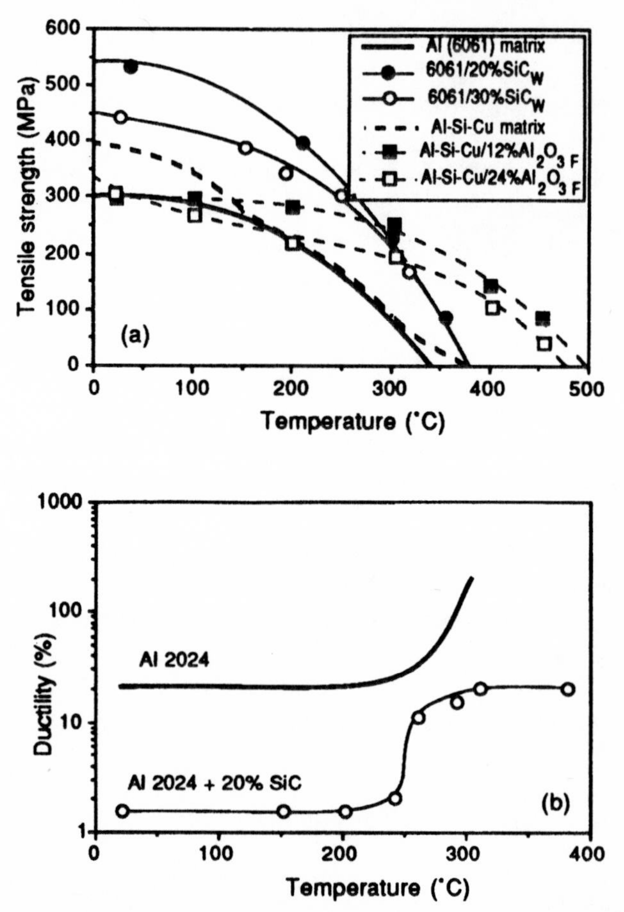

Fig. 7.30 Variations with temperature of (a) ultimate tensile stress[36,96,97] and (b) ductility[98] of reinforced and unreinforced Al. It is interesting to note that the aligned SiC_W (aspect ratio ~ 5) gives the highest degree of strengthening at low to modest temperatures, whereas the longer planar randomly oriented alumina fibres give the best high temperature strength.

propagation can be inhibited by pressure (see Fig. 7.31(c)). For whisker-based systems, Vasudevan *et al.*[100] noted that 5 μm shear bands are observed, in which the whiskers are rotated away from the tensile axis (rather like the long fibres in Fig. 7.32(c)), until a critical point at which shear failure occurs.

An approximate estimate of the retarding effect of the hydrostatic pressure on void growth can be made by re-examining the calculation in §7.2.2 (eqn (7.18)) and simply superposing the external hydrostatic

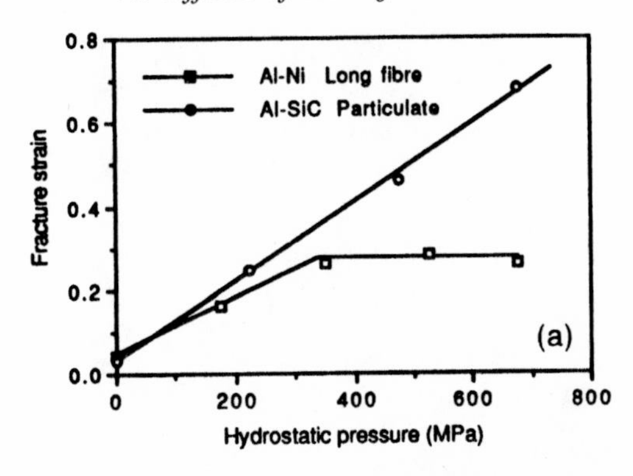

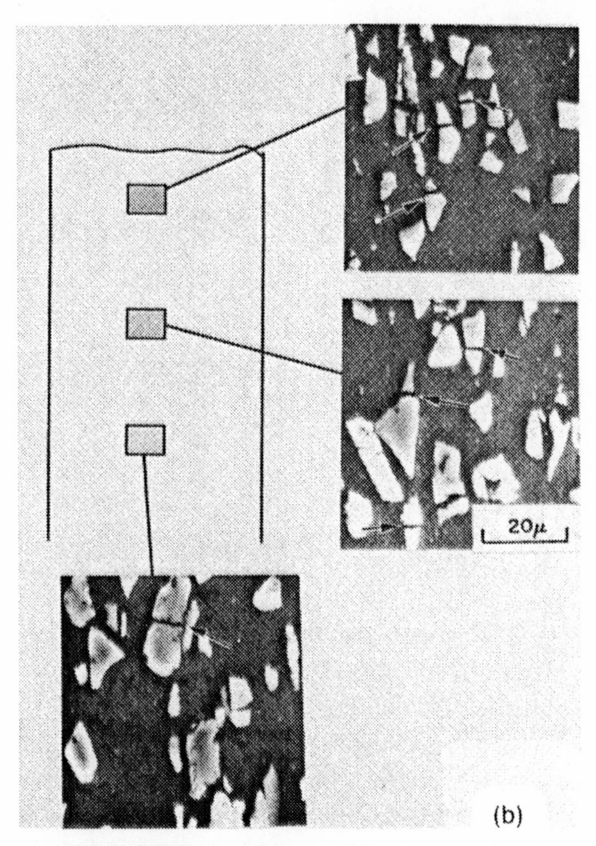

Fig. 7.31 (a) Data showing the influence of pressure on fracture strain for long fibre and discontinuously reinforced systems[100–102]. The suppression of void growth under hydrostatic pressure is illustrated by the inward flow of metal into the elongated voids, seen here for both (b) particulate[101] and (c) [*overleaf*] fibre reinforced systems.[102]

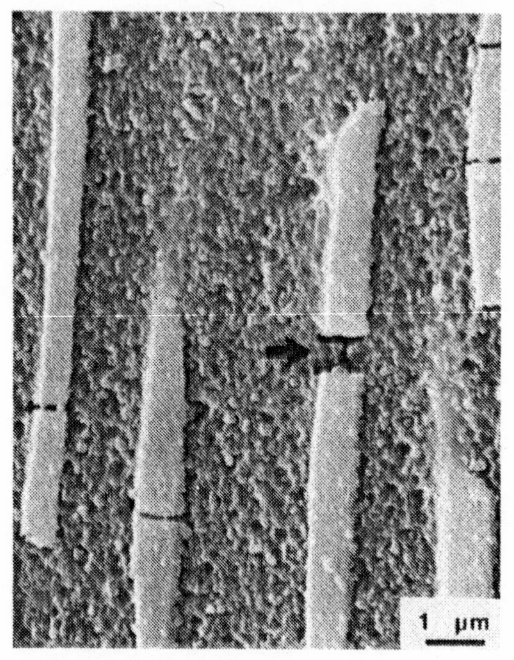

Fig. 7.31 (*cont.*) (c)

pressure (e.g. 800 MPa) directly†

$$3\frac{\dot{d}_0}{d_0} \approx 0.85\dot{\varepsilon}\exp\left(\frac{450 - 3 \times 800}{2 \times 300}\right)$$

$$\approx 0.03\dot{\varepsilon} \qquad (7.19d)$$

so that the rate of growth of a void in the matrix is reduced by a factor of about 60 (cf. eqn (7.19(a)) in §7.2.2) by imposing this pressure.

Stress–strain response

A number of features of the stress–strain response have proven to be pressure-dependent. Increases in the elastic modulus of Al–SiC_W have been reported at high pressure levels[103]. Significant improvements have also been observed in the stress–strain response of Al/SiC composites[101]. This improvement, which is not observed for the unreinforced alloy, is made up of two distinct contributions. A modest increase in the UTS (~20%) is observed, but a larger increase is produced in proof

† In fact the hydrostatic component at the whisker corner would be increased by more than 800 MPa.

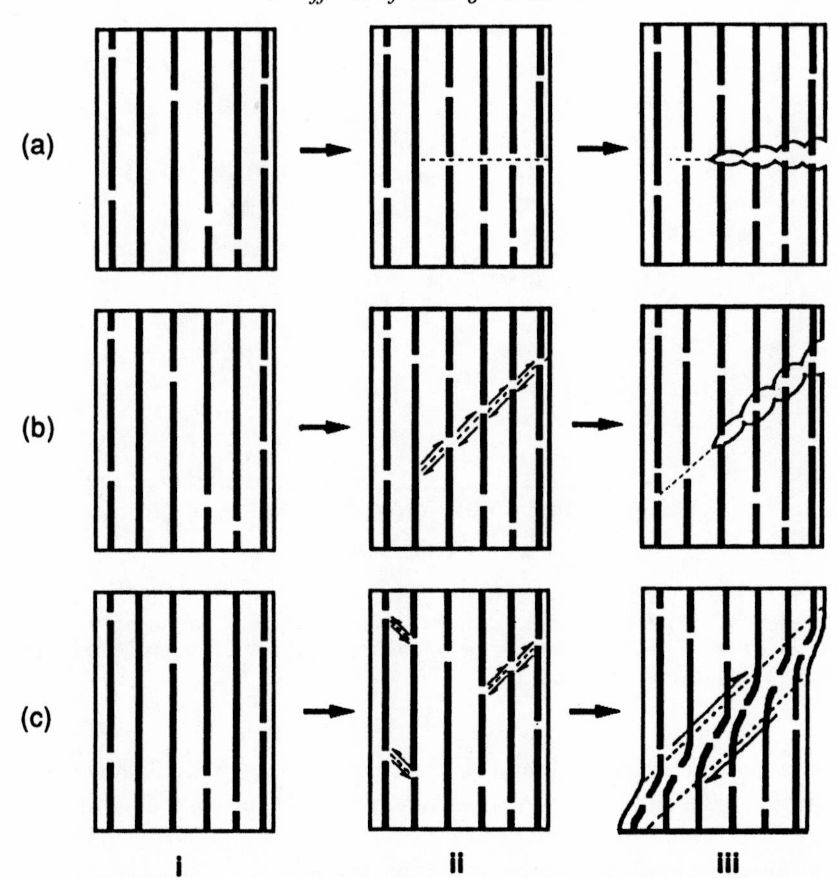

Fig. 7.32 Schematic depiction of the sequence of damage events leading to failure in long fibre composites[102]. In (a), the stress is sufficiently large for fibre fracture to occur and composite failure proceeds by linkage through the matrix. In (b), the matrix yield stress is low so that strain localisation occurs, giving rise to a 45° crack. In (c), the tensile fracture of the matrix is inhibited and the softening effect of damage results in strain localisation aided by rotation of the fibres to form bands. When a compressive hydrostatic stress is superposed, mechanism (c) is favoured.

stress (100%). This is effected merely by pre-pressurising at 300 MPa and subsequently testing at atmospheric pressure. This effect probably originates from the stimulation of dislocations in the matrix near the stiff particles[104]. These act both to reduce the magnitude of the thermal stresses and to work-harden the matrix.

For tests actually carried out under pressure, a significant increase in composite work-hardening is observed with increasing pressure[100]. A

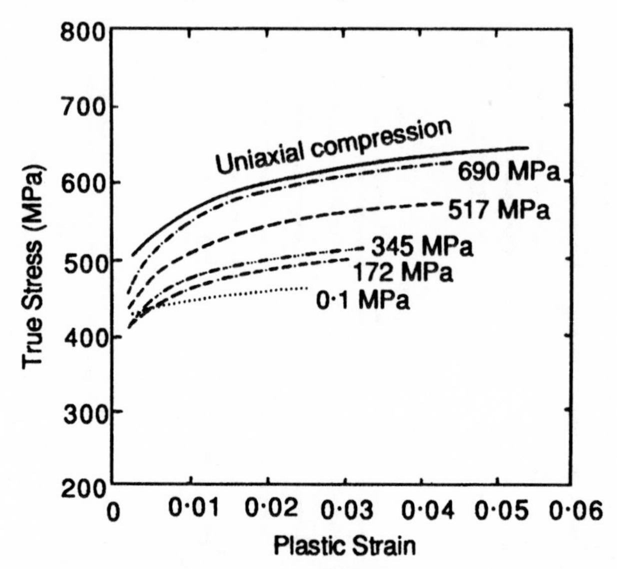

Fig. 7.33 With increasing hydrostatic pressure the tensile flow behaviour of 2014 Al–20 vol% SiC_P approaches the uniaxial compressive response[100]. This suggests that, if void growth can be suppressed, substantial improvements in tensile strength could be achieved.

comparison between tensile and compressive straining (Fig. 7.33) indicates that this arises through an increase in load transfer, brought about by an inhibition of damage accumulation. At large strains, damage can occur, reducing the work-hardening rate, but giving rise to high ductilities because of the decelerated rate of void growth.

References

1. A. Kelly (1966) *Strong Solids*, Oxford University Press.
2. Y. Kawaga and B. H. Choi (1986) in *Composites '86: Recent Advances in Japan and the United States*, K. Kawata (ed.), Jap. Soc. Comp. Mats., pp. 537–43.
3. K. M. Prewo and K. R. Krieder (1972) The Transverse Tensile Properties of Boron Fibre Reinforced Aluminium Matrix Composites, *Metall. Trans.*, **3**, pp. 2201–11.
4. T. Kyono, I. W. Hall and M. Taya (1986) The Effect of Isothermal Exposure on the Transverse Properties of a Continuous Fibre Metal Matrix Composite, *J. Mat. Sci.*, **21**, pp. 4269–80.
5. C. T. Lynch and J. P. Kershaw (1972) *Metal Matrix Composites*, CRC Press, Cleveland, Ohio.
6. D. F. Adams and D. R. Doner (1967) Longitudinal Shear Loading of a Unidirectional Composite, *J. Comp. Mat.*, **1**, pp. 4–17.

7. P. W. Jackson and D. Cratchley (1966) The Effect of Fibre Orientation on the Tensile Strength of Fibre-Reinforced Metals, *J. Mech. Phys. Sol.*, **14**, pp. 49–64.
8. M. J. Pindera (1989) Shear Testing of Fiber Reinforced Metal Matrix Composites, in *Metal Matrix Composites: Testing, Analysis and Failure Modes*, W. S. Johnson (ed.), ASTM STP 1032, Philadelphia, pp. 19–42.
9. R. Hill (1950) *The Mathematical Theory of Plasticity*, Oxford University Press, London.
10. S. W. Tsai (1968) Strength Theories of Filamentary Structures, in *Fundamental Aspects of Fibre Reinforced Plastic Composites*, R. T. Schwartz and H. S. Schwartz (eds.), Wiley, New York, pp. 3–11.
11. R. M. Jones (1975) *Mechanics of Composite Materials*, McGraw-Hill, New York.
12. P. J. Withers, W. M. Stobbs and A. J. Bourdillon (1988) Various TEM Methods for the Study of MMCs, *J. Micros.*, **151**, pp. 159–69.
13. A. F. Whitehouse, R. A. Shahani and T. W. Clyne (1991) Cavitation during Tensile Deformation of Powder Route Particle-Reinforced Aluminium, in *Metal Matrix Composites: Processing, Microstructure and Properties, 12th Risø Int. Symp.*, Roskilde, N. Hansen, D. J. Jensen, T. Leffers, H. Lilholt, T. Lorentzen, A. S. Pedersen, O. B. Pedersen and B. Ralph (eds.), Risø Nat. Lab., Denmark, pp. 741–8.
14. D. Lloyd (1991) Aspects of Particle Fracture in Particulate Reinforced MMCs, *Acta Met. et Mat.*, **39**, pp. 59–72.
15. M. Manoharan and J. J. Lewandowski (1990) Crack Initiation and Growth Toughness of an Al MMC, *Acta Met. et Mat.*, **38**, pp. 489–96.
16. J. R. Fisher and J. Gurland (1981) Void Nucleation in Spheroidized Carbon Steels. Part 1: Experimental, *Metal Sci.*, **15**, pp. 185–202.
17. G. L. Roy, J. D. Embury, G. Edwards and M. F. Ashby (1981) A Model of Ductile Fracture Based on the Nucleation and Growth of Voids, *Acta Metall.*, **29**, pp. 1509–22.
18. A. S. Argon, J. Im and R. Safoglu (1975) Cavity Formation from Inclusions in Ductile Fracture, *Metall. Trans.*, **6A**, pp. 825–37.
19. K. Tanaka, T. Mori and T. Nakamura (1970) Cavity Formation at the Interface of a Spherical Inclusion in a Plastically Deformed Matrix, *Phil. Mag.*, **21**, pp. 267–79.
20. L. M. Brown and W. M. Stobbs (1976) The Work-Hardening of Cu–SiO_2 V. Equilibrium Plastic Relaxation by Secondary Dislocations, *Phil. Mag.*, **34**, pp. 351–72.
21. S. H. Goods and L. M. Brown (1979) The Nucleation of Cavities by Plastic Deformation, *Acta Metall.*, **27**, pp. 1–15.
22. C. Johnson, K. Ono and D. Chellman (1985) Acoustic Emission Behaviour of MMCs, in *Proc. 2nd Int. Conf. on AE*, Nevada, USA, pp. S263–S268.
23. P. M. Mummery, B. Derby, D. J. Buttle and C. B. Scruby (1991) Micromechanisms of Fracture in Particle-Reinforced MMCs: Acoustic Emission and Modulus Retention, in *Proc. 2nd European Conf. on Adv. Mats., Euromat '91*, Cambridge, UK, T. W. Clyne and P. J. Withers (eds.), Inst. of Metals, pp. 441–7.
24. J. D. Evensen and A. S. Verk (1981) The Influence of Particle Cracking on the Fracture Strain of some Al–Si Alloys, *Scripta Met.*, **15**, pp. 1131–3.
25. L. M. Brown and J. D. Embury (1973) The Initiation and Growth of Voids at Second Phase Particles, in *Proc. ICSMA III*, Cambridge, UK, pp. 164–8.

26. P. F. Thomason (1968) A Theory for Ductile Fracture by Internal Necking of Cavities, *J. Inst. Metals*, **96**, pp. 360–4.
27. A. F. Whitehouse and T. W. Clyne (1992) Effects of Reinforcement Content and Shape on Cavitation and Failure in Metal Matrix Composites, *to be published* in *Composites*.
28. A. F. Whitehouse and T. W. Clyne (1992) Cavity Formation during Tensile Straining of Particulate and Short Fibre MMCs, *submitted* to *Acta Met. et Mat.*
29. Y. Flom and R. J. Arsenault (1986) Interfacial Bond Strength in an Al Alloy 6061–SiC Composite, *Mat. Sci. & Eng.*, **A77**, pp. 191–7.
30. T. J. Warner (1989) *Mechanics of Load Transfer in MMCs*, PhD thesis, Univ. of Cambridge.
31. J. R. Rice and D. M. Tracey (1969) On the Ductile Enlargement of Voids in Triaxial Stress Fields, *J. Mech. Phys. Solids*, **17**, pp. 201–17.
32. J. F. Knott (1978) *Fundamentals of Fracture*, Butterworth, London.
33. T. Christman, A. Needleman and S. Suresh (1989) An Experimental and Numerical Study of Deformation in MMCs, *Acta Metall.*, **37**, pp. 3029–50.
34. J. K. Shang, W. Yu and R. O. Ritchie (1988) Role of SiC Particles in Fatigue Crack Growth SiC-Particulate-Reinforced Al Alloy Composites, *Mat. Sci. & Eng.*, **A102**, pp. 181–92.
35. J. J. Lewandowski, C. Liu and W. H. Hunt (1989) Effects of Matrix Microstructure and Particle Distribution on Fracture of an Al MMC, *Mat. Sci. & Eng.*, **A107**, pp. 241–55.
36. D. L. McDanels (1985) Analysis of Stress–Strain, Fracture, and Ductility of Aluminium Matrix Composites Containing Discontinuous Silicon Carbide Reinforcement, *Metall. Trans.*, **16A**, pp. 1105–15.
37. Y. Brechet, J. D. Embury, S. Tao and L. Luo (1991) Damage Initiation in MMCs, *Acta Met. at Mat.*, **39**, pp. 1781–6.
38. A. A. Griffith (1920) The Phenomena of Rupture and Flow in Solids, *Phil. Trans. R. Soc.*, **A221**, pp. 163–97.
39. G. R. Irwin (1948) Fracture Dynamics, in *Fracturing of Metals*, ASM, pp. 147–66.
40. J. R. Rice and M. A. Johnson (1970) The Role of Large Crack Tip Geometry Changes in Plane Strain Fracture, in *Inelastic Behaviour of Solids*, M. F. Kanninen, W. F. Adler, A. R. Rosenfield and R. I. Jaffee (eds.), McGraw-Hill, New York, pp. 641–72.
41. R. M. McMeeking (1977) Finite Deformation Analysis of Crack Tip Opening in Elastic Plastic Materials and Implications for Fracture, *J. Mech. Phys. Solids*, **25**, pp. 357–81.
42. G. T. Hahn and A. R. Rosenfield (1975) Metallurgical Factors Affecting Fracture Toughness of Al Alloys, *Metall. Trans.*, **6A**, pp. 653–70.
43. Y. Flom and R. J. Arsenault (1989) Effect of Particle Size on Fracture Toughness of SiC/Al Composite Material, *Acta Metall.*, **37**, pp. 2413–23.
44. S. Kamat, J. P. Hirth and R. Mehrabian (1989) Mechanical Properties of Particulate Reinforced Al-Matrix Composites, *Acta Metall.*, **37**, pp. 2395–402.
45. R. O. Ritchie and A. W. Thompson (1985) On Macroscopic and Microscopic Analyses for Crack Initiation and Crack Growth Toughness in Ductile Alloys, *Metall. Trans.*, **16A**, pp. 233–48.
46. A. W. Thompson and M. F. Ashby (1984) Fracture Surface Micro-Roughness, *Scripta Met.*, **18**, pp. 127–30.

47. C. R. Crowe, R. A. Gray and D. F. Hasson (1985) Microstructure Controlled Fracture Toughness of SiC/Al Metal Matrix Composites, in *Proc. 5th Int. Conf. Comp. Mats.* (*ICCM V*), San Diego, W. C. Harrigan, J. Strife and A. K. Dhingra (eds.), TMS-AIME, pp. 843–66.
48. F. Vescera, J. P. Keustermans, M. A. Dellis, B. Lips and F. Delannay (1991) Processing and Properties of Zn–Al Alloy Matrix Composites Reinforced by Bidirectional Carbon Tissues, in *Metal Matrix Composites – Processing, Microstructure and Properties, 12th Risø Int. Symp.*, Roskilde, N. Hansen, D. J. Jensen, T. Leffers, H. Lilholt, T. Lorentzen, A. S. Pedersen, O. B. Pedersen and B. Ralph (eds.), Risø Nat. Lab., Denmark, pp. 719–24.
49. D. L. Davidson (1987) Fracture Characteristics of Al–4% Mg Mechanically Alloyed with SiC, *Metall. Trans.*, **18A**, pp. 2115–28.
50. D. L. Davidson (1989) Fracture Surface Roughness as a Gauge of Fracture Toughness: Al-Particulate SiC Composites, *J. Mat. Sci.*, **24**, pp. 681–7.
51. P. Paris and F. Erdogan (1963) A Critical Analysis of Crack Propagation Laws, *J. Basic Engng*, **85**, pp. 528–34.
52. G. J. Dvorak and W. S. Johnson (1980) Fracture of Metal Matrix Composites, *Int. J. Fract.*, **16**, pp. 585–602.
53. G. J. Dvorak and W. S. Johnson (1980) Fatigue Damage Mechanisms in Boron–Aluminium Composite Laminates, in *Advances in Composite Materials*, A. R. Bunsell, C. Bathias, A. Martrenchar, D. Menkes and G. Vercher (eds.), Pergamon, pp. 1117–90.
54. N. Tsangarakis, J. M. Slepetz and J. Nunes (1985) Fatigue Behaviour of Alumina Fiber Reinforced Aluminium Composites, in *Recent Advances in Composites in the United States and Japan*, J. R. Vinson and M. Taya (eds.), ASTM STP 864, Philadelphia, pp. 131–52.
55. J. Nunes, E. S. C. Chin, J. M. Slepetz and N. Tsangarakis (1985) Tensile and Fatigue Behavior of Alumina Fiber Reinforced Magnesium Composites, in *Proc. 5th Int. Conf. on Comp. Mats.* (*ICCM V*), San Diego, W. Harrigan, J. Strife and A. K. Dhingra (eds.), TMS-AIME, pp. 723–45.
56. G. J. Dvorak and J. Q. Tar (1980) Fatigue and Shakedown in Metal Matrix Composites, in *Fatigue of Composite Materials*, J. R. Hancock (ed.), ASTM STP 569, pp. 145–66.
57. W. S. Johnson (1980) Modelling Stiffness Loss in Boron–Aluminium Laminates Below the Fatigue Limit, in *Long Term Behavior of Composites*, T. K. O'Brien (ed.), ASTM STP 813, pp. 160–78.
58. J. L. Rossi, R. Pilkington and R. L. Trumper (1991) Fatigue Damage in a Fibre Reinforced Alloy, in *Proc. 2nd European Conf. on Adv. Mats., Euromat '91*, Cambridge, UK, T. W. Clyne and P. J. Withers (eds.), Inst. of Metals, pp. 448–58.
59. M. D. Sensmeier and P. K. Wright (1990) The Effect of Fiber Bridging on Fatigue Crack Growth in Titanium Matrix Composites, in *Fundamental Relationships Between Microstructures and Mechanical Properties of Metal-Matrix Composites*, P. K. Liaw and M. N. Gungor (eds.), TMS, pp. 441–57.
60. A. R. Ibbotson, C. J. Beevers and P. Bowen (1991) Fatigue Crack Growth in Fibre Reinforced Metal Matrix Composites, in *Proc. 2nd European Conf. on Adv. Mats., Euromat '91*, Cambridge, UK, T. W. Clyne and P. J. Withers (eds.), Inst. of Metals, pp. 469–78.
61. D. B. Marshall, B. N. Cox and A. G. Evans (1985) The Mechanics of Matrix Cracking in a Brittle-Matrix Fiber Composite, *Acta Metall.*, **33**, pp. 2013–21.

62. K. R. Bain and M. L. Gambone (1990) Fatigue Crack Growth of SCS-6/Ti-64 Metal Matrix Composite, in *Fundamental Relationships Between Microstructures and Mechanical Properties of Metal–Matrix Composites*, P. K. Liaw and M. N. Gungor (eds.), TMS, pp. 459–69.
63. A. J. Padkin, M. F. Boreton and W. J. Plumbridge (1987) Fatigue Crack Growth in Two-Phase Alloys, *Mat. Sci. & Tech.*, **3**, pp. 217–23.
64. I. Sinclair and J. F. Knott (1990) Fatigue Crack Growth in Al–Li/SiC Particulate MMCs, in *Proc. Europ. Conf. Frac.* (*ECF 8*), Turin, D. Firrao (ed.), pp. 303–9.
65. D. R. Williams and M. E. Fine (1985) Quantitative Determination of Fatigue Microcrack Growth in SiC_W Reinforced 2124 Al Alloy Composite, in *Proc. 5th Int. Conf. on Comp. Mats.* (*ICCM V*), San Diego, W. C. Harrigan, J. Strife and A. Dhingra (eds.), TMS-AIME, pp. 639–70.
66. T. Christman and S. Suresh (1988) Effects of SiC Reinforcement and Aging Treatment on Fatigue Crack Growth in an Al–SiC Composite, *Mat. Sci. & Eng.*, **102A**, pp. 211–16.
67. D. L. Davidson (1989) The Growth of Fatigue Cracks through Al Alloy–SiC_P, in *Proc. Int. Conf. Frac.* (*ICF7*), Texas, K. Salama, K. Ravi-Chandar, D. M. R. Taplin and P. R. Rao (eds.), pp. 3021–8.
68. S. S. Yau (1983) PhD thesis, North Carolina State Univ.
69. C. R. Crowe and D. F. Hasson (1982) Corrosion Fatigue of SiC/Al MMC in Salt Laden Moist Air, in *Proc. of ICSMA VI*, Melbourne, R. C. Gifkins (ed.), Pergamon, pp. 859–65.
70. D. F. Hasson, C. R. Crowe, J. S. Ahearn and D. C. Cocke (1985) Fatigue and Corrosion Fatigue of SiC/Al MMCs, in *Failure Mech. of High Perf. Mats.*, Cambridge, MA, J. G. Early (ed.), Cambridge Univ. Press, pp. 147–56.
71. C. R. Crowe (1985) Fatigue and Corrosion Fatigue of MMCs, in *Proc. 6th Metal Matrix Composites Tech. Conf.* (*MMCTC*), Santa Barbara, CA, pp. 19.1–19.13.
72. D. F. Hasson and C. R. Crowe (1987) Flexural Fatigue Behaviour of Aramid Reinforced Al 7075 Laminate and Al 7075 Alloy Sheet in Air and in Salt Laden Humid Air, in *Proc. ICCM VI/ECCM 2*, London, F. L. Matthews, N. C. R. Buskell, J. M. Hodgkinson and J. Morton (eds.), Elsevier, pp. 138–45.
73. J. Albrecht, A. W. Thompson and I. M. Bernstein (1979) The Role of Microstructure in Hydrogen-Assisted Fracture of 7075 Aluminium, *Metall. Trans.*, **10A**, pp. 1759–66.
74. D. Nguyen, A. W. Thompson and I. M. Bernstein (1987) Microstructural Effects on Hydrogen Embrittlement in a High Purity 7075 Aluminium Alloy, *Acta Metall.*, **35**, pp. 2417–25.
75. I. M. Bernstein and M. Dollar (1990) The Effect of Trapping on Hydrogen-Induced Plasticity and Fracture in Structural Alloys, in *Hydrogen Effects on Mat. Behaviour*, N. R. Moody and A. W. Thompson (eds.), TMS, pp. 703–15.
76. T. J. Downes, D. M. Knowles and J. E. King (1991) The Effect of Particle Size on Fatigue Crack Growth in an Al-Based MMC, in *Fatigue of Advanced Materials*, Santa Barbara, CA, R. O. Ritchie, R. H. Dauskardt and B. N. Cox (eds.), Mat. & Component Pub. Ltd, Birmingham, pp. 395–407.
77. S. J. Harris and T. E. Wilks (1987) Fatigue Crack Growth in Fibre Reinforced Al, in *Proc. ICCM VI/ECCM 2*, London, F. L. Matthews, N. C. R. Buskell, J. M. Hodgkinson and J. Morton (eds.), Elsevier, pp. 113–27.

78. L. Dignard-Bailey, S. Dionne and S. H. Lo (1989) The Fracture Behaviour of Squeeze-Cast Al203/Zn–Al Composites, in *Fundamental Relationships between Microstructure and Mechanical Properties of MMCs*, Indianapolis, P. K. Liaw and M. N. Gungor (eds.), TMS, pp. 23–35.
79. S. Kumai, J. E. King and J. F. Knott (1990) Short and Long Fatigue Crack Growth in a SiC Reinforced Al Alloy, *Fat. Fract. Engng Mat. Struct.*, **13**, pp. 511–24.
80. C. Liu, S. Pape and J. J. Lewandowski (1988) Effects of Matrix Microstructure and Interfaces on Influencing Monotonic Crack Propagation in SiC/Al Alloy Composites, in *Interfaces in Polymers, Ceramics, and MMCs*, Cleveland, USA, H. Ishida (ed.), Elsevier, pp. 513–24.
81. W. A. Logsdon and P. K. Liaw (1986) Tensile, Fracture Toughness and Fatigue Crack Growth Rate Properties of SiC_W and SiC_P Reinforced Al MMCs, *Eng. Fract. Mech.*, **24**, pp. 737–51.
82. Y. Flom and R. J. Arsenault (1987) Fracture of SiC/Al Composites, in *Proc. ICCM VI/ECCM 2*, London, F. L. Matthews, N. C. R. Buskell, J. M. Hodgkinson and J. Morton (eds.), Elsevier, pp. 189–98.
83. W. H. Hunt (1988) *A Perspective on the Development of MMCs*, Detroit section of TMS (Available from Al Co. of America).
84. T. J. Downes and J. E. King (1991) The Effect of SiC Particle Size on the Fracture Toughness of a MMC, in *Metal Matrix Composites – Processing, Microstructure and Properties, Risø 12th Int. Symp.*, Roskilde, N. Hansen, D. J. Jensen, T. Leffers, H. Lilholt, T. Lorentzen, A. S. Pedersen, O. B. Pedersen and B. Ralph (eds.), Risø Nat. Lab., Denmark, pp. 305–10.
85. J. K. Shang and R. O. Ritchie (1989) Mechanisms Associated with Near Threshold Fatigue Crack Propagation in SiC_P Reinforced Al Composites, in *Proc. ICCM VII*, Guangzou, W. Yunshu, G. Zhenlong and W. Renjie (eds.), Pergamon, New York, pp. 590–4.
86. H. F. Fischmeister, E. Navara and K. E. Easterling (1972) Effects of Alloying on Structural Stability and Cohesion between Phases in Oxide/Metal Composites, *Met. Sci. J.*, **6**, pp. 211–15.
87. W. A. Spitzig, J. F. Kelly and O. Richmond (1985) Quantitative Characterisation of Second Phase Populations, *Metallog.*, **18**, pp. 235–61.
88. A. J. Reeves, H. Dunlop and T. W. Clyne (1991) The Effect of Interfacial Reaction Layer Thickness on Fracture of Ti–SiC_P Composites, *Metall. Trans.*, **23**, pp. 970–81.
89. R. DaSilva, D. Caldemaison and T. Bretheau (1989) Interface Strength Influence on the Mechanical Behaviour of Al–SiCP MMCs, in *Interfacial Phenomena in Composite Materials '89*, Sheffield, F. R. Jones (ed.), Butterworth, pp. 235–41.
90. W. H. Hunt, O. Richmond and R. D. Young (1987) Fracture Initiation in Particle Hardened Materials with High Volume Fraction, in *Proc. ICCM VI/ECCM 2*, London, F. L. Matthews, N. C. R. Buskell, J. M. Hodgkinson and J. Morton (eds.), Elsevier, pp. 209–23.
91. T. F. Klimowicz and K. S. Vecchio (1989) The Influence of Aging Condition on the Fracture Toughness of Al_2O_3-Reinforced Al Composites, in *Fundamental Relationships between Microstructure and Mechanical Properties of MMCs*, Indianapolis, P. K. Liaw and M. N. Gungor (eds.), TMS, pp. 255–67.

92. H. R. Rack and J. W. Mullins (1985) Tensile and Notch Behaviour of SiC_W–2124 Al, in *High Performance Powder Aluminium Alloys – II*, G. Hildeman and M. Koczak (eds.), TMS, Warrendale, Pa, pp. 155–71.
93. S. M. Pickard, B. Derby, J. Harding and M. Taya (1988) Strain Rate Dependence of Failure in 2124 Al/SiC Whisker Composite, *Scripta Met.*, **22**, pp. 601–6.
94. K. Tsuchiya, J. R. Weertman and M. J. Luton (1989) Stress–Strain Behaviour and Creep Properties of Mechanically Alloyed Al–Al_2O_3 Alloys, in *Fundamental Relationships between Microstructure and Mechanical Properties of MMCs*, Indianapolis, P. K. Liaw and M. N. Gungor (eds.), TMS, pp. 565–80.
95. T. Kobayashi and H. Iwanari (1990) Evaluations of Toughness and Mechanical Properties, in *Metal and Cer. Comp., Processing, Modeling and Mech. Behav.*, R. H. Bhagat (ed.), TMS, pp. 227–34.
96. J. Dinwoodie, E. Moore, C. A. J. Landman and W. R. Symes (1985) The Properties and Applications of Short Staple Al_2O_3 Fibre Reinforced Al Alloys, in *Proc. ICCM V*, San Diego, W. C. Harrigan, J. Strife and A. Dhingra (eds.), TMS-AIME, pp. 671–85.
97. A. Sakamoto, H. Hasegawa and Y. Minoda (1985) Mechanical Properties of SiC_W Reinforced Al Composites, in *Proc. ICCM V*, San Diego, W. C. Harrigan, H. Strife and A. Dhingra (eds.), TMS-AIME, pp. 699–707.
98. W. L. Phillips (1978) Elevated Temperature Properties of SiC_W Reinforced Al, in *Proc. ICCM II*, B. Noton, R. Signorelli, K. Street and L. Phillips (eds.), TMS-AIME, Warrendale, Pa, pp. 567–76.
99. J. P. Auger and D. Francois (1977) Variation of Fracture Toughness of a 7075 Al Alloy, *Int. J. Fract.*, **13**, pp. 431–41.
100. A. K. Vasudevan, O. Richmond, F. Zok and J. D. Embury (1989) The Influence of Hydrostatic Pressure on the Ductility of Al–SiC Composites, *Mat. Sci. & Eng.*, **A107**, pp. 63–9.
101. D. S. Liu, B. I. Rickett and J. J. Lewandowski (1989) Effects of Low Levels of Superimposed Hydrostatic Pressure on the Mechanical Behaviour of Al Matrix Composites, in *Fundamental Relationships between Microstructure and Mechanical Properties of MMCs*, Indianapolis, P. K. Liaw and M. N. Gungor (eds.), TMS, pp. 145–58.
102. F. Zok, J. D. Embury, M. F. Ashby and O. Richmond (1988) The Influence of Pressure on Damage Evolution and Fracture in MMCs, in *Mech. and Physical Behav. of Met. and Cer. Composites, Proc. 9th Risø Int. Symp.*, Roskilde, S. I. Anderson, H. Lilholt and O. B. Pedersen (ed.), Risø Nat. Lab., Denmark, pp. 517–26.
103. D. P. Dandekar, J. Frankel and W. J. Korman (1988) Pressure Dependence of the Elastic Constants of SiC/2014 Al Composite, in *Testing Tech. of MMCs*, ASTM, Philadelphia, pp. 79–89.
104. G. Das and S. V. Radcliffe (1969) Pressure-Induced Development of Dislocations at Elastic Discontinuities, *Phil. Mag.*, **20**, pp. 589–609.

8

Transport properties and environmental performance

The previous four chapters have covered the basic deformation and failure behaviour of MMCs under an applied load, including the effects of testing temperature and loading rate. There are, however, other aspects of their performance which are often of considerable importance. For example, MMCs offer scope for high electrical and/or thermal conductivity in combination with good mechanical strength. Good thermal shock resistance, particularly when compared with many competing ceramic-based materials, might thus be expected. Surface properties, and the degradations that might occur in abrasive, corrosive or other aggressive environments, may also be important issues in a wider range of applications. These topics are complex and wide-ranging, but the treatments in this chapter are aimed simply at identifying the ways in which the responses of MMCs in various special situations differ from those of other, more conventional materials.

8.1 Thermal and electrical conduction

There are many applications in which the high electrical and/or thermal conductivities of metals are exploited. A problem commonly arises in situations where this needs to be combined with good mechanical properties, in that conventional strengthening by alloying leads to sharp reductions in these conductivities. Since the use of ceramic inclusions can bring about significant strengthening, there is interest in predicting how the conductivity of the composite will vary with reinforcement properties, volume fraction, aspect ratio, interfacial structure etc. Mathematically, this is a rather simpler problem than that of mechanical behaviour, but a very brief introduction to the mechanisms of conduction (covered in detail in standard texts[1,2]) will be useful.

8.1.1 Heat transfer via electrons and phonons

Heat flows within a material by the transmission of phonons (lattice vibrations) and free electrons (if present). Both of these carriers have a certain mean free path λ between collisions (energy exchange events) and an average velocity v. The thermal conductivity K is related to these parameters by a simple equation derived from kinetic theory

$$K = \tfrac{1}{3}cv\lambda \tag{8.1}$$

where c is the volume specific heat of the carrier concerned. Both metals (electron-dominated) and non-metals (phonons only) have zero conductivity at a temperature of 0 K (where c becomes zero in both cases), followed by a sharp rise to a peak and then a gradual fall as the temperature is progressively increased. The rise reflects the increase in c towards a plateau value ($\sim 3Nk$, where N is Avogadro's number and k is Boltzmann's constant, with all the vibrational modes of all the molecules active), while the fall is caused by the decreasing λ as the greater amplitude of lattice vibration causes more scattering of the carriers. The maximum (low temperature) value of λ is dictated by atomic scale defects for electrons, but by the physical dimensions of the specimen for phonons.

The average carrier velocity is insensitive to temperature in both cases. The phonon velocity (speed of sound) is high in light, stiff materials. The mean free path of a phonon is structure-sensitive and can be very large in pure specimens of high perfection and large grain size. Single crystals of materials like diamond and SiC can therefore have very high thermal conductivities. With the exception of such cases, metals have the highest conductivities because electrons usually have a much larger mean free path than phonons. This can, however, be substantially reduced by the presence of solute atoms and various defects which cause electron scattering. These trends are all apparent in the data shown in Table 8.1. Note that there may in principle be scope for the enhancement of thermal conductivity by incorporation of ceramic (e.g. SiC in Ti), although the sensitivity to size and microstructure for non-metals and the possibility of an interfacial barrier means that care should be exercised in examining data such as these.

8.1.2 Modelling of heat flow in MMCs

The basic equation of heat flow may be written

$$q = -KT' \tag{8.2}$$

where q is the heat flux ($W\,m^{-2}$) arising from a thermal gradient T'

Table 8.1 *Thermal conductivity data*[2–7] *for a range of materials*

Material	K (W m^{-1} K^{-1}) at 300 K	at 900 K
Diamond	600	—
Graphite (parallel to c axis)	355	—
Graphite (normal to c axis)	89	—
Ag	425	325
Cu	400	340
Cu–2% Ag	~390	~340
Cu–2% Be	~130	—
Cu–35% Zn	~100	~180
Cu–40% Ni	~20	~40
Al	~220	~180
Al–8Si–3Cu	~175	—
Mild steel	~60	~35
Ti	~20	~14
Ti–6Al–4V	~6	~6
SiC (single crystal)	~100	—
SiC (polycrystal)	~4–20	~1–5
Al_2O_3 (single crystal)	~100	~20
Al_2O_3 (polycrystal)	~5–30	~2–8

(K m^{-1}) in a material of thermal conductivity K (W m^{-1} K^{-1}). It is important to distinguish K from the thermal diffusivity a ($=K/c$), the parameter determining the rate at which a material approaches thermal equilibrium, which appears in the unsteady diffusion equation

$$\frac{\partial T}{\partial t} = aT''$$

The effective conductivity of composite structures and materials has been the subject of prolonged study[8–11]. Many of the treatments developed are either specific to a particular geometrical arrangement of the constituents or are limited to a dispersed constituent being present at low volume fraction, so that the disturbances to the matrix thermal field which they generate do not overlap (dilute composite). These limitations do not apply to the Eshelby method of treating conduction, which is outlined below.

Although the Eshelby method was originally developed specifically for situations arising from a stress-free strain (such as a martensitic transformation – see Chapter 3), the approach is a powerful one which turns out to be applicable to a wide range of problems[12]. Treatment of thermal conduction[13–15] (and the entirely analogous electrical conduction) serves to illustrate this. The problem is mathematically simpler than that for internal stresses, but the physical operations which the mathematical steps represent are less obvious. Nevertheless, a visualisation such as that shown in Fig. 8.1 can be helpful. This shows isotherms in the (a) real and (b) equivalent homogeneous systems, for a case where the conductivity of the inclusion is greater than that of the matrix. The diagram is analogous to Fig. 3.5, and the correspondence between the governing equations can be seen on writing them in tensor form

$$q_i = -K_{ij}T'_j \tag{8.2a}$$

$$\sigma_{ij} = C_{ijkl}\varepsilon_{kl} \tag{8.2b}$$

Attention is now concentrated on thermal gradients (i.e. the spacings between the isotherms shown in Fig. 8.1), rather than on strains.

Here the equivalent of the 'stress-free strain' in Chapter 3 is a uniform thermal gradient generated in the inclusion. This is the '***transformation thermal gradient***', T'^{T}. It can be imagined as being due to a set of distributed heat sources and sinks within the inclusion, but it is important to recognise that there is no heat flow down this thermal gradient. (This is analogous to a strain produced otherwise than by a stress.) No physical phenomenon corresponding to this (analogous to a martensitic transformation or a temperature change for a stress-free strain) comes readily to mind[†]. When the inclusion is now replaced in the matrix, a distribution of thermal gradients (i.e. a temperature field) results.

It only remains now to superimpose these thermal gradients with the uniform gradient associated with the applied heat flux for this (thermally) homogeneous composite, to give the same distributions of thermal gradient and heat flux as in the real composite. Derivation of the expression for the conductivity of the composite now closely mirrors the stiffness treatment and the corresponding equation numbers in Chapter 3 are given below to emphasise the analogy.

$$q_{\mathrm{I}} = K_{\mathrm{M}}(T'^{\mathrm{C}} - T'^{\mathrm{T}}) \qquad (8.3) \quad \textit{cf.}\ (3.1)$$

$$T'^{\mathrm{C}} = ST'^{\mathrm{T}} \qquad (8.4) \quad \textit{cf.}\ (3.2)$$

[†] A near analogue might be a distribution of dopant along the length of a semiconducting inclusion, giving a gradient of Fermi level for the electrical conduction case.

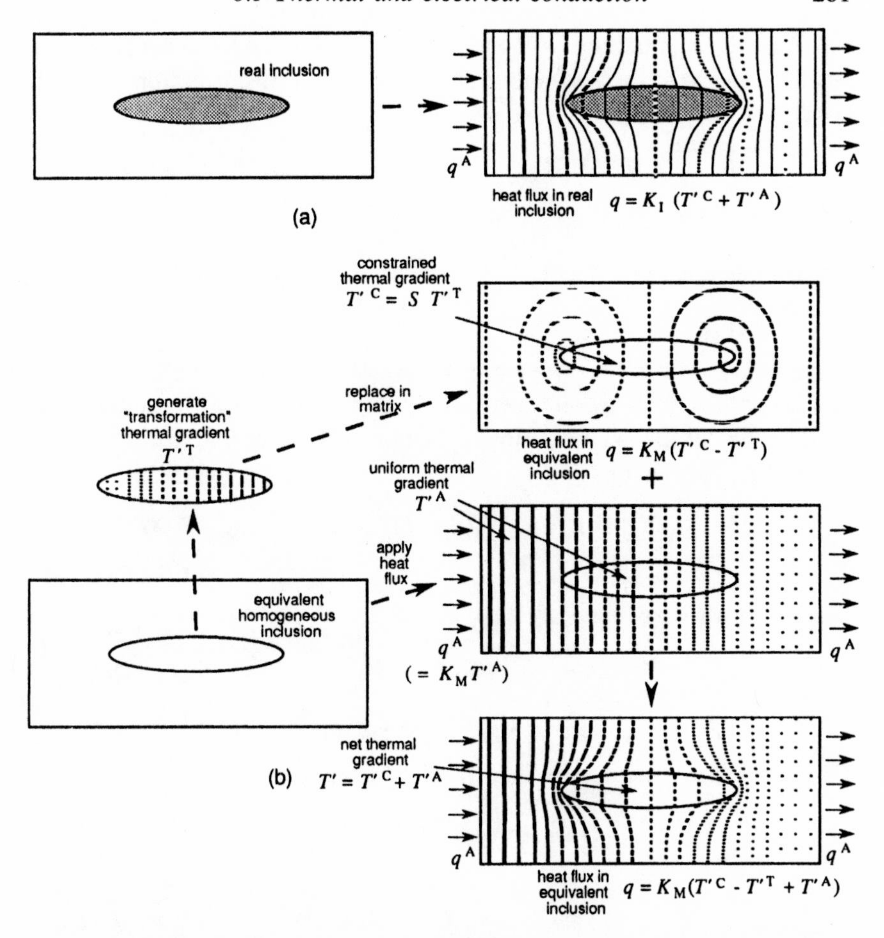

Fig. 8.1 Schematic illustration[15] of the Eshelby method for heat conduction, showing thermal fields for (a) the real composite ($K_I > K_M$) and (b) the equivalent homogeneous inclusion, with and without an applied heat flux. Thermal gradients can be deduced from the spacings of the isotherms, the line density of which increases from right to left, indicating increasing temperature. Note that the thermal gradient is uniform in the inclusion, but variable in the matrix, as for the strain distribution in Chapter 3.

where S is the Eshelby tensor, which is here a second rank tensor (Appendix III), dependent only on inclusion aspect ratio

$$q_I = K_M(S - I)T'^T \qquad (8.5) \quad \textit{cf.}\ (3.3)$$

$$q_I + q_A = K_M(S - I)T'^T + K_M T'^A \qquad (8.6) \quad \textit{cf.}\ (3.9)$$

$$q_I + q_A = K_I(T'^C + T'^A) \qquad (8.7) \quad \textit{cf.}\ (3.12)$$

Equating the expressions for the heat flux in equivalent homogeneous and real inclusions, eqns (8.6) and (8.7), allows the transformation thermal gradient to be evaluated

$$T'^{\mathrm{T}} = -[(K_{\mathrm{I}} - K_{\mathrm{M}})S + K_{\mathrm{M}}]^{-1}(K_{\mathrm{I}} - K_{\mathrm{M}})T'^{\mathrm{A}} \quad (8.8) \quad \textit{cf.}\ (3.13)$$

so that the heat flux in the inclusion is given by

$$q_{\mathrm{I}} + q^{\mathrm{A}} = -K_{\mathrm{M}}(S - I)[(K_{\mathrm{I}} - K_{\mathrm{M}})S + K_{\mathrm{M}}]^{-1}(K_{\mathrm{I}} - K_{\mathrm{M}})T'^{\mathrm{A}} + K_{\mathrm{M}}T'^{\mathrm{A}} \quad (8.9) \quad \textit{cf.}\ (3.14)$$

The non-dilute case is treated in an exactly analogous manner to that of §3.5, with a 'background heat flux' being generated in the matrix by the presence of the neighbouring inclusions. It follows that, for a volume fraction f of inclusions

$$T'^{\mathrm{T}} = -\{(K_{\mathrm{M}} - K_{\mathrm{I}})[S - f(S - I)] - K_{\mathrm{M}}\}^{-1}(K_{\mathrm{M}} - K_{\mathrm{I}})T'^{\mathrm{A}} \quad (8.10) \quad \textit{cf.}\ (3.23)$$

and the mean heat fluxes are

$$\langle q \rangle_{\mathrm{M}} = -fK_{\mathrm{M}}(S - I)T'^{\mathrm{T}} \quad (8.11) \quad \textit{cf.}\ (3.24)$$

$$\langle q \rangle_{\mathrm{I}} = (1 - f)K_{\mathrm{M}}(S - I)T'^{\mathrm{T}} \quad (8.12) \quad \textit{cf.}\ (3.25)$$

The composite conductivity is now found from

$$q^{\mathrm{A}}\ (= K_{\mathrm{M}}T'^{\mathrm{A}}) = K_{\mathrm{C}}\overline{T'^{\mathrm{A}}_{\mathrm{C}}} \quad (8.13) \quad \textit{cf.}\ (3.26)$$

in which

$$\overline{T'^{\mathrm{A}}_{\mathrm{C}}} = fT'^{\mathrm{T}} + T'^{\mathrm{A}} \quad (8.14) \quad \textit{cf.}\ (3.30)$$

so that, making the appropriate substitutions, the composite conductivity is given by

$$K_{\mathrm{C}} = [\![K_{\mathrm{M}}^{-1} + f\{(K_{\mathrm{M}} - K_{\mathrm{I}})[S - f(S - I)] - K_{\mathrm{M}}\}^{-1}(K_{\mathrm{I}} - K_{\mathrm{M}})K_{\mathrm{M}}^{-1}]\!]^{-1} \quad (8.15) \quad \textit{cf.}\ (3.32)$$

This expression can be used to explore basic features of the dependence of K_{C} on fibre content and aspect ratio. Some predicted curves are shown in Fig. 8.2, with K_{C} plotted as a ratio to K_{M}. Fig. 8.2(a) refers to insulating inclusions ($K_{\mathrm{I}} = 0$), which is effectively the case with many MMC systems. The longitudinal conductivity of a continuous fibre composite ($s = 100$ axial plot) is given by a simple weighted mean rule of mixtures (Voigt) line. Reducing the fibre aspect ratio gives a lower conductivity, but it can be seen that the $s = 3$ curve is quite close to that for $s = 100$, so

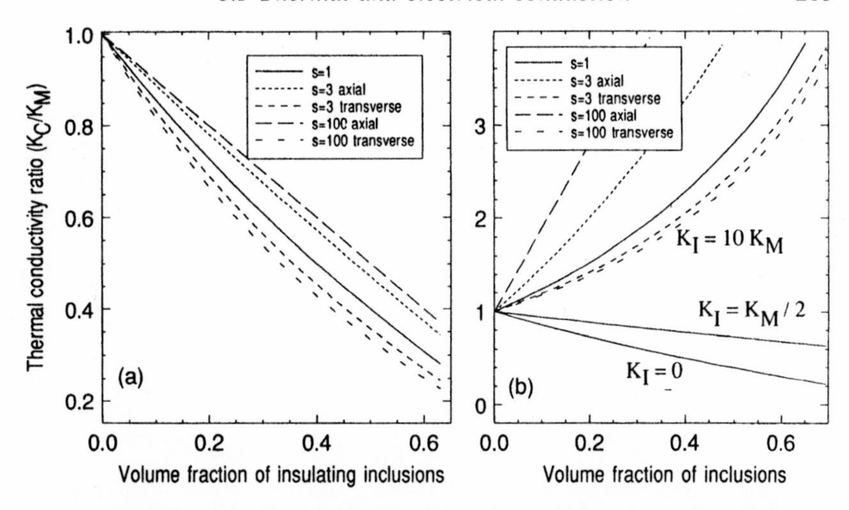

Fig. 8.2 Theoretical predictions obtained by the Eshelby method for the thermal conductivity of composites as a function of fibre content and aspect ratio (a) for insulating inclusions and (b) highly conductive inclusions ($K_I \approx 10K_M$ corresponds approximately to Ti–SiC – see Table 8.1). Also shown in (b) are plots for less conductive equiaxed inclusions.

that (well-aligned) short fibre MMCs have rather similar conductivities to those of the corresponding continuous composites. A particulate composite ($s = 1$), on the other hand, has a significantly lower conductivity. Transverse conductivities are lower still, but it is important to note that a Reuss expression ($1/K_C = (1 - f)/K_M + f/K_I$) will give a gross underestimate ($K_C = 0$) for all f, in cases where $K_I \approx 0$. It may in general be noted that, even if the reinforcement is in effect non-conducting, it will tend to depress the conductivity somewhat less than an alloying operation designed to bring about the same degree of strengthening (see Table 8.1).

Data are shown in Fig. 8.2(b) for cases where the conductivity of the inclusion is significant compared with that of the matrix. Again the axial conductivity of a long fibre composite is given by a linear rule of mixtures, representing an upper bound. Both the equiaxed and transverse cases give appreciably lower values. Hatta and Taya have shown[14] that for a long fibre composite the transverse conductivity, K_{1C}, predicted by the Eshelby method reduces to the expression

$$K_{1C} = K_M + \frac{K_M(K_I - K_M)f}{K_M + (1 - f)(K_I - K_M)/2} \qquad (8.16)$$

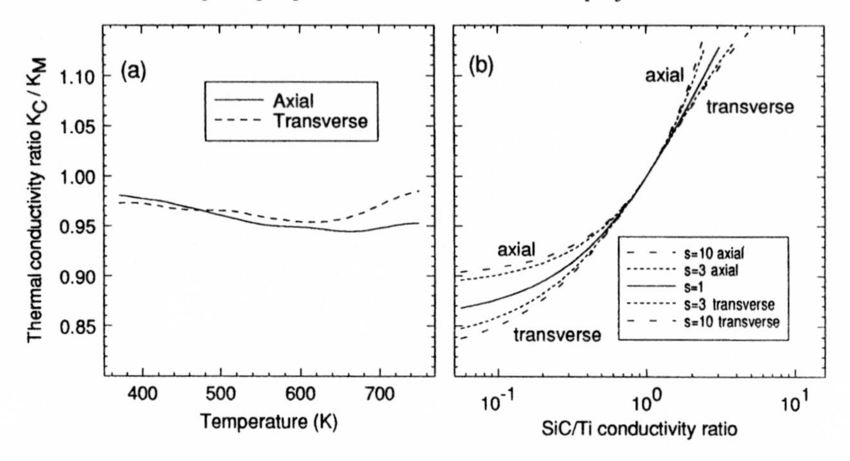

Fig. 8.3 Experimental data[15] and predictions from the Eshelby model for the thermal conductivity of Ti–SiC fibre extruded composites, using the measured mean fibre aspect ratio value of three (see Fig. 8.4). (a) Experimental values obtained using the laser flash method[16], smoothed and expressed as composite/matrix conductivity ratios and (b) predicted values for three fibre aspect ratios, as a function of the fibre/matrix conductivity ratio.

This applies for all K_I values and in the limit of $K_I = 0$ it reduces to

$$K_{1C} = K_M\left(\frac{1 - f}{1 + f}\right) \tag{8.17}$$

It can be seen from Fig. 8.2(b) that eqn (8.16), giving a curve very close to that of the $s = 100$ transverse case, can be used to estimate transverse conductivities across the whole range of aspect ratios, but the axial conductivity is too sensitive to aspect ratio for the simple Voigt rule of mixtures expression to be acceptable for low aspect ratio fibres (especially when these are carrying an appreciable heat flux).

The Ti–SiC system is of particular interest here because, while it appears to constitute a case where the presence of the reinforcement should enhance the matrix conductivity, it is also a combination prone to interfacial chemical reaction (see §6.3). Experimental conductivity data[15,16] are shown in Fig. 8.3(a) for Ti reinforced by 10% of aligned short SiC fibres. The microstructure of the composite material can be seen in Fig. 8.4. Both axial and transverse composite conductivities are close to that of the matrix over the range of temperature studied. These data may be compared with the predictions shown in Fig. 8.3(b), obtained using the Eshelby model. The experimental values indicate that the effective conductivity of the fibres was about 60–70% of that of the matrix.

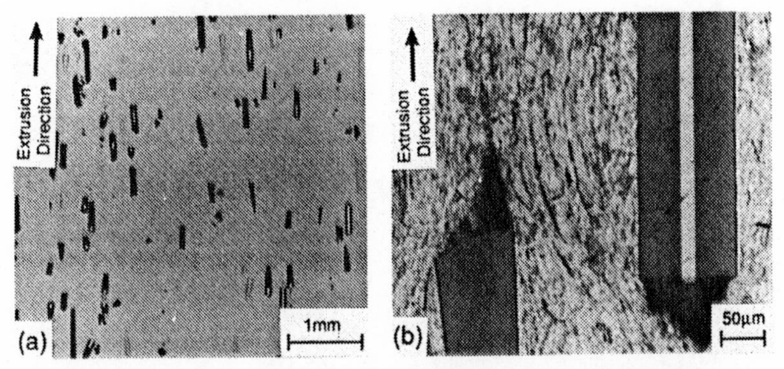

Fig. 8.4 Microstructure of a Ti–10 vol% SiC (W-cored) monofilament composite produced by powder blending and extrusion[15]. (a) SEM micrograph and (b) optical micrograph of a polished longitudinal section. The porosity level was measured by densitometry to be less than 1%. However, it can be seen in (b) that much of this is localised around the fibre ends (although the pores shown are rather extreme examples).

This is rather lower than most handbook values, but it is not unreasonable in view of the very fine grain size in the SiC monofilaments and the possibility of some interfacial thermal resistance (see §8.1.3 below). Evidently the fibres are carrying some heat flux, as considerably lower conductivities would be expected with insulating inclusions, particularly in the transverse direction. An unexpected feature of the experimental data is that the axial conductivity is no greater than the transverse value, and indeed appears to be somewhat lower at higher temperatures. However, the probable explanation for this is apparent in Fig. 8.4(b), where it can be seen that the small amount of porosity present is concentrated at the fibre ends, where it will inhibit axial heat flow in the fibres.

8.1.3 Interfacial thermal resistance and reaction layers

As the above example illustrates, for systems where the inclusions are conducting there may be a thermal resistance across the inclusion/matrix interface, particularly if there is interfacial porosity or a reaction layer. Such a thermal resistance can be characterised by an interfacial heat transfer coefficient or thermal conductance, h (W m^{-2} K^{-1}), defined as the proportionality constant between the heat flux through the boundary and the temperature drop across it

$$q_i = h\,\Delta T_i \tag{8.18}$$

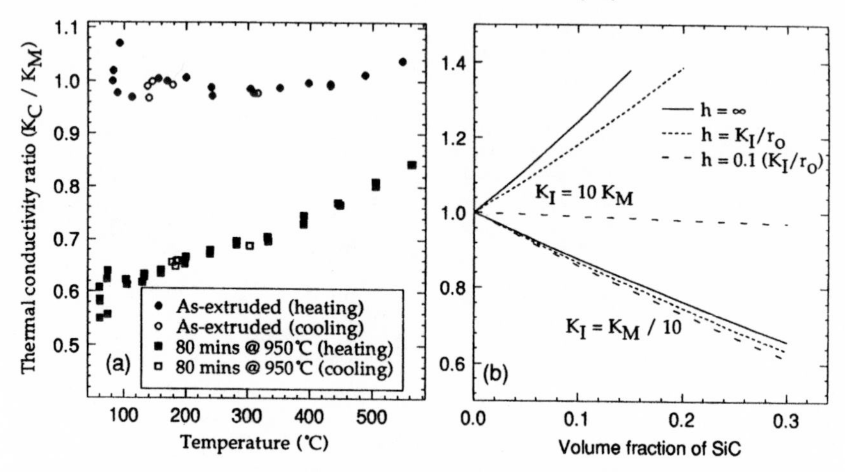

Fig. 8.5 (a) Experimental data[17] for the thermal conductivity of Ti–10 vol% SiC_P composites as a function of temperature, with and without prior heat treatments which increased the thickness of the interfacial reaction layer. It can be seen that the conductivity of the as-extruded composite differed little from that of the matrix, but the heat treated specimens exhibited much lower values – although the difference was reduced at higher temperatures. (b) Theoretical predictions[10] for the thermal conductivity as a function of the volume fraction of spherical inclusions, for two inclusion conductivities and three interfacial thermal conductances.

The Eshelby method can be adapted to treat this case[12,14] by the provision of a coating on the inclusions, having a conductivity and a thickness such that its conductance (i.e. the ratio of its conductivity to its thickness) is equal to h.

Experimental data[17] shown in Fig. 8.5, for Ti–10 vol% SiC_P composites, demonstrate the significance of the reaction layer. For the case of equiaxed inclusions and a relatively dilute composite, an analytical expression given by Hasselman and Johnson[10] can be used to predict the conductivity

$$K_C = K_M \frac{\left[2f\left(\frac{K_I}{K_M} - \frac{K_I}{r_0 h} - 1\right) + \frac{K_I}{K_M} + 2\frac{K_I}{r_0 h} + 2\right]}{\left[f\left(1 - \frac{K_I}{K_M} + \frac{K_I}{r_0 h}\right) + \frac{K_I}{K_M} + 2\frac{K_I}{r_0 h} + 2\right]} \tag{8.19}$$

where r_0 is the radius of the spherical inclusions. Predictions from eqn (8.19) are shown in Fig. 8.5(b) for the inclusion conductivity being much greater and much less than that of the matrix: in each case plots are shown for the interfacial conductance being infinite, equal to that of the inclusions, and small compared with that of the inclusions. The latter case gives a conductivity ratio of about 1.0, in agreement with experiment for the as-extruded composite; the corresponding value of h ($\sim 10^6$ W m^{-2} K^{-1}) actually represents quite good thermal contact and might well arise even with a very thin reaction layer. An idea of the reaction layer thickness for the as-extruded and heat treated composites is given by the microstructures shown in Fig. 8.6. As it is very difficult to avoid a reaction layer of at least the thickness shown in Fig. 8.6(a), it may be concluded that significant enhancement of the conductivity above that of the Ti matrix is not possible with SiC (although it can be achieved with TiB_2 – see §6.3.3).

Of more concern, however, is the sharp reduction effected as the reaction layer gets thicker. The low h, low K_I curve in Fig. 8.5(b) approximates to the insulating inclusions case (cf. Fig. 8.2(a)). It follows that the measured values must correspond not only to the SiC particles carrying no heat flux, but to their effective volume fraction being appreciably above 10%, particularly at lower temperature. This is partly due to the volume of the SiC plus reaction product exceeding that of the unreacted SiC, but it has been proposed[17] that in the heavily reacted case there is a network of fine cracks, caused by stresses arising from the volume reduction associated with the reaction, which raises the effective volume in which heat flow is obstructed. This may account for the reduced effect at higher temperature, because differential thermal expansion stresses will tend to close up the radial cracks. The proposed effect is illustrated in Fig. 8.6(c). It is in any event clear that excessive interfacial reaction is likely to reduce the thermal conductivity.

8.1.4 Electrical resistivity

The passage of electric current through a conductor conforms to the same mathematical laws as does heat flux. The basic equation, which is just a form of Ohm's law, may be written

$$j = \sigma E \tag{8.20}$$

where j is the current density (A m^{-2}), σ is the conductivity (Ω^{-1} m^{-1}) and E is the electric field (V m^{-1}). It is common to deal with the resistivity ρ (Ω m), which is the reciprocal of the conductivity. As with thermal conduction, the conductivity is proportional to the concentration of carriers, and to their mobility. Only electrons or ions can act as carriers of electric charge: because electrons are much more mobile than ions, materials with free electrons (i.e. metals) have vastly greater conductivities than other materials. Hence, MMCs can normally be treated as being

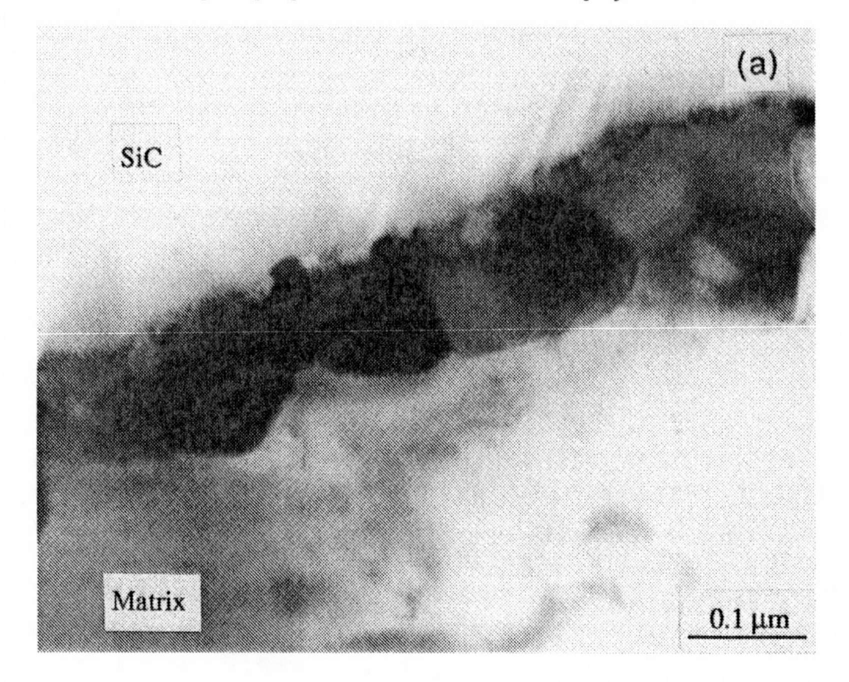

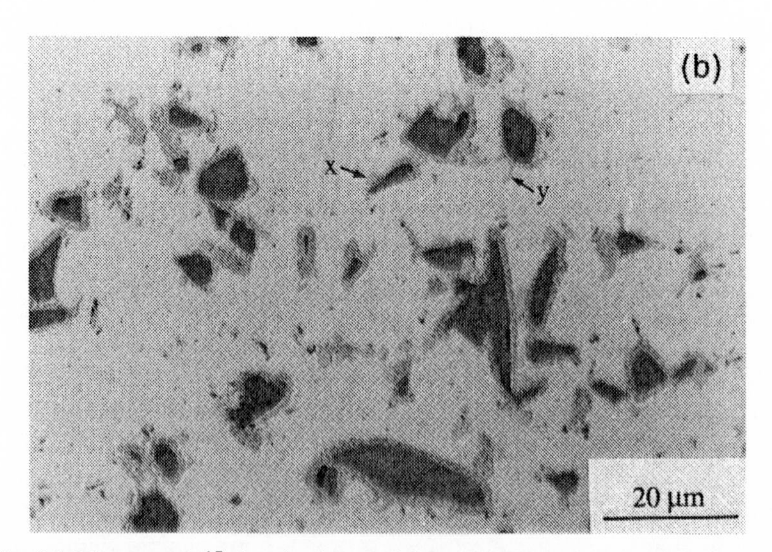

Fig. 8.6 Micrographs[17] showing interfacial reaction layer thickness in Ti–10 vol% SiC_P composites (a) as extruded (TEM) and (b) after 80 minutes at 950 °C (optical). (*cont.*)

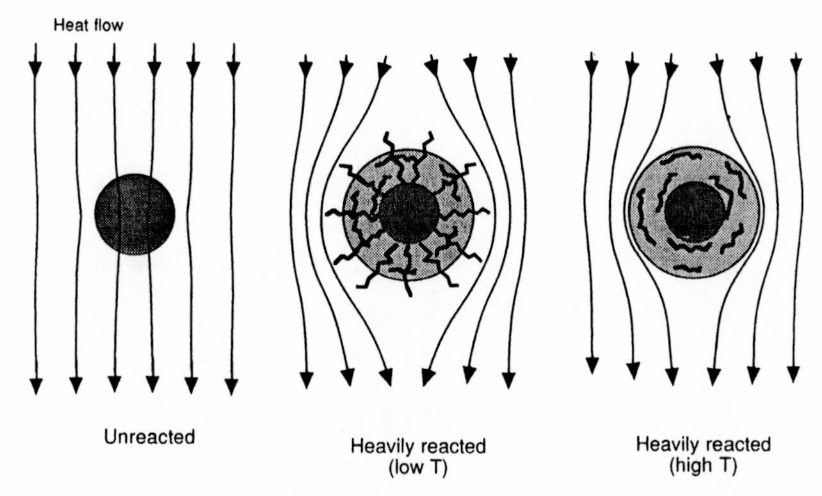

Fig. 8.6 (*cont.*) (c) Schematic illustration of the heat flow patterns around the SiC inclusions for as-extruded, reacted (low temperature) and reacted (high temperature) cases.

Table 8.2 *Electrical resistivity data*[2–5] *for a range of metallic materials*

	ρ (μΩ cm)			ρ (μΩ cm)	
Material	at 300 K	at 900 K	Material	at 300 K	at 900 K
Cu	2.0	5.5	Ag	1.8	5.5
Cu–2% Ag	2.3	—	Al	2.8	10.4
Cu–2% Be	6	—	Mg	5.1	23
Cu–2% Sn	7.6	—	Mild steel	16	85
Cu–30% Zn	7.2	18	Ni–32% Cu	50	150
Cu–40% Ni	52.5	—	Ti	80	250

reinforced with insulating inclusions†, so that interfacial electrical resistance is rarely of any significance. Some resistivity values are shown in Table 8.2. It can be seen that impurities or alloying elements tend to cause a sharp reduction in conductivity (rise in ρ), as for the thermal case. This is, of course, again due to scattering of electrons by atomic scale disturbances to the potential field of the lattice. There is therefore scope for designing composites based on pure metals to give improved combinations of mechanical and electrical properties in the same way as for mechanical

† There are a small number of exceptions to this, such as Ti reinforced with W.

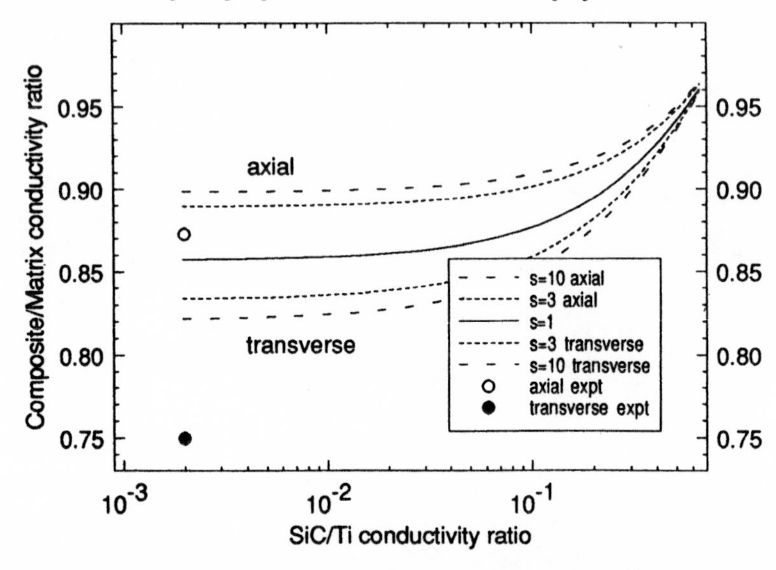

Fig. 8.7 Experimental and predicted data[15] for the electrical conductivity of the Ti–10% SiC short fibre composite shown in Fig. 8.4. The experimental values indicate that the fibres were carrying no electrical current and this is as expected in view of the electrical resistivities of the two constituents (and, indeed, of virtually all MMC combinations).

and thermal properties. Curves such as those shown in Fig. 8.2 are equally applicable to predictions of electrical conductivity as to thermal conductivity.

One point to note, which could in fact also be relevant in the thermal case, is that the electrical resistivity of the matrix in a composite may differ from that of the unreinforced material. For example, the increased dislocation density caused by differential thermal contraction stresses can raise the *in situ* resistivity of the matrix. (Obviously, this effect cannot be avoided by annealing treatments.) Consider the experimental data and predicted curves[15] shown in Fig. 8.7, referring to the Ti–SiC short fibre composite shown in Fig. 8.4. It is clear that the fibres are now making no contribution to the composite conductivity, as the experimental values are actually below those for zero fibre conductivity with the appropriate fibre aspect ratio ($s = 3$). The axial and transverse values are, however, now in approximately the correct proportion, as interfacial porosity makes no difference when the fibres are effectively insulators. The data can be fully explained if the matrix conductivity is slighly lower in the composite than when unreinforced and this may be at least partly a consequence of higher dislocation densities.

In general, provided the matrix microstructure is well characterised, this type of modelling will give reliable predictions. Abukay *et al.*[18] proposed that electron scattering at interfaces, a difficult process to model, was responsible for anomalously high transverse resistivities in Al reinforced with continuous boron fibres, but this is unlikely to be significant in view of the short mean free path for electron scattering compared with inter-fibre distances; their data would also appear to have been influenced by anisotropy of the effective matrix resistivity, probably due to aligned oxide films. Evidently, attention must be paid to processing procedures and microstructural effects if the apparent potential for retaining a low resistivity at moderate fibre contents is to be realised.

8.1.5 Thermal shock resistance

An important property for many applications involving use at elevated temperatures is the resistance to thermal shock. Sudden temperature changes will always tend to set up stresses and, depending on thermal conditions, component geometry, component constraint, etc., these may be sufficient to cause cracking or complete failure. In order to identify a material parameter characterising the resistance to thermal shock, it is conventional to simplify the origin and nature of these thermal stresses in some way. For example, Fig. 8.8 shows how stresses will arise on

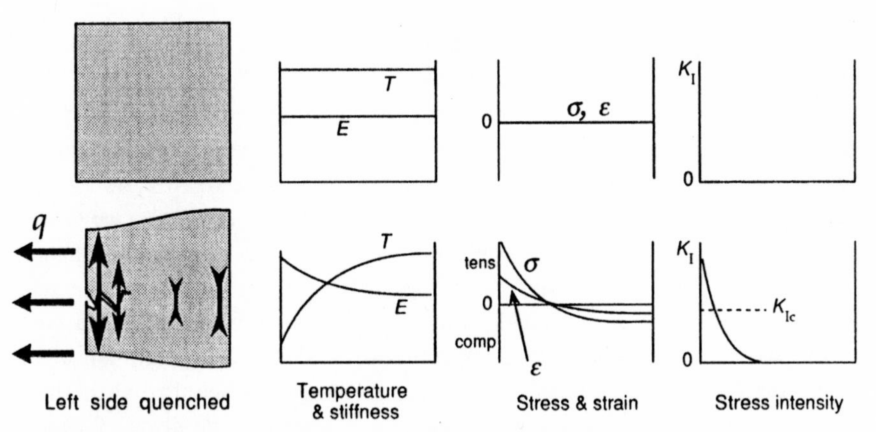

Fig. 8.8 Schematic illustration of the stresses which arise in a material as a consequence of the production of a thermal gradient within it. Of course, in practice there may be multi-dimensional heat flow, specimen constraint, etc., which might lead to more complex stress fields than that shown here.

directionally cooling a homogeneous material. Neglecting the effect of temperature on modulus and taking all the gradients as linear, the condition for fracture can be written

$$K_{max} \approx E\alpha\Delta T\sqrt{\pi a} \geq K_c \qquad (8.21)$$

where K_c is the mode I fracture toughness and a is the size of a pre-existing flaw. The temperature difference ΔT generated depends on cooling (or heating) conditions and on thermal properties of the material. For a linear thermal gradient (steady state), we would have

$$\Delta T = q\frac{\Delta x}{K} \qquad (8.22)$$

where q is the heat flux, K is the thermal conductivity and Δx is the thickness in the heat flow direction. It follows that the condition for avoidance of fracture, with a fixed flaw size, can be written

$$\left(\frac{K_c K}{E\alpha}\right) = R \geq \text{constant} \qquad (8.23a)$$

so that this expression for R is a figure of merit for the material, representing the resistance to thermal shock. In practice, it is often the external heat flux, rather than the thermal profile, which is imposed and in this case the maximum value of ΔT will be inversely proportional to the thermal diffusivity (a) rather than the thermal conductivity (K), giving a slightly different figure of merit expression

$$\left(\frac{K_c a}{E\alpha}\right) = R' \geq \text{constant} \qquad (8.23b)$$

It should be noted that, for anisotropic materials, such as most MMCs, the relevant value of K (or a) is that parallel to the heat flow direction, while the other parameters should be taken transverse to this. As all of these parameters are sensitive to orientation for fibrous MMCs (see §3.6, 5.1, 7.1.2, 7.4.2 and 8.1.2), calculation is necessary to predict the effect of heat flow direction. Results of some such calculations are shown in Table 8.3. It can be seen that the thermal shock resistance of Ti and Ni alloys, for example, is predicted to increase on incorporation of long fibres when the heat flow (and hence the crack propagation) is normal to the fibre direction, but to decrease for heat flow parallel to the fibres. This is primarily a consequence of the poor fracture toughness for cracks running axially (see §7.3), but it can be seen that the variations in the other parameters are also significant.

Table 8.3 *Data for thermal shock resistance*

Matrix	Fibre	f (%)	K (W/m K)		E (GPa)		α ($\mu\varepsilon$/K)		K_c (MPa$\sqrt{}$m)		R (kW/$\sqrt{}$m)	
			∥	⊥	∥	⊥	∥	⊥	∥	⊥	∥	⊥
Al (6XXX)		0	180	180	70	70	23	23	40	40	4.4	4.4
	SiC_P ($s = 1$)	20	138	138	94.7	94.7	18.3	18.3	20	20	1.59	1.59
	SiC_W ($s = 5$)	20	147	132	121	93.4	13.8	20.2	22	14	1.09	1.74
Ti–6Al–4V		0	6	6	115	115	8	8	57	57	0.37	0.37
	SiC fibre ($s = \infty$)	40	11.6	9.3	250	189	5.2	6.7	120	10	0.09	0.86
Ni superalloy		0	11	11	214	214	12	12	150	150	0.64	0.64
	W wire ($s = \infty$)	40	14.6	13.9	309	287	7.8	9.2	~200	~20	0.11	1.15

Note: The parameter R is a figure of merit, given by eqn (8.23a). The parallel (∥) and normal (⊥) values are given relative to the fibre direction; R and K refer to the heat flow direction, while E, α and K_{Ic} refer to the direction of imposed stress or strain. The values of K, E and α for the composites have been calculated by the Eshelby method, using values of K, E and α for SiC and W of 20 and 120, 450 and 410 and 4.0 and 4.3, respectively. The K_c data are all experimentally observed[19–22] or estimated and the R values have been calculated from the other parameters.

There is a shortage of experimental data on thermal shock resistance of MMCs, but some work has been done. Yuen *et al.*[23] used a rastered electron beam to heat small sections of panels by about 800 °C in around half a second, recording the number of thermal cycles required to produce visible cracking as a measure of the thermal shock resistance. They studied both unreinforced superalloys and various alloys reinforced with 40% of aligned tungsten wires. For the composites, heat flow was transverse to the fibre direction. It was observed that considerably more thermal cycles were required to cause visible damage with the composites than with the unreinforced alloys, which is consistent with the relative values of R in Table 8.3 (1.15 for the composite versus 0.64 for the alloy).

8.2 Tribological behaviour

It became clear at a very early stage in the development of MMCs that they offer considerable potential for enhanced wear resistance (§12.1.4). However, understanding of their wear characteristics is still far from complete. This is due in part to the inherent complexity of many wear processes, but the problem is compounded by a possible interplay with microstructural variables in MMCs, such as fibre content, size, orientation, bond strength, etc., which can influence wear behaviour. It is in any event necessary first to consider some basic points about wear.

8.2.1 Fundamentals of wear

Wear involves relative motion between a surface and some other body. One of the most common situations is that of ***sliding wear***, when two surfaces slide over one another, bringing asperities into repeated contact and possibly tearing some of them away from the parent body to form fragments of wear debris. The basic relation describing the rate at which material is removed from the body being considered is the Archard equation

$$Q = \frac{BW}{H} \tag{8.24}$$

where Q (m^2) is the volume removed per unit sliding distance, W (N) is the applied load, H (Pa) is the hardness of the body and B is a constant (the ***dimensionless wear coefficient*** – not to be confused with the coefficient of sliding friction). The wear rate thus increases as the contact load is raised and as the hardness of the material falls. There is therefore an immediate expectation that MMCs will resist wear better than the corresponding unreinforced matrix, because incorporation of the reinforcement would normally raise the hardness. In practice, B is often observed to fall as well (see §8.2.2 below).

The value of B is dependent primarily on local stresses, local temperature and surface chemical effects such as oxidation. For metal/metal sliding, it will in general be small ($\sim 10^{-6}$–10^{-9}) when only oxide wear debris is being removed (mild wear), but can become large ($\sim 10^{-2}$–10^{-5}) when metal starts to become excavated (severe wear). However, it is important to recognise that wear resistance (represented by B or by some other parameter) is not a material property. It can, for example, change with applied load and sliding velocity. Wear mechanism maps (e.g. see Lim and Ashby[24]) can be useful in identifying regimes in which a particular mechanism will operate and hence in which B should be approximately constant. Furthermore, the presence of lubricants, or a change in the nature of the other surface, can have a profound effect on the wear rate.

When material loss is induced by dragging hard, angular particles over the surface, this is usually termed ***abrasive wear***. This can be subdivided into cases where the abrading particles are (a) fixed to, or part of, the other surface (two-body abrasion) and (b) free to move between the two sliding surfaces (three-body abrasion). If, on the other hand, the abrading particles strike the surface freely, the process is termed ***erosion***. In all cases, the wear rate tends to rise sharply if the hardness of the abrading particles is at least 20% greater than that of the surface being abraded. Large, angular particles cause more wear than small, rounded ones.

Essentially, the mechanisms of wear are plastic flow and fracture, both of which can result in excavation and detachment of material. Raising the material hardness will reduce plastic flow, but may increase the danger of fracture. The fracture toughness should then figure in any prognosis about wear rates, and some compromise between high hardness and high fracture toughness may give the best wear resistance. The behaviour tends, however, to be a function of lubrication conditions and abrasive size. A point to note in the context of lubrication is that the wear rate does not necessarily correlate well with the ***coefficient of sliding friction***, μ. Although low friction will often result in reduced wear, increases in μ need not result in the material under test being worn away more quickly. A rise in μ will increase the rate of energy dissipation, but this might simply raise the interfacial temperature without enhancing the wear rate, or it might result in greater wear of the other body only (which may or may not be a problem).

8.2.2 Wear of MMCs with hard reinforcement

A number of studies[25–28] have demonstrated an improved wear resistance for aluminium, steel, etc., in both lubricated and dry conditions, after incorporation of SiC or Al_2O_3 particles or fibres. For example, Hosking *et al.*[25] showed that 2014 Al with 20% of 16 μm SiC gave a wear rate (during dry sliding on steel) which was substantially lower than that of unreinforced 2014 Al, corresponding to a decrease in B from about 2×10^{-3} to 10^{-4}. These workers also found that the size of the reinforcement had a strong effect. It can be seen from the data shown in Fig. 8.9 that higher ceramic contents and larger ceramic particles give greater wear resistance (lower values of B). The reduction in wear rates is even larger because of the effect of increased hardness as the ceramic content is raised (see eqn (8.24)).

These reductions in wear rate are not due to a reduction in the coefficient of friction. Indeed, some workers[29] comparing MMCs with the unreinforced matrices have recorded a decrease in wear rate accompanied by an increase in μ. The improvement can be attributed to the hard particles or fibres remaining intact when struck by asperities or abrading particles, under conditions such that material would have been ploughed out in the unreinforced material. The limited data currently available[26,27,30] suggest that fibres do not offer dramatically greater

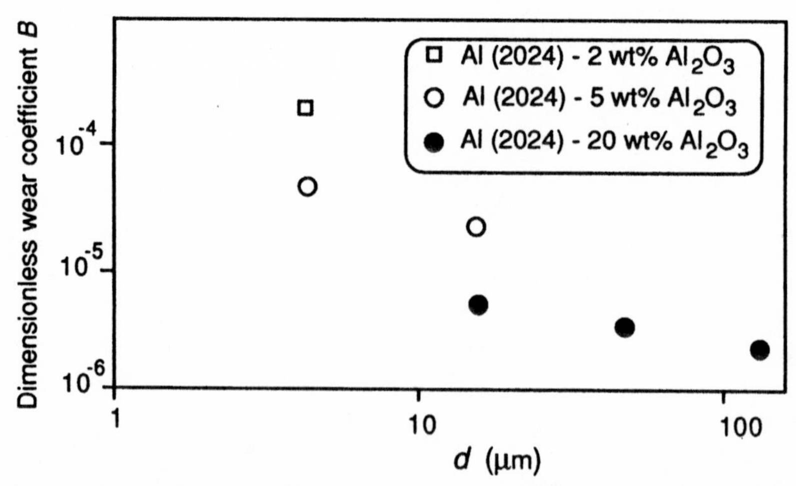

Fig. 8.9 Dependence of the dimensionless wear coefficient on the diameter of the reinforcing particles, during dry sliding of 2024 Al–Al_2O_3 particulate composites against steel. After Hosking *et al.*[25]

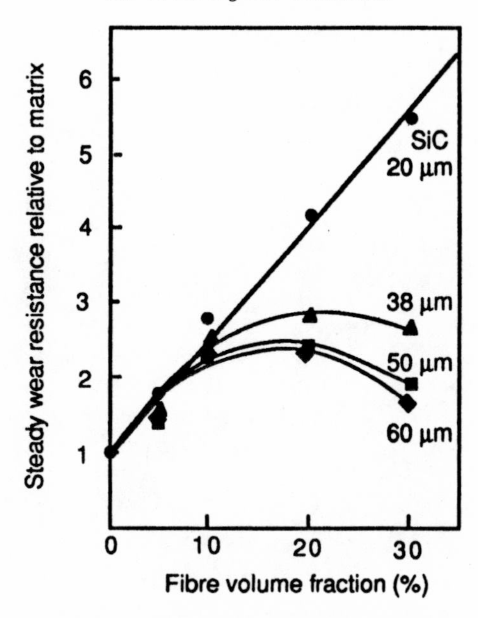

Fig. 8.10 Dependence of the wear resistance ratio on the fibre content for 6061 Al–Saffil® (δ-alumina) composites subjected to abrasive wear[30]. These results are from pin-on-disc experiments in which Al–Saffil® fibre (3 μm diameter) specimens are moving against abrasive paper. The wear resistance is plotted as a ratio of the wear rate of the matrix to that of the composite concerned, with no account taken of hardness variations. Curves are shown for four grades of paper, having SiC particles of different average size.

improvement in wear resistance than do particles of the same material and diameter. Consistent with this is the observation[30] that composites having a planar random distribution of fibres exhibit similar wear behaviour on surfaces normal and parallel to the fibre plane. A high interfacial bond strength might be expected to improve the wear performance as a consequence of enhanced resistance to fibre fracture and excavation, but firm evidence for this effect remains to be collected.

The diameter (smallest dimension) of the reinforcement, relative to the size of the abrading particles, is evidently of considerable importance[28]. This is illustrated by the data of Wang and Hutchings[30] shown in Fig. 8.10. Improvements in wear resistance are clear-cut for fine abrading particles, but less obvious when they are larger. An insight into this behaviour can be obtained from the micrographs of abraded surfaces shown in Figs. 8.11 and 8.12. Fig. 8.11(a) shows unreinforced matrix after abrasion with the finer particles, with regular furrows caused by the ploughing process

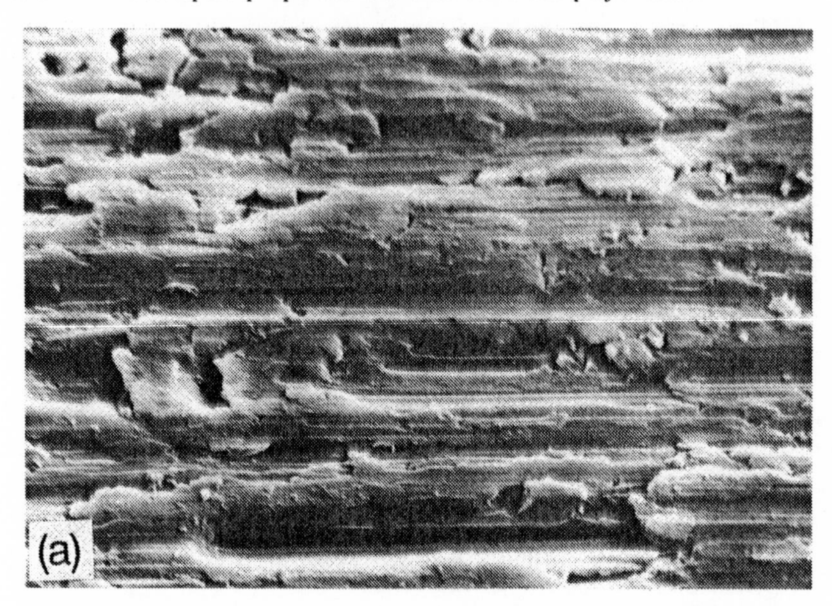

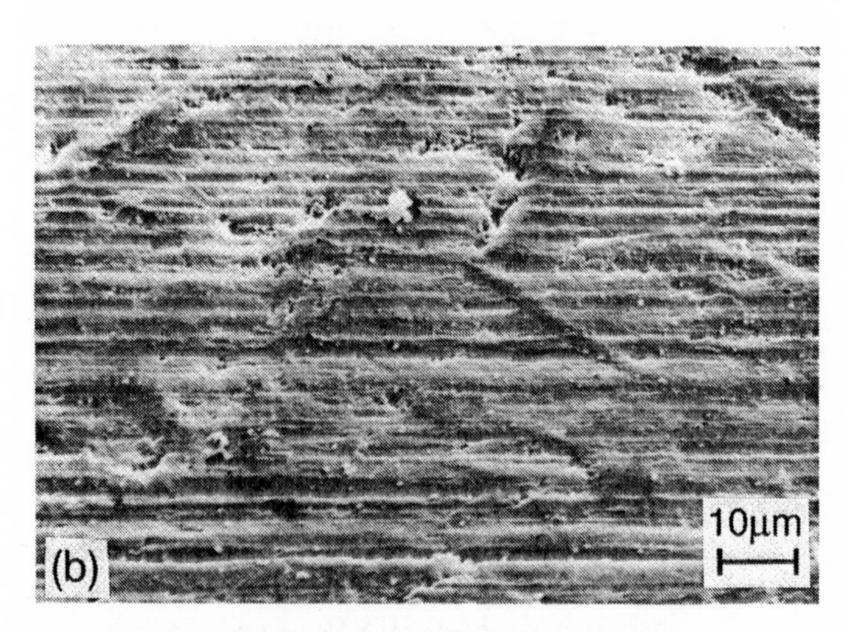

Fig. 8.11 Scanning electron micrographs[30] of the surfaces of (a) unreinforced matrix (6061 Al) and (b) 30 vol% Saffil® composite, after abrasion with 20 μm SiC particles.

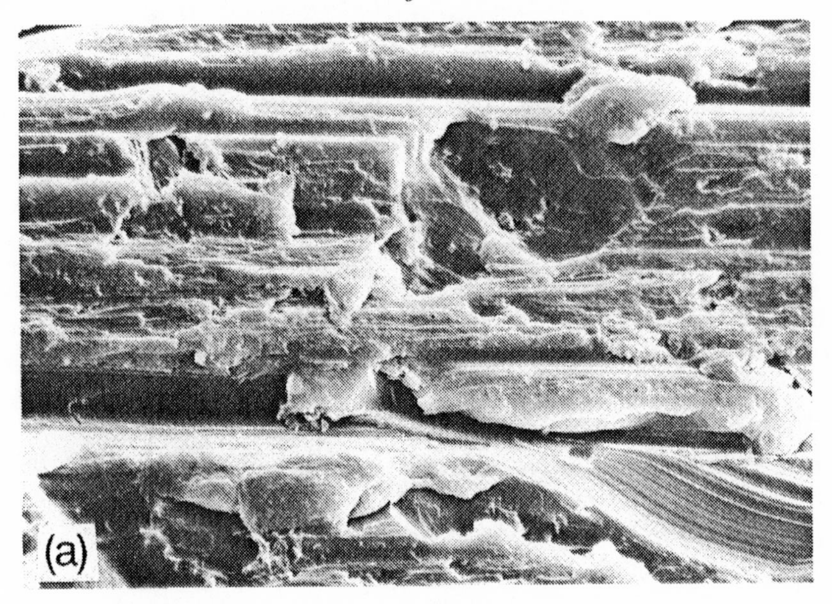

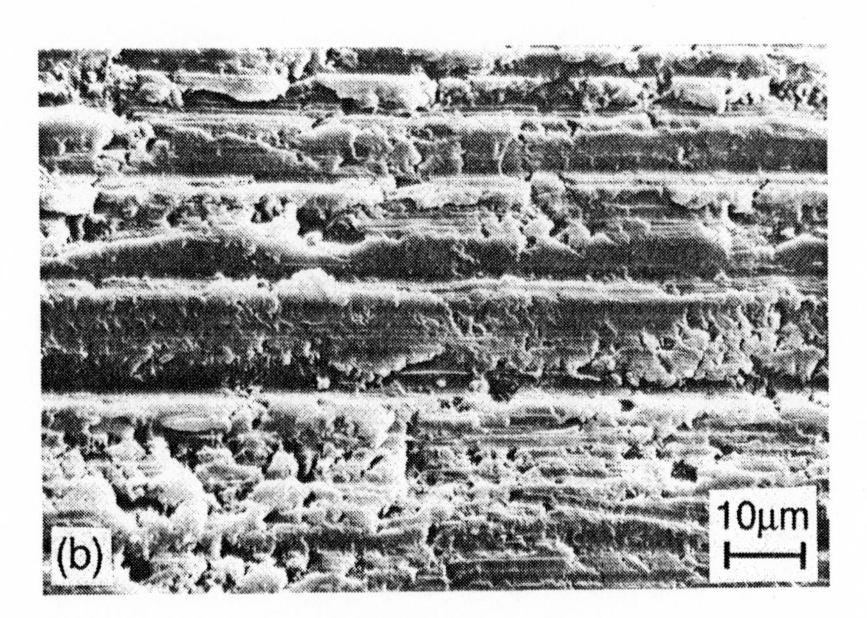

Fig. 8.12 Scanning electron micrographs[30] of the surfaces of (a) unreinforced matrix (6061 Al) and (b) 30 vol% Saffil® composite, after abrasion with 60 μm SiC particles.

clearly evident. This ploughing has been virtually eliminated by the presence of the fibres, which can be seen to be intact just below the surface in Fig. 8.11(b). With the larger abrading particles, however, both matrix and composite become heavily ploughed – see Fig. 8.12(a) and (b). Evidently the fibres become fractured under the higher loads to which they are subjected with larger abrading particles, becoming excavated along with physically deformed matrix.

Wang and Hutchings[30] have proposed a model to quantify this size effect. It can be shown[31] that the critical point load F_* to cause fracture of a long brittle fibre supported in a ductile matrix is given by

$$F_* = \tfrac{1}{2}d^2(3\pi\sigma_{I*}H_M)^{1/2} \tag{8.25}$$

where d is the fibre diameter, σ_{I*} is its tensile strength and H_M is the hardness of the matrix. The mean force transmitted by a single abrading particle (or asperity) can be written as the applied stress, σ^A, divided by the number of contacts per unit area, n, which in turn can be expressed as

$$n = \frac{k}{D^2}\left(\frac{\sigma^A}{H_s}\right)^{1/2} \tag{8.26}$$

where H_s is the hardness of the specimen, D is the mean diameter of the abrading particles and k is a geometrical constant dependent on the packing of the particles, which can be determined experimentally. Larger particles will give a smaller n, raising the actual value of F given by

$$F = \frac{\sigma}{n} \tag{8.27}$$

and hence favouring more fibre fracture. By setting this expression for F equal to that given by eqn (8.25) as being required to fracture fibres of a given diameter, we can write an expression for the critical (minimum) abrading particle diameter to cause fracture

$$D_* = \left(\frac{3\pi}{4}\right)^{3/4} k^{1/2}\left(\frac{\sigma_f}{\sigma^A}\right)^{1/4}\left(\frac{H_M}{H_s}\right)^{1/4} d \tag{8.28}$$

Using data for the present case, including an experimentally determined value for k of 0.62, gives $D_* \sim 30$ to 35 μm. This is consistent with the observation that all the particle sizes above this gave much poorer wear resistance than the one below (Fig. 8.10).

8.2.3 Wear of MMCs with soft reinforcement

The incorporation of a soft constituent into a harder matrix has long been used in materials such as bearing alloys. Al–Sn alloys, for example, have long been used in this way. Superior sliding wear performance is obtained as a result of improved lubrication, either due to shearing of the soft phase located between the sliding surfaces or by improved retention of an added lubricant in surface depressions left in the harder matrix. (The latter effect can also be achieved by simply making the matrix porous.) When improvements to the wear resistance are obtained in this way, it is normally accompanied by a reduction in the coefficient of friction.

A number of materials conventionally regarded as MMCs apparently exhibit improved wear resistance as a result of this effect. The factors identified in the last section as likely to raise the wear resistance, *viz* large reinforcement size, high reinforcement strength (fracture toughness), strong interfacial bonding, etc., will not be relevant in this case. Rather, the critical factors are likely to be those dictating the efficiency of lubrication. It should be noted that these may be highly specific both to the imposed conditions and to the microstructure of the soft constituent. As an example of this, Rohatgi *et al.*[32] found that the presence of flake graphite in an Al alloy depressed both the coefficient of friction and the wear rate during dry sliding against steel, while the same level of a microcrystalline form of graphite had little effect on either parameter. Their data are summarised in Fig. 8.13. These observations are certainly consistent with the known efficiency of flake graphite as a lubricant, associated with the ease of sliding between basal planes.

It should, however, be noted that other workers, e.g. Jha *et al.*[33], have observed the presence of flake graphite in Al to accelerate the wear rates. Evidently the design of soft particle-containing MMCs for wear resistance must be carefully correlated with the end use conditions. This point is highlighted by the data of Kuhlmann-Wilsdorf *et al.*[34], who observed the tribological behaviour of silver–graphite MMCs to be sensitive to the presence of water and hence to the wear surface temperature. These workers found that the presence of adsorbed water on the graphite facilitated basal shear, giving a low μ ($\sim$0.15–0.2) and good wear resistance. If the water is desorbed, for example as a result of the local temperature rising above about 160 °C to 180 °C, then the graphite shears much less readily, μ rises to about 0.35 and the wear rate becomes much higher.

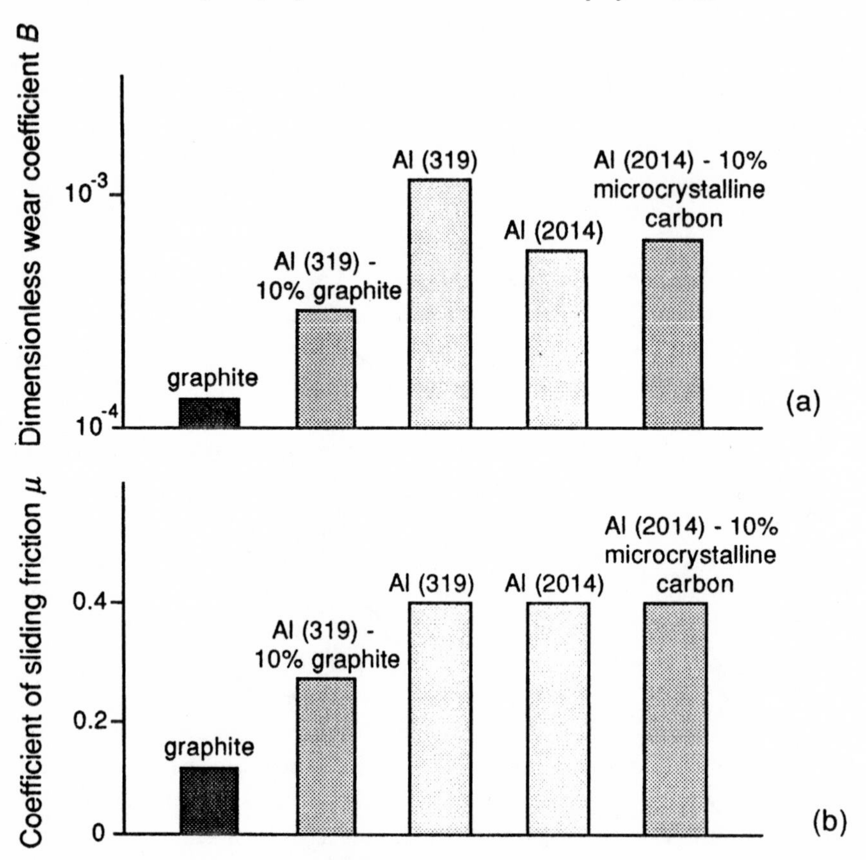

Fig. 8.13 Data[30] for (a) the dimensionless wear coefficient and (b) the coefficient of friction during dry sliding against steel of unreinforced aluminium alloys, composites containing two forms of particulate graphite, and bulk graphite. The load was 7.65 N and the total sliding distance 0.25 km.

8.3 Mechanical damping properties

8.3.1 The origin and measurement of damping capacity

In many practical applications, the rate at which elastic waves, i.e. periodically changing local strains, become attenuated in a material can be of considerable importance. For example, use of components made from materials with a high damping capacity will lead to reduced noise and vibration in machinery. In particular, the amplitude of vibration under resonance conditions will be relatively low, reducing the danger of fatigue failure. Metals vary considerably in their damping capacity, with grey cast irons, for example, well known for their high values. Since grey cast irons are in effect composite materials, in which the interface is thought to play a role in the energy absorption, it is clear that MMCs can be designed to exhibit a high damping capacity.

Damping capacity is a manifestation of ***anelastic behaviour***, i.e. a time-dependent response to an applied load. There are various mechanisms by which energy can be absorbed within a stressed material, other than by the elastic stretching of interatomic bonds. These include grain boundary sliding, dislocation motion (short- or long-range), stress-induced atomic ordering, etc., in metals. In composite materials, additional mechanisms (e.g. interfacial sliding) may become important. The essential requirement is that the mechanism must operate under the local stresses within the timescale for which the external load is applied (before reversal). Consequently, at any given temperature, each mechanism will only be able to respond up to a certain maximum frequency (minimum response time) and down to a certain minimum strain amplitude (local stress level).

These energy-absorbing processes are often said to contribute to ***internal friction*** in the material. Precise measurement of the damping effect under small, carefully controlled strains is a sensitive method of exploring the detailed structure of the material[35,36]. Internal friction is usually characterised by the ***logarithmic decrement***, δ, governing the reduction in amplitude of vibration, A, of a freely decaying system during one cycle

$$\delta = \ln\left(\frac{A_n}{A_{n+1}}\right) \tag{8.29}$$

A common method[37,38] of measuring δ simply involves setting a mechanical system, such as a cantilever or a torsional pendulum, into vibrational motion and monitoring the freely decaying amplitude. An alternative experimental approach[39–41] to the measurement of δ is to maintain the specimen in a condition of forced vibration at constant amplitude, when δ is given by

$$\delta = \frac{\Delta U}{2U} \tag{8.30}$$

where ΔU is the energy absorbed per cycle and the total vibrational energy in one cycle is U. One problem with the test geometries often used in both types of method is that there are variations in the maximum strain amplitude experienced by different parts of the specimen. This can make it difficult to interpret behaviour in regimes near the critical strain amplitude for a particular mechanism.

Certain other terms are sometimes used to represent the damping behaviour of a material. These include the ***loss factor*** η and the ***quality factor*** Q^{-1}, both of which are numerically equal to δ/π. A parameter termed the ***specific damping capacity***, taken to have a magnitude of 2δ, is also used occasionally.

8.3.2 Damping in MMCs

As illustrated in Fig. 8.14, a number of possible damping mechanisms can be identified in MMCs. The local movement of dislocations as they escape from pinning points and/or relax into lower energy configurations (for the local stress state) can occur in much the same manner as in unreinforced metals – where it often constitutes the most significant damping mechanism. The main differences are that the dislocation density

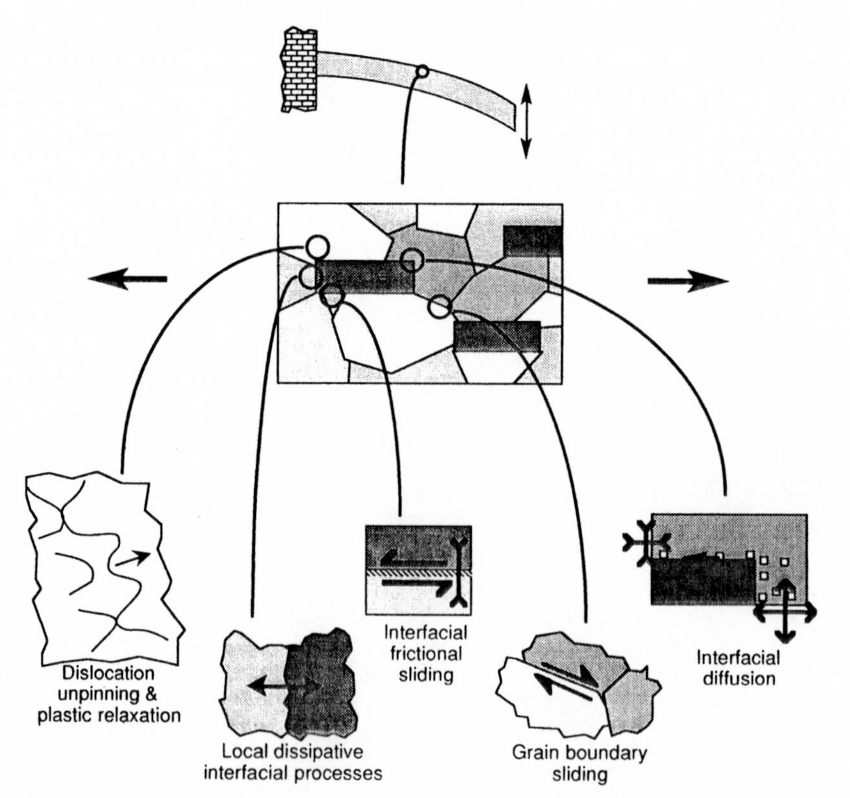

Fig. 8.14 Schematic depiction of candidate micro-mechanisms for damping in MMCs.

is expected to be higher in MMCs and that they may be more efficiently pinned or tangled as a consequence of the plastic flow and work-hardening that has taken place during differential thermal contraction. Similarly, grain boundary sliding can occur in the same way as in unreinforced metals, but the presence of a finer grain size in most MMCs means that there may be rather more scope for this process to occur. On the other hand, it is evident that, for ceramic reinforcement (with little probability of internal dislocation motion or grain boundary sliding), the damping effect from these mechanisms will be reduced in proportion to the remaining volume fraction of matrix in an MMC.

Of rather more interest are the interfacial damping mechanisms that may arise in MMCs. As shown in Fig. 8.14, these include frictional sliding under the influence of an interfacial shear stress and vacancy diffusion under a gradient of hydrostatic stress. These are both potentially efficient as energy-absorbing processes, although diffusion is expected to require relatively high temperatures and low frequencies, while sliding may need high strain amplitudes. On the other hand, there may be scope for energy absorption in an interface by other local processes with a wider response range, particularly if the bonding is poor. Such processes commonly occur when an elastic wave strikes an imperfectly bonded interface and normally leads to interfacial heating. The details of possible processes, which include interfacial microplasticity and interfacial thermal currents, have been discussed by Updike *et al.*[41]

The rather limited data currently available on the damping characteristics of MMCs, which are at first sight somewhat contradictory, are best interpreted in terms of the probable degree of dislocation motion expected under the test conditions, with the possibility of energy absorption at the interface, or possibly within the reinforcement in special cases, being borne in mind. It has been shown[35] that interfacial diffusion can absorb energy in copper containing small particles of silica, but this was at very low frequencies (less than 1 Hz) and at temperatures of at least about 500 K ($\sim 0.4T_{mp}$) for the effect to become noticeable. Moreover, the relaxation time has been found[36] to rise as the third power of the particle size, making the response slower still for the relatively large reinforcements in MMCs. Consequently, it seems likely that this mechanism can be neglected for MMCs because under these conditions various creep processes become operative giving rise to strong damping.

Early work on conventional MMCs[42–44] indicated that the introduction of fibre reinforcement into aluminium raised the damping capacity appreciably. Baker[42] derived the following expression for the log decrement by treating the energy absorption processes as essentially the same plastic flow phenomena as those in the unreinforced matrix and simply evaluating average matrix strains

$$\delta \approx \frac{\sigma_{\mathrm{YM}}(1-f)\,\Delta\varepsilon_{\mathrm{M}}^{\mathrm{P}}}{\left(\frac{fE_{\mathrm{I}}}{2}\right)(\Delta\varepsilon_{\mathrm{M}}^{\mathrm{P}}+\varepsilon_{\mathrm{YM}})^2+2(1-f)\varepsilon_{\mathrm{YM}}\sigma_{\mathrm{YM}}} \tag{8.31}$$

where $\Delta\varepsilon_{\mathrm{M}}^{\mathrm{P}}$ is the plastic strain range experienced by the matrix. This suggests that, for a given volume fraction of stiff inclusions, the damping depends only on the imposed strain range and tends to rise with it. However, the equation predicts that δ should decrease as the volume fraction is raised, which appears to be inconsistent with the experimental findings.

In any event, these observations refer to cases in which high loads, sufficient to cause considerable matrix plasticity, were being imposed at relatively low frequency. Under these circumstances, the presence of a high dislocation density may account for the improved damping capacity, although it seems likely that some interfacial process may also have been contributing. In many practical situations, the imposed strains are insufficient to allow dislocations to move within the dense tangled networks expected in MMCs. For example, consider the data shown in Fig. 8.15 which shows quality factor values, measured[42,45] at 80 kHz, as

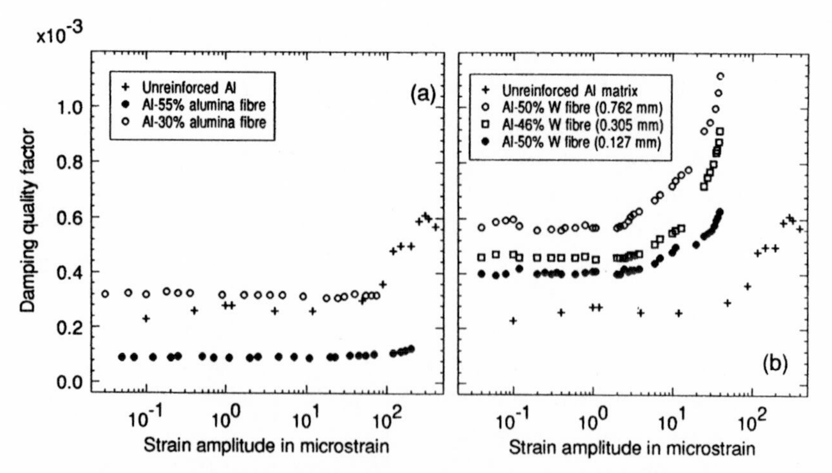

Fig. 8.15 Experimental data[45] giving the damping capacity as a function of the amplitude of imposed strain for (a) Al–Al_2O_3 fibres and (b) Al–W fibres. All measurements were made at 80 kHz.

a function of strain amplitude for pure Al and for several Al-based composites. For pure Al, the value rises from about 2×10^{-4} to about 6×10^{-4} as the strain amplitude becomes sufficient to unpin dislocations (Fig. 8.15(a)). When Al_2O_3 fibres are present, the Q^{-1} values are similar at low strains, but do not rise with strain amplitude, at least up to the limit studied of about 2×10^{-4}. These observations are probably consistent with there being more, but less mobile, dislocations present in the composites. The reduced damping for the higher fibre content confirms that the energy absorption is taking place in the matrix. In contrast, the data in Fig. 8.15(b), which refer to composites containing W fibre, show an increase in damping at higher strain amplitudes, up to maximum levels above those in Al. This is probably due to damping within the metallic fibres. The observation that damping increases with fibre diameter suggests that interfacial energy absorbing processes were not significant.

The above observations notwithstanding, there have been various reports indicating strongly that in certain circumstances interfacial effects can raise the damping capacity of MMCs. For example, Bhagat *et al.*[46] reported large increases in damping on introducing 25–35 vol% of carbon fibres into Al. This is a system in which the interface is often weak[47,48] and a poor bond may well encourage dissipative processes at this boundary (see Fig. 8.14). There have also been reports[49,50] of a progressive enhancement of damping capacity on raising the volume fraction of mica and graphite flakes in Al; these changes are probably also due to interfacial effects, as the constituents themselves do not have inherently high damping capacities. Similarly, Crawley and van Schoor[51] found high damping in Mg–carbon fibre composites, which they attributed to energy absorption at the fibre/matrix interface. Rawal and Misra[52] found strong damping at low frequency (24–38 Hz) in carbon fibre reinforced Al above a strain amplitude of about 3×10^{-5}, which may have been due to interfacial sliding. Updike *et al.*[41] observed enhanced damping in carbon fibre reinforced Al at high frequency (kHz range) and low strain amplitude ($\sim 10^{-8}$–10^{-9}), with no strain amplitude dependence. It is very unlikely that interfacial sliding could be operative in this regime and some faster interfacial process was presumably responsible; possible mechanisms discussed by Updike *et al.*[41] include interfacial microplasticity or microcracking, and thermal currents across the interface arising from a differential thermoelastic effect[53] caused by the mismatch in thermal expansivity.

In summary, there may be scope for considerable enhancement of the damping capacity of MMCs by suitable design of the microstructure. The processes of dislocation motion and grain boundary sliding common in unreinforced metals can also operate in MMCs and it is at least possible that these might operate even more efficiently than in metals as a result of the greater concentration of these defects. However, the limited data available to date suggest that very high damping capacities can result only if the interface can be tailored so as to absorb energy when the material is stressed. This has been exploited[54] in polymer-based composites to develop energy-absorbing components, although only at the expense of destroying the material. Present indications are that damping is enhanced by a weak interface, which may of course impair other mechanical properties, such as stiffness, toughness, strength and wear resistance. Improved understanding of the processes involved may allow more precise design of the microstructure for specific damping characteristics.

8.4 Oxidation and corrosion resistance

8.4.1 High temperature surface degradation of MMCs

There are potential applications for MMCs in which the resistance to oxidation, or some other hot corrosion process, may be of considerable importance. A particular example is provided by the reinforcement of titanium, which significantly improves the creep resistance and stiffness at elevated temperature. However, these gains will be of limited value if oxidation problems then replace stiffness and strength as the factors dictating the upper use temperature – particularly if the presence of the reinforcement degrades the oxidation resistance. A starting point for the discussion of oxidation behaviour is to consider how the reinforcement might induce changes in the oxidation resistance of the matrix.

Significant improvement in oxidation resistance as a result of the presence of the reinforcement has only been reported in a small number of cases. An example of this is provided by the observation[55] that the presence of more than about 10% of fine (3 μm) SiC particles in a Ni matrix has been found to reduce the rate at which it is oxidised in pure oxygen at 1100 °C. This was apparently because the particles impede the transport of Ni^{2+} ions through the growing NiO layer. This effect was not, however, a very strong one and in many cases there will be a danger of the oxidation rate being enhanced, either because the interface can serve as a fast diffusion path or because of some disruption to the formation of a protective film which would normally form on the matrix. Surface-related interfacial problems are likely to be more severe with continuous

Fig. 8.16 Low cycle fatigue fracture surface[56] from a Ti–6Al–4V alloy reinforced with SCS-6 SiC monofilament, after prolonged annealing in air at 500 °C. Cracking of the interfacial reaction zone can be seen.

fibres, because when the interface emerges at the free surface it can serve as a long-range path for transport of oxygen into the interior. For example, several workers[56,57] have reported that titanium alloys reinforced with SiC monofilaments suffer progressive interfacial oxidation after prolonged periods at relatively modest temperatures (300–600 °C). These temperatures are too low to cause significant internal fibre/matrix attack in this case (see §6.3), but oxygen ingress from the free surface is rapid and causes degradation of the interfacial structure and hence of the mechanical properties[56,57]. Fig. 8.16 shows cracking in the interfacial reaction zone after fatigue testing of a specimen annealed in air. These results were obtained with material containing fibres having carbon-rich surface layers, which may encourage oxygen penetration. This may represent a disadvantage for this type of fibre coating, although in view of the affinity of titanium for oxygen, and the fast diffusion path presented by the interface, it is probable that such penetration would occur in any event.

Study[56] of Al–SiC monofilament composites during tests at up to 400 °C revealed no significant enhancement of oxidation rates due to the presence of the fibres. However, there have been one or two reports of some impairment of the oxidation response of Al when SiC is present. For example, the oxidation rate of 6061 alloy was reported[58] to increase when SiC particles were present, particularly if a surface coating of NaCl was first introduced – when formation of a protective spinel layer on the free surface was apparently inhibited. However, some surface blistering was observed with the composite, apparently due to entrapped gas, and the preferential exposure of extra surface area in this way may have been partly responsible for the observed enhancement of the oxidation rate. Transmission electron microscopy studies[59] indicated no strong effects on the oxidation characteristics due to the presence of SiC particles for commercial purity and Mg-containing Al matrices, although it was noted that internal oxidation and void formation appeared to be promoted by these particles in an Al–3.5 wt% Li alloy. Here matrix/particle interfaces appear to act as preferential sites for void nucleation. Voiding may also be encouraged by internal tensile stresses arising from the constraint effect of the SiC particles opposing the volume changes associated with internal oxidation and Li loss to the free surface. With most Al alloys, however, such effects seem to be much less pronounced.

As to degradation of the reinforcement, most ceramic reinforcements are fairly resistant to oxidation. For example, the silica layer on SiC will not grow beyond a few atomic layers in thickness[60] at temperatures below 1000 °C. There are, however, some fibrous reinforcements which should not be exposed to oxygen at high temperatures; these include[61] tungsten, boron and carbon. It is in any event good practice with long fibre MMCs to ensure that the fibre and the fibre/matrix interface do not emerge at the free surface when this is exposed to oxygen at high temperature. Similar problems can also occur in the presence of other aggressive gases such as hydrogen and hydrogen sulphide[62]. A variety of protective coatings, such as plasma-sprayed ceramic layers[63], have been developed for metallic components and these may be required for certain MMCs.

8.4.2 Aqueous corrosion of MMCs

The resistance of MMCs to corrosion in environments containing potential electrolytes is, as in the high temperature oxidation case, unlikely to improve as a result of the presence of the reinforcement, except possibly on the basis of a reduced surface area exposure. This happens, for example, in the case of tungsten fibres in a uranium alloy[64], where the fibres remain unattacked in a salt solution while the alloy corrodes at the same rate as when unreinforced. Such an effect is, however, unlikely to be of any significant benefit in the majority of cases and the main concern is whether the presence of the reinforcement impairs the corrosion resistance of the matrix to a significant degree. General corrosive attack of matrix or fibre is possible, and this can sometimes occur to a different degree when the two constituents are combined in a composite compared with their behaviour in isolation. The problem usually centres on enhanced localised corrosion, which may arise in MMCs as a result of (a) galvanic effects arising from the potential between matrix and reinforcement, (b) crevice corrosion at the matrix/reinforcement interface, or (c) pitting attack at interfacial reaction products.

There are well documented examples where a strong galvanic effect occurs – leading to marked acceleration of the corrosion process. Galvanic corrosion is driven by the electrode potential between the two constituents in the environment concerned. It has been observed on exposure of both Al[65] and Mg[66] reinforced with carbon fibre to sodium chloride solution. The same phenomenon was noted[67] in Al containing a small volume fraction of graphite particulate. That the effect is a galvanic one is confirmed by the anodic and cathodic polarisation curves[67] for corrosion in sea water shown in Fig. 8.17. This will tend to occur in the presence of any suitable electrolyte which can allow the galvanic current to flow and it clearly constitutes a potential obstacle to the use of carbon reinforcement in metals such as Al and Mg in aggressive environments.

Unlike graphite, most reinforcements used in MMCs do not appear to contribute to a large galvanic effect. Measurements[68,69] made on various Al alloys indicated that the corrosion potential was little affected by the introduction of SiC whiskers or Al_2O_3 fibres. Reports of some enhancement of corrosion rates in Al–SiC_W composites seem to be attributable to defects produced during manufacture[70]. However, there have been reports of crevice corrosion at the fibre/matrix interface in some systems. Crevice corrosion is localised attack as a result of restricted convection between the crevice and the external solution, causing depletion of oxygen ion concentration and enrichment of metal, H^+ and Cl^- ions. These conditions often favour dissolution of metals which normally have a protective oxide film. For example, such localised corrosion has been observed[71,72] in Al reinforced with B monofilaments. This was attributed to preferred sites for anodic attack being created by defects or compositional changes at the interface. In general, there will be more danger of crevice or pitting corrosion if some interfacial chemical reaction has already taken place (during manufacture of the composite).

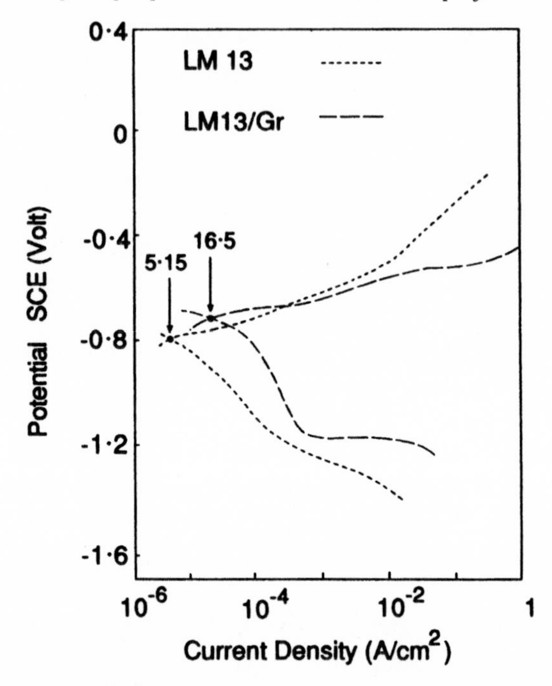

Fig. 8.17 Anodic and cathodic polarisation curves for the aluminium alloy LM13 in sea water, with and without 3 wt% of particulate graphite reinforcement[67]. The corrosion currents (arrowed) are in microamps/cm^2. The cathodic curve (hydrogen evolution) is quite strongly polarised (moved to higher potential) by the presence of the graphite, more than doubling the corrosion current density. The anodic reaction (dissolution of Al) therefore proceeds at a higher rate in the presence of the graphite.

In general, however, the presence of reinforcements such as SiC or Al_2O_3 does not appear to affect the corrosion behaviour in aqueous media to any very marked degree. In fact, the reinforcement can cause a small increase in resistance to pitting corrosion, for example by promotion of matrix precipitation, thus removing from solid solution elements which raise the pitting potential[68]. On the other hand, the higher dislocation densities resulting from differential thermal contraction would be expected to raise the potential of the matrix slightly. In fact, it seems clear[70,73] that the interface is not a preferred location for pitting corrosion in Al–SiC, the important sites being secondary phases and compositional inhomogeneities. The presence of the SiC is apparently relevant only insofar as it affects the incidence of these features. A number of studies[65,74,75] have shown that conventional corrosion protection measures, such as anodising, passivation or polymeric coatings work just as efficiently for MMCs, including those containing carbon.

References

1. R. Berman (1976) *Thermal Conduction in Solids*, Clarendon, Oxford.
2. M. C. Lovell, A. J. Avery and M. W. Vernon (1976) *Physical Properties of Materials*, Van Nostrand Reinhold, New York.
3. A. Goldsmith, T. E. Waterman and H. J. Hirschhorn (1961) *Handbook of Thermophysical Properties of Solid Materials*, Macmillan, New York.
4. E. G. West (1982) *Copper and its Alloys*, Ellis Horwood, New York.
5. D. D. Pollock (1982) *Physical Properties of Materials for Engineers*, CRC Press, New York.
6. J. F. Nye (1985) *Physical Properties of Crystals – Their Representation by Tensors and Matrices*, Clarendon, Oxford.
7. G. H. Geiger and D. R. Poirier (1973) *Transport Phenomena in Metallurgy*, Addison-Wesley, New York.
8. D. K. Hale (1976) The Physical Properties of Composite Materials, *J. Mat. Sci.*, **11**, pp. 2105–41.
9. L. E. Nielsen (1974) The Thermal and Electrical Conductivity of Two-Phase Systems, *Ind. Eng. Chem. Fundam.*, **13**, pp. 17–20.
10. D. P. H. Hasselman and L. F. Johnson (1987) Effective Thermal Conductivity of Composites with Interfacial Thermal Barrier Resistance, *J. Comp. Mats.*, **21**, pp. 508–15.
11. Y. Benveniste (1987) Effective Thermal Conductivity of Composites with a Thermal Contact Resistance between the Constituents: Non-Dilute Case, *J. Appl. Phys.*, **61**, pp. 2840–3.
12. M. Taya (1988) Modelling of Physical Properties of Metallic and Ceramic Composites: Generalized Eshelby Model, in *Mechanical and Physical Behaviour of Metallic and Ceramic Composites, 9th Risø Int. Symp.*, Roskilde, S. I. Andersen, H. Lilholt and O. B. Pedersen (eds.), Risø Nat. Lab., Denmark, pp. 201–31.
13. H. Hatta and M. Taya (1986) Equivalent Inclusion Method for Steady State Heat Conduction in Composites, *Int. J. Eng. Sci.*, **24**, pp. 1159–72.
14. H. Hatta and M. Taya (1986) Thermal Conductivity of Coated Filler Composites, *J. Appl. Phys.*, **59**, pp. 1851–60.
15. F. H. Gordon and T. W. Clyne (1991) Transport Properties of Short Fibre SiC-Reinforced Ti, in *Metal Matrix Composites – Processing, Microstructure and Properties, 12th Risø Int. Symp.*, Roskilde, N. Hansen, D. J. Jensen, T. Leffers, H. Lilholt, T. Lorentzen, A. S. Pedersen, O. B. Pedersen and B. Ralph (eds.), Risø Nat. Lab., Denmark, pp. 361–6.
16. R. Taylor (1980) Construction of Apparatus for Heat Pulse Thermal Diffusivity Measurements from 300–3000 K, *J. Phys. E. Sci. Instr.*, **13**, pp. 1193–6.
17. A. J. Reeves, R. Taylor and T. W. Clyne (1991) The Effect of Interfacial Reaction on Thermal Properties of Titanium Reinforced with Particulate SiC, *Mat. Sci. & Eng.*, **141**, pp. 129–38.
18. D. Abukay, K. V. Rao and S. Arajs (1977) Electrical Resistivity of Aluminium–Boron Composites between 78 K and 400 K, *Fibre Sci. and Tech.*, **10**, pp. 313–18.

19. Y. Flom and R. J. Arsenault (1987) Fracture of SiC/Al Composites, in *Proc. ICCM VI/ECCM 2*, London, F. L. Matthews, N. C. R. Buskell, J. M. Hodgkinson and J. Morton (eds.), Elsevier, pp. 189–98.
20. C. R. Crowe, R. A. Gray and D. F. Hasson (1985) Microstructure Controlled Fracture Toughness of SiC/Al Metal Matrix Composites, in *Proc. 5th Int. Conf. Comp. Mats. (ICCM V)*, San Diego, W. C. Harrigan, J. Strife and A. K. Dhingra (eds.), TMS-AIME, pp. 843–66.
21. K. R. Bain and M. L. Gambone (1990) Fatigue Crack Growth of SCS-6/Ti–6–4 MMC, in *Fundamental Relationships between Microstructure and Mechanical Properties of Metal Matrix Composites*, Indianapolis, M. N. Gungor and P. K. Liaw (eds.), TMS, pp. 459–70.
22. S. M. Soudani and M. L. Gambone (1990) Strain Controlled Fatigue Testing of SCS-6/Ti–6Al–4V MMC, *ibid.*, pp. 669–704.
23. J. L. Yuen, D. W. Petrasek and G. D. Schnittgrund (1990) Effect of Thermal Shock on Fibre-Reinforced Superalloy Composites, in *Fundamental Relationships between Microstructure and Mechanical Properties of Metal Matrix Composites*, Indianapolis, M. N. Gungor and P. K. Liaw (eds.), TMS, pp. 727–44.
24. S. C. Lim and M. F. Ashby (1987) Wear Mechanism Maps, *Acta Metall.*, **35**, pp. 1–24.
25. F. M. Hosking, F. Folgar-Portillo, R. Wunderlin and R. Mehrabian (1982) Composites of Aluminium Alloys: Fabrication and Wear Behaviour, *J. Mat. Sci.*, **17**, pp. 477–98.
26. K. J. Bhansali and R. Mehrabian (1982) Abrasive Wear of Aluminium Matrix Composites, *J. Metals*, **32**, pp. 30–4.
27. Y. M. Pan, M. E. Fine and H. S. Cheng (1990) Wear Mechanisms of Aluminium-Based Metal Matrix Composites under Rolling and Sliding Contact, in *Tribology of Composite Materials*, P. K. Rohatgi, B. G. Blau and C. S. Yust (eds.), ASM, pp. 93–101.
28. I. M. Hutchings (1991) Abrasive and Erosive Wear of Metal Matrix Composites, in *Proc. 2nd European Conf. on Adv. Mat., Euromat '91*, Cambridge, UK, T. W. Clyne and P. J. Withers (eds.), Inst. of Metals, pp. 56–64.
29. K. Akutagawa, H. Ohtsuki, J. Hasegawa and M. Miyazaki (1987) *Reduction of the Friction Coefficient of MMC under Dry Conditions*, SAE Tech. Report 870441.
30. A. G. Wang and I. M. Hutchings (1989) Wear of Alumina Fibre–Aluminium Matrix Composites by Two-Body Abrasion, *Mat. Sci. Tech.*, **5**, pp. 71–5.
31. I. M. Hutchings and A. G. Wang (1988) Fracture under Point Loading of Brittle Fibres in a Ductile Matrix, *Phil. Mag.*, **A57**, pp. 197–206.
32. P. K. Rohatgi, N. B. Dahotre, Y. Liu, M. Lin and T. L. Barr (1988) Tribological Behaviour of Al Alloy–Graphite and Al Alloy–Microcrystalline Carbon Particulate Composites, in *Cast Reinforced Metal Composites*, Chicago, S. G. Fishman and A. K. Dhingra (eds.), ASM, pp. 367–74.
33. A. K. Jha, S. V. Prasad and G. S. Upadhyaya (1988) Dry Sliding Wear of Sintered 6061 Al-Alloy Based Particulate Containing Solid Lubricants, in *Wear Resistance of Metals and Alloys*, G. R. Kingsbury (ed.), ASM, pp. 73–80.
34. D. Kuhlmann-Wilsdorf and D. D. Makel (1988) Friction, Wear and Interfacial Temperatures in Metal–Graphite Composites, in *Cast Reinforced*

Composites, Chicago, S. G. Fishman and A. K. Dhingra (eds.), ASM, pp. 347–60.
35. W. M. Stobbs (1973) The Work-Hardening of Cu–SiO_2 – III. Diffusional Stress Relaxation, *Phil. Mag.*, **27**, pp. 1073–92.
36. R. Monzen, K. Suzuki, A. Sato and T. Mori (1983) Internal Friction Caused by Diffusion Around a Second Phase Particle, *Acta Metall.*, **31**, pp. 519–24.
37. F. J. Guild and P. D. Adams (1981) A New Technique for the Measurement of Specific Damping Capacity of Beams in Flexure, *J. Phys. E. Sci. Instr.*, **14**, pp. 355–63.
38. R. B. Bhagat, M. F. Amateau and E. C. Smith (1988) Logarithmic Decrement Measurements on Mechanically Alloyed Al and SiC Particulate Reinforced Al Matrix Composites, in *Cast Reinforced Metal Composites*, Chicago, S. G. Fishman and A. K. Dhingra (eds.), ASM, pp. 399–405.
39. J. A. DiCarlo and J. E. Maisel (1979) Measurement of the Time Temperature Dependent Dynamic Mechanical Properties of Boron/Al Composites, in ASTM STP 674, pp. 201–27.
40. A. Wolfenden and J. M. Wolla (1988) Damping and Dynamic Modulus Measurements in Metal Matrix Composites with the PUCOT, in *Mechanical and Physical Behaviour of Metallic and Ceramic Composites, 9th Risø Inst. Symp.*, Roskilde, S. I. Andersen, H. Lilholt and O. B. Pedersen (eds.), Risø Nat. Lab., Denmark, pp. 511–16.
41. C. A. Updike, R. B. Bhagat, M. J. Pechersky and M. F. Amateau (1990) The Damping Performance of Aluminium Based Composites, *J. Metals*, **42**, pp. 44–6.
42. A. A. Baker (1968) The Fatigue of Fibre Reinforced Aluminium, *J. Mat. Sci.*, **3**, pp. 412–23.
43. A. Varshavsky (1972) The Matrix Fatigue Behaviour of Fibre Composite Subjected to Repeated Tensile Loads, *J. Mat. Sci.*, **7**, pp. 159–67.
44. H. J. Weiss (1977) The Effect of Axial Residual Stresses on the Mechanical Behaviour of Composites, *J. Mat. Sci.*, **12**, pp. 797–809.
45. A. Wolfenden and J. M. Wolla (1989) Mechanical Damping and Dynamic Modulus Measurements in Alumina and Tungsten Fibre Reinforced Aluminium, *J. Mat. Sci.*, **24**, pp. 3205–12.
46. R. B. Bhagat, M. F. Amateau and E. C. Smith (1988) Damping Behaviour of Squeeze Cast Planar Random Carbon Fibre Reinforced 6061 Aluminium Matrix Composites, in *Cast Reinforced Metal Composites*, S. G. Fishman and A. K. Dhingra (eds.), ASM, pp. 407–13.
47. E. G. Kendall (1974) High Modulus Graphite Fibres, in *Composite Materials*, L. J. Broutman and R. H. Krock (eds.), Academic Press, pp. 341–71.
48. A. A. Baker (1975) Carbon Fibre Reinforced Metals – A Review of the Current Technology, *Mat. Sci. Eng.*, **7**, pp. 177–208.
49. P. K. Rohatgi, N. Murali, H. R. Shetty and R. Chandrashekhar (1976) Improved Damping Capacity and Machinability of Graphite Particle–Aluminium Alloy, *Mat. Sci. & Eng.*, **26**, pp. 115–22.
50. D. Nath, R. Narayan and P. K. Rohatgi (1981) Damping Capacity, Resistivity, Thermal Expansion and Machinability of Aluminium Alloy–Mica Composites, *J. Mat. Sci.*, **16**, pp. 3025–32.
51. E. F. Crawley and M. C. van Schoor (1987) Material Damping in Aluminium and Metal Matrix Composites, *Comp. Mats.*, **21**, pp. 553–68.
52. S. P. Rawal, J. H. Armstrong, M. S. Misra, A. K. Ray and V. K. Kinra

(1986) Damping Measurements of Graphite–Al Composites, in *Proc. 1986 Soc. Exp. Mech. Spring Conf.*, Soc. Exp. Mechanics, Bethel, pp. 999–1008.
53. G. E. Dieter (1986) *Mechanical Metallurgy*, McGraw-Hill, New York.
54. D. Hull (1984) Axial Crushing of Fibre Reinforced Composite Tubes, in *Structural Crashworthiness*, N. Jones and T. Wierzbicki (eds.), Butterworth, Sevenoaks, pp. 118–35.
55. F. H. Stott and D. J. Ashby (1978) The Oxidation Characteristics of Electrodeposited Nickel Composites Containing Silicon Carbide Particles at High Temperature, *Corros. Sci.*, **18**, pp. 183–98.
56. W. Wei (1990) The Effect of Long Term Thermal Exposure on the Interfacial Chemistry and Mechanical Properties of Metal Matrix Composites, in *Fundamental Relationships between Microstructure and Mechanical Properties of Metal Matrix Composites*, Indianapolis, M. N. Gungor and P. K. Liaw (eds.), TMS, pp. 353–70.
57. M. V. Hartley (1988) The Effect of Isothermal Exposure on the Mechanical Properties of a Continuous Fibre Reinforced Titanium Matrix Composite, in *Mechanical and Physical Behaviour of Metallic and Ceramic Composites, 9th Risø Int. Symp.*, Roskilde, S. I. Andersen, H. Lilholt and O. B. Pedersen (eds.), Risø Nat. Lab., Denmark, pp. 383–90.
58. J. A. Little, D. McCracken and N. Simms (1988) High Temperature Oxidation of a Particulate SiC–Aluminium Metal Matrix Composite, *J. Mat. Sci. Letts.*, **7**, pp. 1037–9.
59. P. B. Prangnell and W. M. Stobbs (1990) The Effect of SiC Particulate Reinforcement on the Ageing Behaviour of Aluminium-Based Matrix Alloys, in *Proc. 7th Int. Conf. on Comp. Mats. (ICCM VII)*, Guangzou, China, W. Yunshu, G. Zhenlong and W. Renjie (eds.), Pergamon, pp. 573–8.
60. J. A. Costello and R. E. Tressler (1981) Oxidation Kinetics of Hot Pressed and Sintered α-SiC, *J. Amer. Ceram. Soc.*, **64**, pp. 327–36.
61. I. Kvernes and P. Kofstad (1973) High Temperature Stability of Ni/W Fibre Composites in Oxygen Atmosphere, *Scand. J. Metall.*, **2**, pp. 291–7.
62. M. E. El-Dahshan, D. P. Whittle and J. Stringer (1975) The Oxidation and Hot Corrosion Behaviour of W Fibre Reinforced Composites, *Oxidation of Metals*, **9**, pp. 45–67.
63. J. H. Zaat (1983) A Quarter of a Century of Plasma Spraying, *Ann. Rev. Mat. Sci.*, **13**, pp. 9–42.
64. P. P. Trzaskoma (1982) Corrosion Rates and Electrochemical Studies of a Depleted Uranium Alloy Tungsten Fibre Metal Matrix Composite, *J. Electrochem. Soc.*, **129**, pp. 1398–402.
65. D. M. Aylor and R. M. Kain (1985) Assessing the Corrosion Resistance of Metal Matrix Composite Materials in Marine Environments, in *Recent Advances in Composites in the United States and Japan*, ASTM STP 864, J. R. Vinson and M. Taya (eds.), pp. 632–47.
66. P. P. Trzaskoma (1986) Corrosion Behaviour of a Graphite Fibre/Magnesium Metal Matrix Composite in Aqueous Chloride Solution, *Corrosion*, **42**, pp. 609–13.
67. M. Saxena, O. P. Modi, A. H. Yegneswaren and P. K. Rohatgi (1987) Corrosion Characteristics of Cast Aluminium Alloy–3 wt% Graphite Particulate Composites in Different Environments, *Corros. Sci.*, **27**, pp. 249–56.
68. P. P. Trzaskoma, E. McCafferty and C. R. Crowe (1983) Corrosion Behaviour of SiC/Al Metal Matrix Composites, *J. Electrochem. Soc.*, **130**, pp. 1804–9.

69. T. Otani, B. McEnaney and V. D. Scott (1988) Corrosion of Metal Matrix Composites, in *Cast Reinforced Metal Composites*, S. G. Fishman and A. K. Dhingra (eds.), ASM, pp. 383–90.
70. R. C. Paciej and V. S. Agarwala (1986) Metallurgical Variables Influencing the Corrosion Susceptibility of a Powder Metallurgy SiC_W/Al Composite, *Corrosion*, **42**, pp. 718–29.
71. A. J. Sedriks, J. S. Green and D. L. Novak (1971) Corrosion Behaviour of Al–B Composites in Aqueous Chloride Solution, *Metall. Trans.*, **2**, pp. 871–5.
72. S. L. Pohlman (1978) Corrosion and Electrical Behaviour of B/Al Composite, *Corrosion*, **34**, pp. 156–61.
73. P. P. Trzaskoma (1991) Corrosion, in *Metal Matrix Composites: Mechanisms and Properties*, R. K. Everett and R. J. Arsenault (eds.), Academic Press, New York, pp. 383–404.
74. F. Mansfeld and S. L. Jeanjaquet (1986) The Evaluation of Corrosion Protection Measures for Metal Matrix Composites, *Corros. Sci.*, **26**, pp. 727–34.
75. F. Mansfeld, S. Lin, S. Kim and H. Shih (1988) Electrochemical Impedance Spectroscopy as a Monitoring Tool for Passivation and Localized Corrosion of Aluminium Alloys, *Werkstoffe u. Korrosion*, **39**, pp. 487–92.

9
Fabrication processes

A variety of processes have been and are being developed for the manufacture of MMCs. These may be divided into primary material production, and secondary consolidation or forming operations. A further important distinction can be drawn for the primary processes depending on whether the matrix becomes liquid at any stage. Each technique has its own limitations in terms of component size and shape, and imposes certain micro-structural features on the product. Table 9.1 lists the processing routes discussed in this chapter as well as their applicability to the production of the different composite types[†]*. In the final section, some observations are made with respect to machining and joining of MMCs.*

As can be seen in Table 9.1, many fabrication routes are now available by which a reinforcement can be incorporated into a metal matrix. It is important to note from the outset that making the right choice of fabrication procedure is just as important in terms of the microstructure and performance of a component, as it is for its commercial viability. However, before looking in detail at the various processing options, it is worthwhile dwelling for a moment on selection of the reinforcement. Clearly, the size, shape and strength of the reinforcing particles or fibres is of central importance. Often the choice between the continuous and discontinuous options is relatively straightforward, both in terms of performance and processing cost. However, within each category there exist wide variations in reinforcement size and morphology.

As an example, consider particulate reinforcement. The most convenient form is SiC powder of about 3–30 μm diameter, which is cheap (largely

[†] Directional freezing of eutectics (§9.1.4) and reactive processing are omitted from this Table, as the reinforcement is produced *in situ* and cannot be classified in the normal way.

Table 9.1 *MMC fabrication procedures*

	Type of reinforcement				
	Continuous		Discontinuous		
Processing route	Mono-filament	Multi-filament	Staple fibre	Whisker	Particu-late
Squeeze infiltrate preform (§9.1.1)	(√)	√	√	√	(√)
Spray coat or co-deposit (§9.1.2)	√	√	×	×	√
Stir mixing and casting (§9.1.3)	×	×	(√)	(√)	√
Powder premix/extrude (§9.2.1)	×	×	√	√	√
Slurry coat/hot press (§9.2.1)	(√)	√	×	×	×
Interleave/diffusion bond (§9.2.2)	√	×	×	×	×

× – Not practicable.
(√) – Not common.
√ – Current practice.

because of the mature market for its use as an abrasive) and relatively easy to handle. Extensive research is currently being undertaken on the influence of particle size on properties, especially with respect to failure processes (§7.4.3). This work may point towards a preference for a much tighter size range than is current practice. Furthermore, a change from the common angular form (Fig. 9.1(a)) to spherical particles (Fig. 9.1(b), (c)) might also be expected to offer performance improvements (see §7.4.2) and may lead to superior flow behaviour during handling prior to MMC manufacture. There are also grounds for supposing that platelets or ribbons may have certain advantages over equiaxed particles. Alumina has been produced[1] for MMC use in the form of monocrystalline platelets (Fig. 9.1(d)) of aspect ratio about 5–25 and diameter of the order of 10 μm. These can readily be assembled into preforms for melt infiltration, often with better control over volume fraction than is possible with equiaxed reinforcement. However, given the pressures to manufacture MMCs under economically attractive conditions, particularly for Al-based particulate MMCs aimed at bulk markets, the benefits of changing the type or nature of the reinforcement must first be clearly established and understood if

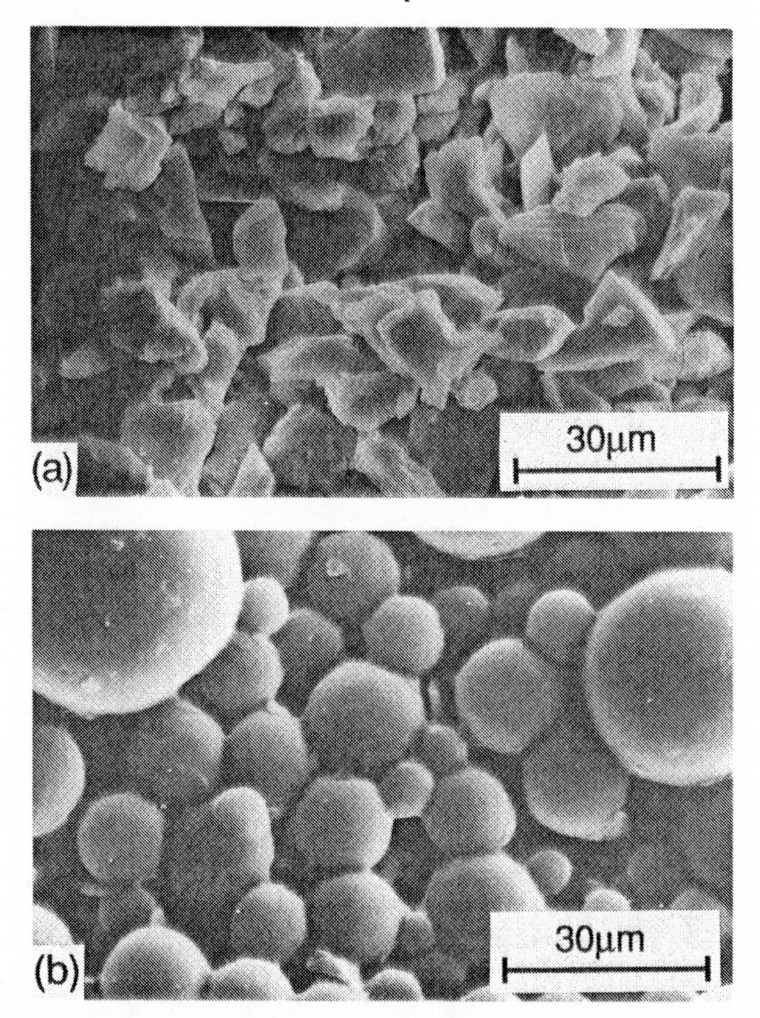

Fig. 9.1 SEM micrographs of ceramic particulate for use in MMCs. (a) Standard angular SiC grit particles, (b) spherical alumina produced by particle melting and in-flight refreezing during plasma spraying – see §9.1.2 (*cont.*).

such a change is accompanied by increased processing or raw materials costs.

9.1 Primary liquid phase processing

Various techniques have been developed which involve the matrix becoming at least partially molten as it is brought into contact with the

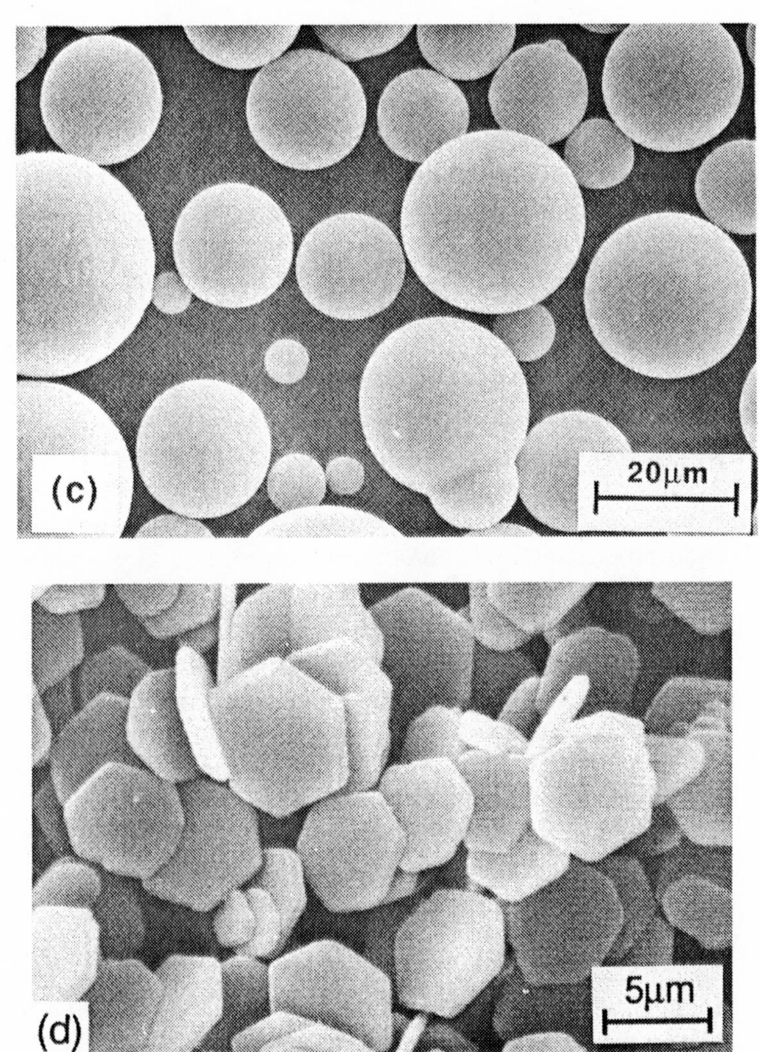

Fig. 9.1 (*cont.*) SEM micrographs of ceramic particulate for use in MMCs. (c) Spherical alumina produced by a sol-gel route (courtesy Dr D. Double, Alcan International, Banbury, UK) and (d) monocrystalline hexagonal α-alumina platelets[1].

ceramic reinforcement. This generally favours intimate interfacial contact and hence a stronger bond, but it can also lead to the formation of a brittle interfacial reaction layer. These features are sensitive to a number of factors such as liquid/ceramic contact time and pressure, which vary widely between the different processes described below.

9.1.1 Squeeze casting and squeeze infiltration

The term ***squeeze casting*** has come to be applied to various processes in which pressure is imposed on a solidifying system, usually via a single hydraulically activated ram. The technique has certain general characteristics, such as a tendency towards fine microstructures (as a result of the rapid cooling induced by good thermal contact with a massive metal mould) and low porosity levels encouraged by efficient liquid feeding[2,3]. The primary distinction between this and conventional pressure die casting is that the ram continues to move during solidification, deforming the growing dendrite array and compensating for the freezing contraction (which is typically about 5%). In addition to this, the ram movement is usually slower, and the applied pressure often greater, than is typical of die casting. Squeeze casting can be applied to MMCs, as reheated powder mixtures or as stir-cast material, but the most common pressure-assisted solidification process for MMC production is properly termed ***squeeze infiltration***. This involves the injection of liquid metal into the interstices of an assembly of short fibres, usually called a '***preform***'†.

Two variants of the process can be identified, depending on whether the preform simply fills a die cavity having a simple shape, such as a cylinder (with the composite being subjected to further shaping operations), or is designed with a specific shape to form an integral part of a finished product in the as-cast form. (The latter type of selectively reinforced casting is now of considerable industrial significance for applications such as pistons – see §12.2.1.) A number of important process characteristics are common to both variants of the process and these are now examined below.

Preform preparation

While preforms can be prepared in various ways, they are commonly fabricated by sedimentation of short fibres from liquid suspension. This procedure has, for example, often been carried out with short alumina fibres, such as 'Saffil®'. It is common to compress the preform while the liquid is being drained off, sometimes with simultaneous gentle heating. A typical experimental procedure would involve decanting the liquid, with fibres in suspension, into an open-topped cylindrical die having fine drainage holes around the base. The liquid is removed by suction while the residue is compressed by a ram. The applied pressure will influence the fibre volume fraction (usually in the range 5–30%) and the degree of

† Processes in which this occurs are also quite widely referred to as squeeze casting.

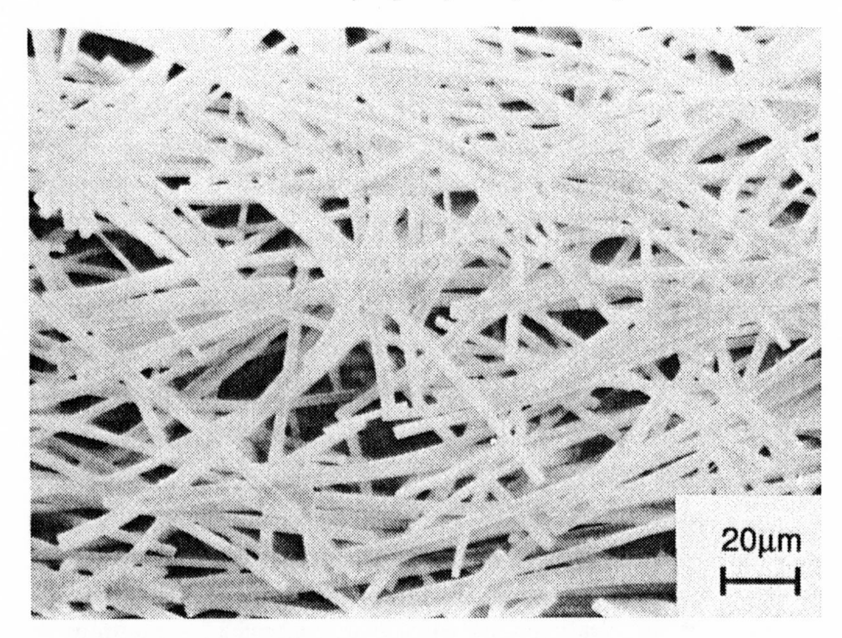

Fig. 9.2 A scanning electron micrograph of a 'Saffil®' (δ-alumina) fibre preform[14]. The fibre volume fraction is about 15%. The sedimentation direction was vertically downwards. While there is a clear tendency for the fibres to align in the plane normal to this direction, it can be seen that many fibres are inclined at an appreciable angle to this plane.

fibre fracture. The fibre distribution in a typical preform is shown in Fig. 9.2. Variants of the above procedure, such as the formation of hollow tube preforms by suction of suspended fibres onto a cylindrical former immersed in the slurry[5], are used for particular shapes. (Preforms can also be prepared from particulate or aligned long fibre reinforcements. The main problem in these cases is usually that the volume fraction of reinforcement becomes very high when pressure is applied, which may be undesirable in itself and can also cause problems of premature solidification and a very high pressure needed for infiltration. One solution for long fibres is to construct a 'hybrid' preform in which they are separated by interleaved planar random mats[6], or some similar arrangement. Long fibres can also be held in place by filament winding onto a mandrel before infiltration[7].)

In order for the preform to retain its integrity and shape, it is often necessary for a binder to be used, particularly if a relatively high volume fraction of fibre ($\geq 10\%$) is required (as unbound short fibres tend to pack naturally with a very low volume fraction). Either fugitive binders, which set at relatively low temperatures, or agents requiring a high temperature firing operation, or some combination of the two types, may be used. Various silica- and alumina-based mixtures have been popular as high temperature binders. The binding agent is normally introduced via the suspension liquid, so that it deposits or precipitates out on the fibres, often forming preferentially at fibre contact points, where it serves to lock the fibre array into a strong network. Typically, a binder might be present in the preform at levels of about 5–10% by weight. There has been interest in the chemical effects induced by the presence of the binder. For example, it has been shown[8,9] that a silica binder is rapidly attacked by Mg in the melt during squeeze infiltration, affecting both interfacial properties and matrix age-hardening characteristics[10].

Squeeze infiltration has been particularly popular for very fine fibres, such as SiC whiskers, which are difficult to blend with metallic powders because of the size difference and the tendency for tenacious fibre agglomerates to be formed, which also makes stir mixing into a melt impractical. Furthermore, many handling procedures lead to fine fibres becoming airborne, creating a risk of inhalation, which is hazardous in the case of whiskers[11]. The preform preparation procedures described above must be carried out within a glove box when whiskers are involved.

Process analysis

Infiltration is usually carried out with equipment of the type shown in Fig. 9.3, although the molten charge is often introduced from the side rather than along the axis of ram travel. The important features are the onset/progression of infiltration, the movement/fracture of the fibres and the heat flow/solidification characteristics[12–18].

The ***infiltration*** and ***fibre movement*** aspects are best considered in terms of the pressure profile through the system, illustrated schematically in Fig. 9.4. In order to carry out any calculations about melt entry and deformation of individual fibres and hence of the preform, a model is needed for the assembly of fibres. Clyne and co-workers[12–14] have shown that, if a certain fraction f_{CL} of the fibres are taken as forming 'crosslinks' between square arrays of 'in-plane' fibres, then expressions can be derived for the preform stiffness and for the applied pressure needed for various phenomena. For example, the applied pressure for the onset of fibre fracture can be written[14]

$$P^{A}_{CL*} = \frac{\sigma_{I*} f_{CL} \phi^2 f}{4} \left[\frac{f(1 - f_{CL})}{2\pi\xi(1 - \phi^2)} \right]^{1/2} \quad (9.1)$$

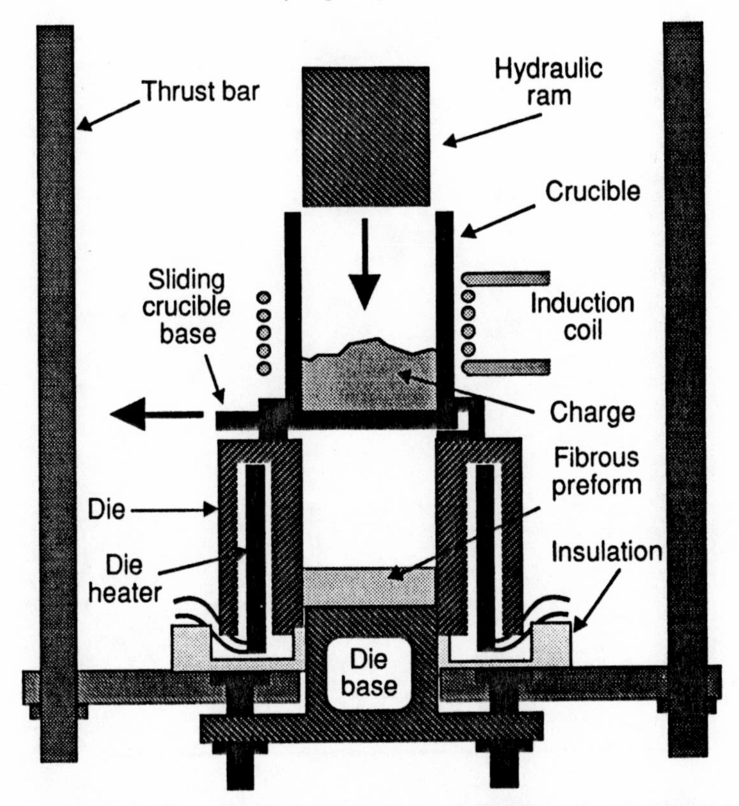

Fig. 9.3 Schematic illustration of an apparatus used for squeeze infiltration. After melting of the charge, it is allowed to fall into the die cavity by withdrawing the sliding base of the crucible, after which the ram is brought down so as to pressurise the melt and force it into the preform. Air escape paths are provided here in the form of a suitable clearance between the die and its base.

where f is the volume fraction of the preform occupied by the fibres and ϕ, ξ are, respectively, the ratios of the 'interplanar spacing' to the length of a crosslinking fibre and to the in-plane fibre spacing.

The pressure necessary for the onset of ***melt entry*** can also be calculated using the Kelvin equation

$$\Delta P_{\mathrm{i}} = \gamma\left(\frac{1}{r_1} + \frac{1}{r_2}\right) \tag{9.2}$$

where γ is the surface tension of the melt and r_1, r_2 are the principal radii of curvature at the front. A lower limit for these radii (which are the same for a square array) is half the interfibre spacing,

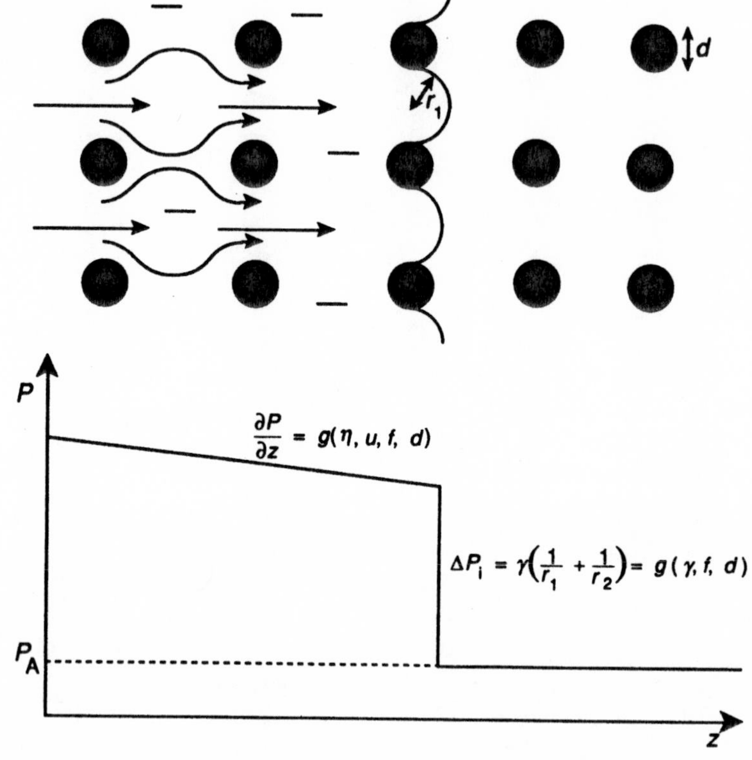

Fig. 9.4 Schematic illustration of the pressure profile through the system during infiltration. A pressure increment is needed at the infiltration front in order to provide the necessary meniscus curvature, while a pressure gradient is necessary behind the front to overcome viscous resistance to flow.

leading[14] to

$$\Delta P_i = \frac{4\gamma}{d\left[\left(\frac{\pi}{2\xi f(1 - f_{CL})}\right)^{1/2} - 1\right]} \tag{9.3}$$

where d is the fibre diameter.

While this geometrical model is necessarily oversimplistic and some difficulty is encountered in specifying appropriate parameters, a number of broad features of the process can be explored using these equations. For example, Fig. 9.5 shows the predicted infiltration pressure for an aluminium melt given by eqn (9.3), as a function of fibre volume fraction, for three fibre diameters. This demonstrates that, while vacuum infiltration ($\Delta P \sim 0.1$ MPa) is possible with relatively coarse fibres ($d \geq 20$ μm),

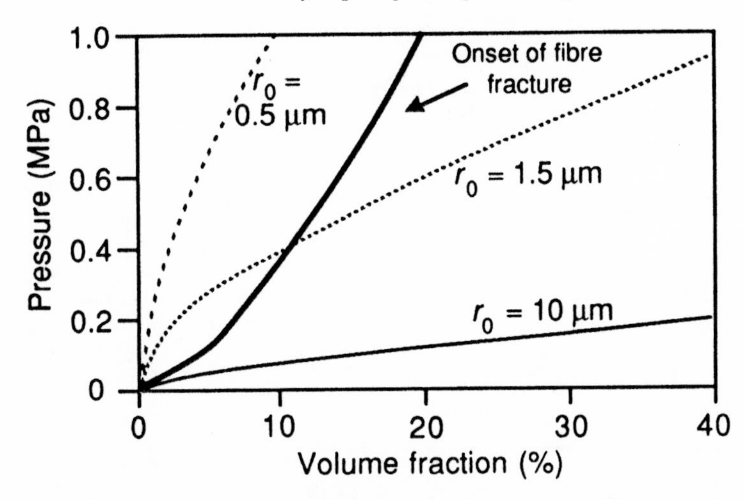

Fig. 9.5 Predicted[14] pressures needed for the onset of melt entry, based on meniscus curvature requirements, as a function of fibre volume fraction in the preform, for three fibre diameters. Also shown for the 3 μm diameter fibre is the predicted pressure needed for fracture of ('crosslinking') fibres.

whiskers are likely to require substantial applied pressures (~1–3 MPa). These plots suggest that extensive fibre fracture is unlikely, as infiltration will take place at similar applied pressures and fibres will not break once infiltration has occurred. This is broadly consistent with experimental observations.

The most significant simplification involved into the above modelling concerns the neglect of wetting (i.e. a contact angle, $\theta = 180°$). Wetting will lower ΔP_i, by allowing reduced curvature[19], and the extent of this reduction has been fully quantified[20]. It may often be the case that this can be neglected, in view of the need for a reduced contact angle to be established extremely quickly if it is to be set up repeatedly at many successive fibre layers during squeeze infiltration. Nevertheless there is considerable interest[21,22] in wetting characteristics and in adding agents which can dramatically encourage wetting[23], allowing infiltration under reduced pressure or pressureless infiltration[22,24] (which is likely to take place relatively slowly). A particular system which shows promise is the use of a nitrogen atmosphere for Al melts containing Mg, which leads to spontaneous slow infiltration of preforms made from various ceramics[22,24]. The mechanism appears to be complex, but probably involves transient formation of AlN. This type of infiltration has been studied as part of the family of reactive processing techniques being

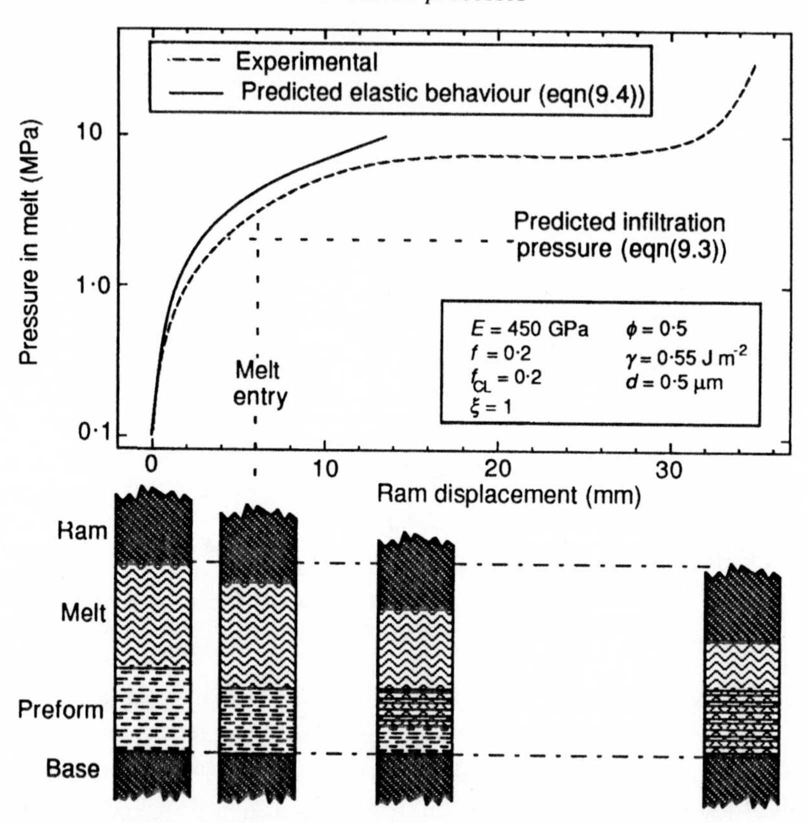

Fig. 9.6 Experimental[25] pressure–ram displacement data during infiltration of a SiC whisker preform with a Mg–Li alloy melt. The pressure data were obtained from a strain gauge attached to a thrust bar (Fig. 9.3). The ram displacement at the point of infiltration (~6 mm) was established from the measured compression of the preform in the casting. (It is an upper limit in view of the possibility of post-infiltration relaxation.) The pressure continues to rise somewhat after infiltration as a result of the resistance to viscous flow through the preform, possibly enhanced by some melt solidification. Also shown are predictions for elastic deformation of the preform and the melt entry pressure.

developed by the Lanxide Corporation (see §9.1.4). This approach offers promise for the fabrication of MMCs with very high ceramic contents, for which pressurised infiltration tends to be difficult.

In any event, quantitative data have been obtained indicating that with $\theta = 180°$ the simple model is broadly reliable for rapid infiltration of a normal preform. For example, the pressure–ram displacement history[25] shown in Fig. 9.6 indicates a melt entry pressure of about 3 MPa, compared with a predicted value using eqn (9.3) of 2.1 MPa, while the elastic deformation of the preform prior to infiltration agrees quite well with the prediction based on bending of the fibres[14]

$$\varepsilon = \frac{8\pi\xi(1-\phi^2)P^A}{3E_I f^2(1-f_{CL})f_{CL}\phi^4} \quad (9.4)$$

where E_I is the Young's modulus of the fibres. In practice, high pressures of at least several tens of MPa are commonly applied, which may be beneficial in terms of helping to reduce porosity arising from the freezing contraction (discussed below under 'Technical variants').

Among other aspects of the process which have been analysed[14] is the pressure gradient needed to overcome the resistance of the melt to viscous flow through the channels between the fibres. A broad conclusion from this analysis is that, largely as a consequence of the low viscosity of liquid metals, infiltration will take place relatively quickly (in less than a few seconds†) with modest pressure differences across the preform of ~1 MPa. It may also be noted that flow is not expected to become turbulent during the passage of the melt through the preform. This means that air entrainment at the front should be minimal, although it is important that air escape paths remain uninterrupted until infiltration is complete.

Finally, another area of major interest concerns the ***heat flow*** and ***solidification*** characteristics of the system. Infiltration normally takes place very quickly. This means that, since the fibres are in most cases relatively fine, local heat exchange with the melt is effectively instantaneous. On this basis, it is a simple matter to estimate the minimum melt superheat needed to avoid premature solidification, using a global heat balance requirement

$$\Delta T = \frac{(T_{mp} - T_p)c_I f}{c_M(1-f)} \quad (9.5)$$

where c_I and c_M are volume specific heats of fibre and matrix, T_{mp} is the melt fusion (liquidus) temperature and T_p is the initial preform temperature. For a preform preheated to, say, 350 °C (a typical die temperature), this equation indicates that an Al melt must be superheated by at least about 150 °C. In practice, higher values are often necessary to avoid excessive solidification immediately behind the advancing melt front, which is a common cause of incomplete infiltration. Among the

† In some cases the duration of melt passage through the preform may be determined, not by the pressure gradient in the liquid, but by the maximum velocity of the ram, particularly when this is relatively slow.

measures put forward to counteract this problem is the use of fibre coatings which release heat on reacting with the melt[18,26]. An alternative approach is to heavily preheat the preform, which is then placed in the die immediately before casting. This gives some control over the final microstructure (see below), although delaying solidification can increase the danger of melt penetration into the air escape paths. Attempts have been made[6] to identify operating regimes of preform and melt temperatures, taking account of the dangers of incomplete infiltration and excessive fibre/matrix reaction. Numerical modelling[14,17,18] of the heat flow is necessary if the solidification characteristics are to be explored in detail.

Composite microstructure

When squeeze infiltration is carried out correctly, the fibre distribution in the composite closely mirrors that in the preform. It is possible to avoid significant microporosity, macrovoids, fibre breakage and local variations in fibre volume fraction, although these and other defects can occur extensively under adverse conditions. Furthermore, preform shape and location within the die can be chosen so as to give selective reinforcement. However, there are other features of the microstructure which are of interest and certain general points can be identified concerning these, although details will obviously vary between different systems. For example, the matrix grain size tends to be largely controlled by heat flow effects, with a coarse columnar structure favoured by substantial preheating of the preform (Fig. 9.7).

The above observations suggest that the fibres do not act as preferential crystal nucleation sites and this seems to be the case for most common systems[14,15,26]. A consequence of this is that during dendritic solidification (which occurs with most matrices of interest), the last liquid to freeze, which is normally solute-enriched, tends to be located around the fibres[9,13–15,27,28]. An example of this can be seen in Fig. 9.8. This relatively prolonged fibre/melt contact, often under high hydrostatic pressure and with solute enrichment, tends to favour formation of a strong interfacial bond, in many cases promoted by a localised chemical reaction. Also worth noting in this context is the question of oxide film formation. It might be expected, at least with metals such as aluminium, that an oxide film would repeatedly form at the infiltration front and become deposited on the fibres. In fact, at least with relatively fine fibres, it can be shown[14] that films of significant thickness (>atomic monolayer) cannot form because of the limited oxygen availability in the system, and this is

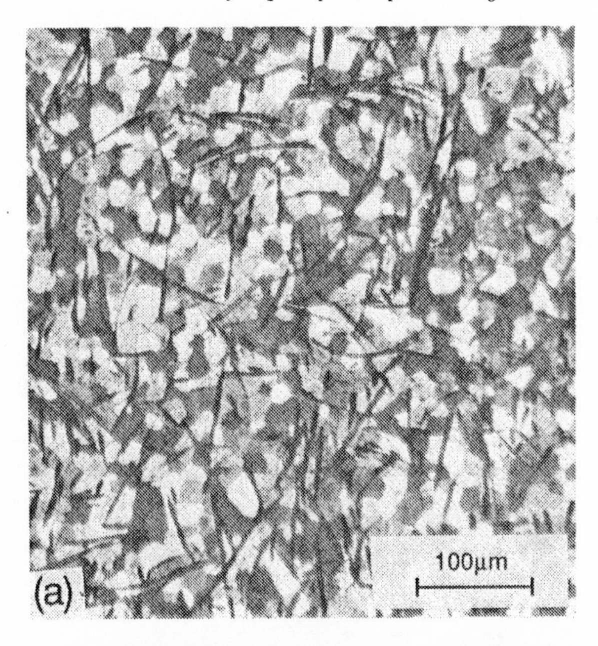

Fig. 9.7 Optical micrographs[14] showing the grain structures of Al–Mg/Saffil® composites, produced with preforms preheated to (a) about 350 °C (transverse section) and (b) about 900 °C (longitudinal section). The strong chilling effect of fibres which are cooler than the melt freezing temperature leads to a finer and less directional grain structure.

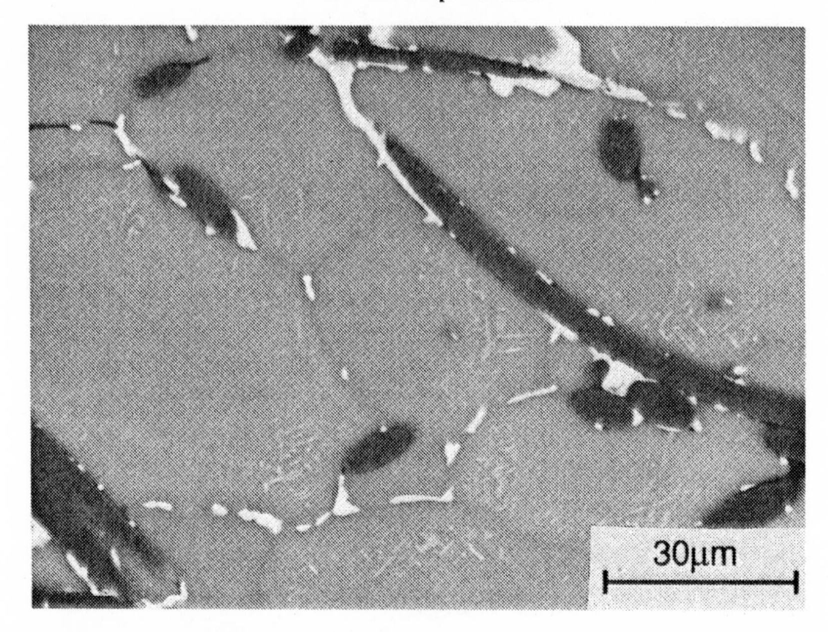

Fig. 9.8 Back-scattered SEM micrograph of an Al–4.5 wt% Cu/Saffil® composite, showing copper-rich (light) regions around the fibres (and at grain boundaries). These solute-enriched regions indicate the locations of the last-solidifying liquid, for solutes (such as copper in aluminium) which have a partition coefficient less than unity.

consistent with microanalysis[8]. The absence of oxide films is a noteworthy feature of squeeze cast material, which probably contributes to the high interfacial bond strengths commonly observed.

Chemical interactions, for example between melt and fibre or melt and binder, can lead to both macro- and micro-segregation effects. This has been observed in various systems[9,10,29,30] and may lead to property variations from one part of the composite to another. The factors influencing heat and mass transport during casting are important in this context, in addition to the importance of melt, fibre and binder compositions.

Technical variants

Various designs have been proposed and developed for squeeze infiltration equipment. While it is not appropriate here to consider these in detail, mention should be made of a few technical points. For example, a distinction can be drawn[6] between 'direct' casting, in which entry of the

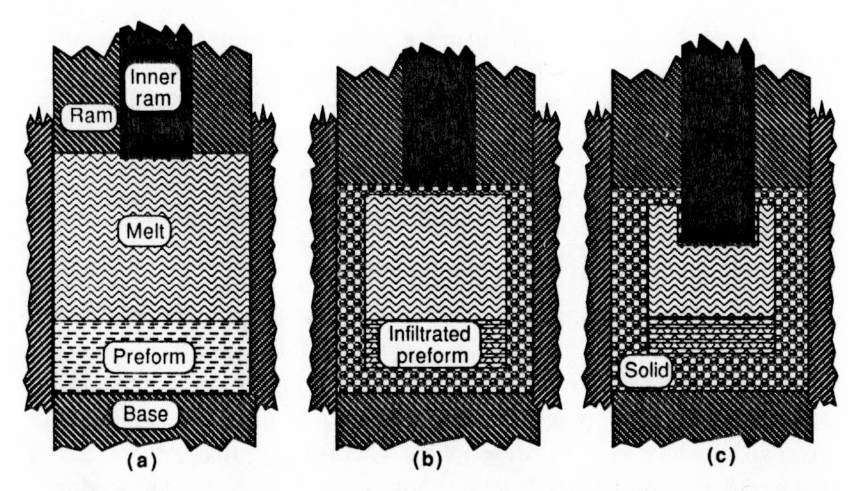

Fig. 9.9 Schematic illustration of a double ram device designed to ensure complete infiltration and low porosity. After initial infiltration (a), solidification progresses inwards forming an envelope, (b), which inhibits transmission of the applied pressure to the central reservoir. (This problem is likely to be worse for alloys freezing with a narrow mushy zone, such as pure metals and eutectic alloys, which form a strong envelope more quickly.) At this stage, the inner ram is pressurised (c) so as to break through the envelope and to re-establish a pressure to counteract the continuing solidification shrinkage.

charge into the die and melt pressurisation are separate operations, and 'indirect' casting, in which entry and pressurisation are effected by a single piston stroke. The latter arrangement, which is similar to conventional pressure die casting, allows thinner sections to be cast and is better suited to die evacuation, but melt entry is usually more turbulent and large section ingates are needed if the melt pressure is to be maintained during infiltration and to feed shrinkage.

In fact, the problem of maintaining the melt pressure is central to the avoidance of porosity and incomplete infiltration, even with direct casting, and some attention has been devoted to this aspect. Transmission of pressure from the ram becomes difficult after a solidified envelope has formed. Arrangements have been designed to overcome this using direct casting, such as the set-up shown in Fig. 9.9. After a solid envelope has formed, and the main ram has ceased to move, high pressure is applied to an inner ram, which breaks through the envelope and ensures that feeding of the solidification shrinkage continues and porosity is eliminated.

Indirect squeeze casting can also be carried out by injecting a metal/fibre slurry into a die. For example, a technique has been described[31] in which such a slurry has been produced by a powder blending and reheating procedure. In this case, very large ingates are needed in view of the high viscosity of the slurry, although having a viscous charge actually brings advantages in terms of reduced turbulence and air entrainment on entry into the die.

9.1.2 Spray deposition

A number of processes have evolved in which a stream of metallic droplets impinges on a substrate in such a way as to build up a composite deposit. The reinforcement can be fed into the spray, if particulate, or introduced onto the substrate in some way, if fibrous. The techniques employed fall into two distinct classes, depending whether the droplet stream is produced from a molten bath, or by continuous feeding of cold metal into a zone of rapid heat injection. In general, spray deposition methods are characterised by rapid solidification, low oxide contents, significant porosity levels and difficulties in obtaining homogeneous distributions of reinforcement.

The Osprey process (melt atomisation)

Spray deposition was developed commercially in the late 1970s and throughout the 1980s by Osprey Ltd (Neath, UK) as a method of building up bulk material by atomising a molten stream of metal with jets of cold gas, with the effect that most such processes are covered by their patents or licences and are now generally referred to as 'Osprey Processes'[32–34]. The potential for adapting the procedure to particulate MMC production by injection of ceramic powder into the spray was recognised at an early stage and has been developed by a number of primary metal producers[35,36]. A typical apparatus, designed in this case for the production of cylindrical ingots, is illustrated in Fig. 9.10. Very high melt feed rates, reaching the $kg\,s^{-1}$ range, can be developed, allowing ingots measured in tonnes to be produced during relatively brief runs. The ceramic particulate is normally injected into the gas stream, rather than the melt stream, and does not normally become significantly heated in flight. Although the ceramic content can in principle be controlled by the relative feed rates of melt and ceramic, in practice the overspray loss rate for the ceramic gets higher as the feed ratio is increased and so 20–25% of particulate is the upper limit for successful incorporation.

The metal droplet size depends on nozzle design and flow rates of metal and atomising gas. A fairly reliable prediction of the mean size of atomised

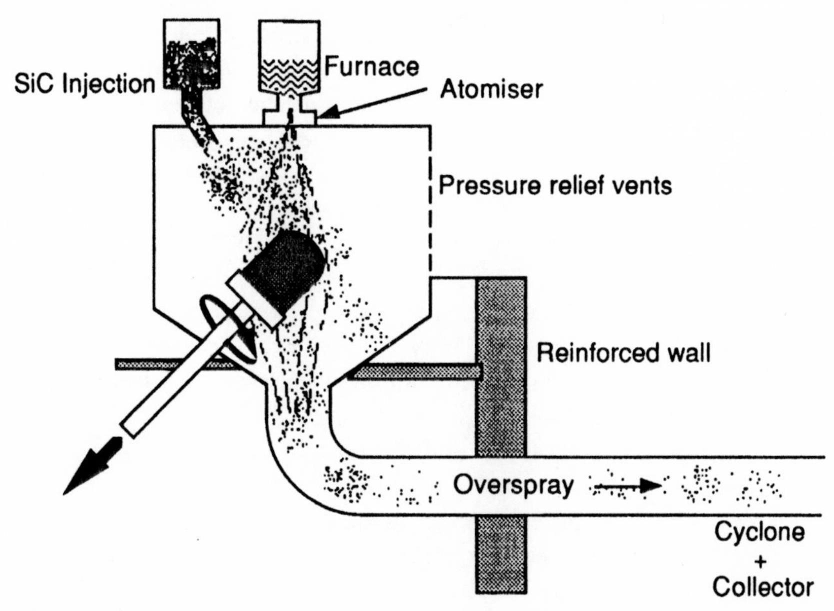

Fig. 9.10 Schematic illustration[35] of an Osprey spray casting arrangement for composite manufacture. Ceramic particles are injected into the spray cone and become incorporated into the deposited ingot. The substrate is continuously rotated and withdrawn, keeping the flight distance approximately constant.

droplets can be obtained from the semi-empirical Lubanska correlation[37]

$$\bar{d} \approx 50 d_{\text{melt}} \left[\frac{\eta_{\text{melt}}}{\eta_{\text{gas}} N_{\text{We}}} \left(1 + \frac{j_{\text{melt}}}{j_{\text{gas}}} \right) \right]^{1/2} \tag{9.6}$$

in which η is the kinematic viscosity, j is the mass flow rate and the Weber number is given by

$$N_{\text{We}} = \frac{u^2 \rho_{\text{melt}} d_{\text{melt}}}{\gamma_{\text{melt}}} \tag{9.7}$$

where u is the droplet velocity and γ is the surface tension. Use of this equation suggests a mean diameter of about 100–200 μm for the atomisation of Al with argon in a typical Osprey process, although variation over a wide range is undoubtedly possible. As the ceramic particles might be between 3 and 20 μm in diameter, this implies a fairly substantial mass difference between metal and ceramic particles, giving very different flight times and raising the probability of in-flight collisions. The droplet velocities typically average[38] about 10–40 m s^{-1}, giving a typical flight time of a few tens of ms. As the freezing time would be somewhat greater than that for droplets of this size[39], it follows that most droplets will not freeze in flight. In fact, as a result of this and the high droplet arrival flux, it seems likely that a thin layer of liquid, or more probably semi-solid, is often present on the top of the ingot as it forms. The solidification rate, although high, is therefore likely to be considerably lower than in many 'Rapid Solidification Processes'.

A feature of much MMC material produced by the Osprey route is a tendency towards inhomogeneous distribution of the ceramic particles. It is common to observe ceramic-rich layers approximately normal to the overall growth direction. This may be the result of hydrodynamic instabilities in the powder injection and flight patterns or possibly to the repeated pushing of particles by the advancing solidification front (§9.1.3) in the liquid or semi-solid layer, until the ceramic content is too high for this to continue. In any event, it is usually possible to minimise these inhomogeneities by empirical changes to the spraying conditions and they can then often be reduced or eliminated by subsequent secondary processing such as extrusion (§9.3.1). Among the other notable microstructural features of Osprey MMC material are a strong interfacial bond, little or no interfacial reaction layer and very low oxide content. Porosity in the as-sprayed state is typically about 5%, but this is normally eliminated by secondary processing. A number of commercial alloys have been explored for use in Osprey route MMCs[40,41].

Thermal spraying

Thermal spraying involves the feeding of powder, or in some cases wire, into the hot zone of a torch, where it is heated and accelerated as separate particles by the rapid expansion of a gas. The heating is usually generated by an electric arc or by gas combustion. Particularly good process control is possible with plasma spraying, which is now a mature technology[42] with considerable potential for future development[43]. A typical design for a plasma spray torch is shown in Fig. 9.11. Another promising technique[44,45] for spraying of MMCs is the use of high velocity oxy-fuel (HVOF) guns. Such processes are in established industrial use for the production of surface coatings. There has recently been interest in composite deposits, which are readily produced with twin powder feeding arrangements. In addition to the production of bulk MMC material in this way[46–48], there is also considerable interest[49–51] in graded composition MMC coatings, an example of which is shown in Fig. 9.12, exhibiting both good wear resistance (ceramic-rich region) and strong adhesion to a metallic substrate (metal-rich region).

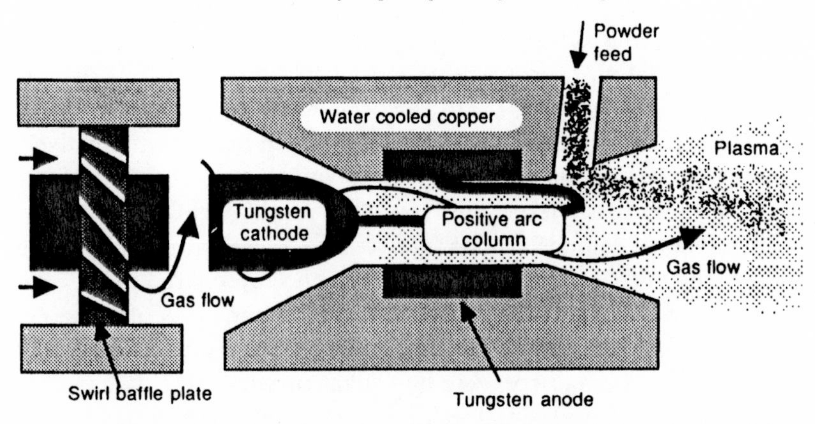

Fig. 9.11 Schematic section through a Plasma Spray Torch. Powders are injected close to the core of the plasma, where temperatures are of the order of 10 000–20 000 K. Although the dwell time in this region is normally less than 1 ms, most materials can readily be melted and sprayed with this process.

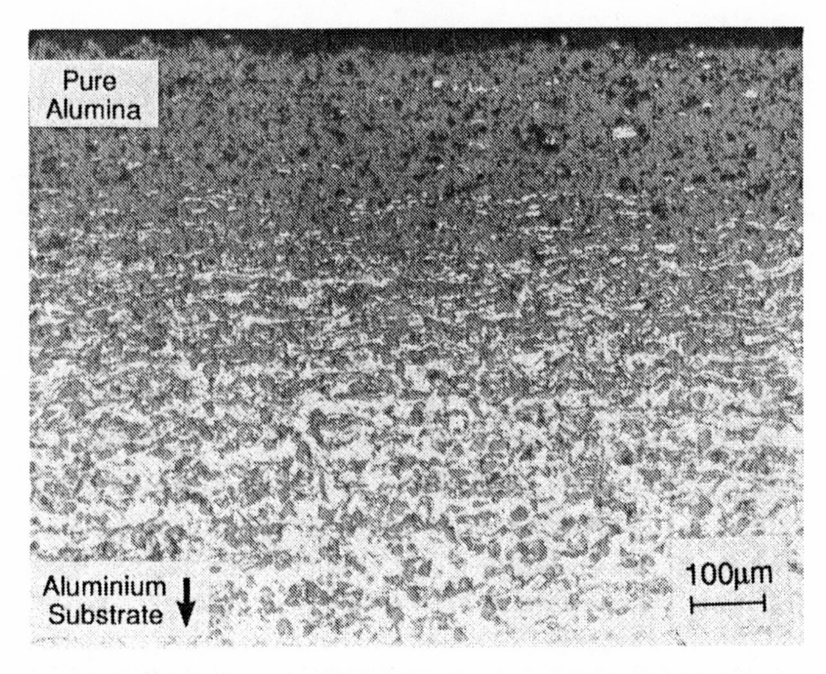

Fig. 9.12 A graded composition MMC coating produced by plasma co-spraying. This structure was produced by co-spraying of alumina and aluminium powders, using independently variable feed rates. The plasma power was progressively increased so that melting of the alumina particles became complete as the volume fraction became large.

Thermal spraying differs in several respects from melt atomisation processes. Deposition rates (usually $< 1\ \mathrm{g\,s^{-1}}$) are slower, but particle velocities (~ 200–$700\ \mathrm{m\,s^{-1}}$) are higher[52], particularly for HVOF guns[44,45]. In general, the substrate and deposit tend to remain relatively cool and the quenching rates for each individual splat are very high[53] ($\geq 10^6\ \mathrm{K\,s^{-1}}$). Porosity levels are typically several %, but can be below 1%. Other types of composite material have also been produced by plasma spraying† and include laminated metal/ceramic layer structures[54] and fibre-reinforced laminae produced by spraying onto an array of monofilaments[55] (Fig. 9.13). There is also interest in producing long fibre MMCs by simultaneous thermal spraying and filament winding[56]. It has been suggested[57] that a RF plasma torch may be better suited to this type of composite production than a DC system, as droplet impact velocities are lower so that there is less damage to the fibres.

9.1.3 Slurry casting ('compocasting')

Arguably the simplest and most economically attractive method of MMC manufacture is to simply stir mix the liquid metal with solid ceramic particles and then allow the mixture to solidify. The slurry can be continuously agitated while the ceramic is progressively added. This can in principle be done using fairly conventional processing equipment and can be carried out on a continuous or semi-continuous basis. This type of processing is now in commercial use for Al–SiC_p composites[58,59] and the material produced is suitable for further operations such as pressure die casting[60], but details of the conditions employed during the stir casting have not been published. Nevertheless, the main attractions and difficulties are fairly clear. The problems may be divided into three main areas – forming difficulties, microstructural inhomogeneities and interfacial chemical reactions.

Viscous resistance to forming operations

The increase in viscosity on adding solid to liquid is well documented[61,62] and for fibres is so severe that fibrous MMCs containing more than a few % of ceramic cannot be made in this way. Interest in slurry processing was stimulated by the development of ***rheocasting***[63,64], in

† Ceramic particles are readily melted by plasma spraying, in which the core temperatures are $\sim 10\,000$–$20\,000$ K, but this is often difficult with HVOF guns, in which the maximum flame temperature is usually less than 3000 K and the particle dwell times are very short.

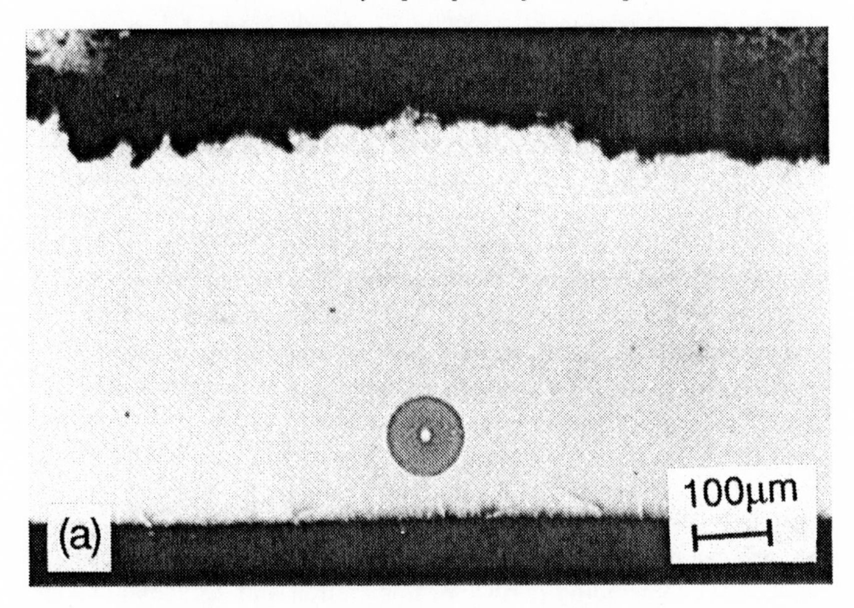

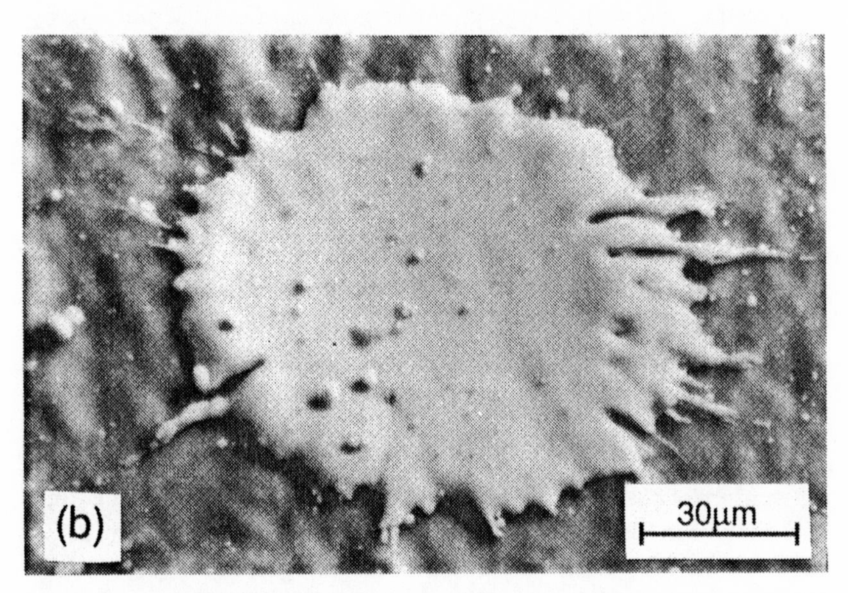

Fig. 9.13 (a) A SiC monofilament surrounded by a Ti matrix deposited by Vacuum Plasma Spraying. The fibre is enveloped by lateral spreading of droplets after impact. Typically, a 50 μm diameter droplet arriving at about 300 m s^{-1} will spread laterally in all directions by about 50–70 μm in forming a splat 3 μm thick (b). This readily allows monofilaments of 100–150 μm diameter to be enveloped, provided they are not very closely spaced.

which a solidifying liquid metal slurry can be made to remain relatively fluid with up to about 40% solid, provided it is continuously agitated so as to break up the dendritic structure to form spherical particles. This led readily to the concept of ***compocasting***[65,66], in which a similar operation is carried out with a suspension of ceramic particles. In this case agitation is necessary, not to break up dendrites, but to encourage initial particle wetting, discourage particle agglomeration and counter particle sedimentation. Broadly, the viscosity[67,68] will not increase by more than a factor of two or so with up to about 25 vol% particulate, provided the particles remain well-dispersed in suspension. This viscosity is sufficiently low to allow fairly conventional casting operations to be carried out. It can, however, be difficult to ensure a good dispersion[69] and problems can be encountered in the initial incorporation of the particles into the melt[70,71].

Microstructural homogeneity

There are several possible sources of microstructural inhomogeneity in stir cast MMCs, including particle agglomeration and sedimentation in the melt, gas bubble entrapment, porosity from inadequate liquid feeding during casting and particle segregation as a result of particle pushing by an advancing solidification front. The last of these is probably the least well understood and the most difficult to eliminate. A typical microstructure illustrating the effect of particle pushing during solidification is shown in Fig. 9.14. The effect can be aggravated by post-casting treatments in the semi-solid range[73,74], which tend to coarsen the microstructure. The situation is likely to be improved by ensuring that solidification is rapid, both as a result of refinement in the scale of the structure and because a critical growth velocity is predicted, above which solid particles should be enveloped rather than pushed[75–79].

Basically, the pushing effect arises from a 'disjointing force' exerted on the particle by the advancing front if the total interfacial energy would be raised by formation of a solid/particle interface. This is a simple thermodynamic effect, but modelling of the pushing process is complicated by the variety of phenomena which might influence the net retarding force acting on the particle. Prominent among these is the requirement for liquid to flow around the particle and into the narrow channel between particle and solidification front, which gives rise to a viscous drag force. Among other effects of possible significance are the disturbance to the heat flow and solute redistribution processes and the presence of buoyancy forces. Further complications arise from the presence of neighbouring particles, which is expected to oppose pushing by inhibiting the required liquid

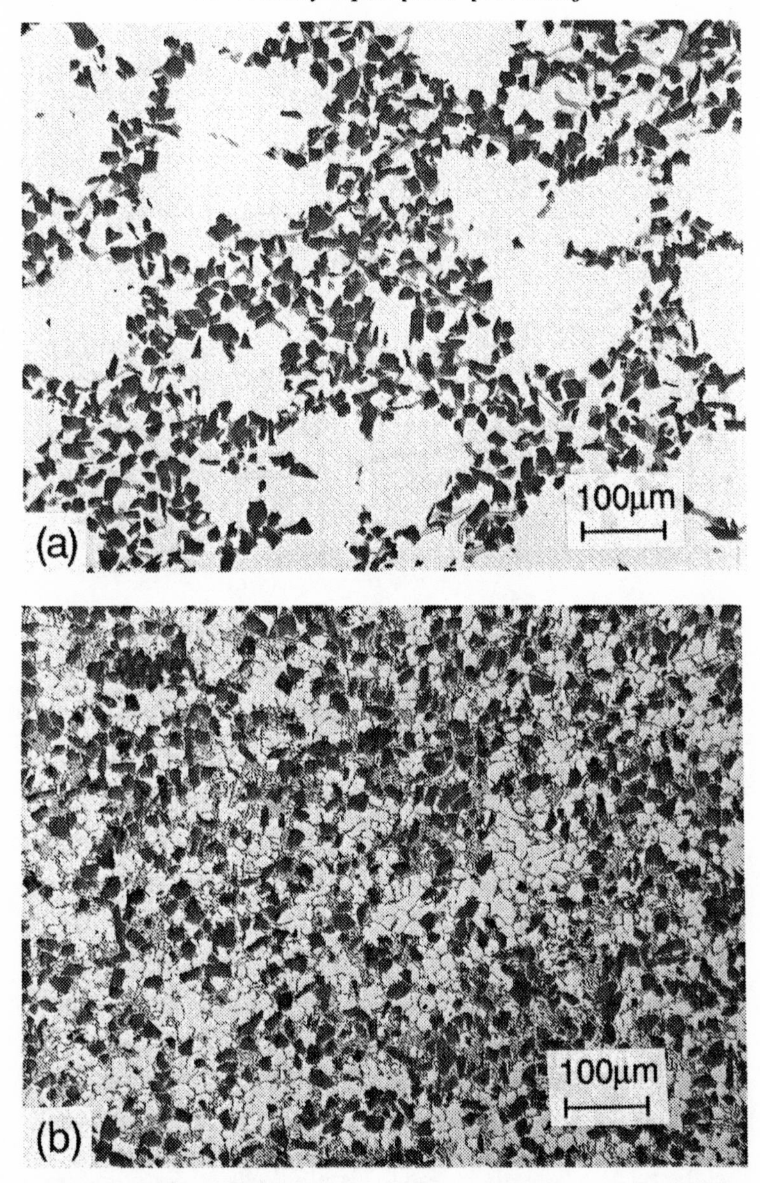

Fig. 9.14 Microstructures[72] of Al–7 wt% Si/20 vol% SiC_P, (a) investment cast (slow cooling) and (b) pressure die cast (rapid cooling). At the slower cooling rate, it is clear that the SiC particles have been pushed into the interdendritic regions by the growing dendrites, causing severe clustering. For the more rapid cooling, it seems likely that the growing dendrites have at least partially engulfed the particles, so that less pushing has occurred; in any event, because the scale of the dendrite arm spacing is similar to that of a typical particle diameter, pushing can only occur over short distances and does not lead to pronounced clustering.

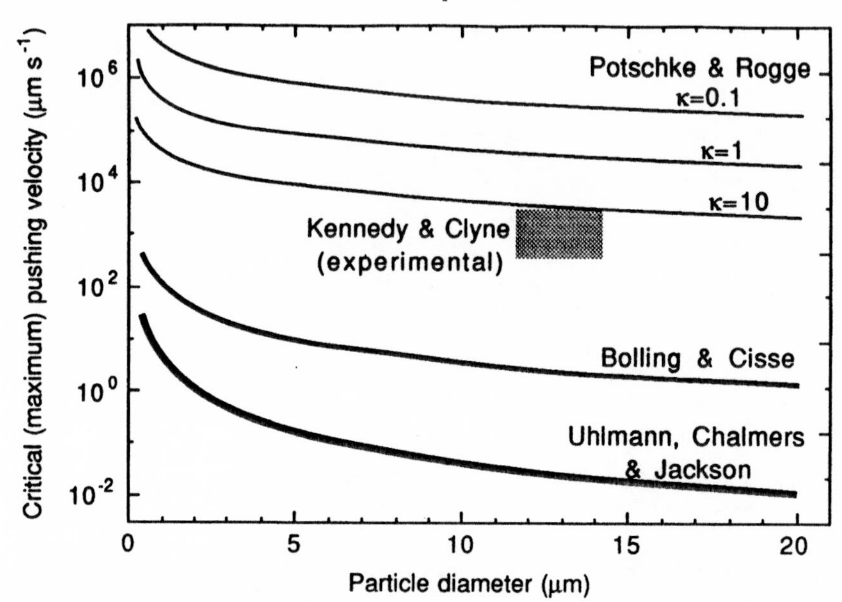

Fig. 9.15 Predicted dependence of the critical velocity for particle pushing on the particle size, as given by the models of Uhlmann *et al.*[75], Bolling and Cisse[76], and Potschke and Rogge[79]. The data used are those appropriate for the Al–Al_2O_3 system, although some of the values are not accurately established. The thermal conductivity ratio κ, which is only used in the Potschke and Rogge model, has a value of around 0.1. An experimental datum point is shown from the work of Kennedy and Clyne[80], indicating that with Al–Al_2O_3 particles the maximum pushing velocity can be lower-bounded at about 500 μm s^{-1}. This is consistent with the predictions of the Potschke and Rogge model, but not with those of the other two models.

flow. A simple correction to the liquid viscosity in the viscous drag term may, however, be inappropriate, depending on whether it is the supply of liquid to the general area of the front or flow within the particle/solid channel which is the critical process. Constitutional undercooling of the growth front, leading to morphological instability and the formation of cells or dendrites, is also expected to favour engulfment, although particles may then be pushed laterally by the thickening cells or dendrite arms (see Fig. 9.14).

Several analytical expressions have been presented[75–79] for the critical (maximum) pushing velocity. Predictions from three of these models are presented in Fig. 9.15. The Uhlmann, Chalmers and Jackson model[75] is based on the thermodynamics of the thin liquid film, while that of Bolling and Cisse[76] involves the kinetic undercooling of the front and interface curvature effects. These models apparently underestimate the maximum pushing velocity, as the experimental datum shown for Al_2O_3 particles in aluminium was obtained[80] in the presence of a buoyancy-driven body force and hence represents a lower bound. The Potschke and Rogge model[79], which is apparently the most complete treatment currently available, is consistent with this experimental result, particularly when it is recognised that neighbouring particles were present – the average volume fraction was 5%. It also predicts a (rather weak) dependence on the ratio, κ, of the thermal conductivity of the particle to that of the matrix. An insulating particle is predicted to be more easily pushed as a result of the local delay in solidification. (A model proposed by Stefanescu *et al.*[78] gives a plot very close to the Potschke and Rogge curve for a κ value of unity and also predicts a higher critical velocity as κ falls; however, the model breaks down as κ approaches 2, above which predicted critical velocities are negative.)

In any event, ceramic particles in Al can apparently be pushed at surprisingly high velocities, up to the mm s^{-1} range, in broad agreement with the Potschke and Rogge model. Other experimental work[81–83] is broadly consistent with the conclusion that pushing takes place readily in Al-based systems. While interfacial energy values are obviously relevant, it seems probable that pushing will occur quite readily in particulate MMCs based on other metals. Combinations in which the ceramic is a better thermal conductor than the matrix (e.g. SiC in Ti – see §8.1.2) are predicted to have lower maximum pushing velocities, but the values are still relatively high. Also, it is both predicted and observed[84] that finer particles are more readily pushed. (This is in contrast to sedimentation-driven segregation prior to casting, which is less pronounced for finer particles.) In practice, it is difficult to cast MMCs so as to eliminate pushing entirely, but relatively fast cooling is generally desirable as it will both discourage pushing and limit the distances over which pushing can occur. Centrifugal casting[82,85] offers some potential for counteracting the pushing effect with a buoyancy force. There is, of course, also some scope for homogenising the structure during secondary processing (§9.3).

Interfacial reaction

Stir casting involves prolonged liquid/ceramic contact, which can cause excessive interfacial reaction. This has been studied in detail[58,86] for Al–SiC, in which the formation of Al_4C_3 and Si can be extensive. This both degrades the final properties of the composite and raises the viscosity of the slurry substantially, making subsequent casting difficult. The rate of reaction is reduced, and can become zero, if the melt is Si-rich, either by prior alloying or as a result of the reaction. The reaction kinetics and Si levels needed to eliminate it are such that it has been concluded[86] that compocasting of Al–SiC_P is suited to conventional (high Si) casting alloys, but not to most wrought alloys. There are few systematic data available on reaction during stir casting with other systems, but evidently problems would arise in many cases. At the present time, commercially cast aluminium matrix composites usually contain Al_2O_{3P} reinforcement, in cases where the application is not suited to an alloy rich in Si.

9.1.4 Reactive processing ('in situ *composites*')

Directional solidification of eutectics

Directional solidification has long been used to produce anisotropic material, often with a high degree of microstructural regularity and perfection. In certain special cases, this procedure can be used to produce castings which are in effect metal matrix composites. A binary alloy melt of exactly the eutectic composition normally freezes at a congruent temperature to form an aligned two phase structure. If the volume fraction of one of the phases is sufficiently low, it is energetically favourable for it to solidify in the form of fibres, rather than the more commonly observed lamellae. The critical volume fraction depends on interfacial energy and solute diffusion characteristics, but it is typically about 5%. An example of such a composite is shown in Fig. 9.16. The fibres may be intermetallic compounds of high strength and stiffness, with a strong fibre/matrix interface. Furthermore, by selecting the growth conditions, the fibre diameter and spacing can be controlled to some extent. A considerable amount of research has been carried out[87–90] into the production and properties of such directionally solidified eutectics, particularly those based on cobalt and nickel alloys. Promising materials have been made in this way, with particularly good creep resistance. The level of interest has dropped somewhat in recent years, partly in view of the very slow growth rates often needed to ensure good microstructural control, but more significantly because of inherent limitations in the nature and volume fraction of the reinforcement. Furthermore, although there is no problem of interfacial chemical reaction (as the two phases should be in thermodynamic equilibrium), the microstructures were often found to be morphologically unstable, particularly in the presence of thermal gradients[91–93].

Fig. 9.16 SEM image[88] of a Ni–Cr/TaC composite after deep etching to reveal the fibres of TaC. These only constitute about 6% by volume; if the reinforcing phase is present at a much higher volume fraction than this, then it forms with a lamellar morphology.

Directed metal oxidation

There has been considerable interest recently in the family of processes developed by the Lanxide Corporation, in which metal is directionally oxidised so as to produce a near net shape component containing metal and ceramic[94,95]. Although full details are often unavailable, a typical process would involve raising an aluminium melt to high temperature, with additions such as Mg, so that the alumina skin becomes unstable and the metal moves by capillary action into an array of ceramic particles, such as alumina. Many other metal/ceramic systems have been explored[96]. Similar principles are employed in the efforts to develop pressureless infiltration processes (see §9.1.1).

Exothermic reaction processes

A similar approach to directed metal oxidation has been developed by the Martin Marietta Corporation in their XD ('exothermic dispersion') process, which involves heating various mixtures to high temperatures, such that a self-propagating exothermic reaction takes place. Typically, a very fine dispersion of some stable ceramic phase is created. This process has been extensively studied for reinforced intermetallic systems such as TiB_2 in titanium aluminides[97] and in nickel aluminide[98]. Little has been published about porosity levels in the as-reacted materials, but consolidation, for example by hot isostatic pressing (HIPing), can be carried out and there is in any event scope for controlling the pore content via choice of reactions and conditions. As with directionally solidified eutectics, it is expected that the products of these *in situ* reactions should be thermodynamically stable. However, they may also suffer from morphological instability in a similar way during service at very high temperature[96].

9.2 Primary solid state processing

It is possible to produce metal matrix composites without the matrix ever becoming even partially liquid while in contact with the ceramic, although this may result in less intimate interfacial contact. One of the main problems with this approach is that material handling procedures tend to be cumbersome and hence expensive.

9.2.1 Powder blending and pressing

Mixing of metallic powder and ceramic fibres or particulate is a convenient and versatile technique for MMC production, offering, for example, excellent control over the ceramic content across the complete range. The blending can be carried out dry or in liquid suspension. This is usually followed by cold compaction, canning, evacuation (degassing) and a high temperature consolidation stage such as HIPing or extrusion. Some information has been published concerning these procedures[99–101], with at least one report[102] that HIPing gives rise to a higher interfacial bond strength and hence improved properties. It can be difficult to achieve a homogeneous mixture during blending[103,104], particularly with fibres (and especially whiskers), which tend to persist in the form of tangled agglomerates with interstitial spaces far too small for the penetration of matrix particles. The relative size of metal and ceramic particles has also been identified[105] as significant during blending of powder route particulate MMCs. Another notable feature of much powder route material is the presence of fine oxide particles†, usually present in Al–MMCs in the form of plate-like particles a few tens of nm thick, constituting about 0.05–0.5 vol%, depending on powder history and processing conditions[106,107] (see Fig. 10.11).

9.2.2 Diffusion bonding of foils

A technique in commercial use for titanium reinforced with long fibres involves the placement of arrays of fibres between thin metallic foils, often by means of a filment winding operation, followed by a hot pressing operation. The plastic flow processes taking place during pressing have been analysed by Chang and Scala[108], who showed both experimentally and theoretically that the pressure required for consolidation is independent of the fibre volume fraction. They also predicted that the pressure required rises with increasing coefficient of friction between the contacting surfaces. In practice, the matrix creep characteristics and the factors affecting surface oxide film durability are also likely to be important. The procedure is attractive for titanium MMCs because: (a) reinforcement of Ti with continuous fibres is particularly attractive (in view of the scope for significant improvements in creep resistance and stiffness), (b) routes involving liquid Ti suffer from rapid interfacial chemical reaction[109,110] and (c) Ti is well suited to diffusion bonding operations because it dissolves its own oxide[111] at temperatures above about 700 °C. Suitable conditions for composite production have been established[112,113] and typically involve a few hours at around 900 °C. Changes in alloy composition have their advantages and disadvantages; for example, additions of Al, Mo or V slow the kinetics of interfacial reaction[113], but also tend to make the rolling of thin foils more difficult. Whilst the addition of nickel allows reduced diffusion bonding temperatures[114], it also leads to formation of coarse intermetallics such as Ti_2Ni, which

† For titanium composites, the native oxide layers on the Ti particles dissolve in the matrix during consolidation and the oxygen normally remains in solution after processing.

impairs the mechanical properties. The overall procedure is slow and cumbersome and there can be difficulties in obtaining very high fibre volume fractions and homogeneous fibre distributions. Furthermore, a significant interfacial reaction layer (~1 μm) usually forms during the heat treatment necessary for consolidation. A microstructure is shown in Fig. 9.17(a). There is considerable interest in fibre coatings (§6.4) designed to reduce these problems of interfacial attack. Composites based on superplastic aluminium alloys have also been made by the hot pressing of foils[115].

9.2.3 Physical vapour deposition (PVD)

Several PVD processes have been used in fabrication of MMCs[116]. All PVD processes are relatively slow, but the fastest is evaporation – involving thermal vaporisation of the target species in a relatively high vacuum. An evaporation process used[117] for fabrication of long fibre reinforced titanium involves continuous passage of fibre through a region having a high partial vapour pressure of the metal to be deposited, where

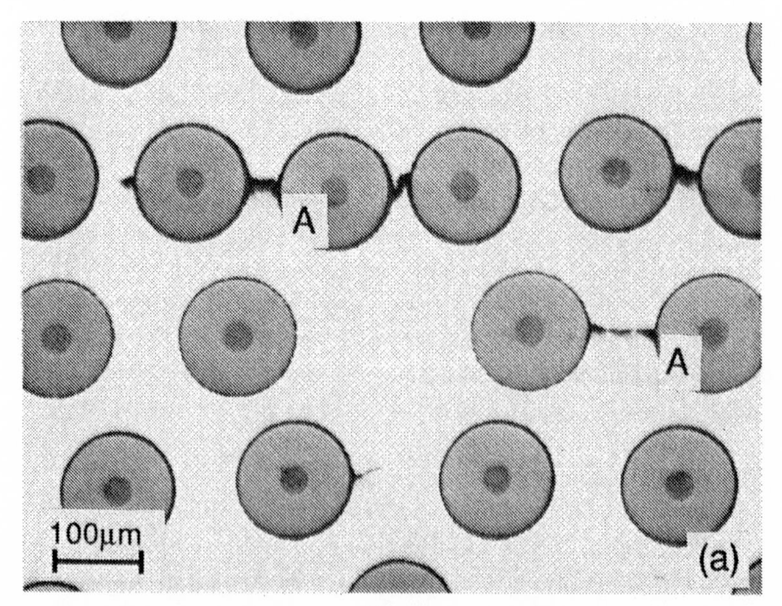

Fig. 9.17 Long fibre reinforced Ti-based MMCs. (a) Diffusion bonded Ti–6Al–4V with SiC monofilaments[112], showing residual porosity at points marked'A' (triple points where two foils meet a fibre). (*cont.*)

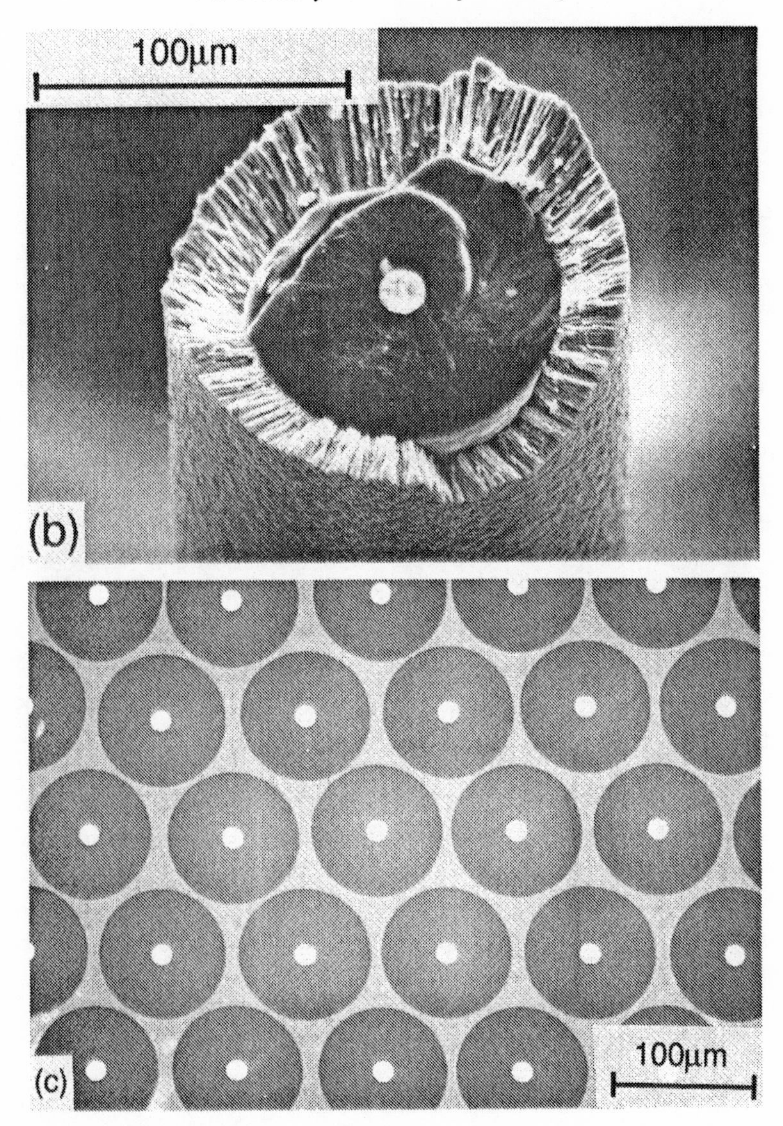

Fig. 9.17 (*cont.*) (b) A SiC monofilament with a 35 μm vapour-deposited layer of Ti–5Al–5V and (c) a Ti–5Al–5V/80 vol% SiC composite produced by HIPing of a bundle of monofilaments with 8 μm thick coatings[117].

condensation takes place so as to produce a relatively thick coating on the fibre. The vapour is produced by directing a high power (~10 kW) electron beam onto the end of a solid bar feedstock.

One advantage[117] of the technique is that a wide range of alloy compositions can be used; differences in evaporation rate between different solutes become compensated by changes in composition of the molten pool formed on the end of the bar until a steady state is reached in which the alloy content of the deposit is the same as that of the feedstock†. Titanium and aluminium alloys and Ti–Al intermetallic compounds have been deposited[117]. A further point worthy of note is that there is little or no mechanical disturbance of the interfacial region; this may be of significance when the fibres have a diffusion barrier layer, or a tailored surface chemistry, which might be degraded by the droplet impacts in spray deposition (§9.1.2) or the frictional motion in diffusion bonding (§9.2.2). Typical deposition rates are ~5–10 μm min^{-1}. Composite fabrication is usually completed by assembling the coated fibres into a bundle and consolidating this in a hot pressing or HIPing operation. A very uniform distribution of fibres is produced in this way, with fibre contents of up to about 80%. A typical coated fibre and consolidated composite are shown in Fig. 9.17(b). The fibre volume fraction can be accurately controlled via the thickness of the deposited coatings.

Other PVD processes have also been used in MMC manufacture. Ion plating involves passing the vapour through a gas (usually Ar) glow discharge around the substrate, leading to bombardment by ions of both Ar and the coating material. This gives a denser deposit than evaporation, but a slower deposition rate. The process has been applied to Al deposition on C fibres[118,119]. Sputter deposition, in which a target of the coating material is bombarded by ions of a working gas (Ar) and atoms are ejected towards the substrate, is slower still, but has the advantage of being applicable to virtually all materials – even those with very low vapour pressures. It is not a promising technique for matrix deposition, but may have potential for thin fibre coatings or to introduce small quantities of elements with low vapour pressures during evaporation.

9.3 Secondary processing

A number of secondary processes can be applied to MMC material, usually with the objectives of consolidation (porosity elimination), generating fibre alignment and/or forming into a required shape. These involve high temperatures and large strain deformations, and several of

† This becomes impractical for some elements having a very low vapour pressure at typical pool temperatures, e.g. Nb, as sufficiently high concentrations would be difficult or impossible to achieve. There may, however, be scope for the incorporation of such elements by simultaneous sputtering from a solid target nearby.

the issues covered in Chapter 5 on this topic are relevant here. In this section, attention is concentrated on the nature of the deformation involved in the different processes and of the microstructural features of the product.

9.3.1 Extrusion and drawing

Extrusion may be carried out on discontinuous MMCs produced in various ways, commonly by squeeze infiltration or by powder blending. Generally, extrusion pressures required are higher than for unreinforced material and heating more rapid, leading to greater limitations on extrusion speed to avoid liquation and surface tearing[120]. A primary interest centres on the microstructural changes induced during the process, particularly when fibres are present. There is scope for alignment of fibres parallel to the extrusion axis, but at the expense of progressive fibre fragmentation. It has been shown[121] that the degree of fibre fracture decreases with increasing temperature and decreasing total strain rate. This can be explained in terms of the kinetics of the stress relaxation processes, which act to limit the build-up of tensile stress within the fibre. An example of data confirming the significance of local strain rate is shown in Fig. 9.18. The most effective way to minimise the peak strain rate (for a given extrusion ratio) is to use tapered or streamlined die shapes and these are now widely employed for MMC extrusion[122]. The maintenance of a high billet temperature is also an advantage in terms of the retention of a high fibre aspect ratio and the process can be carried out while the matrix is in a partially liquid state[123]. This reduces fibre fracture substantially and the problem of defects induced by the presence of liquid can be largely eliminated by use of a cladding which will remain solid during extrusion[123]. The main difficulties lie in the selection of suitable die materials and the control of heat flow during extrusion so as to maintain the required temperature field. Fibre strength is also relevant in determining the final aspect ratio and, for example, whiskers will normally exhibit higher values than staple fibres extruded under the same conditions[121].

Other microstructural features of extruded MMCs have attracted attention, notably the formation of ceramic-enriched ‘bands’ parallel to the extrusion axis. Examples of these are shown in Fig. 9.19(a). There has been some uncertainty as to whether these can form spontaneously in homogeneous material or require some inhomogeneity in the starting material. This is difficult to establish definitely, but it seems

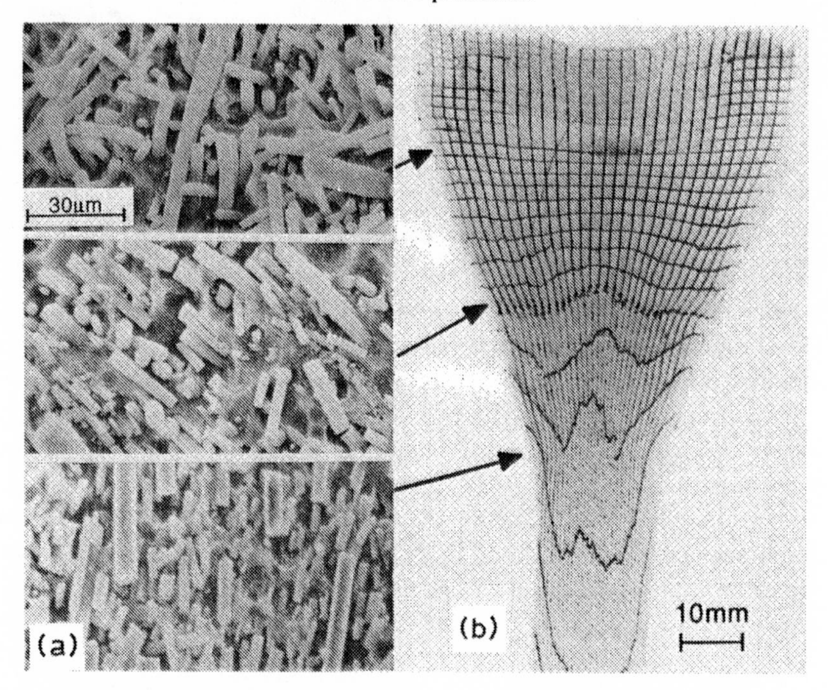

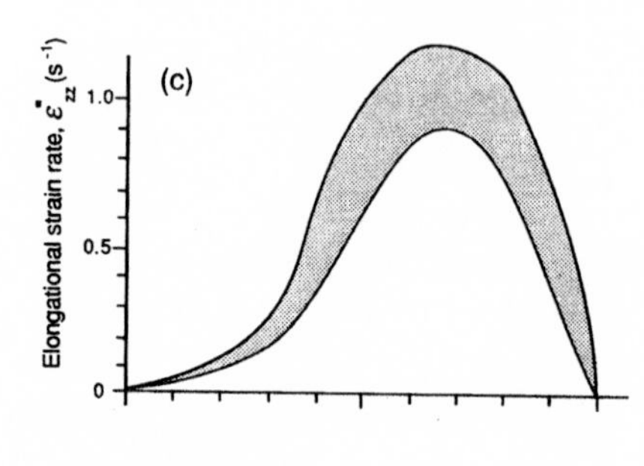

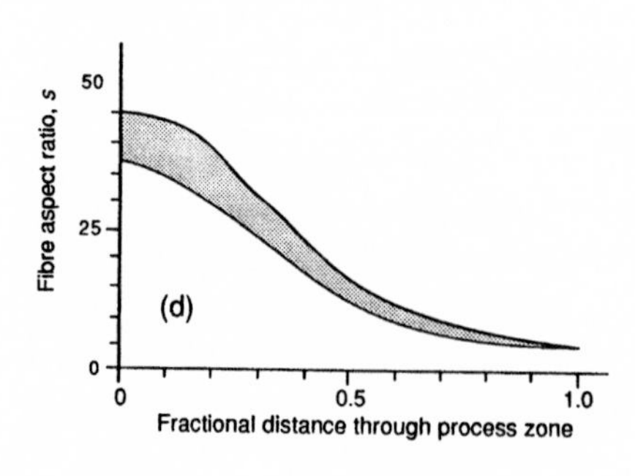

Fig. 9.18 Progressive fibre fragmentation during extension. (a) SEM micrographs showing fibre alignment and aspect ratio changes through the process zone of an interrupted extrusion run; (b) the strain field can be seen in the accompanying X-ray radiograph of a section through the extrudate, which reveals a set of gold wires embedded in the billet, originally in the form of an orthogonal mesh. (c) Elongational strain rate, obtained from an X-ray radiograph of the type shown in (b), and (d) measured fibre aspect ratio, as a function of distance traversed through the process zone during extrusion[121].

probable that the extrusion conditions and aspect ratio of the reinforcement are the most important factors and that severe initial distribution inhomogeneity is not necessary. The mechanism of band formation is still unclear, but it appears to involve the concentration of shear strain in regions where ceramic particles or fibres accumulate. A possible mechanism[125] is illustrated in Fig. 9.19(d). Minimisation of peak strain rates and strain rate gradients appears to reduce the effect so that, for example, it is more pronounced with a shear face die than with tapered or streamlined dies[121]; banding also appears to be more prevalent with fibres than with particulate reinforcement, which is probably a consequence of their greater interaction with each other during gross plastic deformation of the composite.

Drawing involves a rather similar strain field to extrusion, but the stress state in the process zone has a smaller compressive hydrostatic component. In addition, it is normally carried out at a lower temperature. One consequence of this is that there is a much greater risk of internal cavitation during drawing, particularly if the interfacial bond strength is low. Surface finish, on the other hand, is often superior.

9.3.2 Rolling, forging and hot isostatic pressing

In general the microstructural effects observed for MMCs after various conventional forming processes have been applied are explicable in terms of knowledge about local temperatures, stresses and strain rates, along similar lines to those outlined above for extrusion. Processes such as rolling and forging generally involve high deviatoric strains being imposed relatively quickly[126], and hence can readily cause damage such as cavitation, fibre fracture and macroscopic cracking, particularly at low temperature. Rolling in particular involves high local strain rates, often with sharp reductions in temperature caused by contact with cold rolls, and is generally unsuitable for consolidation operations. Forging is more easily carried out at relatively low strain rates and elevated temperatures. Very high temperatures, and the possibility of matrix liquation, on the other hand, lead quickly to the generation of macroscopic defects such as hot tearing or hot shortness. In contrast to these forming processes, Hot Isostatic Pressing (HIPing) generates no (volume-averaged) deviatoric stresses and so is unlikely to give rise to either microstructural or macroscopic defects. It is an attractive method for removing residual porosity, which can include surface-connected porosity, as long as some form of encapsulation is provided. It has been quite widely applied to MMCs[127–129]. However, it should be noted that it can be very difficult to

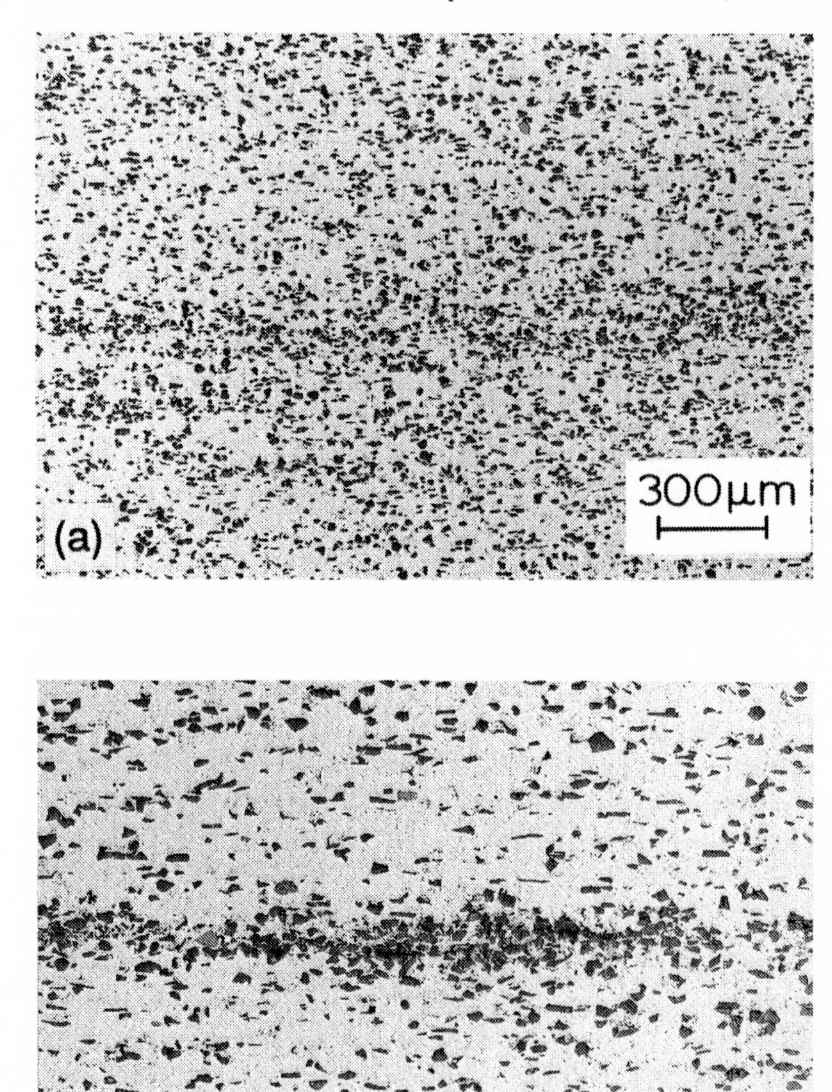

Fig. 9.19 Formation of ceramic-rich bands during extrusion. (a) Slight or (b) moderate banding is commonly observed in particulate MMCs[124], while (c) the effect can be severe in short fibre reinforced material[121]. Cavities can form in ceramic-rich bands. The asymmetry of the ceramic content distribution on either side of the band, apparent in (c), can be explained in terms of superimposed gradients of temperature and strain rate across the section of the extrudate (*cont.*).

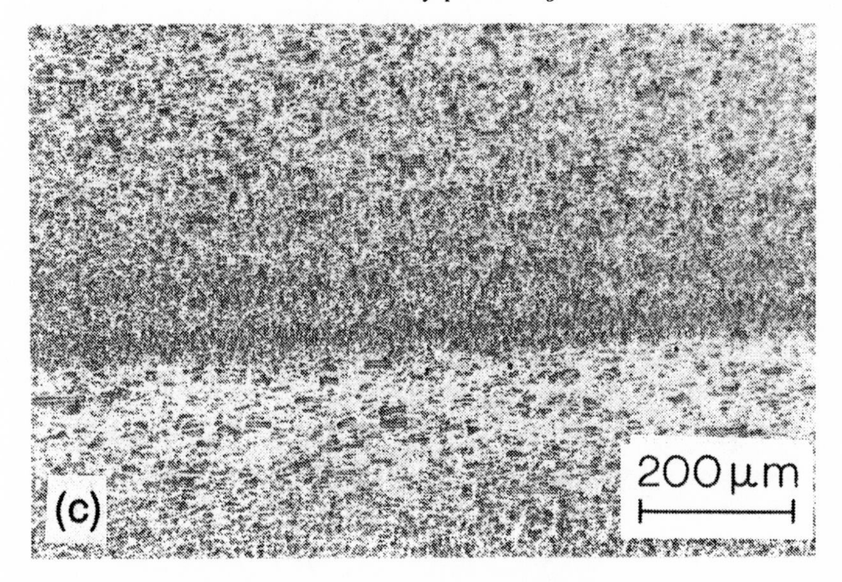

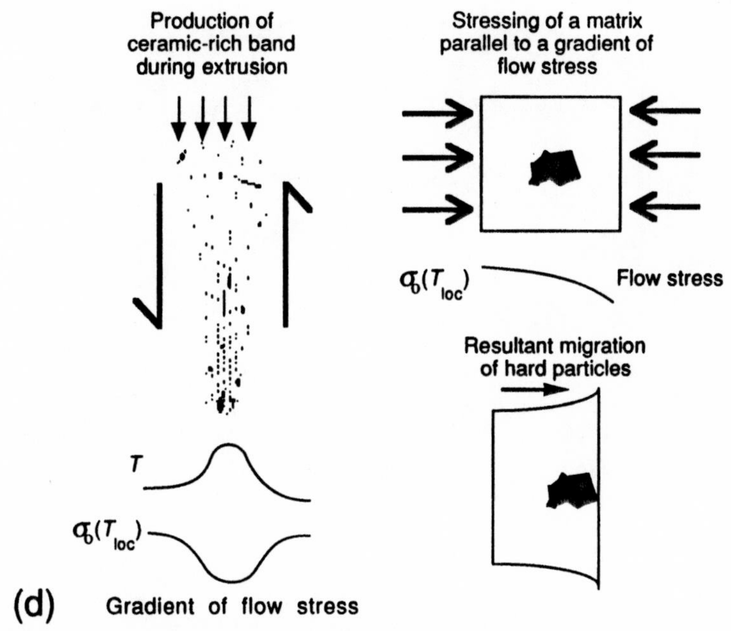

Fig. 9.19 (*cont.*) A proposed mechanism (d), after Ehrstrom and Kool[125], involves ceramic particle migration down a gradient of matrix flow stress associated with a thermal gradient, caused by the strain localisation. Fibres appear to be more prone to the effect than particles, possibly because they rotate, and hence interact more with each other, during the process.

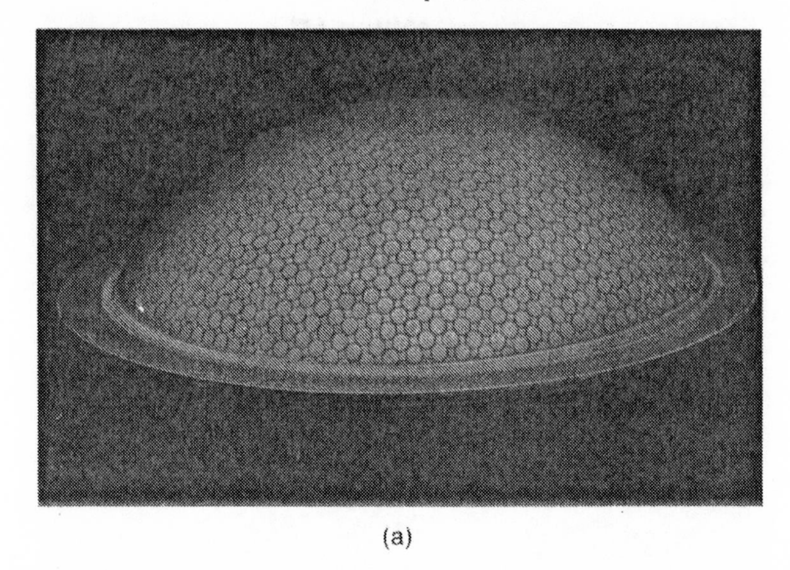

(a)

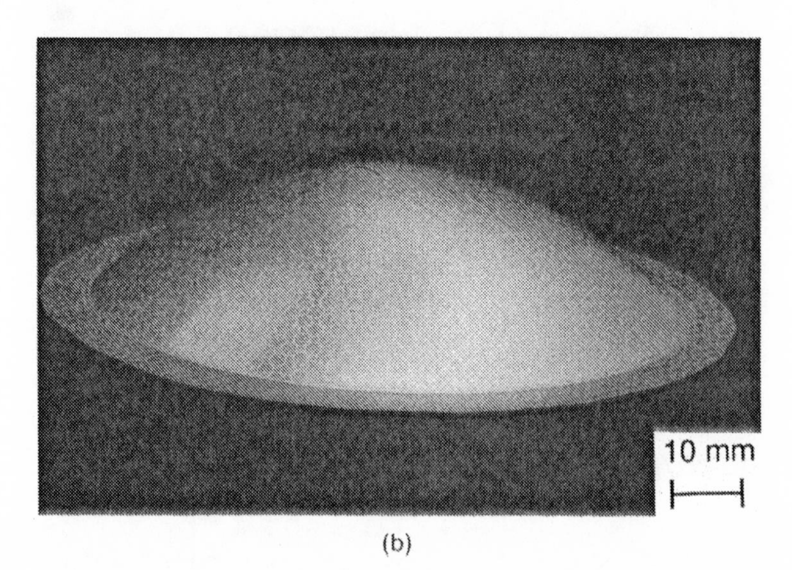

(b)

Fig. 9.20 Sheet forming of MMC by thermal cycling induced superplasticity[132]. Bulge tested specimens of (a) 6061 Al–10 vol% SiC_P and (b) unreinforced 1100 Al alloy. The composite was thermally cycled between 125 °C and 450 °C, with a 6 minute cycle, while stressed by a bulge pressure of about 8 atmospheres. The through-thickness strain in the centre after 1140 cycles was about 120%, with no sign of imminent failure. The 1100 Al alloy was tested isothermally at 350 °C (close to the 'diffusional mean' temperature of the cycle in (a) – see §5.3.2, and failed at a strain of less than 50%.

remove residual porosity in regions of very high ceramic content, such as within particle clusters, and the absence of any shear stresses means that such clusters cannot be dispersed during HIPing.

9.3.3 Superplastic processing and sheet forming

There has been considerable interest in the processing of MMCs in the superplastic state, as this offers promise for carrying out forming operations involving large plastic strains, without inducing microstructural damage. A conventional superplastic Al alloy has been used[115] for making long fibre MMCs by diffusion bonding, exploiting good flow characteristics to ensure low porosity after consolidation. This type of superplasticity will not allow such a composite to be shaped, as the fibres inhibit flow. However, it is now well-established that the presence of the reinforcement can be used to promote superplastic behaviour by the generation of internal stresses during thermal cycling. The details of this effect were outlined in Chapter 5. This type of superplastic behaviour should be clearly distinguished from that dependent on a fine stable grain size and a high strain rate sensitivity exponent[130], and has been termed internal stress superplasticity[131] and mismatch-induced superplasticity[132]. The dramatic increase in maximum plastic strain which is achieved by the superimposition of thermal cycling during deformation is illustrated in Fig. 9.20, which shows a thin sheet of an MMC (with a normal tensile ductility of 7% at 350 °C) after bulge testing (without failure) to a strain of 120% by cycling between 125 °C and 450 °C while applying a pressure of about 8 atmospheres[132]. Such a procedure offers considerable promise as a method of forming particulate and short fibre MMC components into complex and convoluted shapes.

9.4 Machining and joining

The development for MMCs of what might be termed 'tertiary' processing operations, such as finish machining and final assembly joining, is still in its infancy, although it is attracting increasing interest as more applications are being explored. In this section a very brief summary is given of factors to be taken into account when applying such processes to MMCs. General surveys of current machining techniques and criteria for their selection are available in the literature[133,134].

9.4.1 Mechanical cutting

Conventional cutting, turning, milling and grinding operations can usually be applied to MMCs, but there is often a problem of excessive tool wear.

In general, such problems become more significant with increasing reinforcement volume fraction and size. The volume fraction effect is evidently attributable to the tool encountering more hard material, while larger particles or fibres resist excavation more strongly and hence stress the tooling more highly. Diamond-tipped or -impregnated tools are therefore usually necessary for monofilament-reinforced MMCs, while tungsten carbide, or even high speed steel, tooling may be adequate for short fibre and particulate material. The strength of the reinforcement is, however, also relevant; aluminium reinforced with SiC whiskers has, for example, been found[135] to be more difficult to machine than other composites containing weaker fibres. For most MMCs, the best results are obtained with sharp tooling, an appropriate cutting speed, copious cooling/ lubrication and a high material feed rate. These aspects have, for example, been emphasised[136] with regard to Al–SiC_p composites. Diamond tooling has been found[136,137] to give better performance than cemented carbides and ceramics and to allow relatively high tool speeds. In contrast, if carbide tooling is used, then a low cutting speed will increase the tool life[136,137]. Further general features of MMC machining have also been identified[138–140]. Wire saws (in which a diamond-impregnated wire is translated along its axis while in abrasive contact with the workpiece) can be useful in cutting MMCs, although cutting rates are usually relatively slow and only straight cuts can normally be produced.

9.4.2 Electrical cutting

Several cutting processes involving an electric field between tool and workpiece are potentially useful for MMCs. Electrochemical machining involves removal of material by anodic dissolution, using a shaped cathode to determine the geometry of the cut. An ionic electrolyte is flushed through the cathode/workpiece gap and this serves to carry away debris such as undissolved fibres. There would be difficulties in cutting and removing long fibres, although this could be done for some cutting geometries by combining the electrochemical action with mechanical grinding, using an abrasive moving cathode. Normal electrochemical machining, however, involves no mechanical contact between electrode and workpiece, so that very little damage is induced in the remaining material.

Conventional electrical discharge machining (spark erosion) involves cutting with a moving wire electrode bathed in a stream of dielectric fluid. Material removal results from the high local temperatures and liquid pressure pulses generated on the workpiece surface, which does not come into mechanical contact with the wire. As cutting is a thermal and mechanical, rather than a chemical, process, ceramic fibres can be cut quite readily. The process can, however, be rather slow and damage can be relatively severe.

9.4.3 High energy beam and fluid jet cutting

MMCs can be successfully cut using various high energy beams. For cutting along the fibre axis, or for discontinuous composites, the fibres do not need to be fractured and cutting, melting or volatilisation of the matrix is all that is required. This can be achieved using lasers, electron beams, etc., as with unreinforced material, although the effect of the reinforcement on the thermal conductivity may influence the response of the material. At sufficiently high beam power[141], fibres can be melted/volatilised or, more probably, fractured under the various mechanical and thermal stresses induced by the beam. There tends, however, to be a high level of microstructural damage induced during the cutting process, in the form of cracking, interfacial debonding and heat affected microstructures. Laser cutting is also being applied[142] to engineering ceramics and CMCs.

Low damage levels can be combined with high cutting rates by using a concentrated jet of high velocity fluid, usually water, containing abrasive particles in suspension. This technique has been found to be applicable to both continuous[143] and discontinuous[144] MMCs. As there is no significant temperature rise, damage is only of mechanical origin and tends to be localised close to the machined surface[145]. The fluid jet technique is considered[143] to offer the best combination of cleanliness and maximum speed of the cut. A related process is abrasive flow machining, which is used to generate a good surface finish. A gel containing abrasive particles flows over workpiece surfaces under pressure. This can be useful when polished surfaces are needed, particularly on components of complex shape.

9.4.4 Joining processes

There have been few investigations to date concerning methods of joining MMCs. Conventional welding processes are likely to be unsatisfactory, particularly in fibrous MMCs, because the distribution of reinforcement will tend to be strongly disturbed in the fusion zone and may be entirely absent. Even in particulate MMCs, using MMC filler rods, pushing effect in the weld metal (see §9.1.3) will tend to generate marked inhomogeneities. Furthermore, these problems of reinforcement distribution cannot be remedied by post-welding heat treatments in the manner that is often employed to modify the fusion and heat affected zone microstructures in unreinforced material. These marked inhomogeneities may make the joint area prone to strain localisation and failure. It is therefore preferable to consider processes in which the joint produced is relatively narrow, such as brazing, diffusion bonding, adhesive bonding, friction welding, laser or electron beam welding and mechanical fastening.

Diffusion bonding is commonly applied to Ti alloys (see §9.2.2) and the process is suitable for application to Ti-based MMCs, provided the thermal history is controlled so as to limit the matrix/reinforcement interfacial reaction[146]. More problematic is the use of diffusion bonding for Al alloys, although it has been shown[147] that Al-based MMCs can be diffusion bonded with the use of suitable interlayers. Friction welding has been successfully used[148] for Al–SiC_P composites, with few problems in terms of particulate distribution and of post-welding control of the matrix microstructure. There is evidently scope for controlling constitutional effects during the bonding. For example, Suganuma *et al.*[149] brazed 6061 Al–10% Al_2O_3 short fibre composite with 150 μm thick sheets of Al–10% Si and of Al–1% Mn with thin (~5 μm) outer layers Al–10% Si. Brazing was carried out at temperatures above the liquation temperature of Al–10% Si, but below that of Al–1% Mn (and of 6061 Al). Joint strength was found[149] to be substantially higher for the duplex Al–Mn/Al–Si brazing sheet, than for the Al–Si sheet, reflecting the value of keeping the liquation zone as narrow as possible.

It may also be noted that the scope for tailoring the thermal expansivity of MMCs, by control over fibre content and orientation distribution, can be of importance in making joints, particularly between dissimilar materials; homogeneous or graded composition MMCs may prove useful as expansivity matching layers (see Chapter 12).

References

1. V. Massardier, D. Dafir, M. Rmili, P. Merle and A. Faure (1991) 6061 Processing Alloy–Al_2O_3 Platelets Aluminium Composite and Mechanical Properties, in *Metal Matrix Composites – Processing, Microstructure and Properties, 12th Risø Int. Symp.*, Risø, N. Hansen, D. J. Jensen, T. Leffers, H. Lilholt, T. Lorentzen, A. S. Pedersen, O. B. Pedersen and B. Ralph (ed.), Risø Nat. Lab., Denmark, pp. 503–8.
2. S. Rajagopal (1981) Squeeze Casting: A Review and Update, *J. Appl. Metalworking*, **1**, pp. 3–14.
3. S. K. Verma and J. L. Dorcic (1988) Manufacturing of Composites by Squeeze Casting, in *Cast Reinforced Metal Composites*, Chicago, S. G. Fishman and A. K. Dhingra (eds.), ASM, pp. 115–26.
4. M. H. Stacey, M. D. Taylor and A. M. Walker (1987) A New Alumina Fibre for Advanced Composites, in *Proc. ICCM VI/ECCM 2*, London,

F. L. Matthews, N. C. R. Buskell, J. M. Hodgkinson and J. Morton (eds.), Elsevier, pp. 5.371–5.381.
5. R. J. Sample, R. B. Bhagat and M. F. Amateau (1988) High Pressure Squeeze Casting of Unidirectional Graphite Fiber Reinforced Aluminium Matrix Composites, in *Cast Reinforced Metal Composites*, Chicago, S. G. Fishman and A. K. Dhingra (eds.), ASM, pp. 179–83.
6. H. Fukununga (1988) Squeeze Casting Processes for Fiber Reinforced Metals and their Mechanical Properties, *ibid.*, pp. 101–7.
7. J. Charbonnier, S. Dermarkar, M. Santarini, J. Fages and M. Sabatie (1988) High Performance Metal Matrix Composites Manufactured by Squeeze Casting, *ibid.*, pp. 127–32.
8. G. R. Cappelman, J. F. Watts and T. W. Clyne (1985) The Interface Region in Squeeze-Infiltrated Composites containing δ-Alumina Fibre in an Aluminium Matrix, *J. Mat. Sci.*, **20**, pp. 2159–68.
9. C. H. Li, L. Nyborg, S. Bengtsson, R. Warren and I. Olefjord (1989) Reactions between SiO_2 Binder and Matrix in δ-Al_2O_3/Al–Mg Composites, in *Interfacial Phenomena in Composite Materials 1989*, Sheffield, F. R. Jones (ed.), Butterworth, pp. 253–7.
10. C. M. Friend, I. Horsefall, S. D. Luxton and R. J. Young (1988) The Effect of Fibre/Matrix Interfaces on the Age-Hardening Characteristics of δ-Alumina Fibre Reinforced AA6061, in *Cast Reinforced Metal Composites*, Chicago, S. G. Fishman and A. K. Dhingra (eds.), ASM, pp. 309–15.
11. J. D. Birchall, D. R. Stanley, M. J. Mockford, G. H. Pigott and P. J. Pinto (1988) The Toxicity of Silicon Carbide Whiskers, *J. Mat. Sci. Letts.*, **7**, pp. 350–2.
12. T. W. Clyne and M. G. Bader (1985) Analysis of a Squeeze Infiltration Process for Fabrication of Metal Matrix Composites, in *Proc. ICCM V*, San Diego, W. Harrigan, J. Strife and A. K. Dhingra (eds.), TMS-AIME, pp. 755–71.
13. T. W. Clyne (1987) Fabrication and Microstructure of Metal Matrix Composites, in *Proc. ICCM VI/ECCM2*, London, F. L. Matthews, N. C. R. Buskell, J. M. Hodgkinson and J. Morton (ed.), Elsevier, pp. 2.275–2.286.
14. T. W. Clyne and J. F. Mason (1987) The Squeeze Infiltration Process for Fabrication of Metal Matrix Composites, *Metall. Trans.*, **18A**, pp. 1519–30.
15. A. Mortensen and J. A. Cornie (1987) On the Infiltration of Metal Matrix Composites, *Metall. Trans.*, **18A**, pp. 1160–3.
16. A. Mortensen, V. J. Michaud, J. A. Cornie, M. C. Flemings and L. J. Masur (1988) Kinetics of Fibre Preform Infiltration, in *Cast Reinforced Metal Composites*, Chicago, S. G. Fishman and A. K. Dhingra (eds.), ASM, pp. 7–14.
17. A. Mortensen and V. J. Michaud (1990) Infiltration of Fibrous Preform by a Binary Alloy: Part I – Theory, *Metall. Trans.*, **21A**, pp. 2059–72.
18. J. M. Quenisset, R. Fedou, F. Girot and Y. Lepetitcorps (1988) Effect of Squeeze Casting Conditions on Infiltration of Ceramic Preforms, in *Cast Reinforced Metal Composites*, Chicago, S. G. Fishman and A. K. Dhingra (eds.), ASM, pp. 133–8.
19. F. Delannay, L. Froyen and A. Deruyterre (1988) Wetting of Solids by Liquid Metals in Relation to Squeeze Casting of MMCs, *ibid.*, pp. 81–84.
20. S. Nourbakhsh, F. Liang and H. Margolin (1989) Calculation of

Minimum Pressure for Liquid Metal Infiltration of a Fiber Array, *Metall. Trans.*, **20A**, pp. 1861–6.
21. S. Schamm, J. P. Rocher and R. Naslain (1989) Physicochemical Aspects of the K_2ZrF_6 Process Allowing the Spontaneous Infiltration of SiC (or C) Preforms by Liquid Aluminium, in *Proc. 3rd European Conf. on Comp. Mats. (ECCM3)*, Bordeaux, A. R. Bunsell, P. Lamicq and A. Massiah (eds.), Elsevier, pp. 157–61.
22. M. K. Aghajanian, M. A. Rocazella, J. T. Burke and S. D. Keck (1991) The Fabrication of Metal Matrix Composites by a Pressureless Infiltration Technique, *J. Mat. Sci.*, **26**, pp. 447–54.
23. H. Scholz and P. Greil (1991) Nitridation Reactions of Molten Al–(Mg, Si) Alloys, *J. Mat. Sci.*, **26**, pp. 669–77.
24. M. K. Aghajanian, J. T. Burke, D. R. White and A. S. Nagelberg (1989) A New Infiltration Process for the Fabrication of Metal Matrix Composites, *SAMPE Quarterly*, **34**, pp. 43–6.
25. J. F. Mason (1990) *The Fabrication and Mechanical Properties of Mg–Li Alloys Reinforced with SiC Whiskers*, PhD thesis, Univ. of Cambridge.
26. J. P. Rocher, J. M. Quenisset and R. Naslain (1985) A New Casting Process for Carbon (or SiC-based) Fibre–Aluminium Low Cost Composite Materials, *J. Mat. Sci. Letts.*, **4**, pp. 1527–9.
27. A. P. Diwanji and I. W. Hall (1988) Effect of Manufacturing Variables on the Structure and Properties of Squeeze Cast C/Al MMCs, in *Cast Reinforced Metal Composites*, Chicago, S. G. Fishman and A. K. Dhingra (eds.), ASM, pp. 225–30.
28. A. A. Das, A. J. Clegg, B. Zantout and M. M. Yakoub (1988) Solidification under Pressure: Aluminium and Zinc Alloys containing Discontinuous SiC Fibre, *ibid.*, pp. 139–47.
29. Y. Le Petitcorps, J. M. Quenisset, G. Leborgne and M. Barthole (1991) Segregation of Magnesium in Squeeze Cast Aluminium Matrix Composites Reinforced with Alumina Fibres, *Mat. Sci. & Eng.*, **A135**, pp. 37–40.
30. J. F. Mason, C. M. Warwick, P. Smith, J. A. Charles and T. W. Clyne (1989) Magnesium–Lithium Alloys in Metal Matrix Composites – A Preliminary Report, *J. Mat. Sci.*, **24**, pp. 3934–46.
31. R. M. K. Young and T. W. Clyne (1986) A Powder-Based Approach to Semisolid Processing of Metals for Fabrication of Die Castings and Composites, *J. Mat. Sci.*, **21**, pp. 1057–69.
32. R. W. Evans, A. G. Leatham and R. G. Brooks (1985) The Osprey Preform Process, *Powder Metallurgy*, **28**, pp. 13–19.
33. A. G. Leatham, A. Ogilvey, P. F. Chesney and J. V. Wood (1989) Osprey Process – Production Flexibility in Materials Manufacture, *Metals & Materials*, **5**, pp. 140–3.
34. P. F. Chesney, A. G. Leatham, R. Pratt and D. Zebrowski (1989) The Osprey Process – A Versatile Manufacturing Technology for the Production of Solid and Hollow Rounds and Clad (Compound) Billets, in *Proc. 1st European Conf. on Adv. Mat. and Processes, Euromat '89*, Aachen, H. E. Exner and V. Schumacher (eds.), DGM, pp. 247–54.
35. T. C. Willis (1988) Spray Deposition Process for Metal Matrix Composite Manufacture, *Metals & Materials*, **4**, pp. 485–8.
36. W. Kahl and J. Leupp (1989) Spray Deposition of High Performance Al Alloys via the Osprey Process, in *Proc. 1st European Conf. on Adv. Mats.*

and Processes, Euromat '89, Aachen, H. E. Exner and V. Schumacher (eds.), DGM, pp. 261–6.
37. H. Lubanska (1970) Correlation of Spray Data for Gas Atomization of Liquid Metals, *J. Metals*, **22**, pp. 45–9.
38. P. S. Grant and B. Cantor (1989) Modelling of Spray Forming, *Cast Metals*, **4**, pp. 140–51.
39. T. W. Clyne, R. A. Ricks and P. J. Goodhew (1984) The Production of Rapidly Solidified Powder by Ultrasonic Gas Atomization, *Int. J. Rap. Solidif.*, **1**, pp. 59–84.
40. J. White and T. C. Willis (1989) The Production of Metal Matrix Composites by Spray Deposition, *Mater. Des.*, **10**, pp. 121–7.
41. J. White, N. A. Darby, I. R. Hughes, R. M. Jordan and T. C. Willis (1990) Metal Matrix Composites Produced by Spray Deposition, *Adv. Mater. Technol. Int.*, pp. 96–9.
42. J. H. Zaat (1983) A Quarter of a Century of Plasma Spraying, *Ann. Rev. Mat. Sci.*, **13**, pp. 9–42.
43. E. Lugscheider (1988) The Family of Plasma Spray Processes – Present Status and Future Prospects, in *1st Plasma Technik Symp.*, Lucerne, Switzerland, H. Eschenauer, P. Huber, A. R. Nicoll and S. Sandmeier (eds.), Plasma Technik, pp. 23–48.
44. X. Guo, P. Howard, G. Milidantri and H. Zhang (1990) Jet Kote Sprayed Composite Coating for High Temperature Tribological Applications, in *Thermal Spray Research and Applications*, Long Beach, CA, T. F. Bernecki (ed.), ASM, pp. 599–604.
45. K. A. Kowalski, D. R. Marantz, M. F. Smith and W. L. Oberkampf (1990) HVOF: Particle, Flame Diagnostics and Coating Characteristics, *ibid.*, pp. 587–592.
46. P. A. Dearnley, K. A. Roberts and T. W. Clyne (1989) Some Observations on the Microstructure of Boron Carbide Reinforced Titanium Composites Produced by Spray Co-deposition, in *12th Int. Plansee Seminar*, H. Bildstein and H. M. Orter (eds.), Reutte, Austria, pp. 523–38.
47. B. Gudmundsson, B. E. Jacobson, L. Berglin, L. L'Estrade and H. Gruner (1988) Microstructure and Erosion Resistance of Vacuum Plasma Sprayed CoNiCrAlY/Al_2O_3 Composite Coatings, in *1st Plasma Technik Symp.*, Lucerne, Switzerland, H. Eschnauer, P. Huber, A. R. Nicoll and S. Sandmeier (eds.), Plasma Technik, pp. 105–14.
48. M. Vedani and G. Piatti (1991) Mechanical Properties of Al–SiC Particulate Composites Produced by a Vacuum Plasma Spray Co-deposition Technique, in *Metal Matrix Composites – Processing, Microstructure and Properties, 12th Risø Int. Symp.*, Risø, N. Hansen, D. J. Jensen, T. Leffers, H. Lilholt, T. Lorentzen, A. S. Pedersen, O. B. Pedersen and B. Ralph (eds.), Risø Nat. Lab., Denmark, pp. 713–8.
49. F. Barbalat, C. Blain, P. Luquet, T. C. Lu and M. Jeandin (1989) Study of Plasma Sprayed Graded W–Cu Composites using Glow Discharge Spectrometry Compared to Electron Probe and Quantitative Image Analysis, in *Proc. 1st European Conf. on Adv. Mats. and Processes, Euromat '89*, Aachen, H. E. Exner and V. Schumacher (eds.), DGM, pp. 267–72.
50. T. Fukushima, S. Kuroda and S. Kitahara (1990) Gradient Coatings Formed by Plasma Twin Torches and their Properties, in *Proc. 1st Int. Symp. on Functionally Gradient Materials*, Sendai, Japan, M. Yamanouchi,

M. Koizumi, T. Hirai and I. Shiota (eds.), Soc. of Non-Traditional Technology, Tokyo, Japan, pp. 145–50.
51. K. A. Roberts and T. W. Clyne (1991) The Mechanical Behaviour of Graded Composition Coatings Produced by Plasma Spraying, in *2nd Plasma Technik Symp.*, Lucerne, Switzerland, H. Eschenauer, P. Huber, A. R. Nicoll and S. Sandmeier (eds.), Plasma Technik, pp. 267–78.
52. M. F. Smith and R. C. Dykhuizer (1988) Effect of Chamber Pressure on Particle Velocities in Low Pressure Plasma Spray Deposition, *Surf. & Coat. Techn.*, **34**, pp. 25–31.
53. R. McPherson (1981) The Relationship between the Mechanism of Formation, Microstructure and Properties of Plasma-Sprayed Coatings, *Thin Solid Films*, **83**, pp. 297–310.
54. P. A. Siemers, M. J. Jackson, R. L. Mehan and J. R. Rairden (1985) Production of Composite Structures by Low Pressure Plasma Deposition, *Ceram. Eng. Sci. Proc.*, **6**, pp. 896–907.
55. R. R. Kieschke and T. W. Clyne (1989) Plasma Processing of Titanium-Based Composites, in *6th World Conference on Titanium*, Cannes, P. Lacombe, R. Tricot and G. Beranger (eds.), Soc. Franc. Metall., pp. 1789–94.
56. L. J. Westfall (1987) Composite Monolayer Fabrication by an Arc Spraying Process, in *Nat. Thermal Spray Conf. – Thermal Spray: Advances in Coating Technology*, Orlando, Florida, D. L. Houck (ed.), ASM, pp. 417–26.
57. R. W. Smith (1991) Plasma Spray Processing – The State of the Art and Future – From a Surface to a Materials Processing Technology, in *2nd Plasma Technik Symp.*, Lucerne, Switzerland, S. Blum-Sandmeier, H. Eschnauer, P. Huber and A. R. Nicoll (eds.), Plasma Technik, pp. 17–38.
58. M. Skibo, P. L. Morris and D. J. Lloyd (1988) Structure and Properties of Liquid Metal Processed SiC Reinforced Aluminium, in *Cast Reinforced Metal Composites*, Chicago, S. G. Fishman and A. K. Dhingra (eds.), ASM, pp. 257–61.
59. D. G. Evans, P. L. Morris, R. W. Hains, C. Jowett and P. Achim (1989) *Production Extrusion of AA 6061–SiC Metal Matrix Composites*, Alcan, Kingston.
60. W. R. Hoover (1991) Die Casting of Duralcan Composites, in *Metal Matrix Composites – Processing, Microstructure and Properties, 12th Risø Int. Symp.*, Risø, N. Hansen, D. J. Jensen, T. Leffers, H. Lilholt, T. Lorentzen, A. S. Pedersen and B. Ralph (eds.), Risø Nat. Lab., Denmark, pp. 387–92.
61. D. G. Thomas (1965) Transport Characterisation of Suspension: a Note on the Viscosity of Newtonian Suspensions of Uniform Spherical Particles, *J. Colloidal Sci.*, **20**, pp. 267–77.
62. R. J. Farris (1968) Predictions of the Viscosity of Multimodal Suspensions from Unimodal Viscosity Data, *Trans. Soc. Rheology*, **12**, pp. 281–301.
63. D. B. Spencer, R. Mehrabian and M. C. Flemings (1972) Rheological Behaviour of Sn–15% Pb in the Crystallization Range, *Metall. Trans.*, **3**, pp. 1925–32.
64. P. A. Joly and R. Mehrabian (1976) The Rheology of a Partially Solid Alloy, *J. Mat. Sci.*, **11**, pp. 1393–418.
65. R. Mehrabian, R. G. Riek and M. C. Flemings (1974) Preparation and

Casting of Metal-Particulate Non-Metal Composites, *Metall. Trans.*, **5**, pp. 1899–905.
66. C. G. Levi, G. S. Abbaschian and R. Mehrabian (1978) Interface Interactions During Fabrication of Al Alloy–Alumina Fibre Composites, *Metall. Trans.*, **9A**, pp. 697–711.
67. D. J. Lloyd (1988) Properties of Shape Cast Al–SiC Metal Matrix Composites, in *Cast Reinforced Metal Composites*, Chicago, S. G. Fishman and A. K. Dhingra (eds.), ASM, pp. 263–9.
68. T. Z. Kattamis and J. A. Cornie (1988) Solidification of Particulate Ceramic–Aluminium Alloy Composites, *ibid.*, pp. 47–52.
69. P. K. Rohatgi, R. Asthana, M. A. Khan and P. Ostermier (1988) Mixing Quality Modelling in the Manufacture of Cast Metal Matrix Particulate Composites, *ibid.*, pp. 85–92.
70. S. Ray (1988) Porosity in Foundry Composites Prepared by the Vortex Method, *ibid.*, pp. 77–80.
71. P. K. Rohatgi and R. Asthana (1988) Transfer of Particles and Fibres from Gas to Liquid during Solidification Processing of Composites, *ibid.*, pp. 61–6.
72. D. J. Lloyd (1991) Factors Influencing the Properties of Particulate Reinforced Composites Produced by Molten Metal Mixing, in *Metal Matrix Composites – Processing, Microstructure and Properties, 12th Risø Int. Symp.*, Risø, N. Hansen, D. J. Jensen, T. Leffers, H. Lilholt, T. Lorentzen, A. S. Pedersen and B. Ralph (ed.), Risø Nat. Lab., Denmark, pp. 81–99.
73. M. A. Bayoumi and M. Sueri (1988) Partial Remelting and Forming of Al–Si/SiC Composites in their Mushy Zone, in *Cast Reinforced Metal Composites*, Chicago, S. G. Fishman and A. K. Dhingra (eds.), ASM, pp. 167–72.
74. S. Skolianos and T. Z. Kattamis (1987) Microstructural Coarsening in Al–4.5Cu–2Mn Alloy, in Al–4.5Cu–2Mn Alloy, in *Solidification Processing* '87, Sheffield, H. Jones (ed.), Inst. of Metals, pp. 207–10.
75. D. R. Uhlmann, B. Chalmers and K. A. Jackson (1964) Interactions between Particles and Solid–Liquid Interface, *J. Appl. Phys.*, **35**, pp. 2986–93.
76. G. F. Bolling and J. Cisse (1971) A Theory for the Injection of Particles with a Solidifiying Front, *J. Cryst. Growth*, **10**, pp. 56–66.
77. A. A. Chernov, D. E. Temkin and A. M. Melnikova (1976) Theory of the Capture of Solid Inclusions during the Growth of Crystals from the Melt, *Sov. Phys. Crystallog.*, **21**, pp. 369–74.
78. D. M. Stefanescu, B. K. Dhindaw, S. A. Kacar and A. Moitra (1988) Behaviour of Ceramic Particles at the Solid–Liquid Metal Interface in MMCs, *Metall. Trans.*, **19A**, pp. 2847–55.
79. J. Potschke and V. Rogge (1989) On the Behaviour of Foreign Particles at an Advancing Solid–Liquid Interface, *J. Cryst. Growth*, **94**, pp. 726–38.
80. A. R. Kennedy and T. W. Clyne (1991) Particle Pushing during the Solidification of Metal Matrix Composites, *Cast Metals J.*, **4**, pp. 160–4.
81. P. K. Rohatgi, F. M. Yarandi and Y. Liu (1988) Influence of Solidification Conditions on Segregation of Aluminium–Silicon Carbide Composites, in *Cast Reinforced Metal Composites*, Chicago, S. G. Fishman and A. K. Dhingra (eds.), ASM, pp. 249–55.
82. L. Lajoye and M. Suery (1988) Modelling of Particle Segregation during Centrifugal Casting of Al-Matrix Composites, *ibid.*, pp. 15–20.

83. D. M. Stefanescu, A. Moitra, A. S. Kacar and B. K. Dhindaw (1990) The Influence of Buoyant Forces and Volume Fraction of Particles on the Particle Pushing/Entrapment Transition During DS of Al/SiC and Al/Graphite Composites, *Metall. Trans.*, **21A**, pp. 231–9.
84. J. W. McCoy and F. E. Wawner (1988) Dendritic Solidification in Particle-Reinforced Cast Aluminium Composites, in *Cast Reinforced Metal Composites*, Chicago, S. G. Fishman and A. K. Dhingra (eds.), ASM, pp. 237–42.
85. A. R. Kennedy and T. W. Clyne (1992) A Study of the Engulfment of Ceramic Particles by an Advancing Solidification Front in the Presence of Body Forces Induced by Centrifuging, in *2nd European Conf. on Adv. Mats. and Processes, Euromat '91*, Cambridge, UK, T. W. Clyne and P. J. Withers (eds.), Inst. of Materials, pp. 198–204.
86. D. J. Lloyd (1989) The Solidification Microstructures of Particulate Reinforced Al/SiC Composites, *Comp. Sci. & Tech.*, **35**, pp. 159–80.
87. F. R. Mollard and M. C. Flemings (1967) Growth of Composites from the Melt, *Trans. Met. Soc. AIME*, **239**, pp. 1526–46.
88. M. Rabinovitch, J. F. Stohr, T. Khan and H. Bibring (1983) Directionally Solidified Composites for Application at High Temperatures, in *Handbook of Composites, Vol. 4 – Fabrication of Composites*, A. Kelly and S. T. Mileiko (eds.), Elsevier, Amsterdam, pp. 295–372.
89. K. Dannemann, N. S. Stoloff and D. J. Duquette (1987) High Temperature Fatigue of Three Nickel-Based Eutectic Composites, *Mat. Sci. & Eng.*, **95**, pp. 63–71.
90. J. C. Gerdeen, J. C. Predebon and P. M. Schwab (1987) Elastic-Plastic Analysis of Directionally Solidified Lamellar Eutectic Composites, *J. Eng. Mater. Technol.* (*Trans. ASME*), **109**, pp. 53–8.
91. H. E. Cline (1971) Shape Instabilities of Eutectic Composites at Elevated Temperatures, *Acta Metall.*, **19**, pp. 481–90.
92. G. Staniek, K. Fritscher and G. Wirth (1979) Effect of Transverse Thermal Gradient on the Microstructure of a Directionally Solidified 73C Eutectic Alloy, *Zeit. Werkstofftech.*, **10**, pp. 49–57.
93. G. Smolka, A. Maciejny and A. Dytkowicz (1988) High Temperature Stability of the Ni(Cr, Al)–TiC *in situ* Composite, in *Mechanical and Physical Behaviour of Metallic and Ceramic Composites*, 9th Int. Risø Symp., S. I. Andersen, H. Lilholt and O. B. Pedersen (eds.), Risø Nat. Lab., Denmark, pp. 475–8.
94. M. S. Newkirk, A. W. Urquhart, H. R. Zwicker and E. Breval (1986) Formation of Lanxide Ceramic Composite Materials, *J. Mat. Res.*, **1**, pp. 81–9.
95. M. S. Newkirk, H. D. Lesher, D. R. White, C. R. Kennedy, A. W. Urquhart and T. D. Claar (1987) Preparation of Lanxide Ceramic Composite Materials: Matrix Formation by the Directed Oxidation of Molten Metals, *Ceram. Eng. Sci. Proc.*, **8**, pp. 879–85.
96. D. Lewis (1991) *In Situ* Reinforcement of Metal Matrix Composites, in *Metal Matrix Composites: Processing and Interfaces*, R. K. Everett and R. J. Arsenault (eds.), Academic Press, Boston, pp. 121–50.
97. L. Christodoulou, P. A. Parrish and C. R. Crowe (1988) XD™ Titanium Aluminide Composites, in *High Temperature/High Performance Composites*, Reno, Nevada, F. D. Lemkey, S. G. Fishman, A. G. Evans and J. R. Strife (eds.), MRS, Pittsburgh, pp. 29–34.
98. R. K. Visanwadham, J. D. Whittenberger, S. K. Mannan and B. Sprissler

(1988) Elevated Temperature Slow Plastic Deformation of NiAl/TiB_2 Particulate Composites, *ibid.*, pp. 89–94.

99. A. D. McLeod, C. Gabryel, D. J. Lloyd and P. Morris (1989) Particle Incorporation Studies in Support of the Dural Process, in *Processing of Ceramic and Metal Matrix Composites*, Halifax, Nova Scotia, Pergamon, pp. 228–35.
100. A. K. Jha, S. V. Prasad and G. S. Upadhyaya (1989) Preparation and Properties of 6061 Aluminium Alloy/Graphite Composites by PM Route, *Powder Metall.*, **32**, pp. 309–12.
101. G. S. Upadhyaya (1989) Powder Metallurgy Metal Matrix Composites: an Overview, *Met. Mater. Process*, **1**, pp. 217–28.
102. A. Niklas, L. Froyen, L. Delaey and L. Buekenhout (1991) Comparative Evaluation of Extrusion and Hot Isostatic Pressing as Fabrication Techniques for Al–SiC Composites, *Mat. Sci. & Eng.*, **A135**, pp. 225–9.
103. J. H. T Haar and J. Duszczyk (1991) Mixing of Powder Metallurgical Fibre-Reinforced Aluminium Composites, *Mat. Sci. & Eng.*, **A135**, pp. 65–72.
104. H. J. Rack (1991) Processing of Metal Matrix Composites, in *Metal Matrix Composites: Processing and Interfaces*, R. K. Everett and R. J. Arsenault (eds.), Academic, Boston, pp. 83–101.
105. W. H. Hunt, O. Richmond and R. D. Young (1987) Fracture Initiation in Particle Hardened Materials with High Volume Fraction, in *Proc. ICCM VI/ECCM2*, London, F. L. Matthews, N. C. R. Buskell, J. M. Hodgkinson and J. Morton (eds.), Elsevier, pp. 209–23.
106. N. Hansen and D. J. Jensen (1990) Recrystallization of Metals Containing Particles and Fibres, in *Recrystallization '90*, T. Chandra (ed.), TMS-AIME, Warrendale, PA, pp. 79–88.
107. A. F. Whitehouse, R. A. Shahani and T. W. Clyne (1991) Cavitation during Tensile Deformation of Powder Route Particle-Reinforced Aluminium, in *Metal Matrix Composites: Processing, Microstructure and Properties, 12th Risø Int. Symp.*, Roskilde, N. Hansen, D. J. Jensen, T. Leffers, N. Lilholt, T. Lorentzen, A. S. Pedersen, O. B. Pedersen and B. Ralph (eds.), Risø Nat. Lab., Denmark, pp. 741–8.
108. M. Chang and E. Scala (1974) Plastic Deformation Processing Compressive Failure Mechanism in Al Composite Materials, in *Composite Materials: Testing and Design*, ASTM STP 546, pp. 561–79.
109. P. Martineau, M. Lahaye, R. Pailler, R. Naslain, M. Couzi and F. Creuge (1984) SiC Filament/Titanium Matrix Composites Regarded as Model Composites: Part 2. Fibre/Matrix Chemical Interactions at High Temperatures, *J. Mat. Sci.*, **19**, pp. 2749–70.
110. W. J. Wheatley and F. W. Wawner (1985) Kinetics of the Reaction between SiC (SCS-6) Filaments and Ti–6Al–4V Matrix, *J. Mat. Sci. Letts.*, **4**, pp. 173–5.
111. N. F. Kasakov (1985) *Diffusion Bonding of Materials*, Pergamon.
112. P. G. Partridge and C. M. Ward-Close (1989) Diffusion Bonding of Advanced Materials, *Metals & Materials*, **4**, pp. 334–9.
113. P. R. Smith and F. H. Froes (1984) Developments in Ti MMCs, *J. Metals*, **36**, pp. 19–26.
114. C. G. Rhodes, A. G. Gosh and R. A. Spurling (1987) Ti–6Al–4V–2Ni as a Matrix for a SiC Reinforced Composite, *Metall. Trans.*, **18A**, pp. 2151–6.
115. D. J. Lloyd (1984) Fabrication of Fibre Composites Using an Aluminium Superplastic Alloy as Matrix, *J. Mat. Sci.*, **19**, pp. 2488–92.

116. R. K. Everett (1991) Deposition Technologies for MMC Fabrication, in *Metal Matrix Composites: Processing and Interfaces*, R. K. Everett and R. J. Arsenault (eds.), Academic Press, Boston, pp. 103–19.
117. C. M. Ward-Close and P. G. Partridge (1990) A Fibre Coating Process for Advanced Metal Matrix Composites, *J. Mat. Sci.*, **25**, pp. 4315–23.
118. T. Ohsaki, M. Yoshida, Y. Fukube and K. Nakamura (1977) The Properties of C Fibre Reinforced Al Composites Formed by Ion-Plating Process and Hot Pressing, *Thin Solid Films*, **45**, pp. 563–8.
119. R. W. Gardiner and M. C. McConnell (1987) The Properties of Advanced Al Alloys by Vapour Deposition, *Metals & Materials*, **3**, pp. 254–8.
120. S. Brusethaug, O. Reiso and W. Ruch (1990) Extrusion of Particulate-Reinforced Aluminium Billets Made by DC Casting, in *Fabrication of Particulates Reinforced Metal Composites*, J. Masounave and F. G. Hamel (eds.), ASM, pp. 173–80.
121. C. A. Stanford-Beale and T. W. Clyne (1989) Extrusion and High Temperature Deformation of Fibre-Reinforced Aluminium, *Comp. Sci. & Tech.*, **35**, pp. 121–57.
122. H. J. Rack and P. W. Niskaenen (1984) Extrusion of Discontinuous Metal Matrix Composites, *Light Metal Age*, **42**, pp. 9–12.
123. K. Suganuma, T. Fujita, K. Niihara, T. Okamoto, M. Koizuma and N. Suzuki (1989) Hot extrusion of AA 7178 Reinforced with Alumina Short Fibres, *Mat. Sci. & Tech.*, **5**, pp. 249–54.
124. R. A. Shahani (1991) *Microstructural Development During Thermomechanical Processing of Aluminium-Based Composites*, PhD thesis, Univ. of Cambridge.
125. J. C. Ehrstrom and W. H. Kool (1988) Migration of Particles During Extrusion of Metal Matrix Composites, *J. Mat. Sci. Letts.*, **7**, pp. 578–80.
126. G. E. Dieter (1986) *Mechanical Metallurgy*, McGraw-Hill, New York.
127. S. Ito, K. Miyazaka, N. Yoneda and K. Asaka (1989) Preparation of SiC–Fe Composite using HIP, *J. Jap. Soc. Powder Metall.*, **36**, pp. 831–6.
128. H. Morimoto and K. Ohuchi (1989) Effects of HIP Consolidating Temperature on the Mechanical Properties of SiC Whisker Reinforced Aluminium Alloy Composites, *J. Iron & Steel Inst. Jap.*, **75**, pp. 1541–8.
129. R. Moritz (1989) Steel Matrix – Titanium Carbide Composite Materials by Hot Isostatic Pressing (HIP), in *12th Int. Plansee Seminar*, Garmischpartenkirchen, Austria, H. Bildstein and H. M. Orter (eds.), Metallwerk Plansee GmbH, pp. 847–62.
130. K. A. Padmannabhab and G. J. Davies (1980) *Superplasticity*, Springer-Verlag, Berlin.
131. M. Y. Wu and O. D. Sherby (1984) Superplasticity in a Silicon Carbide Whisker Reinforced Aluminium Alloy, *Scripta Met.*, **18**, pp. 773–6.
132. Y. C. Chen, G. S. Daehn and R. H. Wagoner (1990) The Potential for Forming Metal Matrix Composite Components via Thermal Cycling, *Scripta Met.*, **24**, pp. 2157–62.
133. H. S. Ismail and K. K. B. Hon (1988) An Expert System for the Selection of Machining Processes, in *Expert Systems in Engineering*, D. T. Pham (ed.), IFS Publications.
134. R. F. Firestone (1988) Guide to Newer Methods for Machining Ceramics, *Ceramic Industry*, pp. 17–18.
135. H. Matsubara, Y. Nishida, M. Yamada, I. Shirayanagi and T. Imai (1987)

Si_3N_4 Whisker Reinforced Al Alloy Composite, *J. Mat. Sci. Letts.*, **6**, pp. 1313–15.
136. C. T. Lane (1990) Machining Characteristics of Particulate-Reinforced Aluminium, in *Fabrication of Particulates Reinforced Metal Composites*, J. Masounave and F. G. Hamels (eds.), ASM, pp. 195–202.
137. A. R. Chambers and S. E. Stephens (1991) Machining of Al–5Mg Reinforced with 5 vol% Saffil® and 5 vol% SiC, *Mat. Sci. & Eng.*, **A135**, pp. 287–90.
138. G. A. Chadwick and P. J. Heath (1990) Machining Metal Matrix Composites, *Metals & Materials*, **5**, pp. 73–6.
139. J. B. Pond (1990) Rough Cut, *Cutting Tool Eng.*, **42**, pp. 20–31.
140. M. K. Brun, M. Lee and F. Gorsler (1985) Wear Characteristics of Various Hard Materials for Machining SiC-Reinforced Al-Alloy, *Wear*, **104**, pp. 21–9.
141. S. Utsunomiya, Y. Kagawa and Y. Kogo (1986) in *Composites '86: Recent Advances in Japan and the United States*, K. Kawata (ed.), Jap. Soc. Comp. Mats., pp. 589–95.
142. M. Naeem, M. E. Preston and J. R. Tyrer (1990) Machining of Engineering Ceramics with a High Power CO_2 Laser, *Adv. Manuf. Eng.*, **2**, pp. 27–36.
143. M. Taya and R. J. Arsenault (1989) *Metal Matrix Composites: Thermomechanical Behaviour*, Pergamon, Oxford.
144. E. Savrun and M. Taya (1988) Surface Characterisation of SiC_W/2124 Al and Al_2O_3 Composites Machined by Abrasive Water Jet, *J. Mat. Sci.*, **23**, pp. 1453–8.
145. H. Ho-Cheng (1990) A Failure Analysis of Water Jet Drilling in Composites, *Int. J. Mach. Tools Manufac.*, **30**, pp. 423–30.
146. K. Kuriyama and E. R. Wallach (1991) Diffusion Bonding of SiC–Ti Metal Matrix Composites, in *Interfacial Phenomena in Composite Materials '91*, Leuven, I. Verpoest and F. R. Jones (eds.), Butterworth, pp. 282–3.
147. R. S. Bushby, V. D. Scott and R. L. Trumper (1991) Diffusion Bonding of Metal Matrix Composites, in *Metal Matrix Composites – Processing, Microstructure and Properties, 12th Risø Int. Symp.*, Risø, N. Hansen, D. J. Jensen, T. Leffers, H. Lilholt, T. Lorentzen, A. S. Pedersen, O. B. Pedersen and B. Ralph (eds.), Risø Nat. Lab., Denmark, pp. 271–6.
148. O. T. Midling, O. Grong and M. Camping (1991) A First Report on the Microstructural Integrity and Mechanical Performance of Friction Welded Al–SiC Composites, *ibid.*, pp. 529–34.
149. K. Suganuma, T. Okamoto and N. Suzuki (1987) Joining of Alumina Short Fibre Reinforced AA6061 Alloy to AA6061 Alloy and to Itself, *J. Mat. Sci.*, **23**, pp. 1580–4.

10

Development of matrix microstructure

Initially, suitable matrix alloy compositions and heat treatments were proposed solely on the basis of experience gained with unreinforced alloys. However, as this chapter demonstrates, the incorporation of a reinforcing phase can have a pronounced effect on the development of matrix microstructure. Because of their sensitivity to matrix microstructure, this is an especially important consideration for discontinuously reinforced systems.

10.1 Dislocation structure and behaviour

Plastic deformation in the matrix of a composite is never completely homogeneous. The reinforcement interrupts flow, giving rise to distinctive dislocation structures. As well as having immediate implications for the flow properties of the matrix, these structures also indirectly influence flow behaviour via changes in the precipitation and aging response.

10.1.1 The influence of thermal stress on dislocation structure

In many MMCs, thermal stresses can give rise to dislocation densities which are 10–100 times greater than those for comparable unreinforced alloys. The conditions under which the creation of thermally stimulated dislocations is energetically favourable were first studied by Ashby and Johnson[1], who found that, for incoherent particles, the critical misfit decreases with increasing particle size. For short fibres and particles, the simplest mechanism of stress relief is to punch out a dislocation loop (a disc of vacancies or interstitials) into the matrix (Fig. 10.1(a,b)). This mechanism, illustrated schematically in Fig. 4.16, is well understood[4] in terms of a relaxation in the local stress field. A number of quantitative

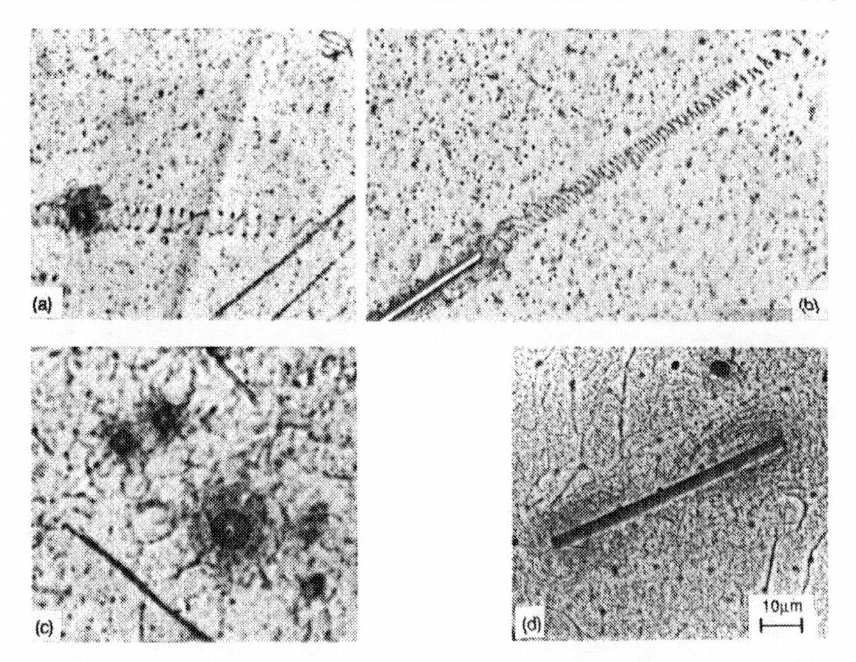

Fig. 10.1 Optical micrographs taken from transparent AgCl matrix/glass particle-containing analogue systems[2,3]. These illustrate for (a) spherical and (b) fibrous reinforcement how prismatic punching of dislocations (decorated by dissociated Ag) is driven by thermal misfit stresses. The dislocation tangling mechanism[2] shown in (c) and (d) is an alternative means of misfit strain relief.

models have been proposed (see §4.4.2) to describe the extent of punching and its relationship to particle aspect ratio[5–7]. As exemplified by the prismatic loops shown in Fig. 10.2, there is usually a specific relationship between the crystal orientation of the matrix host grains and the available punching directions. If the orientations of neighbouring grains are unfavourable in terms of the resolved stresses for punching, then a complex tangle of secondary dislocations is often generated instead of, or in addition to, the loops (Fig. 10.1(c,d)). Irrespective of the precise dislocation mechanism, the heavily dislocated zone surrounding each particle is expected to increase in size with decreasing matrix friction stress[†].

A number of experimental investigations have focused on the effect of thermal stresses on dislocation density in discontinuous MMCs. TEM

[†] I.e. the stress required to move a dislocation through the matrix.

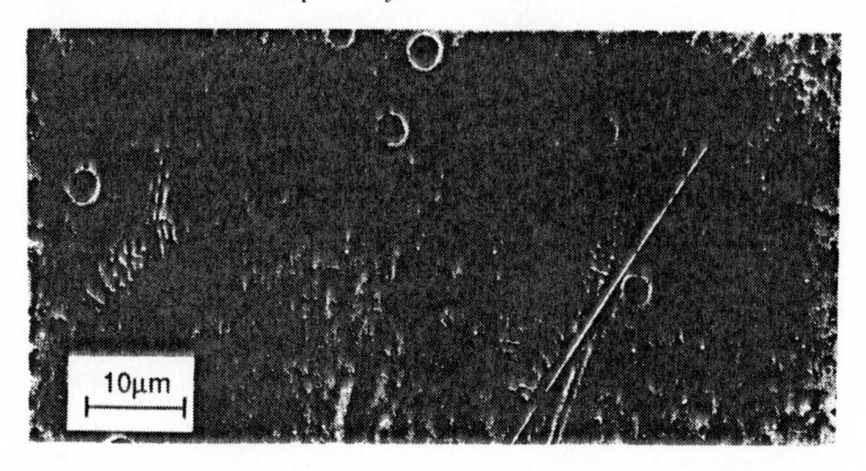

(a)

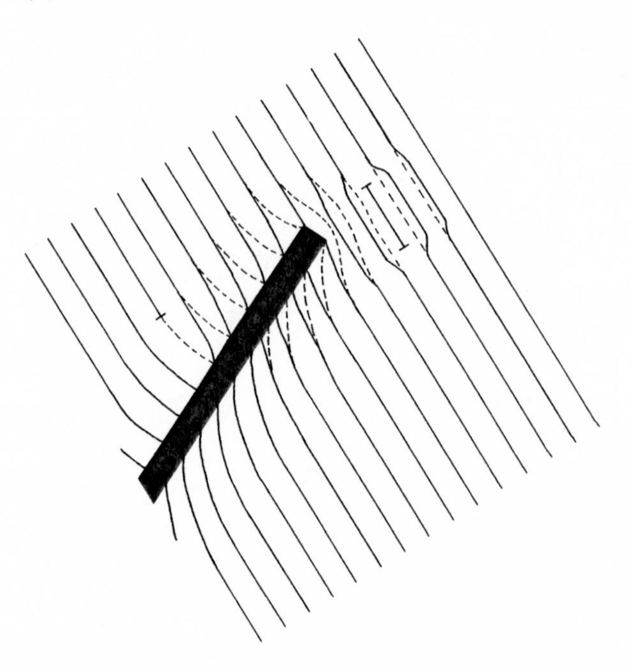

(b)

Fig. 10.2 (a) Micrograph and (b) corresponding schematic[2,3] illustrating how the punching of prismatic interstitial (RHS of fibre) and vacancy (LHS of fibre) loops can lower the thermal misfit strain (from the dashed to the continuous lines) when the fibre lies at an angle to the lattice planes associated with the punching direction.

observations† for commercially pure Al–2 vol% SiC_W confirm[9] the expected enhancement of dislocation density, although, somewhat surprisingly, no spatial density variations were observed prior to straining in this case (see Fig. 10.5(a)). This result could be explained in terms of the distance over which punched dislocations might travel into a commercially pure Al matrix[10]. (Certainly it appears that in Fig. 10.5(a) dislocations punched from the fibre have become entangled in neighbouring oxide stringers.) In contrast, for 2124 Al/SiC_P composite, both direct dislocation observation[11] and θ' precipitate labelling[12] (§10.2) suggest that the thermally generated dislocation density does indeed decrease with distance from the reinforcement. For the smallest particles, however, the decrease in dislocation density was not as evident as for larger particles and this may explain the absence of a spatial variation for fine whisker-containing 2124 Al material[10].

It is possible to calculate the extent of plastic strain (ε^P) as a function of the radial distance from a spherical particle (r) on the basis of ideally plastic behaviour[13]

$$\varepsilon_r^P = -2\varepsilon_\theta^P = \frac{\sigma_{YM}(1-\nu_M)}{G(1+\nu_M)}\left[1-\left(\frac{r^P}{r}\right)^3\right] \tag{10.1a}$$

$$U^P(r) = \frac{\sigma_{YM}^2(1-\nu_M)}{G(1+\nu_M)}\left[\left(\frac{r^P}{r}\right)^3 - 1\right] \tag{10.1b}$$

where the matrix has yield stress σ_{YM}, Poisson's ratio ν_M and a plastic zone size r^P. The spatial variation in dislocation density can then be deduced‡ in terms of the plastic work $U^P(r)$. As shown in Fig. 10.3, this variation has been confirmed experimentally by direct[11] and indirect methods[12]. For angular particles, precipitate observations (Fig. 10.4) suggest that the dislocation density is greatest at sharp corners[12], in agreement with expectations.

For continuous fibre reinforced composites, geometric considerations limit the number of available dislocation mechanisms for thermal stress relief. However, etch pit measurements of dislocation density in Cu containing a low volume fraction of W fibres show a similar radial distribution to the particulate results described above (see Fig. 10.6(b)). In a similar

† There is always concern when viewing dislocation structures in thin foils about changes either induced by the foil preparation method itself or simply as a result of the introduction of free surfaces[8] (§11.6). Apparent dislocation densities also change with the diffraction conditions. Comparative observations can still be meaningful, although quantitative measurements are likely to be very inaccurate.

‡ Assuming the dislocation density $\rho(r) \approx \rho_0 + U^P(r)/Gb^2$.

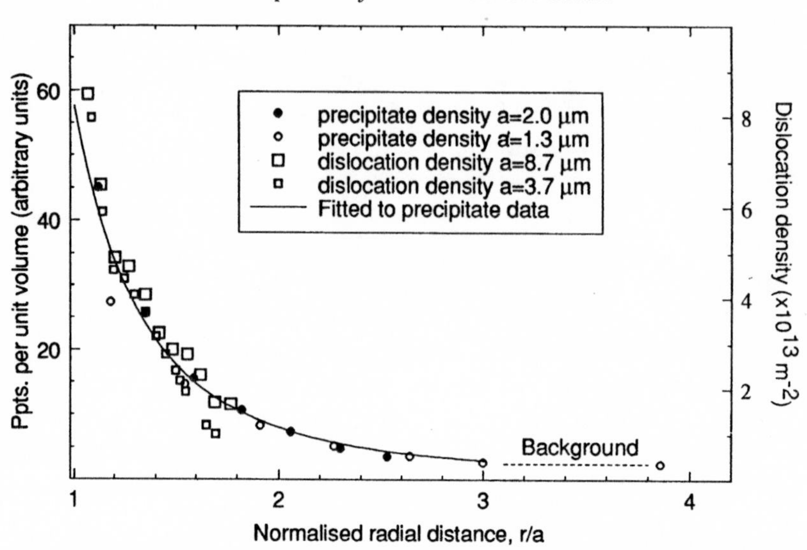

Fig. 10.3 The measured dislocation density profile within a 2124 Al alloy matrix, as a function of the (normalised) distance from the SiC_P reinforcement[11]. The slight decrease in average dislocation density with decreasing particle size, *a*, is in disagreement with the predictions of Arsenault and Shi[14] (dislocation density to be inversely proportional to size) and Dutta and Bourell[15] (independent of size). The θ' precipitate density[12] variation in Al–4 wt% Cu, aged so as to decorate the dislocations present at 190 °C, seems to confirm the direct dislocation measurements and conforms to the simple $1/r^3$ plastic strain relation (continuous line) of eqn (10.1) surprisingly well. While the growth of the differently orientated variants has been shown[12] to be sensitive to stress at the aging temperature, the elastic component of the thermal misfit is not expected to be large.

vein, profuse tangles have been observed near the interface of Al–graphite, while radial dislocations have been observed for Mg–graphite[18]. In contrast to the above results, the incorporation of steel fibres in Al was found not to affect the dislocation density[19].

10.1.2 The influence of plastic straining on dislocation structure

In cubic metals, plastic deformation is normally associated with the generation and movement of dislocations. In Chapter 4 a distinction was drawn between dislocations formed as a geometrical consequence of plastic deformation, i.e. those related directly to the plastically generated misfit between matrix and particle (such as Orowan loops – see Fig. 4.9), and those leading to the relaxation of such a misfit (such as prismatic punching and tangles – §4.4.2). In observing actual dislocation structures,

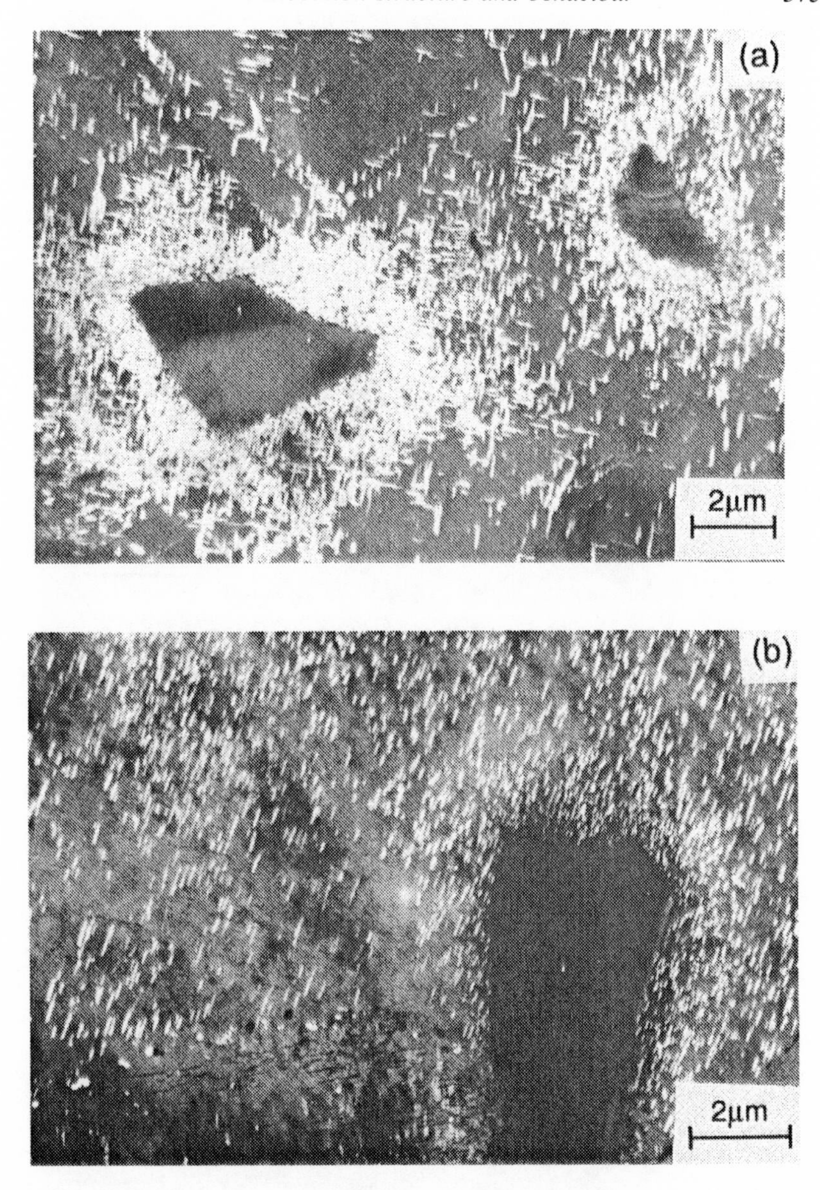

Fig. 10.4 TEM micrographs[12] showing the fall-off in θ' density around a SiC_P in Al–4 wt% Cu after aging at 190 °C. At this temperature θ' nucleates almost exclusively at dislocations (§10.2). In (a) the specimen was quenched prior to aging at 190 °C; this highlights the distribution of thermally nucleated dislocations at room temperature. In (b), the precipitates result from 0.5% plastic strain prior to aging at 190 °C. Clearly, it may be possible to locally age-harden the matrix in regions of high stress, i.e. in the vicinity of the particles, by such a heat treatment.

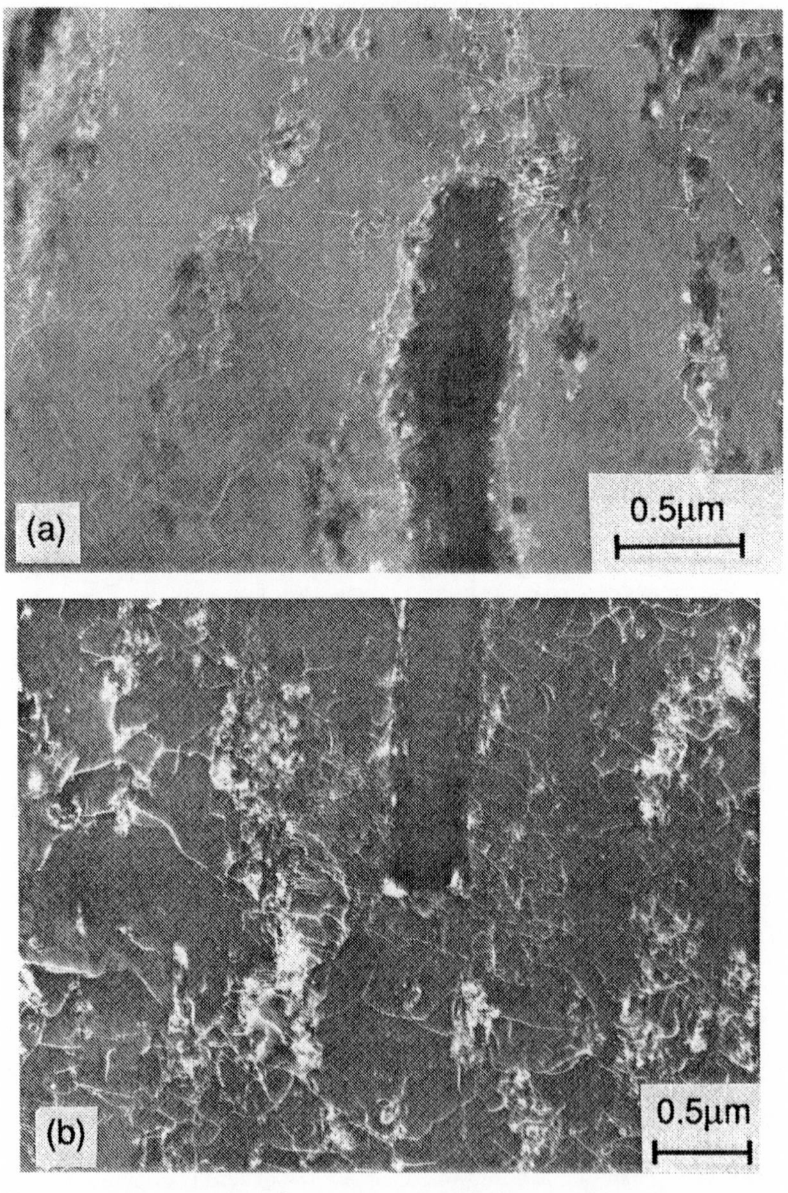

Fig. 10.5 TEM micrographs[9] of powder route CP Al–2 vol% SiC_W, showing the development of dislocation structures with increasing plastic strain. (a) *Prior to straining* – there is a high density of dislocations, but they are rather evenly distributed throughout the matrix; only occasional fibre-related tangles are present. (b) *1–3% strain* – many dislocations are visible in the immediate vicinity of the whisker ends (within 1 μm), mostly in the form of nets and subgrain walls (cf. low dislocation density in the unreinforced Al). (*cont.*)

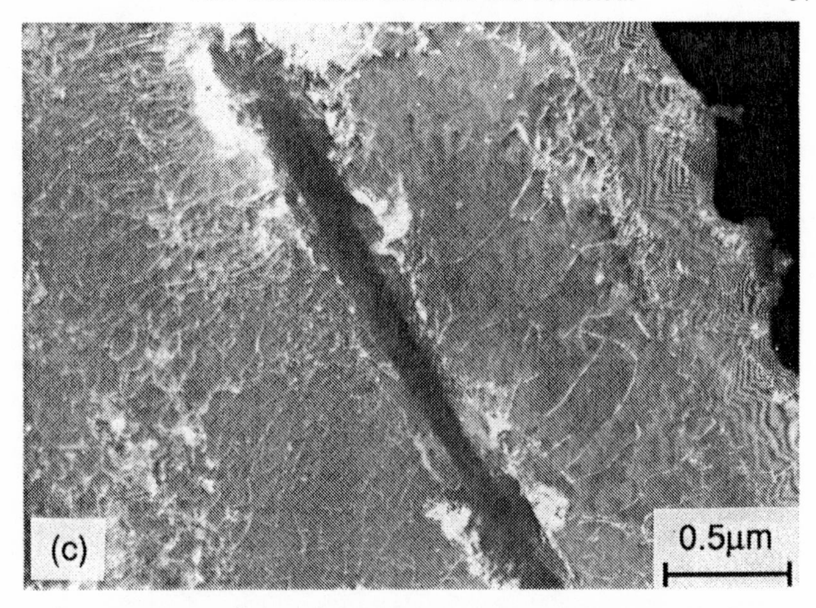

Fig. 10.5 (*cont.*) (c) 5% *strain* – dislocations can be seen all along the whisker length (most intense at ends); subgrain walls (parallel to extrusion direction) are visible, with nets spreading into the matrix. At still higher strains[9], extensive subgrain structures develop; cell and wall dimensions are smallest near whisker ends; occasional fractured whiskers can be seen (cf. diffuse subgrains in unreinforced Al).

it is difficult to make such neat distinctions, although from a load transfer point of view the difference remains valid. Instead, dislocation structures appear complex and chaotic. In an attempt to simplify matters, Barlow and Hansen[9] studied dislocation development in recrystallised CP Al–2 vol% SiC_W material, and their findings are summarised in Fig. 10.5. The propensity for subgrain wall and dislocation network formation has also been observed by others[20,21] and is possibly a consequence of the rearrangement of primary dislocations into lower energy structures (§10.3).

In agreement with the TEM observations, precipitation evidence[12] (Fig. 10.6) also points to an increase in dislocation activity with straining, as well as a tendency for higher dislocation densities to form at the poles of fragments oriented along the loading direction (Fig. 10.4(b)).

Few observations of dislocation structures in continuous fibre composites have been reported as a function of externally applied strain, although the Cu–W system described in Fig. 10.6 is in effect plastically strained axially at high fibre fractions as a result of the internal constraint[16]. An increase in dislocation density with increased straining has been observed for Al–steel, but no variation in density with reinforcement volume fraction or with radial distance from the fibres was evident[19].

10.2 Precipitation behaviour

The presence of a reinforcing phase has both direct implications for the precipitation behaviour, through, for example, interfacial solute segregation, and indirect implications, for example via changes in nucleation and growth kinetics caused by variations in dislocation and vacancy concentrations. In this section, attention is focused on the principles governing different precipitation phenomena rather than the specifics relating to any one particular system, so as to provide a framework for the understanding of aging kinetics and microstructural control generally.

10.2.1 Monitoring of aging kinetics

Of the many different means of monitoring aging, the following three methods are most informative:

- *Differential scanning calorimetry* (*DSC*) for the identification of the precipitation and dissolution temperatures of various phases, via their exotherms and endotherms
- *Transmission electron microscopy* (§11.6) for precipitate identification, population and size evaluation
- *Hardness measurements* as an indicator of the mechanical response to aging.

With respect to hardness measurements, particular care must be taken when choosing the scale of the indentation. Microhardness measurements can suffer from large fluctuations (±10VHN) because of the danger of indenting directly upon a reinforcing particle. Macrohardness measurements, on the other hand, suffer less scatter (±2VHN), but sensitivity to the presence of the reinforcement can obscure the effect of changes in the matrix.

In the following sections, attention will be focused on age-hardening precipitates, because it is their development which most strongly influences mechanical behaviour.

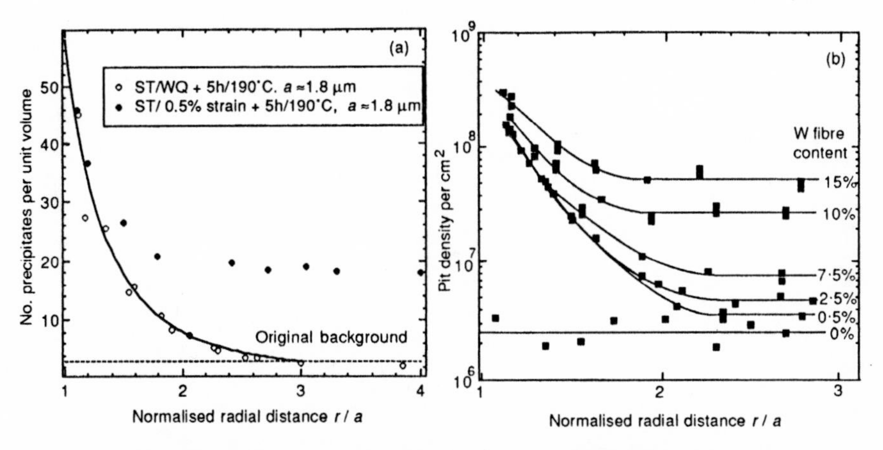

Fig. 10.6 (a) Measured precipitate densities† around a SiC particle after solution treatment (ST) plus water quenching (WQ) and after ST/WQ followed by plastic straining and annealing. The change in θ' density[12] observed in Al–4 wt% Cu/SiC_P upon straining suggests that the dislocation density increases more by an expansion of the heavily dislocated zone outwards into the matrix than by an increase in the peak density at the interface. This observation is in contrast to the behaviour predicted[17] for dispersion hardened alloys (i.e. dislocation forest size approximately independent of strain). (b) The spatial variation in etch pit density observed for continuously reinforced Cu with increasing W fibre fraction[16]. When the matrix contains few fibres, the thermal fibre/matrix misfit is accommodated locally in a manner similar to that for short fibres (Figs. 10.1 and 10.4), perhaps by axial matrix shear locally[16] with a plastic strain field inversely proportional to r^2. At large fibre fractions however, the matrix undergoes self-induced *global* plastic strain axially ($\sim$ 1% at 15%) and, similar to (a), the associated increase in dislocation density appears to be most marked in the plateau region.

10.2.2 Dislocation-nucleated precipitation

Many precipitates nucleate at heterogeneous sites within the matrix. Of these, those which nucleate on dislocations are probably the most important in the context of accelerated aging in MMCs. Good examples are provided by the θ' (Al_2Cu) phase in the Al–Cu binary system and the S′ (Al_2CuMg) phase in commercial 2124 Al alloys. As one would expect, DSC evidence for the Al–4% Cu system (Fig. 10.7) shows that increasing the particle content progressively lowers the temperature at which θ' forms. This *accelerates* the θ'-dominated age-hardening response (above the Gunier–Preston zone (GPZ) solvus – Fig. 10.8(a)).

As discussed in §10.1, the evaluation of dislocation densities by direct observations in the electron microscope is haphazard at best. Using a

† θ' precipitates in 2124 Al (§10.2.2) and etch pits in Cu nucleate on dislocations.

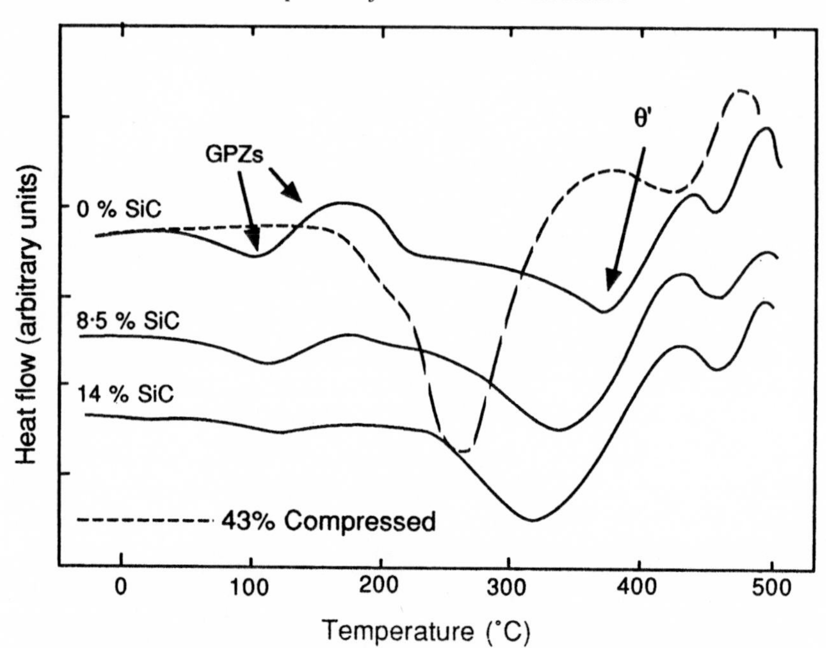

Fig. 10.7 DSC traces for Al–4 wt% Cu with 0, 8.5 and 14 vol% SiC_P. Note the shifting of the θ' exotherm (trough) to a lower temperature effected by incorporation of SiC, signifying a reduction in the activation energy for nucleation. A similar effect is brought about by cold deformation (43% compression) of the unreinforced alloy (dashed line). In contrast, the magnitude of the exo- and endotherms for the formation and dissolution of the GPZs decrease with increasing SiC fraction and are shifted to slightly higher temperatures†.

suitable alloy and heat treatment[12], the precipitation of dislocation-nucleated phases provides a complementary means of mapping dislocation densities, free from the difficulties of direct observation (e.g. Fig. 10.4). Unambiguous interpretation, however, requires a proper consideration of the nucleation and growth process. For Al–4% Cu, while homogeneous nucleation is possible, θ' nucleates almost invariably on dislocations below 200 °C. The initial nucleation rate will thus be proportional to the dislocation density, but will fall as potential nucleation sites are consumed and the solute is depleted locally. Consequently, the local density of precipitates will only be proportional to the dislocation density if the

† Care must be exercised when comparing the magnitude of the exotherm: for simple volume fraction reasons one would expect the exotherm to decrease in magnitude with increasing volume fraction of SiC.

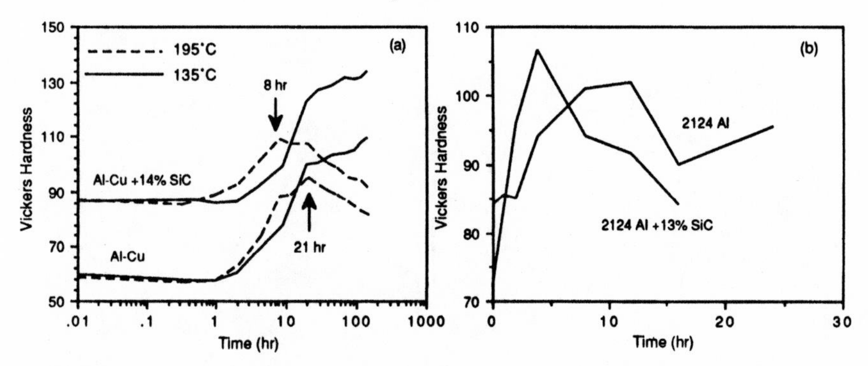

Fig. 10.8 (a) Macrohardness-derived aging curves[22] for Al–4% Cu/14% SiC_P composite and for the control alloy for aging temperatures of 135 °C (i.e. below the GPZ solvus) and 195 °C. At 135 °C it is difficult to discern a retardation in hardening response, corresponding to the suppression of the formation of GPZs observed in Fig. 10.7. At 195 °C on the other hand, the response is clearly accelerated because of the increased dislocation-assisted nucleation of θ'. (b) The aging response of commercial 2124 Al–13% SiC_P (S' dominated) is accelerated[10] at 177 °C for similar reasons. Note that for the macrohardness tests the composite results are affected by the presence of the reinforcing phase as well as by the matrix microstructure.

distribution is observed *before* significant solute depletion has occurred (i.e. at a very early stage).

While the precipitates tend to nucleate upon dislocations below 200 °C, there is some evidence to suggest that the particular variant which precipitates is influenced by any elastic fields which may be present[12]. This effect is exemplified by the change in the relative densities of the two precipitate variants around the circumference of the particle shown in the TEM micrographs of Fig. 10.9.

10.2.3 Vacancy-related precipitation

In certain cases, the nucleation of precipitates is aided by the presence of excess vacancies, which accelerate the kinetics of diffusional processes. A good example of this is provided by the formation of Gunier–Preston zones (GPZs) in Al–Cu. For this particular system, DSC evidence (Fig. 10.7) suggests that the incorporation of SiC_P (or prior cold work) causes the suppression of GPZ formation and its elevation to *higher* temperatures. This has been explained both in terms of the presence of matrix defects at and near the ceramic interface[23] which act as vacancy

Fig. 10.9 The two variants of θ' precipitate in Al–4 wt% Cu around a SiC particle viewed under two dark field conditions. For this precipitate, the variants tend to *grow* preferentially such that the planes of the plates lie normal to a compressive stress[12]. That the density of each variant is largest in regions where the variant is growing radially to the particle is suggestive of a compressive hoop stress and a tensile radial stress. This implies that the stress state at the aging temperature was the reverse of that commonly found at room temperature. Because the *nucleation* process is related primarily to the dislocation density, the precipitate density when the two variants are summed is approximately constant around the particle circumference.

sinks†, and by the mopping up of vacancies by the extra dislocations[22]. A further complication is the competitive interaction between the GPZ formation and the earlier formation of θ'.

10.2.4 Homogeneously-nucleated precipitation

One would expect the precipitation of a homogeneously nucleated phase to be unaffected by the presence of the reinforcement, provided they did not chemically interact. A good example of this is provided by the coherent δ' (Al_3Li) phase in Al–Li, which nucleates independently of structural defects. However, somewhat surprisingly, whilst DSC evidence shows only a slight shift of the δ' precipitation exotherm upon incorporation of SiC reinforcement, the age-hardening response of Al–Li is actually accelerated[22] by SiC particles at 175 °C (as it is by plastic deformation of the unreinforced alloy). Furthermore, TEM evidence shows the size and spatial variation in precipitate density to be unaffected in both cases. Further evidence that the accelerated age hardening is probably related to something other than the δ' phase is given by the fact that in the

† A similar effect is responsible for the formation of precipitate free zones adjacent to grain boundaries in Al–Mg–Zn alloys[24,25].

unreinforced alloy the size of the δ' precipitates at peak hardness is ~30 nm, which is about the size for which Orowan hardening becomes more energetically favoured than shearing[26,27], whereas in the composite the peak hardness occurs at a smaller size (~20 nm). The presence of a zone, denuded of precipitates, at the particle/matrix interface may be responsible (see below).

10.2.5 Interfacial precipitation and precipitate-free zones

Clearly, segregation of solute to the reinforcement/matrix interface could have a marked effect on composite properties. Haaland *et al.*[28] (7091 Al/SiC), and Liu *et al.*[29] (A17XXX/SiC) report that over-aging lowers interface strength (under fatigue and fracture), while Mahon *et al.*[30] report that, for 2124 Al, interface precipitation can improve load transfer. In contrast, precipitate free zones (PFZs) can form near the matrix/reinforcement interface when stable phases at the boundary act as sinks for solute atoms in the matrix, or when the interface depletes the local vacancy concentration. Prangnell[22] has highlighted the growth of PFZs (<90 nm) in Al–Li/SiC_P. However, while there have been a large number of reports of segregation to, or away from, the reinforcement interface[22,31,32], difficulties associated with the measurement of interface properties in discontinuous systems have limited the inferences which can be drawn for mechanical properties.

10.3 Grain structure, texture, recovery and recrystallisation

10.3.1 Deformation-induced features

Grain structure

A non-deforming particle acts as an obstacle to dislocation glide and, as discussed in §4.2.2, can be by-passed only by the formation of an Orowan loop. Each Orowan loop transfers stress to the particle and so, in general, only a small number of dislocations can pass the particle before relaxation mechanisms will act to reduce the particle stress by the stimulation of secondary dislocations. As a result, after deformations typical of most processing routes (e.g. rolling and extrusion), the dislocation density in a zone surrounding each particle is often considerably higher than in the matrix as a whole† (§10.1).

Recovery processes are able to reduce the energy stored in the vicinity of the particles by the removal of point defects and through the

† This is dependent on the processing condition[21]; for example, extrusion at high temperature and with low peak strain rates results in less pronounced deformation zones[33].

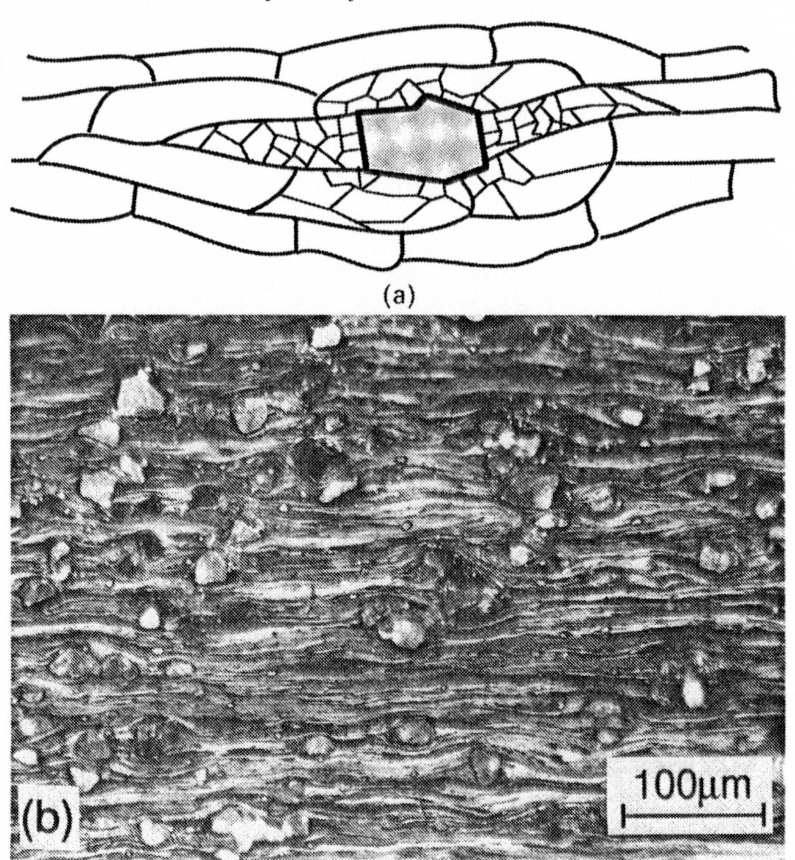

Fig. 10.10 (a) Schematic after Humphreys *et al.*[33] highlighting the localised refinement of the recovered subgrain structure and the deflected appearance of high aspect ratio subgrains around the particle zone. These features are evident in the optical micrograph (b) of an Al 10% SiC_P (d = 20 μm) composite subsequent to cold rolling[34].

rearrangement of dislocation structures, often so as to form subgrain boundaries. Dislocation rearrangement is especially favoured in high stacking fault energy metals (e.g. Al) which cross-slip easily, and this leads to a fine subgrain size locally (see Fig. 10.10). For whisker reinforcement, this region is often approximately whisker shaped[35], and approximately 1–2 times larger than the whiskers themselves, while for particles, evidence suggests that the subgrain structure is finer at angularities and corners[36]. This refinement is often counterbalanced by enhanced recovery-led subgrain growth within the locally deformed zone[37]. Furthermore, when considering composite microstructures, it is extremely important to note that in composites made by the powder route, the subgrain and grain structure is strongly influenced by the fine oxide particles; this is exemplified by the micrographs of Fig. 10.11. On a larger scale, the reinforcing particles also seem to disturb slip resulting in a substructure with a deflected appearance (Fig. 10.10(b)).

Deformation texture

If a single crystal is deformed in a tensile test, the active slip direction rotates towards the tensile axis. During the deformation of a polycrystal the crystallites also rotate, but they exert a mutual constraint on each other which limits the possible shape change. Depending on the number of available slip systems, these constraints limit the possible grain rotations, and give rise to textures characteristic of the crystal structure as well as the mode of deformation. The relaxation of deformation-induced primary dislocation structures around the particles gives rise to local rotations which are contrary to the sense of the matrix as a whole[36,39]. Consequently, the presence of large particles tends to reduce the deformation texture by an amount dependent on their volume fraction[35] (Fig. 10.12).

The extent of rotation within the subgrains is greater for larger particles[41] and decreases with distance from the particle (see Fig. 10.13). In the case of whisker reinforcement, the subgrain misorientation has been observed to be most marked near the whisker end region[21,35]. As with fracture behaviour (§7.4.5), particle clusters can behave in two ways, depending on whether they move individually or collectively. For example, Liu *et al.*[35] found that when the deformation zones for a number of SiC whiskers overlapped, large misorientations resulted (Table 10.1). However, when a cluster moves collectively, the deformation of the

Table 10.1 *The percentage of whiskers and whisker clusters with associated subgrain rotations*[35]

Single whisker		Clusters	
Rotation angle (°)	Percentage	Rotation angle (°)	Percentage
>15	70	>50	75
10–15	20	30–50	10
5–10	10	15–30	15

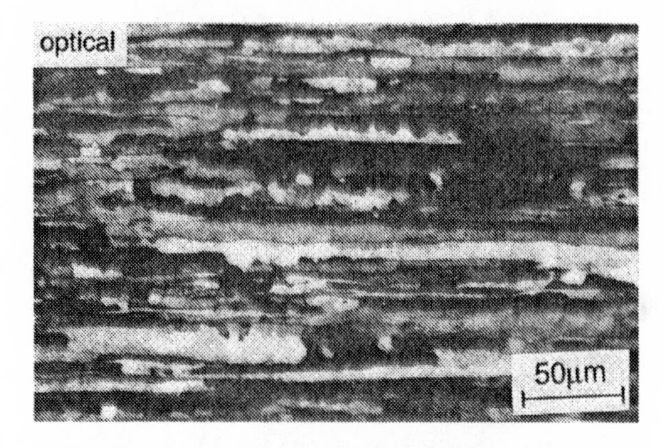

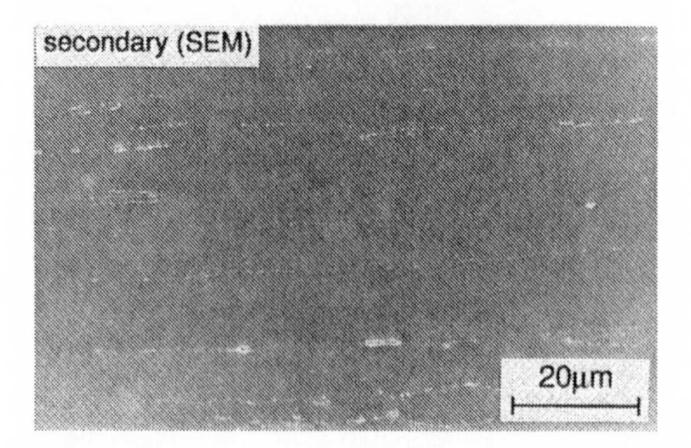

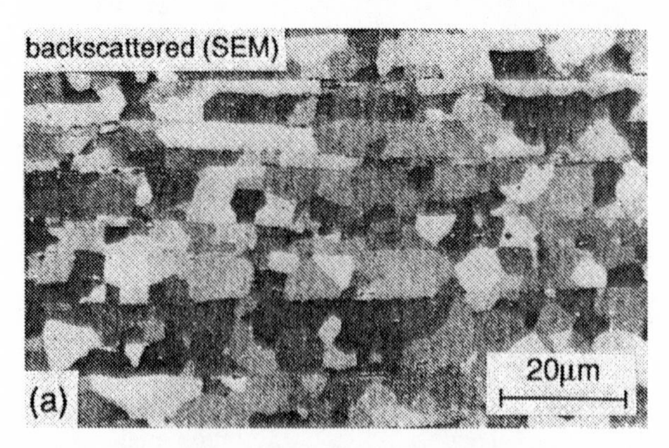

Fig. 10.11 Grain and subgrain structures of as-extruded powder route CP aluminium[38] (a) unreinforced and (b, opposite) reinforced with 5% of spherical Al_2O_3 particles. The fine oxide particles, apparent in the secondary electron micrographs, are broken up into long, aligned stringers, revealing the flow pattern around the reinforcing spheres. (*cont.*)

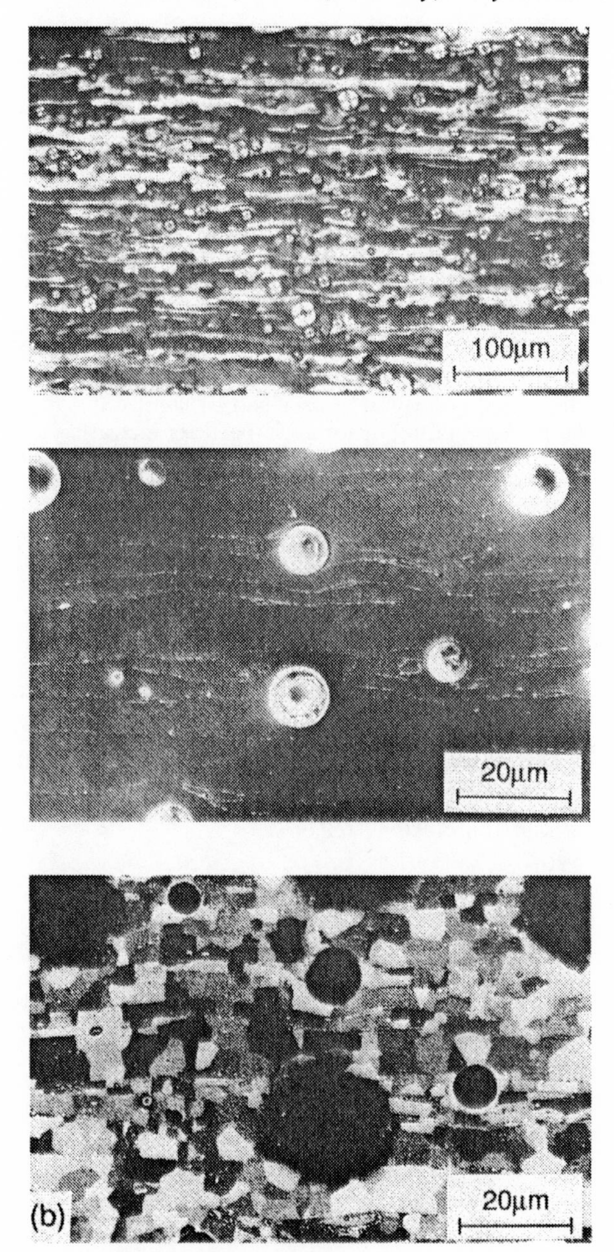

Fig. 10.11 (*cont.*) These fine particles inhibit recrystallisation by Zener pinning of subgrain boundaries, ensuring that the grain structure reflects the distortion of the original Al powder. The subgrain structure, visible from channelling contrast in the back-scattered images is somewhat finer in the composite as a result of more localised deformation, but is largely determined by the Zener pinning effect.

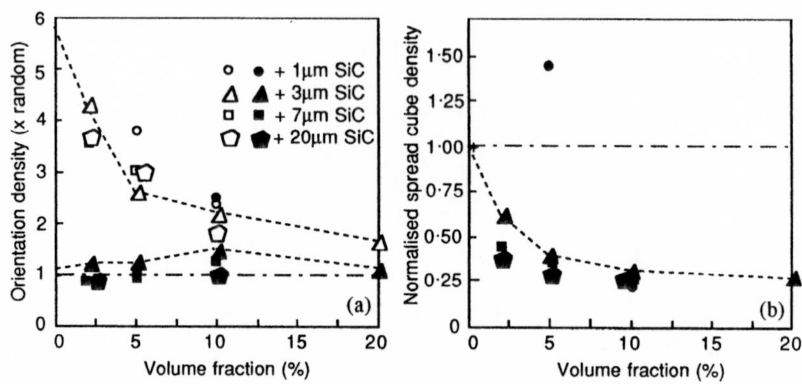

Fig. 10.12 (a) Plot of matrix texture[34] versus particle content for Al-based alloys showing how the presence of non-deformable particles increasingly smears out the deformation texture of the matrix (open symbols). This is especially marked in the volume fraction range $0 < f < 2.5\%$. The closed symbols show the recrystallised intensity. (b) Plot illustrating the weakening of the spread cube texture relative to the unreinforced alloy in recrystallised specimens. Such plots must be treated carefully since they tend to smear out the existence of distinct components, e.g. $\{100\}\langle 013\rangle$[40].

intermediate region can be extremely low while the deformation zone surrounding the cluster is large and corresponds to that around a single larger particle. This behaviour is exemplified by the micrographs in Fig. 10.14; while the stored energy in the zone around the cluster is large, there is little *within* the cluster itself.

10.3.2 Recrystallisation

It is now well established that, according to size, particles can have two effects on recrystallisation. These are:

- *particles can pin*[41] *high angle grain boundaries if the particles are small* ($f/d > 0.1$ μ*m*$^{-1}$)
- *particles can stimulate formation of viable recrystallisation nuclei*[41,43] *if they are large* ($> \sim 1$ μ*m*)

Zener pinning

This occurs because the boundary experiences a drag effect caused by the reduction in grain boundary area when the particle lies on the boundary. The drag pressure, for a random array of spherical particles, is given

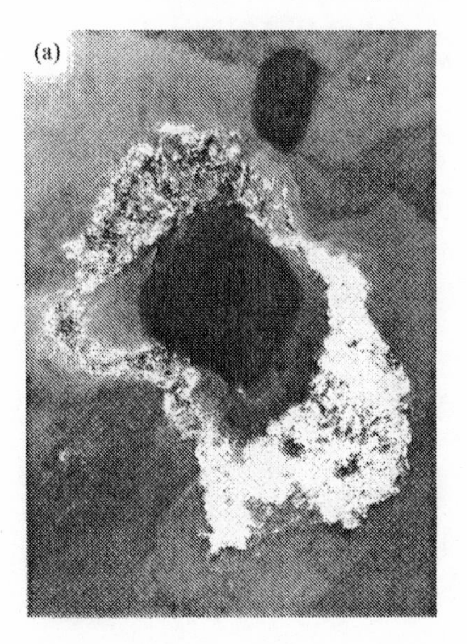

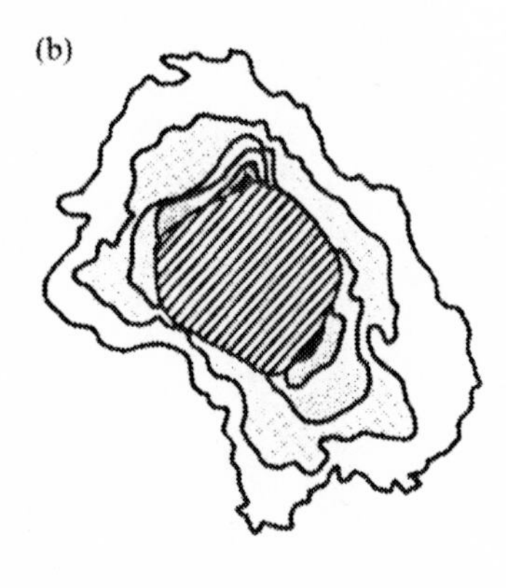

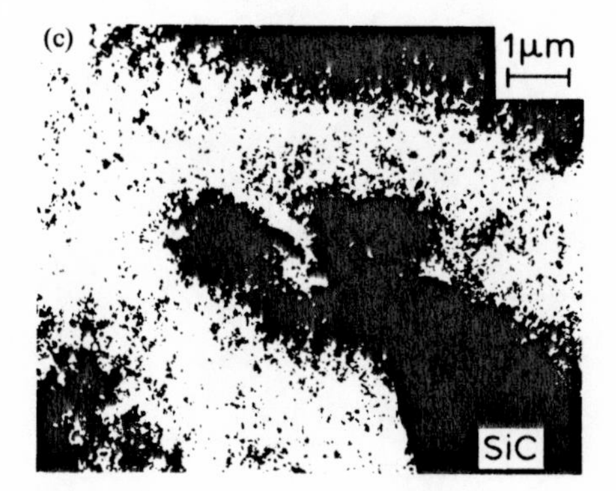

Fig. 10.13 Humphreys[41] has used (a) bright field TEM (here the 2° misorientation) to reveal the deformation zone around a 2 μm Si particle in an Al matrix deformed to a shear strain of 40%. His results (b) illustrate the dramatic increase in misorientation as the reinforcing Si particle is approached (the contours show 2, 5, 8, 11, 16 and 20° misorientation). An Al (100) weak beam image, (c), demonstrates that during deformation of Al–4 wt% Cu/SiC_P highly misoriented regions can occur locally at particle angularities[42].

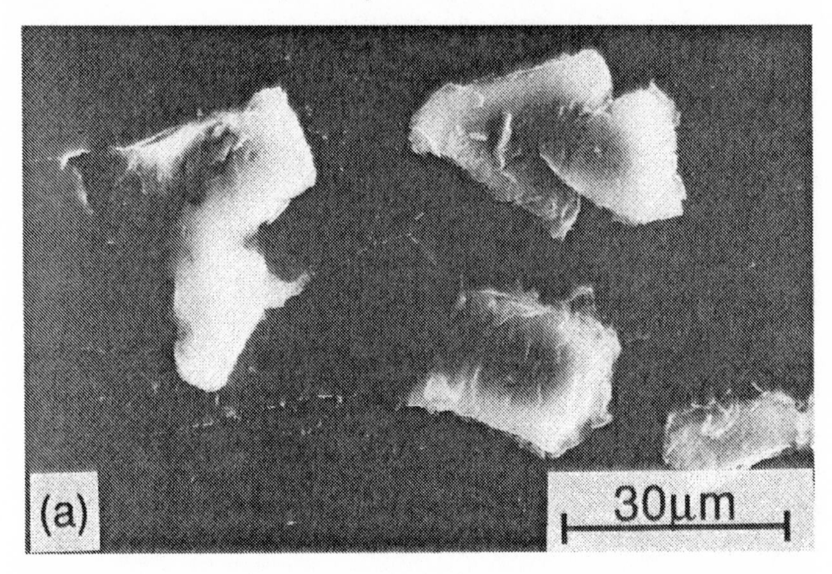

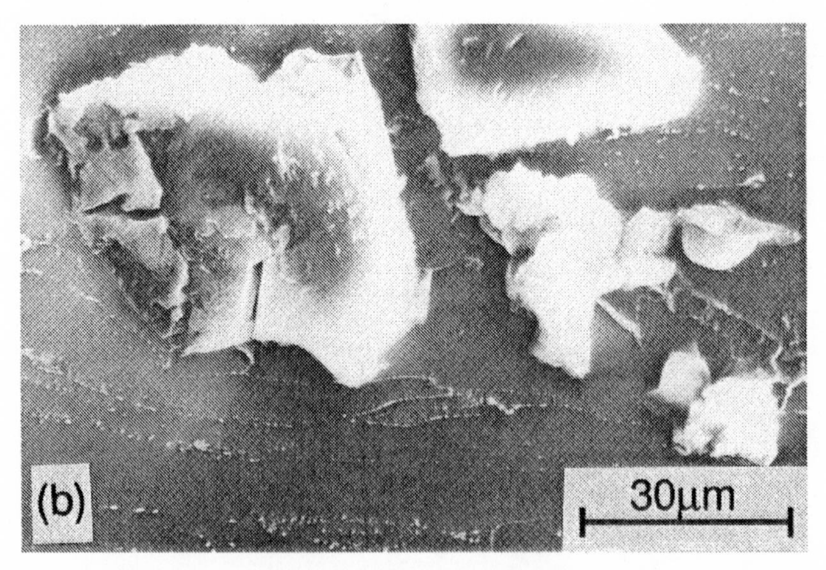

Fig. 10.14 SEM micrographs of as-extruded powder route CP Al–10 vol% SiC_P, in which the fine oxide stringers reveal features of the matrix flow field during deformation[38]. These micrographs demonstrate that the matrix has largely maintained its original grain structure within the region containing the cluster of SiC particles and hence that it is shielded from deformation during processing. In effect the three particles behave as a single larger particle with an associated deformation pattern outside the cluster.

approximately[44] by

$$P_{\mathrm{Zener}} \approx \frac{2\gamma f}{d} \tag{10.2}$$

This can be compared with the driving pressure for recrystallisation, which may be estimated in terms of the reduction in dislocation density ($\Delta\rho$)

$$P_{\mathrm{recryst}} \approx U \approx Gb^2 \, \Delta\rho \tag{10.3}$$

This is evidently an overestimate, as some of the dislocations will annihilate spontaneously or assemble into lower energy arrays before the passage of the recrystallisation front. In any event, this indicates that for a typical drop in dislocation density of $10^{15}\,\mathrm{m}^{-2}$, $P_{\mathrm{recryst}} \sim 3$ MPa. Substitution into eqn (10.2) of values appropriate for reinforcements in MMCs demonstrates that these could never act so as to pin a recrystallisation front or inhibit formation of a recrystallisation nucleus. For an array of fine particles, such as included oxide from powder route Al, use of eqn (10.2) suggests that significant drag pressures ($\sim$0.1 MPa) are possible, but that these are still rather small. In this context it should be remembered that the particles are not randomly distributed and that a bowed front will experience many more particles than the planar front used to derive eqn (10.2). These effects will tend to enhance the pinning efficiency: in practice it is established[21,33,38] that inhibition of both nucleation and growth of recrystallisation fronts by Zener drag is considerable in Al MMCs containing oxide dispersions, but negligible in their absence.

Particle stimulated nucleation (PSN)

This occurs provided that the matrix has sufficient stored energy in the vicinity of the particle for the nucleus to grow into the matrix from the deformed zone. The viability of this process is affected by subgrain mobility.

Given that MMCs contain particles which are greater than 1 μm in size, one expects and indeed finds that recrystallisation can be promoted by the reinforcement at relatively low volume fractions. A low nucleation efficiency has been observed[35,38] because of pinning of high angle grain boundaries by SiC and low angle grain boundaries by Al_2O_3 dispersoids which restrict the growth of nuclei. Consequently, the grain size is usually greater than that predicted on the basis of one grain per particle[45] (§4.2.2). The effect of increased reinforcement content in promoting a finer recrystallised grain size is illustrated by Fig. 10.15. These data

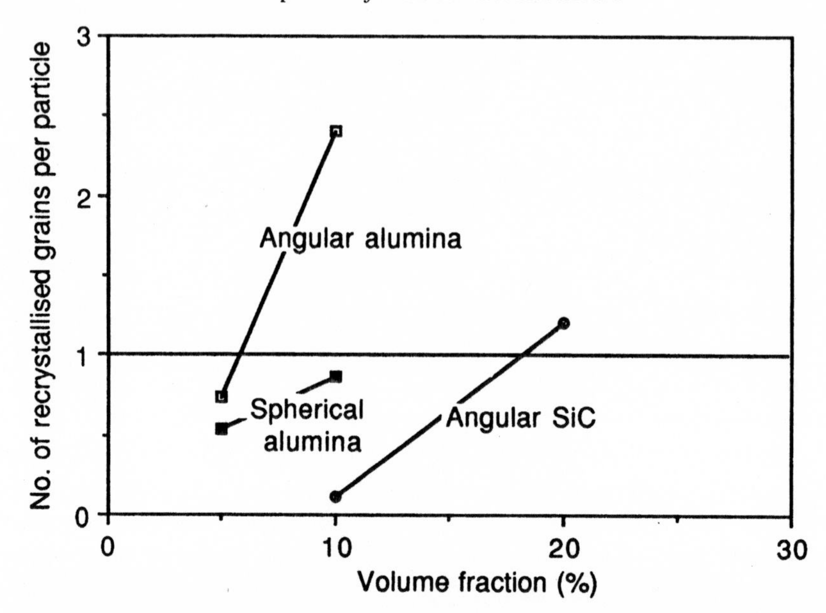

Fig. 10.15 The spread in the nucleation efficiency for several different Al-based reinforcing systems, after cold working and annealing[38]. The SiC-containing system had a much higher oxide concentration than the Al_2O_3 composites, and this is reflected in the lower recrystallisation efficiency. The grain size becomes finer as the reinforcement volume fraction is raised, at least over this range. However, it is very sensitive to fine oxide content as this can affect both the nucleation and growth of recrystallising grains.

demonstrate that particle angularities, and the associated strain localisation, appear to favour the promotion of more viable nucleation sites. However, they also indicate the importance of factors affecting boundary mobility, notably the presence of fine oxide particles.

The sensitivity of recrystallisation to local volume fraction is exemplified by Fig. 10.16, which shows an association between fine grain size and high volume fraction regions. However, as Fig. 10.17 indicates, at very high reinforcement contents, recrystallisation is sometimes suppressed. It has been proposed that in such cases subgrain growth is impeded so that the composite can recover (i.e. anneal out defects), but not recrystallise: again the presence of oxide particles is critical in promoting this effect. That their role in determining the competition between recrystallisation and recovery is complex is highlighted by recent evidence which suggests that dynamic recovery during extensive cold deformation is aided by the presence of a low oxide concentration, while

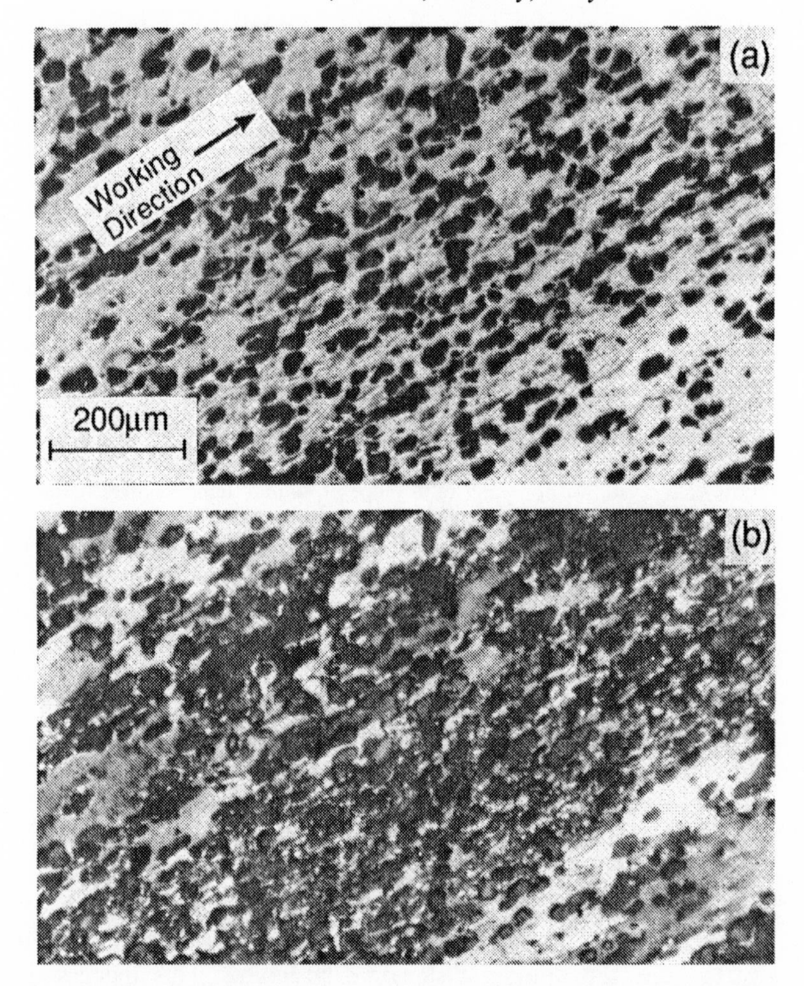

Fig. 10.16 Recrystallisation is sensitive to the particle content on a local scale. In this micrograph local volume fraction variations in nominally Al–10 vol% SiC_P have given rise to significant variations in recrystallised grain size[38]. The distribution of ceramic particles can be clearly seen in (a) a conventional optical micrograph, while (b) taken in polarised light, shows that the grain size is substantially smaller in the regions of higher ceramic content.

a higher concentration will obstruct dislocation rearrangement thereby encouraging recrystallisation[47]. Finally, Liu *et al.*[35] found that, while single SiC whiskers are not very powerful as nucleation sites, whisker clusters are effective sites for recrystallisation. This is probably a combination of the increased size and stored energy associated with the deformation zone within the cluster (Table 10.1).

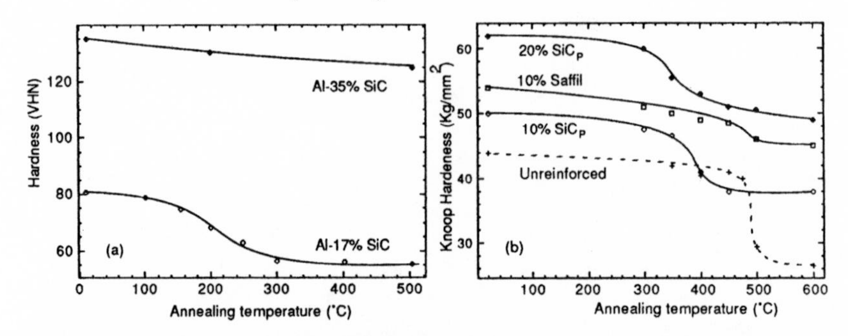

Fig. 10.17 Hardness changes during recrystallisation of powder route SiC_P MMCs (a) 3 μm particles[33] and (b) 15 μm particles[46] in Al. Higher particle contents tend to promote local matrix deformation, giving rise to larger stored energies and thus favouring the formation of recrystallisation nuclei. Growth of recrystallised grains, however, may be somewhat inhibited by the reinforcing particles, so that completion of recrystallisation may take place at similar temperatures to more dilute composites – see (b). At very high reinforcement contents, extended recovery may take place instead of recrystallisation – see (a). Note that absolute values of both hardness and recrystallisation temperatures are very sensitive to the nature of the fine oxide dispersion present.

In view of the increase in stored energy of an MMC compared with similarly deformed unreinforced alloys, it is not surprising that recrystallisation temperatures are generally lower (Fig. 10.18) and recrystallisation times shorter. Provided the particle size is not too small, this effect is more pronounced the smaller the reinforcing particles (at constant volume fraction), through an increase in the number of nucleation sites. At high temperatures, there is no dislocation accumulation at the non-deforming phase, because diffusion and dislocation climb are able to relax the primary dislocation structures. This reduces the stored energy within the vicinity of the particles considerably and removes the driving force for particle-stimulated recrystallisation.

Recrystallisation textures

Naturally, the texture after recrystallisation reflects the orientation of regions which are preferred as nucleation sites. For discontinuous MMCs, these regions are the highly misoriented zones in the vicinity of the particles, so it is not surprising that recrystallisation usually results in a sharp fall in the strength of any texture originally present (Fig. 10.12). (Retention of texture is often a useful indicator that extended recovery rather than recrystallisation has occurred.) In cases where the unreinforced matrix would ordinarily exhibit a strong recrystallisation texture,

Table 10.2 *Experimentally and theoretically calculated weakening of the ⟨111⟩ deformation texture brought about by increasing the volume fraction of differently sized reinforcing particles*[40]. *The predictions assume that each particle develops a deformation zone misoriented with respect to the ⟨111⟩, the size of which has been estimated by direct observations, and that these zones act as nucleation sites.*

	Whiskers			Particles (0.8 μm)			Particles (3.0 μm)		
Vol% →	4	10	20	4	10	20	4	10	20
Experimental	8%	18%	—	5%	12%	—	10%	22%	39%
Calculated	10%	30%	30%	5%	14%	14%	6%	23%	34%

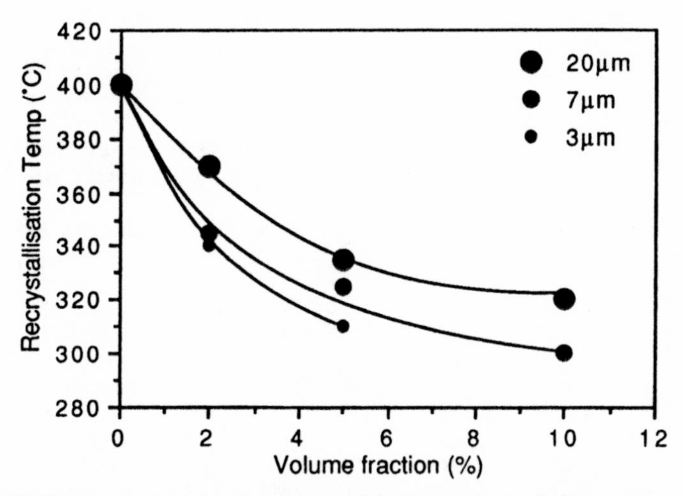

Fig. 10.18 Increases in stored energy with increasing reinforcement content combine with increases in the number of nucleation sites with decreasing particle diameter to progressively lower the temperature at which 50% of a 60% deformed powder route Al–SiC_P has recrystallised[33].

one would expect this to be much weaker when reinforcement is present[48] (Fig. 4.12). The extent of the weakening is related to the size and volume fraction of the reinforcement as shown in Table 10.2. Certain textures have, however, been specifically attributed to the effect of the presence of reinforcing particles through a combination of favoured misorientation angles[36] and of the accelerated rate of growth of the favourably oriented grains[49].

References

1. M. F. Ashby and L. Johnson (1969) On the Generation of Dislocations at Misfitting Particles in a Ductile Matrix, *Phil. Mag.*, **19**, pp. 1009–22.
2. D. C. Dunand and A. Mortensen (1991) On Plastic Relaxation of Thermal Stresses in Reinforced Metals, *Acta Met. et Mat.*, **39**, pp. 127–39.
3. D. C. Dunand and A. Mortensen (1991) Dislocation Emission at Fibres – II. Experiments and Microstructure of Thermal Punching, *Acta Met. et Mat.*, **39**, pp. 1417–29.
4. F. J. Humphreys (1983) Deformation Mechanisms and Microstructures in Particle-Hardened Alloys, in *Deformation of Multi-Phase and Particle Containing Materials*, J. B. Bilde-Sørensen, N. Hansen, A. Horsewell, T. Leffers and H. Lilholt (eds.), Risø Nat. Lab., Denmark, pp. 41–52.
5. M. Taya and T. Mori (1987) Dislocations Punched-Out Around a Short Fibre in a Short Fibre Metal Matrix Composite Subjected to Uniform Temperature Change, *Acta Metall.*, **35**, pp. 155–62.
6. D. C. Dunand and A. Mortensen (1991) On the Relaxation of a Mismatching Spheroid by Prismatic Loop Punching, *Scripta Met. et Mat.*, **25**, pp. 761–6.
7. D. C. Dunand and A. Mortensen (1991) Dislocation Emission at Fibres – I. Theory of Longitudinal Punching by Thermal Stresses, *Acta Met. et Mat.*, **39**, pp. 1405–16.
8. M. Vogelsang, M. Fisher and R. J. Arsenault (1986) An *in situ* HVEM Study of Dislocation Generation at Al/SiC Interfaces in MMCs, *Metall. Trans.*, **17A**, pp. 379–89.
9. C. Y. Barlow and N. Hansen (1991) Deformation Structures and Flow Stresses in Al Containing Short Whiskers, *Acta Met. et Mat.*, **39**, pp. 1971–80.
10. T. Christman, A. Needleman and S. Suresh (1990) Microstructural Development in an Al-Alloy SiC Whisker Composite, *Acta Metall.*, **37**, pp. 3029–50.
11. C. T. Kim, J. K. Lee and M. R. Plichta (1990) Plastic Relaxation of Thermoelastic Stress in Al/Ceramic Composites, *Metall. Trans.*, **21A**, pp. 673–82.
12. P. Prangnell and W. M. Stobbs (1991) The Effect of Internal Stresses on Precipitation Behaviour in Particulate Reinforced Al Matrix MMCs, in *Metal Matrix Composites – Processing, Microstructure and Properties, 12th Risø Int. Symp.*, Roskilde, N. Hansen, D. J. Jensen, T. Leffers, H. Lilholt, T. Lorentzen, A. S. Pedersen, O. B. Pedersen and B. Ralph (eds.), Risø Nat. Lab., Denmark, pp. 603–10.
13. J. K. Lee, Y. Y. Earmme, H. I. Aaronson and K. C. Russel (1980) Plastic Relaxation of the Transformation Strain Energy of a Misfitting Spherical Precipitate: Ideal Plastic Behaviour, *Metall. Trans.*, **11A**, pp. 1837–47.
14. R. J. Arsenault and N. Shi (1986) Dislocation Generation Due to Differences between Coefficients of Thermal Expansion, *Mat. Sci. & Eng.*, **81**, pp. 175–87.
15. I. Dutta and D. L. Bourell (1988) A Theoretical Investigation of Accelerated Aging in Metal Matrix Composites, *J. Comp. Mat.*, **22**(9), pp. 829–49.
16. K. K. Chawla and M. Metzger (1972) Initial Dislocation Distributions in W Fibre–Cu Composites, *J. Mat. Sci.*, **7**, pp. 34–9.

17. L. M. Brown and W. M. Stobbs (1971) The Work-Hardening of Cu–SiO_2 – II. The Role of Plastic Relaxation, *Phil. Mag.*, **23**, pp. 1201–33.
18. S. P. Rawal, L. F. Allard and M. S. Misra (1987) Characterisation of Interface Structure in Gr/Al and Gr/Mg Composites, in *Proc. ICCM VI/ECCM2*, London, F. L. Matthews *et al.* (eds.), Elsevier, pp. 169–82.
19. M. R. Pinnel and A. Lawley (1970) Correlation of Uniaxial Yielding and Substructure in Al–Stainless Steel Composites, *Metall. Trans.*, **1**, pp. 1337–48.
20. O. D. Sherby, R. H. Klundt and A. K. Miller (1977) Flow Stress, Subgrain Size and Subgrain Stability at Elevated Temperature, *Metall. Trans.*, **8A**, pp. 843–50.
21. R. A. Shahani and T. W. Clyne (1991) Recrystallization in Fibrous and Particulate MMCs, *Mat. Sci. & Eng.*, **A135**, pp. 281–5.
22. P. B. Prangnell and W. M. Stobbs (1990) The Effect of SiC Particulate Reinforcement on the Ageing Behaviour of Aluminium-Based Matrix Alloys, in *Proc. 7th Int. Conf. on Comp. Mats.* (*ICCM7*), Guangzou, China, W. Yunshu, G. Zhenlong and W. Renjie (eds.), Pergamon, pp. 573–8.
23. S. Abis and G. Donzelli (1988) Effects of Reinforcements on the Ageing Process of an Al–Cu/SiO_2, Al_2O_3, *J. Mat. Sci. Letts.*, **7**, pp. 51–2.
24. G. W. Lorimer and R. B. Nicolson (1966) Further Results on the Nucleation of Precipitates in the Al–Zn–Mg System, *Acta Metall.*, **14**, pp. 1009–13.
25. G. W. Lorimer (1967) *The Nucleation of Precipitates in Age Hardening Alloys*, PhD Thesis, Univ. of Cambridge.
26. J. Glazer, T. S. Edgecumbe and J. W. Morris (1986) Theoretical-analysis of Aging Response of Al-Li Alloys Strengthened by Al_3Li Precipitates, in *Proc. Al–Li III*, C. Baker, P. J. Gregson, S. J. Harris and C. J. Peel (eds.), Oxford, Inst. of Metals, pp. 369–75.
27. K. Mahalinagam, B. P. Gu, G. L. Liedel and T. H. Sanders (1987) Coarsening of $\delta'(Al_3Li)$ Precipitates in Binary Al–Li Alloys, *Acta Metall.*, **35**, pp. 483–98.
28. R. S. Haaland, G. M. Michal and G. S. Chottiner (1989) Mechanical Behaviour of 7091Al/SiC Interfaces, in *Fundamental Relationships between Microstructure and Mechanical Properties of MMCs*, M. N. Gungor and P. K. Liaw (eds.), Indianapolis, TMS, pp. 779–92.
29. C. Liu, S. Pape and J. J. Lewandowski (1988) Effects of Matrix Microstructure and Interfaces on Influencing Monotonic Crack Propagation in SiC/Al Alloy Composites, in *Interfaces in Polymer, Cer., & MMCs*, H. Ishida (eds.), Cleveland, USA, Elsevier, pp. 513–24.
30. G. J. Mahon, J. M. Howe and A. Vesudevan (1990) Microstructural Development and the Effect of Interfacial Precipitation on the Tensile Properties of an Al/SiC Composite, *Acta Metall.*, **38**, pp. 1503–12.
31. P. J. Withers, W. M. Stobbs and A. J. Bourdillon (1988) Various TEM Methods for the Study of MMCs, *J. Micros.*, **151**, pp. 159–69.
32. S. R. Nutt and R. W. Carpenter (1985) Non Equilibrium Phase Distribution in an Al–SiC Composite, *Mat. Sci. & Eng.*, **75**, pp. 169–77.
33. F. J. Humphreys, W. S. Miller and M. R. Djazeb (1990) Microstructural Development During Thermomechanical Processing of Particulate Metal Matrix Composites, *Mater. Sci. Tech.*, **6**, pp. 1157–66.
34. A. Bowen, M. Ardakani and F. J. Humphreys (1991) The Effect of Particle Size and Volume Fraction on Deformation and Recrystallisation Textures in Al–SiC Metal Matrix Composites, in *Metal Matrix Composites –*

Processing, Microstructure and Properties, 12th Risø Int. Symp., Roskilde, N. Hansen, D. J. Jensen, T. Leffers, H. Lilholt, T. Lorentzen, A. S. Pedersen, O. B. Pedersen and B. Ralph (eds.), Risø Nat. Lab., Denmark, pp. 241–6.
35. Y. L. Liu, N. Hansen and D. J. Jensen (1989) Recrystallization Microstructure in Cold-Rolled Aluminium Composites Reinforced by Silicon Carbide Whiskers, *Metall. Trans.*, **20A**, pp. 1743–53.
36. R. A. Shahani and T. W. Clyne (1991) The Effect of Reinforcement Shape on Recrystallisation in MMCs, in *Metal Matrix Composites – Processing, Microstructure and Properties, 12th Risø Int. Symp.*, Roskilde, N. Hansen, D. J. Jensen, T. Leffers, H. Lilholt, T. Lorentzen, A. S. Pedersen, O. B. Pedersen and B. Ralph (eds.), Risø Nat. Lab., Denmark, pp. 665–60.
37. F. J. Humphreys (1980) Nucleation of Recrystallization in Metals and Alloys with Large Particles, in *Recrystallization and Grain Growth in Multi-Phase and Particle Containing Materials*, N. Hansen, A. R. Jones and T. Leffers (eds.), Risø Nat. Lab., Roskilde, Denmark, pp. 35–44.
38. R. A. Shahani (1991) *Microstructural Development during Thermomechanical Processing of Aluminium-Based Composites*, PhD Thesis, Univ. of Cambridge.
39. P. F. Chapman and W. M. Stobbs (1969) The Measurement of Local Rotations in the Electron Microscope, *Phil. Mag.*, **19**, pp. 1015–30.
40. D. J. Jensen, Y. L. Liu and N. Hansen (1991) Hot Extrusion of Al–SiC. Texture and Microstructure, in *Metal Matrix Composites – Processing, Microstructure and Properties, 12th Risø Int. Symp.*, Roskilde, N. Hansen, D. J. Jensen, T. Leffers, H. Lilholt, T. Lorentzen, A. S. Pedersen, O. B. Pedersen and B. Ralph (eds.), Risø Nat. Lab., Denmark, pp. 417–22.
41. F. J. Humphreys (1979) Recrystallization Mechanisms in Two-Phase Alloys, *Metal Sci.*, **13**, pp. 136–45.
42. P. B. Prangnell (1991) *Development of Matrix and Interface Microstructures and Effect on Mechanical Behaviour of SiC_P/Al MMCs*, PhD Thesis, Univ. of Cambridge.
43. N. Hansen (1975) Recristallisation Accélérée et Retardée dans les Produits Renforcés par des Dispersions, *Mém. Sci. Rev. Metall.*, **72**, pp. 189–203.
44. E. Nes, N. Ryum and O. Hunderi (1985) On the Zener Drag, *Acta Metall.*, **33**, pp. 11–22.
45. W. S. Miller and F. J. Humphreys (1990) Strengthening Mechanisms in Metal Matrix Composites, in *Fundamental Relationships Between Microstructure and Mechanical Properties of Metal–Matrix Composites*, P. K. Liaw and M. N. Gungor (eds.), TMS, Warrendale, Pa, pp. 517–41.
46. R. A. Shahani and T. W. Clyne (1989) Factors Affecting Recrystallisation in Fibrous and Particulate MMCs, in *Metal Matrix Composites – Property Optimization and Applications*, London, Inst. of Metals, Paper 21.
47. Y. L. Liu, D. J. Jensen and N. Hansen (1992) Recovery and Recrystallisation in Cold Rolled Al-SiC_W Composites, *Metall. Trans.*, **23**, pp. 807–19.
48. F. J. Humphreys (1977) The Nucleation of Recrystallization at Second Phase Particles in Deformed Aluminium, *Acta Metall.*, **25**, pp. 1323–44.
49. D. J. Jensen, N. Hansen and Y. L. Liu (1989) Texture Development During Recrystallization of an Al–SiC Composite, in *Materials Architecture, 10th Risø Symposium*, J. B. Bilde-Sørensen (ed.), Roskilde, Denmark, Danish Atomic Energy, pp. 409–14.

11

Testing and characterisation techniques

Progress in our understanding of metal matrix composites is heavily reliant on the range of available experimental techniques and the correct selection by the researcher of the most appropriate method for a specific task. In this chapter an outline is given of some of the main testing and characterisation techniques which have been used for composite evaluation, looking first at the basic principles and capability of each technique, before going on to discuss application to MMCs. Each section also contains a source of references which provide the necessary technical details.

11.1 Measurement of Young's modulus

11.1.1 Basic principle and capability

The aim is to characterise the relationship between an applied load and a material's elastic (reversible) variation in strain. The available experimental approaches can be split into two broad classes; mechanical or static methods, and ultrasonic or dynamic methods. Their measurement capability in terms of accuracy is similar (± 0.5 GPa).

Of all the different means of characterising stiffness, the following measures are perhaps the most popular.

(1) the tangent to the initial stress–strain slope
(2) the tangent, subsequent to prestraining, of the initial slope on reloading
(3) the tangent, subsequent to prestraining, of the initial slope of the unloading curve
(4) the tangent to the reloading curve, subsequent to prestraining and low stress amplitude cycling about zero until stress/strain hysteresis is negligible

(5) the speed of ultrasonic waves through the medium
(6) the frequency of resonant standing vibrations, which are related to the dimensions of the specimen as well as to the stiffness of the material.

Dynamic methods of stiffness evaluation ((5) and (6)), which utilise the compressions and rarefactions of an ultrasonic wave, inevitably involve much smaller applied loads than conventional static measures.

11.1.2 Application to MMCs

The plethora of test procedures proposed for the measurement of Young's modulus in discontinuously reinforced MMC materials reflects the difficulty of precise measurement in these systems. Truly reversible stress–strain behaviour relies upon the absence of plastic deformation. For discontinuous systems, large point-to-point variations in stress, arising from mismatches in stiffness and thermal expansion coefficient between the phases, often gives rise to plastic deformation at very low applied loads. Nieh and Chellman[1] were the first to point out the impracticality of using the initial portion of the loading curve to assess the stiffness of discontinuously reinforced systems, and to propose an alternative method. In the following subsections the advantages and practical difficulties associated with each of the above methods are considered.

(1) *The tangent to the initial stress–strain slope (Fig. 11.1.1(a))*

Prior to prestraining, *continuous fibre* composites often exhibit a linear elastic response for loading along the alignment direction at low strains[2,3], from which the stiffness can be evaluated[†]. Careful measurements are required, because subsequent to matrix yielding the curve remains linear, though of reduced slope. For discontinuous reinforcement, however, the limited extent of the proportional regime, combined with alignment-related difficulties at very low strains, can give rise to large measurement errors ($\pm \sim 5$ GPa)[1].

(2) *The tangent, subsequent to prestraining, of the initial slope on reloading (Fig. 11.1.1(a))*

The advantage of this technique over (1) is that the proportional regime is extended by the prestraining. However, the unload/reload loop tends

[†] Refs. 2 and 3 report without explanation that the initial stiffness seems to be *higher* than rule of mixtures would suggest. This is contrary to all theoretical expectations.

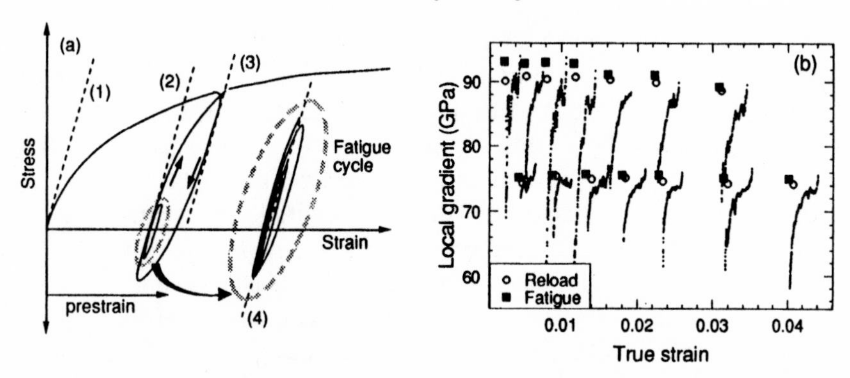

Fig. 11.1.1 (a) Schematic after Prangnell *et al.*,[5] depicting four mechanical means of characterising the Young's modulus. The diagram highlights the characteristically small initial proportional regime (1) and the difficulty in assigning an unloading modulus (3). In (b) the variation in unloading gradient (3) is plotted alongside values calculated by methods (2 – reload) and (4 – fatigue) with increasing tensile plastic strain for solution treated 2618 Al–12.5 vol% SiC_P composite[5]. The results suggest that methods (2) and (4) give reliable measures of stiffness, although their absolute values differ by about 1–3 GPa in the example case. The assignment of an unloading modulus becomes increasingly impractical at large strains. The decrease in composite stiffness with straining is probably related to the accumulation of damage.

to exhibit limited hysteresis (and thus microplasticity) and hence observed gradients may underestimate the stiffness.

(3) *The tangent, subsequent to prestraining, of the initial slope of the unloading curve*[4]

Here the practical difficulty is associated with choosing the point at which to 'draw' the tangent (see Fig. 11.1.1(a)). Two effects cause difficulties; machine backlash, and the occurrence of considerable reverse plastic flow upon unloading. In any event, Fig. 11.1.1(b) indicates that, at least at low plastic strains, a short region of constant slope (i.e. a 'modulus') is commonly observed.

(4) *The tangent to the reloading curve, subsequent to prestraining and low stress amplitude cycling about zero until stress/strain hysteresis is negligible*[5] (*Fig. 11.1.1(a)*)

The principle behind this technique is to reduce dislocation motion through arrangement stabilisation and local work-hardening under low stress fatigue cycling. Consequently, dislocation motion is negligible during the actual stiffness measurement stroke.

Prangnell *et al.*[5] show that considerable difficulty is associated with the application of methods (1) and (3) (error ~3 GPa – Fig. 11.1.1(b)) to discontinuous MMCs, but that reproducible measures of stiffness can be obtained by methods (2) and (4) (error ~0.5–1.5 GPa). Of these, the fatigue method (4) gives the slightly higher values and is probably to be favoured, because the linearity of the low amplitude stress fatigue curve suggests a more truly elastic response. One clear lesson to be learned from this study is the danger of using computer-controlled testing machine derived values of stiffness, without critical analysis of the method of derivation. This is especially important when using stiffness to monitor damage accumulation, where increasing reverse flow with increased straining could cause an overestimation of damage (Chapter 7).

(5) *The speed of ultrasonic waves through the medium*

The elastic constants can be evaluated directly from the velocity of sound (if the density (ρ) is known) via the relations

$$G = \rho u_s^2 \qquad \frac{E(1-\nu)}{(1+\nu)(1-2\nu)} = \rho u_L^2 \tag{11.1.1}$$

where u_L and u_s are the longitudinal and shear wave velocities in isotropic bulk material†. These relations must be modified for anisotropic materials such as directional composites. For example, for uniaxially aligned composites

$$u_L^2 = \frac{C_{C33}}{\bar{\rho}_C} \quad \left(\text{or, when } \lambda \sim D,\ u_L^2 = \frac{E_{C3}}{\bar{\rho}_C}\right) \tag{11.1.2}$$

for which one might predict the results

$$u_L^2 = \frac{(1-f)C_{M33} + fC_{I33}}{(1-f)\rho_M + f\rho_I} \quad \left(\text{or, when } \lambda \sim D,\ u_L^2 = \frac{(1-f)E_{M3} + fE_{I3}}{(1-f)\rho_M + f\rho_I}\right) \tag{11.1.3}$$

using the rule of mixtures for stiffness (C-tensor) and Young's modulus (E_3).

† Strictly the velocity is dependent on specimen shape: if the rod diameter, D, is of the order of, or larger than, the wavelength of the propagating sound ($\lambda \sim D$) then the longitudinal velocity is given by $\sqrt{(E/\rho)}$. (In this case the transverse stresses are zero, whereas in eqn (11.1.1) the transverse *strains* are taken to be zero (i.e. $\lambda \ll D$)).

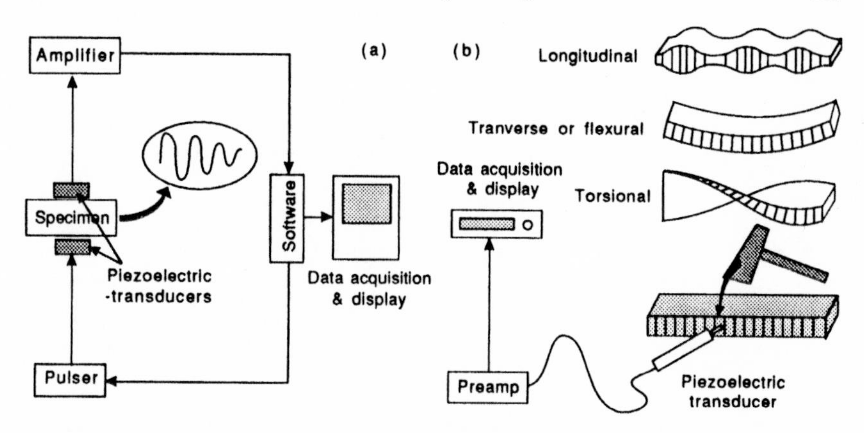

Fig. 11.1.2 Schematic diagram of (a) a typical system for the measurement of ultrasonic velocity (5) and (b) for the measurement of resonant frequency (6).

The primary advantages of ultrasonic stiffness measurement are:

- stresses and strains are of very low amplitude
- high sensitivity of measurement ($\sim \pm 0.7\%$ for Al–SiC[6])
- the ability to measure stiffness quickly on small specimens.

The wave velocity can be determined in a number of ways[7], the simplest being the time of flight of the first pulse, or the time delay between successive face-to-face reflections (see Fig. 11.1.2(a)). Provided the wavelength is long compared with the microstructural detail, good correspondence with static measures of stiffness can be achieved. The transition from composite sampling to individual phase sampling, which occurs as the ultrasonic wavelength approaches microstructural dimensions, is exemplified by work on fine (125 μm diameter B rods in Al) and coarse (2000 μm diameter Fe rods in Lucite) continuously reinforced systems, using 5 MHz ($\lambda \sim 1000$ μm) ultrasonic waves[8]. Displacement measurements at the surface of the fine scale system have shown that, whilst the displacement amplitudes are different for the B fibres and the Al matrix, the time of arrival of the pulse is the same. In contrast, measurements at the surface of the Lucite–Fe system indicate that the signal arrives later in the Lucite than in the stiffer steel rods.

Using a suitable theoretical model for the composite stiffness variation with reinforcement content and aspect ratio (e.g. the Eshelby approach of Chapter 3), velocity measurements can be used to evaluate the reinforcement volume fraction. If the density of the composite is known

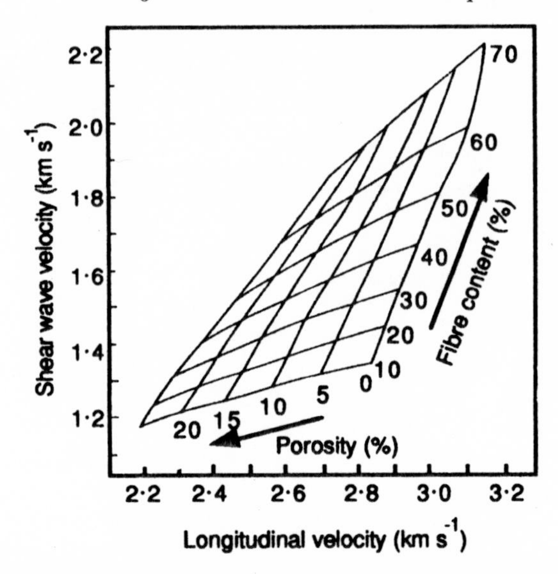

Fig. 11.1.3 Evaluation of composite constitution in terms of the compressive and shear wave velocities propagating normal to the fibre direction in continuously reinforced GRP composites[11]. This can be achieved because, whereas composite velocities parallel to the fibre direction are largely insensitive to the presence of spherical pores, a longitudinal wave passing normal to the fibres is much more sensitive. Shear waves on the other hand, polarised at right angles to the fibres and propagating normal to the fibres, are more sensitive to reinforcement content than porosity. Combining the two measurements thus facilitates simultaneous evaluation of porosity and fibre content.

and a suitable stiffness model is able to account for residual porosity, velocity measurements (i.e. stiffness/density) can be used to estimate porosity content and reinforcement content at the same time[9]. In addition, because the wave velocities of different waves (e.g. longitudinal, shear) are, in certain cases, affected differently by the presence of pores, measurement of two wave-type velocities can also be combined to give simultaneous reinforcement content and porosity information[10] (Fig. 11.1.3).

(6) *The use of resonant vibrations*

Here the principle is to deduce composite stiffness by measuring the fundamental resonance frequencies (v_n) of vibration excited by striking the specimen. The stiffness–frequency relationship is dependent on specimen shape; for cylinders under longitudinal, transverse and torsional resonance, theoretical equations have been derived analytically[11] (Table 11.1.1).

Table 11.1.1 *Theoretically predicted moduli for isotropic bars*[11] *with appropriate correction factors*†

Shape	Longitudinal (E)	Flexural (transverse) (E)	Torsional (G)
Cylinder (L, d)	$(\rho/K_n)(2Lv_n/n)^2$	$1.261\,886\rho L^4 v_n^2 T/d^2$	$\rho(2Lv_n/n)^2$
Prism (L, h, b)	approx. as above with $d^2 = (h^2 + b^2)$	$0.946\,42\rho L^4 v_n^2 T/h^2$	$\rho(2Lv_n/n)^2 R_n$

† The longitudinal correction factor (K_n) for the nth mode is given by

$$K_n = 1 - \pi^2 n^2 \nu^2 d^2/8L^2 \tag{11.1.4}$$

in which ν is Poisson's ratio and $d/\lambda \ll 1$, where the nth mode has a wavelength $\lambda_n = 2L/n$.

The flexural correction factor T is dependent on d/L for cylindrical specimens and h/L for prisms. Ref. 13 lists tables of T, within which T tends to unity as d/L or h/L tends to zero; for prisms of long aspect ratio, T is given approximately by[13]

$$T = 1 + 6.585(1 + 0.0752\nu + 0.8109\nu^2)(h/L)^2 - 0.868(h/L)^4 - \frac{8.340(1 + 0.2023\nu + 2.173\nu^2)(h/L)^4}{1 + 6.338(1 + 0.14081\nu + 1.536\nu^2)(h/L)^2} \tag{11.1.5}$$

The torsional correction factor (R_n) is given by

$$R_n = R_0(1 + n^2(h/L)^3(0.01746 + 0.00148n + 0.00009n^2)) \tag{11.1.6}$$

with

$R_0 = 1.18559$ for a square-ended prism

$R_0 = (1 + b/t)^2/(4 - 2.521(b/t))$ for a flat rod ($b \gg t$)

Relations for bars of rectangular section are approximate (better than 1% error).

Measurement of resonant frequencies can be made using the set-up shown in Fig. 11.1.2(b). Usually, the most easily excited mode is the flexural mode, for which there are nodes at about $L/4$ from each end. Free resonance is thus aided by supports situated at these positions, although excitation with the specimen supported upon a foam rubber mat is also suitable. After excitation, complex vibrations soon die away to leave the fundamental resonant frequency, which can then be measured with a contact transducer or microphone. Like the ultrasonic measurement technique, this test gives a fast, low strain amplitude measure of the modulus. Specimen shape is important, since it determines the ease with which different modes can be excited, as well as the relationship between frequency and modulus. Provided this relationship has been calculated for the relevant specimen geometry and anisotropy (and specimen dimensions can be measured to high precision), then accurate measurements of modulus can be obtained.

In a round-robin series of parallel tests of dynamic methods of Young's modulus evaluation of Inconel®, Wolfenden *et al.*[12] found stiffness evaluation variations of ~1.5% between the various methods. Whilst this is to some extent reassuring, caution should be exercised. Metal matrix composites often have local density and porosity variations as well as significant anisotropy and this, combined with fabrication-related geometry constraints, can lead to serious systematic errors in the wave analysis if these factors are not taken into account.

Examples:

(1) Al–B, Cu–W[2]; Al–B[3]
(2) Al–SiC_P[5]
(3) Al–SiC_P[4,5], Al–$SiC_{W,P}$, Al–B_4C_P[1]
(4) Al–SiC_P[5]
(5) Al–B, Lucite–Fe[8]; Al–SiC[6]; Cu–W[13]; measurements at elevated temperature[14]; pore and reinforcement variation maps[9]
(6) Dental composites[15]; Al–SiC_W[16].

References

1. T. G. Nieh and D. J. Chellman (1984) Modulus Measurements in Discontinuous Reinforced Al Composites, *Scripta Met.*, **18**, pp. 925–8.
2. M. R. Pinnel, D. R. Hay and A. Lawley (1968) Microstrain Behaviour of MMCs, in *Metal Matrix Composites*, ASTM STP 438, pp. 95–107.
3. W. F. Stuhrke (1968) The Mechanical Behaviour of Al–B Composite Material, in *Metal Matrix Composites*, ASTM STP 438, pp. 108–33.
4. D. J. Lloyd (1991) Aspects of Fracture in Particulate Reinforced MMCs, *Acta Met. et Mat.*, **39**, pp. 59–71.
5. P. B. Prangnell, T. J. Warner and W. M. Stobbs (1992) Private Communication.
6. A. Wolfenden, M. R. Harmouche and S. V. Haues (1988) Anelastic and Elastic Measurement in Al-MMCs, in *Testing Tech. of MMCs*, Philadelphia, ASTM STP 964, P. R. DiGiovanni and N. R. Adsit (eds.), pp. 207–15.
7. E. P. Papadakis (1976) Ultrasonic Velocity and Attenuation: Measurement Methods with Scientific and Industrial Applications, in *Physical Acoustics Principles and Methods*, W. P. Mason and R. N. Thurston (eds.), Vol. **XII**, Academic Press, New York, pp. 1285–92.
8. S. Huber, W. R. Scott and R. Sands (1988) Detection of Ultrasonic Waves Propagating in B/Al and Steel/Lucite Composite Materials, in *Review of*

Prog. in Quantitative NDE, D. O. Thompson and D. E. Chimenti (eds.), Vol. **7**, Plenum Press, New York, pp. 1065–73.
9. K. Telschow, J. Walter and D. Kunerth (1988) Ultrasonic Characterisation of Nonuniform Porosity Distributions in SiC Ceramic, in *Review of Prog. in Quantitative NDE*, D. O. Thompson and D. E. Chimenti (eds.), Vol. **7**, Plenum Press, New York, pp. 1285–92.
10. W. N. Reynolds and S. J. Wilkinson (1978) The Analysis of Fibre-Reinforced Porous Composite Materials by the Measurement of Ultrasonic Wave Velocities, *Ultrasonics*, **16**, pp. 159–63.
11. S. Spinner and W. E. Teft (1961) A Method for Determining Mechanical Resonance Frequencies and for Calculating Elastic Moduli from these Frequencies, in *Proc. ASTM*, **61**, pp. 1221–38.
12. A. Wolfenden, M. R. Harmouche, G. V. Blessing, Y. T. Chen, P. Terranova, V. Dayal, V. K. Kinra, J. W. Lemmens, R. R. Phillips, J. S. Smith, P. Mahmoodi and R. J. Wann (1981) Dynamic Young's Modulus Measurements in Metallic Materials: Results of an Interlaboratory Testing Program, *J. Test. and Evaluation*, **17**, pp. 2–13.
13. D. L. McDanels, R. W. Jech and J. W. Weeton (1965) Analysis of Stress–Strain Behaviour of W Fibre Reinforced Cu Composites, *Trans. TMS-AIME*, **23**, pp. 636–42.
14. K. Heritage, C. Frisby and A. Wolfenden (1988) Impulse Excitation Technique for Dynamic Flexural Measurements at Moderate Temperature, *Rev. Sci. Instrum.*, **56**, pp. 973–4.
15. M. Braem, P. Lambrechts, V. Van Doren and G. Vanherle (1988) The Impact of Composite Structure on its Elastic Response, *J. Dental Res.*, **65**, pp. 648–53.
16. R. L. Mehan (1968) Fabrication and Evaluation of Sapphire Whisker Reinforced Al-Composites, in *Metal Matrix Composites*, ASTM STP 438, pp. 29–58.

11.2 Characterisation of plastic strain history

11.2.1 Basic principle and capability

The aim is to characterise the relationship between an applied load and a material's plastic stress–strain response. With careful execution (see the following section), the response can be characterised to within $\pm 0.1\%$ strain, ± 2 MPa stress. There are essentially four different loading configurations in common usage

(1) *Tension*
(2) *Compression*
(3) *Load reversal*
(4) *Shear*

Each configuration involves its own set of experimental difficulties and problems when applied to long and short fibre composites and these are discussed in turn in the next section.

11.2.2 Application to MMCs

(1) *Tension*

Few difficulties have been reported for simple tensile testing and many specimen geometries are acceptable. Guidelines for tensile testing have been compiled by Roebuck *et al.*[1] The primary concern is often that of the gripping arrangement. For example, diffusion bonded Ti–SiC monofilament lay-ups often require the welding or electroplating of tabs to the specimen to facilitate specimen gripping. When testing unidirectional and cross plied continuously reinforced systems care must be taken to ensure that specimen edges are accurately parallel to the 0° oriented fibres[2], while mixed mode loading is difficult to avoid for off-axis loading (item (4)). Further, as with polymeric composites, the specimen must be wide enough to ensure that the failure mode is not influenced by edge effects.

(2) *Compression*

Compression testing can be particularly useful for examining plastic flow and creep properties, in view of the low tensile ductility of many composites. The primary difficulty is the prevention of buckling, which is usually avoided by the use[3,4] of small gauge length/gauge diameter ratios (<2–3). Though this effectively inhibits bending, the gauge length may be very short (particularly for MMCs available only with small cross-section), making strain measurement difficult. Great care must be taken with external means of strain measurement (e.g. Linear Variable Displacement Transducers (LVDTs), laser interferometers or scanners (§11.10) because of difficulties associated with defining a gauge length[3], whilst strain gauges have limitations of maximum surface curvature, maximum measurable strain and inherent compressive–tensile strain asymmetries (item (3)). Clip-gauges are preferred, although attachment to the gauge section can present difficulties. Round[3] and flat[2] coupons have been tested successfully using split collet grips (as illustrated in Fig. 11.2.3(b) in connection with Bauschinger testing). The use of alignment pins between the upper and lower grip sections is sometimes necessary for the compressive loading of stiff specimens, although load transfer to the pins limits the reliability of specimen loading. The problem of specimen misalignment is evident in much of the literature published to date. Fig. 11.2.1 illustrates the kinds of features which can be observed when the alignment is not absolutely correct.

Simple cylinders of aspect ratio 2 have also been tested in compression[5],

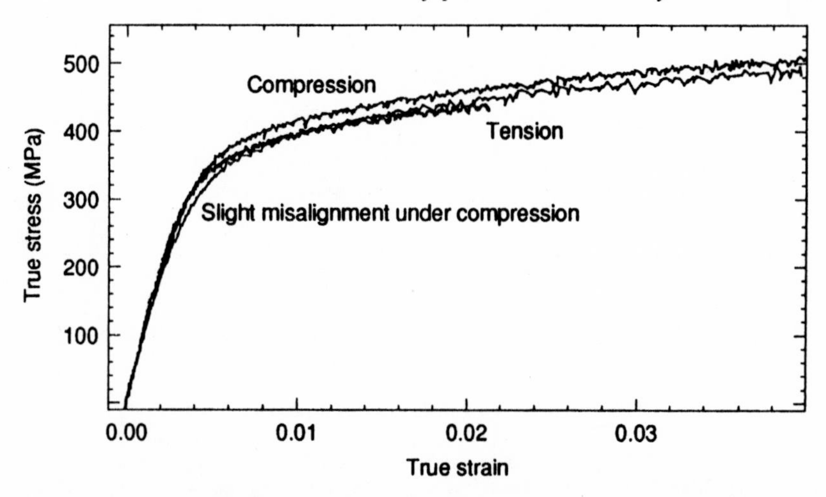

Fig. 11.2.1 A high degree of alignment in compression is crucial for the undertaking of compression tests. The yield point is particularly sensitive to misalignment and this is exemplified by the data shown above for solution treated 8090 Al/17% SiC_P. The initial drop in the compressive curve was observed for specimens which exhibited only a very slight degree of skewness subsequent to testing. Clearly, it is absolutely crucial to minimise this error when interpreting microplasticity variations between tensile and compressive loading in terms of damage accumulation. (Plot courtesy of P. B. Prangnell and T. J. Downes.)

using stabilising collars to maintain their lateral position. Lateral constraint at the specimen end is always a problem for short gauge lengths[5], whether it arises from the tabs, loading shoulders of coupons, or from the stabilising collars in the case of cylindrical specimens. This constraint can give rise to considerable barelling of the specimen and is an especially serious problem for squat specimens (Fig. 11.2.2(a)). Barrelling results in stress–strain non-linearity in the elastic region[4] and has even more serious effects in the plastic regime[6] (Fig. 11.2.2(b)). While a comparison of the tensile and compressive stress–strain behaviour can give useful information about the extent of damage accumulation under tensile straining[7], it is clearly essential that the influence of end constraint is fully accounted[3] for before drawing quantitative conclusions.

Laminates are prone to non-uniform strain through the thickness, especially for angle ply composites and at short gauge lengths[2] (due to a large lateral deformation of the inner plies). A number of long gauge length methods have been proposed for the testing of MMC laminates and have been designed so as to circumvent the problems of delamination and lateral movement, whilst allowing legitimate local failure by ply buckling.

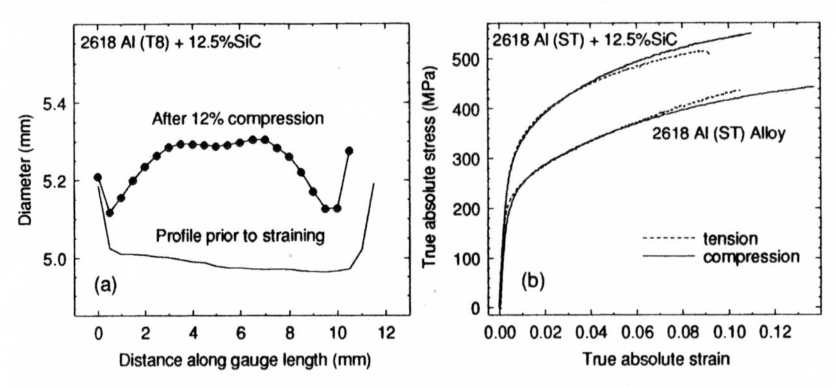

Fig. 11.2.2 (a) A comparison of the variation in gauge length profile under tensile and compressive straining of 2618 Al–12.5 vol% SiC_P[3]. Non-uniform deformation is caused by constraint at the ends of the gauge length as the cross-section increases there. For 2618 Al the extent of barrelling is found to be a function of age hardening treatment, being more severe for the T8 heat treatment than in the solution treated state[3]. (b) As illustrated by the dip in the compressive curve relative to the tensile at strains >6%, even in the solution treated state (aged at R.T. for one week), gauge length constraint can seriously affect the stress–strain response[3]; this is a serious problem for compressive and push–pull testing but is less so for tension testing because of the large gauge length to diameter ratios normally used. Note how damage accumulation with tensile straining means that the tensile response of the particulate composite falls below that of the compressive.

In essence, these methods prevent buckling by sideways support of long flat coupons[2]. One approach achieves this by testing the specimen between two steel plates; whilst this prevents buckling, load transfer to the plates via frictional forces can give rise to an overestimation of the yield stress[2]. To remedy this problem the specimen can be laterally restrained by a honeycomb structure. The honeycomb behaves rather like a bellows parallel to the loading direction (stiffness[4] typically ~1 MPa), but it resists crumpling in the lateral direction and hence helps to delay buckling of the specimen.

(3) *Load reversal* (*Bauschinger testing*)

In 1886 Johann Bauschinger reported that for the steels he examined[8], ‘by loading in tension or compression above the elastic limit, the elastic limit for compression or tension respectively is lowered significantly, the more so, the greater is the initial loading above the limit’. Despite extensive research during the intervening years, this effect, which is

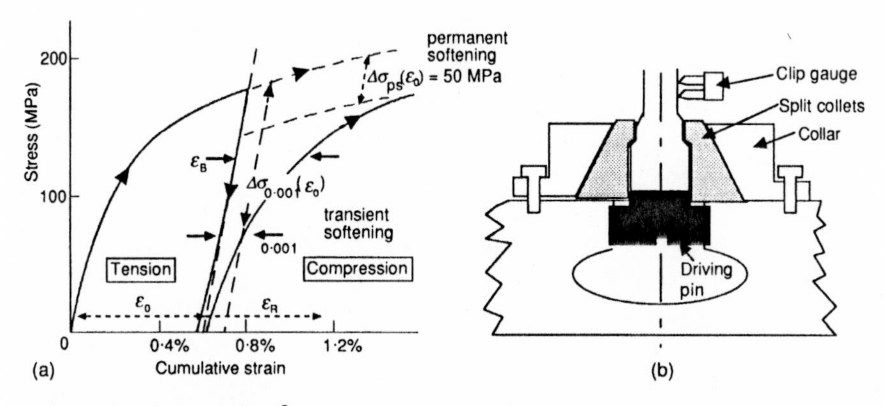

Fig. 11.2.3 (a) A typical[9] plot for a short fibre composite showing the transient softening regime[10], the permanent softening ($\Delta\sigma_{ps}$), the Bauschinger strain (ε_B)[11] and the 0.1% softening ($\Delta\sigma_{0.1\%}$), all of which have been used to characterise the Bauschinger effect. These parameters are all a function of the prestrain ε_0. (b) A typical gripping arrangement for Bauschinger effect testing[3]. The driving pin holds the specimen end in compression throughout the cycle.

common to all inhomogeneous materials (Fig. 11.2.3(a)), is still not well understood. Its importance for MMCs derives from the sensitivity of the stress reversal response to the internal stresses.

The basic explanation proposed by Orowan is now generally accepted, although the details are still debated. He suggested that under (e.g. tensile) load, dislocations sweep through the matrix until they are held up by the dispersed phase, producing Orowan loops and hence internal stresses (§4.3). These plastically generated mean stresses act so as to reduce the average stress in the matrix, i.e. they oppose further forward deformation and hence increase the macroscopic flow stress of the composite. At any time during forward straining, many dislocations are piled up in the vicinity of the particles. This distribution of dislocations has two implications for reverse flow. Firstly, the piled-up dislocations encounter much less resistance to motion when loading is reversed, because they initially sample a random array of obstacles[9], and so travel quite large distances before being blocked by other stiff obstacles. This is evidenced by the large degree of '***transient softening***' (Fig. 11.2.3(a)). The second effect is that on reverse loading the applied (e.g. compressive) stress is of the same sense as the mean matrix stress caused by the Orowan loops from the initial plastic deformation, so that reverse deformation is *easier* than it would otherwise be. This is referred to as '***permanent softening***' ($\Delta\sigma_{ps}$) (Fig. 11.2.3(a)).

Fig. 11.2.3(b) illustrates a suitable experimental arrangement for the elimination of specimen movement during stress reversal. The primary difficulty arises because of the contradictory requirements for the minimisation of the effects of gauge length end constraint and for the avoidance of buckling in compression. Experience[3] indicates that gauge length constraint limits the available plastic strain range to less than ~6%. Once measured, various characteristics of the stress–strain curve can be used to quantify the Bauschinger effect (BE) in MMCs (Fig. 11.2.3(a)). There is still some uncertainty surrounding their exact relationships with the internal stress distribution. This is largely because on plastic straining the associated local stresses soon rise to levels which cannot be maintained, so that relaxation occurs (either by cross-slip, diffusion or debonding, etc.), and the work-hardening rate decreases (§4.4). Relaxation lowers the internal mean stresses by reducing the inclusion–hole misfit, complicating the relationship between internal stress and the Bauschinger effect parameters listed below.

- The most common approach[13] for traditional materials has been to measure the permanent softening $\Delta\sigma_{ps}$ (Fig. 11.2.3(a)) and to relate this to the mean matrix stress introduced during the selected prestrain ε_0. Upon further straining (ε_F) in the direction of the prestrain (e.g. tension) the composite yield stress is given† by (§4.3)

$$\sigma^{A}_{ten}(\varepsilon_F) = \sigma_{YM}(|\varepsilon_0| + |\varepsilon_F|) + [\langle\sigma_M(\varepsilon_0 + \varepsilon_F)\rangle^{P} - \langle\sigma_M\rangle^{\Delta T}] \quad (11.2.1a)$$

Assuming microstructural strengthening (§4.3.2) to be independent of the direction of straining, the yield stress in the reverse direction (e.g. compressive) is lowered by the mean matrix stress generated during the prestrain $[\langle\sigma_M(\varepsilon_0)\rangle^{P}]$ so that

$$\sigma^{A}_{com}(\varepsilon_R) = \sigma_{YM}(|\varepsilon_0| + |\varepsilon_R|) + [\langle\sigma_M(\varepsilon_R)\rangle^{P} - \langle\sigma_M(\varepsilon_0)\rangle^{P} + \langle\sigma_M\rangle^{\Delta T}] \quad (11.2.1b)$$

Ideally, in order to evaluate the mean stress generated by a given prestrain $\langle\sigma_M(\varepsilon_0)\rangle^{P}$, one would like to substract $\sigma^{A}_{ten}(\varepsilon_F = 0)$ and $\sigma^{A}_{com}(\varepsilon_R = 0)$; however, in this region the curve is dominated by transient softening of the intrinsic matrix yield stress σ_{YM}. This difficulty can be circumvented

† The terms within the brackets are actually the difference between the mean stresses parallel and perpendicular to the loading axis, the thermal residual stress term ($\langle\sigma_M\rangle^{\Delta T}$) being added for compressive prestrain and subtracted for a tensile prestrain (§4.1). Note that the thermal stress can be evaluated by simply comparing the initial flow stresses in tension and compression (§4.1).

by extrapolating the forward and reverse curves at larger strains (once they have become parallel) back to $\varepsilon_F = \varepsilon_R = 0$ to give[9]

$$\Delta\sigma_{ps}(\varepsilon_0) = 2[\langle\sigma_M(\varepsilon_0)\rangle^P \pm \langle\sigma_M\rangle^{\Delta T}] \tag{11.2.2}$$

This approach has been successfully applied to metal W fibre reinforced composites[14,15]. However, for discontinuous Al-based composites this method is often unworkable, partly because the forward and reverse curves never become truly parallel†. This is because these MMCs are characterised by extensive transient softening as well as limited tensile ductility[3]. A practical solution has been adopted by Taya *et al.*[16] who have measured the difference in forward and reverse yield stress measured arbitrarily at 0.1% strain after the prestrain (Fig. 11.2.3(a)). While this has the advantage that it does not require ultimate parallelity and limits the effect of differences in mean stress and microstructural strengthening development upon further forward and reverse straining, caution must be exercised because it can be affected by the transient softening which occurs at low strains.

- The ***Bauschinger strain*** ε_B[18] is commonly used to describe the BE when the degree of permanent softening is small. Because the Bauschinger strain depends on both the permanent softening and the work-hardening rate, which has a rather complex nature, it can prove difficult to interpret. In certain cases, the Bauschinger strain can become infinite; this is especially common for systems affected by thermal stresses when the prestrain is of the same sense as the mean thermal matrix stress.

- Atkinson, Brown and Stobbs[11] related the roundedness of the reverse curve to the internal mean stress ($\langle\sigma_M(\varepsilon_0)\rangle^P$). This method has the advantage that only very small reverse strains are required, but it does depend on the existence of a definite relation between the local stresses which determine the transient softening and the long range mean stresses.

- Another parameter of interest is the reverse plastic straining upon unloading. Reverse flow is associated with the inhomogeneity of stress and thus could be used as a measure of the internal stress. However, no formal relationship has yet been proposed.

In summary, whilst the Bauschinger effect has great potential for the identification of internal stress development in MMCs, and has been used

† There is also concern about differences in the rate at which the internal stresses reverse[3,17] and about the rate of increase of matrix work-hardening under reverse straining.

Table 11.2.1 *Evaluation*[20] *of the* (i) *torsion of a hoop-wound tube,* (ii) *biaxial stressing of a tube,* (iii) *off-axis tension,* (iv) *Iosipescu test,* (v) *picture frame,* (vi) *rail shear and* (vii) 45° *tests*

	Test						
Criterion	*i*	*ii*	*iii*	*iv*	*v*	*vi*	*vii*
Can test general laminates	×	×	×	√	√	√	×
Negligible stress concentrations	?	?	√	×	×	×	√
Large uniform shear region	√	√	√	×	×	√	√
Low cost	×	×	√	?	?	?	√

successfully in continuously reinforced metal fibre composites and dispersion strengthened materials, uncertainty of interpretation hinders its widespread adoption for application to discontinuous MMCs. Calibration of the Bauschinger effect as applied to MMCs is required, perhaps by *in situ* identification of the concomitant internal stress development in a similar manner to the X-ray work undertaken by Wilson for dispersion-strengthened materials[17].

(4) *Shear*

The anisotropic nature of composites presents special problems for the design of effective shear test equipment. The main difficulty is the application of a pure shear, and is primarily a result of the strong interaction between normal and shear modes when composites are loaded away from the alignment axis. The most reliable methods of measurements centre on the testing of tubes, either by direct tube torsion (shear stress = applied torque × radius of the tube/moment of inertia), or by a combination of pressure and axial loading (see Table 11.2.1), which allows pure shear in addition to a range of other loading modes. These methods suffer from the drawback that, whilst the stress state is easily analysed, special construction of tubular specimens is required which is often expensive or troublesome for MMCs. A variety of other tests have been proposed (see below). While the off-axis tension test, the ±45° test, and the Iosipescu test have shown up well in comparative studies on unidirectional composites[19], the results are very sensitive to the exact stress state in the test volume.

Shear tests, such as the off-axis tension test (*iii*), can be used to define the shear moduli by a separation of the shear and normal components.

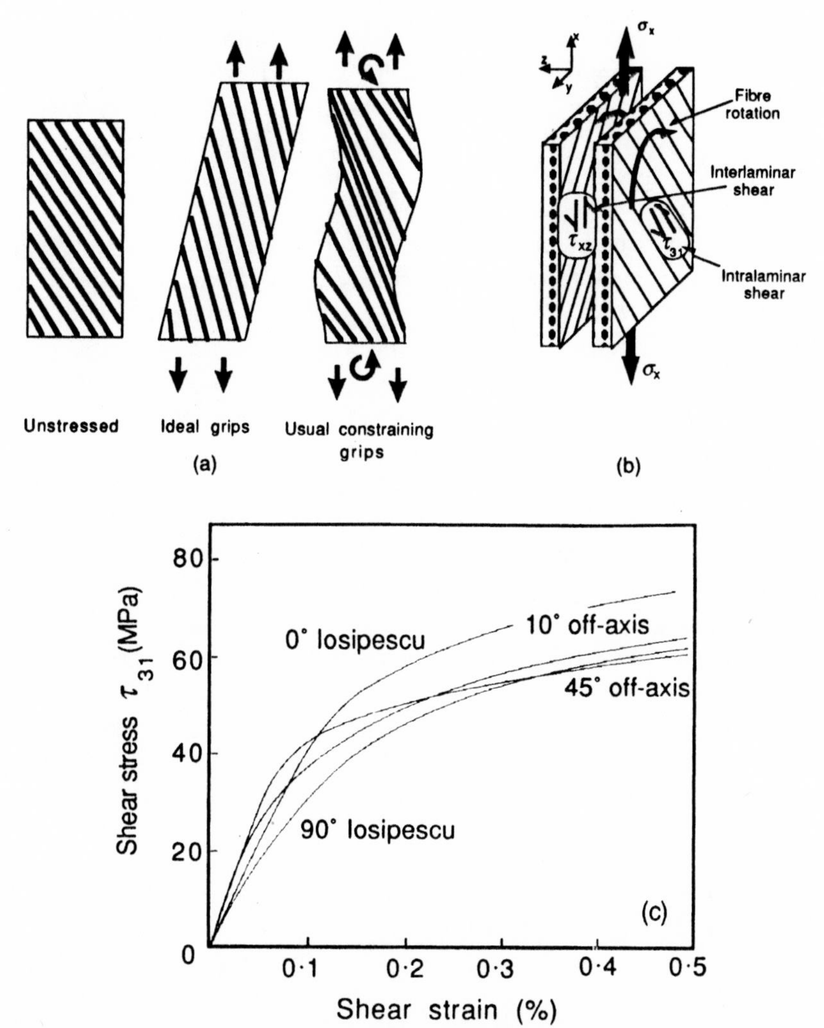

Fig. 11.2.4 (a) End clamping of off-axis specimens can given rise to significant errors in the determination of shear properties[22]. (b) The generation of inter-laminar shear stresses. (c) The shear response of Al–B composites with respect to the principle coordinate system under off-axis and Iosipescu loading[20].

However, upon plastic deformation shear deformation is likely to be dependent on both shear and normal stresses[21] since relevant yield criteria involve more than one component of stress, complicating post-yielding analysis. A further complication is the gripping of the specimen ends; as illustrated in Fig. 11.2.4, simple clamping exerts a considerable constraint which gives rise to a very non-uniform stress field. Interlaminar stresses can also arise in laminates (Fig. 11.2.4(b)). The Iosipescu test (*iv*) was designed as a result of finite element calculations aimed at producing an almost pure shear state in the test section. However, 0° and 90° laminates give rise to different shear stresses in contradiction to theory (Fig. 11.2.4(c)).

Shear test measurements made by the off-axis tension and Iosipescu tests have been compared by Pindera[21]. Aspect ratios of 12–16 were chosen (to limit constraint) for the off-axis tests using 10° and 45° loading configurations. Good agreement between the test measurements of modulus was observed only after correction for constraint in the off-axis test (this is minimal for the 45° orientation) and for non-uniform deformation fields (using finite element analysis) for the Iosipescu test. With regards to shear strength, the 90° Iosipescu test is unreliable due to the early initiation of local failure at the notches which can propagate right across the specimen. The reliability of 0° Iosipescu tests can be increased by optimisation of the notch angle[23] using modulus parameters obtained through preliminary 45° off-axis tests. This is achieved through the attainment of a nearly uniform shear stress across the test section. The off-axis test can only provide a measure of shear yielding in the presence of other loading components. Finally the work by Pindera[21] has highlighted the sensitivity of shear loading experiments to experimental factors such as specimen, grip, and strain gauge misalignment.

Examples:

(2) compression – tests using Celanese rig (Gr–Al, Gr–Mg)[24], tests using cylindrical specimens (Al–FP Al_2O_3)[6]
(3) Bauschinger – continuous fibre (Cu–W[14,15]), whisker (SiC_W[10]), particulate (SiC_P[12,16])
(4) shear – tube torsion test (Al–SiC_W[25], Al–Gr[26]) Iosipescu test (Al–Gr, Al–B[21]), off-axis tension test (Al–Gr, Al–B[21]).

References

1. B. Roebuck, T. A. E. Gorley and L. N. McCarteney (1989) Mechanical Property Test Procedures for MMCs, *Mat. Sci. & Eng.*, **5**, pp. 105–17.
2. J. M. Kennedy (1989) Tension and Compression Testing of MMC Materials, in *Metal Matrix Composites: Testing, Analysis and Failure Modes*, Philadelphia, ASTM STP 1032, W. S. Johnson (ed.), pp. 7–18.
3. P. B. Prangnell, W. M. Stobbs and P. J. Withers (1992) Considerations in the Use of Yield Asymmetries for the Analysis of Internal Stresses in MMCs, *Mat. Sci. & Eng.*, In Proof.

4. P. A. Lagace and A. J. Vizzini (1988) The Sandwich Column as a Compressive Characterisation Specimen for Thin Laminates, in *Composite Materials: Testing and Design*, Philadelphia, ASTM STP 972, J. D. Whitcomb (ed.), pp. 148–60.
5. W. M. Bethoney, W. M. Nunes and J. A. Kidd (1988) Compressive Testing of MMCs, in *Testing Tech. of MMCs*, Philadelphia, ASTM STP 664, P. R. DiGiovanni and N. R. Adsit (eds.), pp. 319–28.
6. N. H. Polakowski and E. J. Ripling (1966) *Strength and Structure of Engineering Materials*, Prentice Hall, New Jersey, p. 314.
7. A. K. Vasudevan, O. Richmond, F. Zok and J. D. Embury (1988) The Influence of Hydrostatic Pressure on the Ductility of Al–SiC Composites, *Mat. Sci. & Eng.*, **A107**, pp. 63–9.
8. J. Bauschinger (1886) Mittheilungen XV: Über die Veränderung der Elasticitätsgrenze und der Festigkeit des Eisens und Stahls durch Strecken und Quetschen, durch Erwärmen und Abkuhlen und durch wiederholte Beanspruchung, *Mitthelungen aus dem Mechanisch-technischen Laboratorium der königlichen Hochschule in München*, **13**, p. 1.
9. L. M. Brown (1979) Precipitation and Dispersion Hardening, in *Strength of Metals and Alloys, Proc. ICSMA 5*, Aachen, P. Haasen, V. Gerold and G. Kostorz (eds.), Pergamon Press, Oxford, pp. 1551–71.
10. P. J. Withers, W. M. Stobbs and O. B. Pedersen (1989) The Application of the Eshelby Method of Internal Stress Determination to Short Fibre MMCs, *Acta Metall.*, **37**, pp. 3061–84.
11. J. D. Atkinson, L. M. Brown and W. M. Stobbs (1974) The Work-Hardening of Cu–SiO_2 – IV. The Bauschinger Effect and Plastic Relaxation, *Phil. Mag.*, **30**, pp. 1247–80.
12. R. J. Arsenault (1987) Unusual Aspects of the Bauschinger Effect in Metals and MMCs, *Mats. Forum*, **10**, pp. 43–53.
13. L. M. Brown and W. M. Stobbs (1976) The Work-Hardening of Cu–SiO_2 – V. Equilibrium Plastic Relaxation by Secondary Dislocations, *Phil. Mag.*, **34**, pp. 351–72.
14. H. Lilholt (1977) Hardening in 2-Phase Materials – I. Strength Contributions in Fibre-Reinforced Cu–W, *Acta Metall.*, **25**, pp. 571–85.
15. O. B. Pedersen (1985) Mean Field Theory and the Bauschinger Effect in Composites, in *Fundamentals of Deformation and Fracture, Proc. IUTAM, Eshelby Mem. Symp., April 1984*, K. J. Miller and B. Bilby (eds.), Cambridge Univ. Press, pp. 129–44.
16. M. Taya, K. E. Lulay, K. Wakashima and D. J. Lloyd (1990) Bauschinger Effect in SiC_P–6061 Al Composite, *Mat. Sci. & Eng.*, **A124**, pp. 103–11.
17. D. V. Wilson (1965) Reversible Work Hardening in Alloys of Cubic Metals, *Acta Metall.*, **13**, pp. 807–14.
18. R. L. Wooley (1953) The Bauschinger Effect in some fcc and bcc Metals, *Phil. Mag.*, **44**, pp. 597–618.
19. S. Lee and M. Munro (1978) Evaluation of In-Plane Shear Test Methods for Composite Materials by the Decision Analysis Technique, *Composites*, **9**, pp. 49–55.
20. C. M. Browne (1988) Alternative Methods for the Determination of Shear Modulus in a Composite Material, in *Testing Tech. of MMCs*, Philadelphia, ASTM STP 964, P. R. DiGiovanni and N. R. Adsit (eds.), pp. 259–74.
21. M.-J. Pindera (1989) Shear Testing of Fiber Reinforced MMCs, in *Metal Matrix Composites: Testing, Analysis and Failure Modes*, Philadelphia, ASTM STP 1032, W. S. Johnson (ed.), pp. 19–42.

22. N. J. Pagano and J. C. Halpin (1968) Influence of End Constraints in the Testing of Anisotropic Bodies, *J. Comp. Mat.*, **2**, pp. 18–31.
23. A. Wang and D. Dasgupta (1986) *Development of Iosipescu-Type Test for Determining In-Plane Shear Properties of Fiber Composite Materials: Critical Analysis and Experiment*, VILU-ENG-86-5021 Report, Univ. of Illinois, Urbana.
24. D. J. Chang, G. L. Steckel, W. D. Hanna and F. Izaguirre (1988) Compressive Properties and Laser Absorptivity of Unidirectional MMCs, in *Testing Tech. of MMCs*, Philadelphia, ASTM STP 964, P. R. DiGiovanni and N. R. Adsit (eds.), pp. 18–30.
25. S.-C. Chou, J. L. Greem and R. A. Swanson (1988) Mechanical Behaviour of SiC/2014 Al Composite, in *Testing Tech. of MMCs*, Philadelphia, ASTM STP 964, P. R. DiGiovanni and N. R. Adsit (eds.), pp. 305–16.
26. R. B. Francini (1988) Characterisation of Thin Wall Gr/Metal Pultruded Tubing, in *Testing Tech. of MMCs*, Philadelphia, ASTM STP 964, P. R. DiGiovanni and N. R. Adsit (eds.), pp. 396–408.

11.3 Internal stress measurement by diffraction

11.3.1 Basic principle and capability

Diffraction exploits the crystal structure of the stressed material, treating each crystallite as a microscopic strain gauge (see below). X-ray and neutron diffraction are both commonly used and have the following characteristics:

Characteristic	*X-rays* ($\lambda = 0.154$ nm)	*Neutrons* ($\lambda = 0.129$ nm)
Strain resolution	± 10 μstrain	± 10 μstrain
Depth penetration[1]	~53 μm Al ~7 μm Ti	~70 mm Al ~15 mm Ti
Sampling volume	~0.5 mm^2 (50 μm depth)	$\gtrsim$1 mm^3

When an external load is applied to a crystal the lattice distorts. The elastic component of this distortion is achieved by a reversible change in the interplanar spacings, the plastic component by a movement of the atoms between atomic sites, with no change in the interplanar spacings. Through the Bragg equation, diffraction provides a measure of the interplanar spacing (d)

$$2d \sin \theta = \lambda \tag{11.3.1}$$

where 2θ is the angle between the incident and diffracted beams and λ the incident wavelength. Effectively the atomic planes within each crystallite act as microscopic strain gauges, so that by comparing the measured d value with the strain free interplanar spacing d_0 the *elastic* strain (ε) can be determined

$$\varepsilon = \frac{d - d_0}{d_0} = \cot \theta_0(\Delta\theta) \tag{11.3.2}$$

11.3.2 Application to MMCs

Diffraction techniques can measure lattice strain, so that some form of analysis based on the stiffness of the planes concerned is required before the stress fields can be calculated. Stress changes can be measured over macroscopic dimensions[2–4], but for composites it is usually more useful to measure the stresses between the phases[5,6]. For MMCs, the 'wavelength' of the local stress fluctuation within the matrix is much smaller than the spatial resolution of the diffraction technique. Consequently, such stress variations cause a broadening of the diffraction peak. However, a non-zero *average stress* in each phase will give rise to a small, but measurable, shift in the position of the peak corresponding to that particular phase (Fig. 11.3.1).

As to whether X-rays or neutrons should be used for the determination of internal mean stresses, the choice is often influenced by the penetrating power required (see above). For X-rays, the measured strain is an average taken over a thin surface layer. This is not always a disadvantage; for example, by choosing X-rays of different penetration one could monitor the variation with depth of matrix stresses in Al–graphite composites[7]. X-rays have also been used to measure residual stresses in barrier coatings

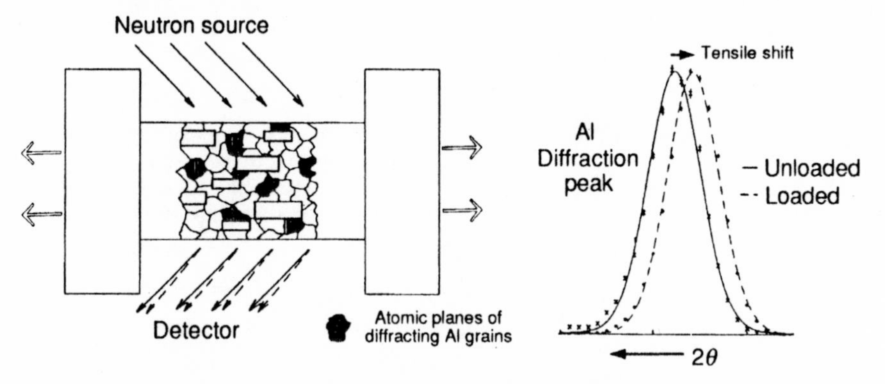

Fig. 11.3.1 A schematic illustrating the application of the neutron diffraction method to MMCs. The average phase strain is deduced in terms of the shift in the position of a particular diffraction peak.

(1 μm thick) on monofilaments, prior to incorporation in MMCs[8]: in this case a limited penetration is an advantage. At the surface of any material, the normal stress component must be zero and the stress state biaxial. With increasing depth the stress state becomes increasingly representative of the bulk state. Hanabusa, Nishioka and Fujiwara[9] found that, as might be expected, a triaxial stress state can be assumed only when the penetration depth is larger than a few 'wavelengths' of the fluctuating local stress system (approximately equal to the interparticle spacing). This condition may be satisfied when using X-rays, although it must be remembered that the sampling is not uniform over the penetration depth but favours the near surface regions and that machining will almost certainly cause significant surface damage.

Examples

- Internal stress measurement by diffraction[2]
- Measurement of residual stresses (X-rays[6,7,10,11,12]; neutrons[5,13,14])
- Measurement of stresses under load (§4.5; X-rays[17]; neutrons[15,16])
- *In situ* measurement of thermal stress generation and relaxation (§5.3.1; neutrons[5])
- Measurement of stresses in fibre coatings (X-rays[8]).

Pros	Cons
General	
√ non-destructive	× difficult to measure stress-free lattice parameter
√ simultaneously give strains in both phases	× sensitive to non-stress-related changes in lattice parameter (e.g. solid solution variations)
X-Rays	
√ good peak shape definition	× prone to surface relaxation effects
√ readily available	× sensitive to surface roughness
√ can carry out near-surface stress *vs.* depth studies	× more penetrating radiation (shorter λ) gives smaller peak shifts
Neutrons	
√ measures bulk internal strains	× few neutron sources
√ can carry out *in situ* experiments on the diffractometer table (e.g. mech. tests[15], thermal cycling[5])	× industrial work can be expensive
√ can carry out stress mapping non-destructively	

References

1. A. D. Krawitz, J. E. Brune and M. J. Schmank (1982) Measurements of Stress in the Interior of Solids with Neutrons, in *Residual Stress and Stress Relaxation*, E. Kula and V. Weiss (eds.), Plenum Press, New York, pp. 139–55.
2. A. J. Allen, M. T. Hutchings, C. G. Windsor and C. Adreani (1981) Measurement of Internal Stress within Bulk Materials using Neutron Diffraction, *NDT International*, (Oct.), pp. 249–54.
3. T. Leffers, T. Lorentzen, D. Juul Jensen and J. K. Kjems (1987) Measurement of Internal Stress by Neutron Diffraction using a Position Sensitive Detector, in *Res. Stresses in Sci. and Tech.*, E. Macherauch and V. Hauk (eds.), DGM Verlag, **1**, pp. 143–50.
4. P. Webster (1992) Spatial Resolution and Strain Scanning in *Neutron Diffraction Measurement of Residual Stresses*, NATO Conference, ASI Series E Applied Sci., vol. 216, M. T. Hutchings (ed.), Oxford, 1991, pp. 235–51.
5. P. J. Withers, D. Juul Jensen, H. Lilholt and W. M. Stobbs (1987) The Evaluation of Internal Stresses in a Short Fibre MMC by Neutron Diffraction, in *Proc. ICCM VI/ECCM 2*, F. L. Matthews *et al.* (eds.), London, Elsevier, **2**, pp. 255–65.
6. R. J. Arsenault and M. Taya (1985) The Effects of Differences in Thermal Coefficients of Expansion in SiC_W–6061 Al Composite, in *Proc. ICCM V*, W. C. Harrington *et al.* (eds.), San Diego, AIME, pp. 21–36.
7. S. D. Tsai, D. Mahulikar, H. L. Marcus, I. C. Noyan and J. B. Cohen (1981) Residual Stress Measurements on Al–Graphite Composites using X-Ray Diffraction, *Mat. Sci. & Eng.*, **47**, pp. 145–9.
8. C. M. Warwick, R. R. Kieschke and T. W. Clyne (1991) Sputter Deposited Barrier Coatings on SiC Monofilaments for Use in Reactive Metal Matrices – Part II. System Stress State, *Acta Met. et Mat.*, **239**, pp. 437–44.
9. T. Hanabusa, K. Nishioka and H. Fujiwara (1983) Criterion for the Triaxial X-Ray Residual Stress Analysis, *Z. Metallkunde*, **74**, pp. 307–13.
10. D. V. Wilson (1965) Reversible Work-Hardening in Alloys of Cubic Metals, *Acta Metall.*, **13**, pp. 807–14.
11. H. M. Ledbetter and M. W. Austin (1987) Internal Strain (Stress) in an SiC_P–Al Reinforced Composite: An X-Ray Diffraction Study, *Mat. Sci. & Eng.*, **89**, pp. 53–61.
12. A. D. Krawitz (1985) The Use of X-Ray Stress Analysis for WC-Base Cermets, *Mat. Sci. & Eng.*, **75**, pp. 29–36.
13. A. D. Krawitz (1992) Stress Measurements in Composites by Neutron Diffraction in *Neutron Diffraction Measurements of Residual Stresses*, NATO Conference, ASI Series E Applied Sci., vol. 216, M. T. Hutchings (ed.), Oxford 1991, pp. 405–20.
14. S.Majumdar, J. P. Singh, D. Kupperman and A. D. Krawitz (1991) Application of Neutron Diffraction to Measure Residual Strains in Various Engineering Composite Materials, *J. Eng. Mat. & Tech.*, **113**, pp. 51–9.
15. A. J. Allen, M. Bourke, S. Dawes, M. T. Hutchings and P. J. Withers (1992) The Analysis of Internal Strains Measured by Neutron Diffraction in Al/SiC MMCs, *Acta Met. et Mat.*, **40**, pp. 2361–73.
16. P. J. Withers and T. Lorentzen (1989) A Study on the Relation Between the Internal Stresses and the External Loading Response in Al/SiC Composites, in *Proc. ICCM VII*, Guangzou, W. Yunshu, G. Zhonlon and W. Renjie (eds.), **2**, pp. 429–34.

17. H. P. Cheskis and R. W. Heckel (1968) *In situ* Measurement of Deformation Behaviour of Individual Phases in Composites by X-Ray Diffraction, in *Metal Matrix Composites*, ASTM STP 438, pp. 76–91.

11.4 Photoelastic determination of internal stress

11.4.1 Basic principle and capability

Photoelastic stress modelling provides an effective practical means of determining internal stresses in two-dimensional systems. This is because fringe patterns seen when the photoelastic test-piece is viewed between polarising and analysing polaroids can be interpreted in terms of the stress field within it. Accurate analysis of the fringe pattern allows the determination of:

- *the difference in in-plane principal stresses, i.e. the maximum shear stress* ($\pm 5\%$)
- *the principal stress directions* ($\pm 2°$)
- *the absolute values of the in-plane principal stresses* ($\pm 20\%$).

The method relies on the fact that photoelastic materials are birefringent under elastic strain. Effectively, this means that an incident plane polarised wave is split into two perpendicular components along the directions of the principal stresses ($\sigma_{a,b}$), which travel at speeds related to the principal stresses. Consequently, one component becomes retarded relative to the other, depending on the stresses, so that when the two components are combined on leaving the medium they generate bright and dark contours of maximum shear stress ($\sigma_a - \sigma_b$) when examined through an analysing polaroid. The standard procedure therefore involves loading a two-dimensional photoelastic model (i.e. the out-of-plane stress equals zero) and viewing through a polariscope.

The approach can be extended to three-dimensional systems either by 'freezing-in' molecular scale strain after loading at a suitable temperature (and subsequently viewing a thin section)[1], or by detecting light scattered from the region of interest within an unsectioned model[2]. The former method, which is often the more convenient, involves heating the model (usually araldite) to the 'stress-freezing' temperature T_s (sometimes termed, rather inappropriately, the critical temperature), which is typically about 20–40 K above the glass transition temperature, T_g. The choice of T_s is made such that Van der Waals bond energies are dominated by available thermal energy and the molecular arrangement is then

Table 11.4.1 *Published*[1,3] *photoelastic data for CT200 resins and HT range hardeners for sodium light* ($\lambda = 590$ *nm*). *The values of* T_g *and* T_s *for these systems are* 100 *and* 130 °*C respectively. The stiffness values at* T_s *tend to be sensitive to the exact composition and curing cycle and should be measured for each batch*

Araldite resin	Modulus (MPa)	Poisson's ratio	Fringe value ($kN\,m^{-1}$)	CTE (10^{-6}/K)
CT200–37 wt% HT907 ($< T_g$)	3000	0.34	15	~50
CT200–37 wt% HT907 ($\sim T_s$)	2	0.48	0.28	~150
CT200–20 wt% HT901 ($\sim T_s$)	8	0.48	0.35	~150

influenced solely by the covalent bonds of the crosslinked network. The model is loaded at this temperature, causing local strains proportional to the local stresses. (It should be noted that in this temperature regime the material has a low stiffness, a high stress sensitivity and a high thermal expansivity, relative to the values exhibited below T_g – Table 11.4.1). The assembly is then cooled at a rate sufficiently low to avoid the generation of significant thermal gradients within the model. As the temperature falls below T_g, Van der Waals bonding is re-established, locking the molecular configuration into the anisotropic form characteristic of the loaded state. Because the stresses exist on the molecular scale, cutting of the model relaxes only a very thin (atomic scale) surface layer, and hence the pattern is to all intents and purposes unchanged upon careful slicing.

Identifying principal stress directions: this can be achieved by viewing the model slice under plane polarised light and plotting the ***isoclinics*** (black fringes connecting points of equal inclination of the principal stress directions to the polariser). These are formed by virtue of the fact that when one of the principal stresses is parallel with the polariser, the polarised light is not split into two components so that the light exits the model unchanged, and is thus stopped by the analyser. By rotating the model with respect to the polariser and analyser it is possible to plot the direction of the principal stresses everywhere (Fig. 11.4.1). In general the isoclinics tend to confuse the isochromatics (see below) and so are removed using two quarter-wave plates.

Determining the principal stress difference: for monochromatic light polarised at a general angle to the principal stress directions, the fast and slow components interfere either constructively or destructively on leaving

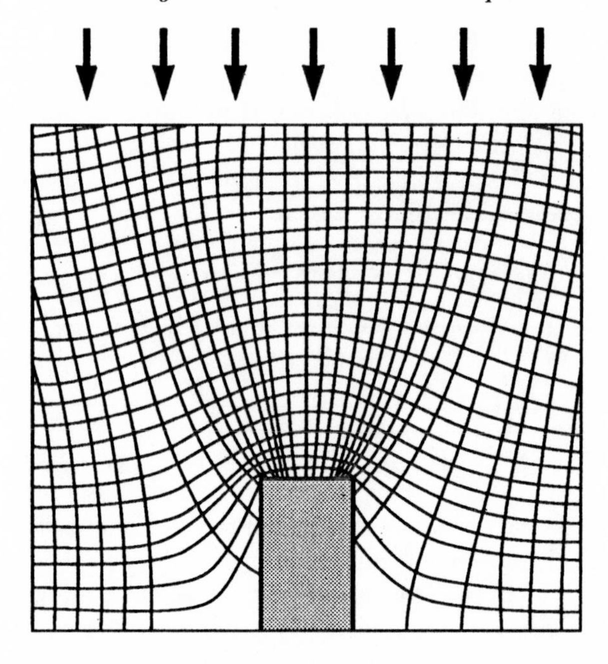

Fig. 11.4.1 A plot of the principal stress directions measured for a slice from a 3-D model containing a cylinder ($s = 3$) for which the stress patterns arising from axial compression have been stress-frozen. Note how the stress trajectories are analogous to the field and equipotential lines of electrostatics.

the model depending on the maximum shear stress ($\sigma_a - \sigma_b$) and the slice thickness (t). This results in fringes of equal maximum shear stress (e.g. Fig. 3.9)

$$\sigma_a - \sigma_b = \frac{nf}{t} \qquad (\sigma_a, \sigma_b = \text{in-plane principal stresses}) \qquad (11.4.1)$$

where n is the order of the fringe and f is the material fringe value. In white light, the fringes form as bands of colour (***isochromatics***), as each wavelength interferes independently. Accurate determination of the principal stress difference can be achieved using the Tardy[4] method for the identification of fractional fringe orders ($\sim$0.2 of a fringe). Alternatively, the fringe pattern can be viewed in white light, and the colour of the point of interest compared with the Michel-Levy chart. The retardation can then be read off from the chart and converted into the fractional number of sodium wavelengths ($\lambda = 590$ nm – Table 11.4.1).

Separating the principal stresses: the isochromatic fringes provide an

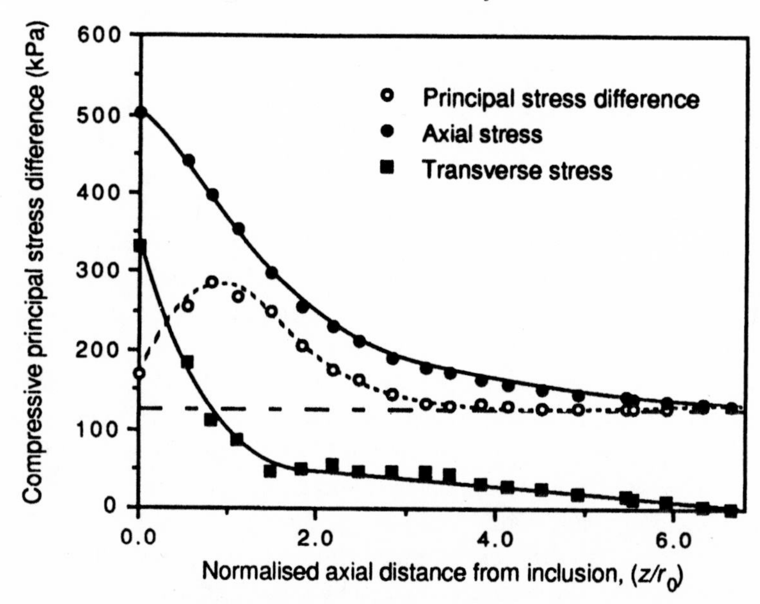

Fig. 11.4.2 The variation in the principal stress difference (taken directly from the fringe order) down the symmetry axis of a composite containing a metal ellipsoid[3] (aspect ratio 3) and loaded to 125 kPa: the principal stresses have been separated using eqn (11.4.2) and the curvature of the transverse stress trajectory along the stress axis. It is important to note that the error in the determination of σ_a is cumulative with distance down the axis towards the inclusion. In spite of the errors involved, it is clear that the transverse stress is responsible for the drop in the stress difference (see Fig. 3.10) and that the hydrostatic stress at the end of the cylinder is very large indeed. The axial stress at the cylinder end is consistent with an average inclusion stress value of 600 kPa (predicted by the Eshelby approach) given that the stress is expected to rise towards the centre of the ellipsoid. The peak transverse stress predicted for the corresponding ellipsoid is ~260 kPa, but lies slightly away from the inclusion.

easy method of determining principal stress differences, but, except at points where one of the principal stresses is already known (e.g. at a model boundary), they do not yield the individual stresses. Jessop[5] proposed an integration technique which enables one to deduce the principal stress by integrating along a line from a known point (i.e. a boundary). This method is tedious and inaccurate for a general point, but can be simplified if carried out along a symmetry axis (e.g. Fig. 11.4.2). The integration procedure relates the change in principal stress $\delta\sigma_a$ to the step size δz and the radius of curvature (ρ) of the orthogonal stress directions (σ_b and σ_c)

$$\delta\sigma_a = \frac{2(\sigma_a - \sigma_b)}{\rho}\,\delta z \qquad (11.4.2)$$

11.4.2 Application to MMCs

Photoelastic techniques have been used successfully for the determination of matrix stressing under loading[1] as well as during interface testing of long fibre composite systems. Care must, however, be exercised when selecting a suitable 'reinforcement' for frozen stress studies. The very low stiffness of araldite resins above the stress-freezing temperature means that use of metal reinforcement represents a system with an effectively infinite stiffness ratio. Further, given the large expansion coefficient of araldite relative to most metals, fringes are generated from thermal stresses, which superimpose upon any load-induced fringes†. However, araldite resins can be produced with different stiffnesses (a factor of ten) through a selective choice of hardener and the hardener/resin ratio. This allows the production of completely photoelastic systems, with stiffness ratios typical of metal matrix composites. This also gives further advantages in reducing the thermal stresses almost to zero, and allowing the stress fields within the reinforcement to be investigated[1].

Examples:

- Scattered light[2]
- Three-dimensional frozen stress: Modelling of short fibre composites[1]
  : Modelling of pull-out tests, push-out tests[6]

Pros	Cons
√ cheap to undertake	× production of models can be time-consuming
√ can determine stresses in both phases	× stiffness of araldite may vary from batch to batch
√ easy to evaluate principal stress differences	× difficult to evaluate the principal stresses accurately
√ good for checking theoretical models or for complex reinforcement geometries	× can only model relatively simple composites
	× can model only in the elastic regime

† In one study[3] the influence of thermal residual stresses has been minimised through the use of Mg–30 wt% Li as reinforcement which has an unusually large expansion coefficient (35 μstrain/K).

References

1. P. J. Withers, G. Cecil and T. W. Clyne (1991) Frozen Stress Photoelastic Determination of the Extent of Fibre Stressing in Short Fibre Composites, in *Proc. of 2nd European Conf. on Adv. Mats., Euromat '91*, Cambridge, T. W. Clyne and P. J. Withers (eds.), Inst. of Metals, pp. 134–40.
2. B. J. Briscoe and D. R. Williams (1988) Developments in Laser Light Scattered Photoelasticity in Observing Interfacial Stress Gradients of Model Composite Interfaces, in *Conf. on Interfaces in Polymer, Ceramic and Metal Matrix Composites*, H. Ishida (ed.), Elsevier, pp. 89–99.
3. P. J. Withers, A. N. Smith, T. W. Clyne and W. M. Stobbs (1990) A Photoelastic Investigation of the Validity of the Eshelby Modelling Approach to MMCs, in *Fundamental Relationships Between Microstructure and Mechanical Properties of MMCs*, Indianapolis, M. N. Gungor and P. K. Liaw (eds.), TMS-AIME, pp. 225–40.
4. H. L. Tardy (1929) Methode Practique d'Examen et de la Birefringence des Verres d'Optique, *Revue d'Optique*, **8**, pp. 59–69.
5. H. T. Jessop (1949) The Determination of the Separate Stresses in Three-Dimensional Stress Investigations by the Frozen Stress Method, *J. Sci. Instr.*, **26**, pp. 27–31.
6. M. C. Watson and T. W. Clyne (1992) The Use of Single Fibre Pushout Testing to Explore Interfacial Mechanics in SiC Monofilament-Reinforced Ti – I. A Photoelastic Study of the Test, *Acta Met. et Mat.*, **40**, pp. 131–9.

11.5 Metallographic preparation

11.5.1 Basic principle and capability

The aim of metallographic preparation is simple; to prepare a surface such that one or more of the following features can be clearly delineated without unnecessary distortion or the introduction of new features not originally present. Most preparation procedures involve cutting and grinding, lapping and polishing, and, if necessary, electropolishing or anodising stages so as to reveal:

- *distributions of reinforcement size, shape, orientation and location*
- *matrix grain structure, presence of precipitates, etc.*
- *information relating to crystallographic texture and distribution of plastic strain.*

Ideally, the surface of a bulk metallographic specimen should be totally flat and representative of any similarly oriented plane within the material. As there is usually little or no difficulty in obtaining contrast between the reinforcement and the matrix when viewed in the optical or scanning electron microscope, it is then a straightforward matter to make deductions about the volume fraction, shape, size, etc., of the reinforcement. In addition, it is often possible to obtain contrast between different phases in the matrix, between separate matrix grains and, in some cases, between regions of the matrix which differ in other respects, such as having variable dislocation densities as a result of differential plastic straining.

11.5.2 Application to MMCs

The objectives of specimen preparation may be divided into the avoidance of (a) surface relief and (b) microstructural distortion. There may then be further objectives concerned with the production of contrast, which are more readily achieved during specimen preparation rather than by adjusting the optical conditions during viewing. The avoidance of relief may be essential in order to see particular microstructural features and this can be especially difficult with MMCs as a result of the large difference in hardness between matrix and reinforcement. Microstructural distortion, particularly in the form of smearing of the matrix within the surface layer, can also be difficult to avoid and this can be misleading when exploring the levels of porosity and reinforcement content. (These particular features may be best characterised via densitometry – §11.8.) There are, however, a range of procedures which can be used to minimise relief and distortion effects.

Cutting and grinding

Most MMCs can readily be sectioned with a rotary abrasive cut-off wheel, typically having a thickness of 1–2 mm. In some cases a powered hacksaw or bandsaw may be suitable. Copious lubrication is often essential, particularly with cut-off wheels. For small specimens or in cases where minimal surface damage is required, it may be preferable to use a higher precision device such as a high speed diamond slitting wheel or a diamond wiresaw. The latter technique (typically using a diamond-impregnated metallic wire about 300 μm in diameter) induces very low damage and is capable of producing a long (~100 mm), deep (~50 mm) planar cut, but it is rather slow. Spark machining has rather similar characteristics but is usually even slower. If grinding is necessary it is usually done after mounting, using 180 grit SiC paper. When polishing SiC_P reinforced systems care must be exercised so as not to confuse original reinforcing particles and grit arising from the abrasive paper.

Lapping and polishing

The polishing procedure is often critical if relief is to be avoided with MMCs. Although the details will vary, certain principles can be identified.

(A number of these have been summarised[1] for aluminium-based MMCs.) An automated polishing device is almost essential in order to achieve the necessary control and reproducibility with regard to contact pressures, lubricant dispensing rates and wheel speeds. Generally, a sequence of 45, 15, 6 and 3 μm diamond abrasive is appropriate, using napless cloths. The use of napped (or worn) cloths tends to leave the reinforcement in strong relief. Napless cloths are, however, more prone to contamination and may need to be flushed clean periodically. It can be important to clean the specimen carefully between wheels, using an ultrasonic bath, particularly if it contains any porosity. Abrasive particles tend to become embedded in the matrix if high contact pressures or low wheel speeds are used, while fine reinforcement debris can accumulate if the pressure is too low. It is often beneficial to carry out a final brief polish using fine colloidal silica or γ-alumina as an abrasive.

Electropolishing and anodising

Electropolishing can be an effective technique for the production of a macroscopically flat surface, while minimising the damage induced during the process. It is in general unnecessary to prepolish the surface to better than 6 μm. A typical arrangement[1] for electropolishing an Al MMC is shown in Fig. 11.5.1. In this case a black film forms as the voltage is raised

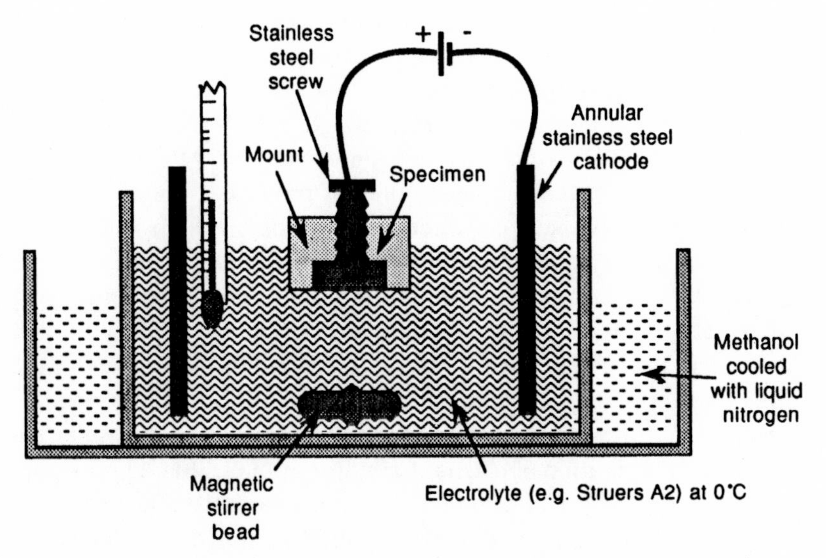

Fig. 11.5.1 Typical arrangement for electropolishing or anodising[1].

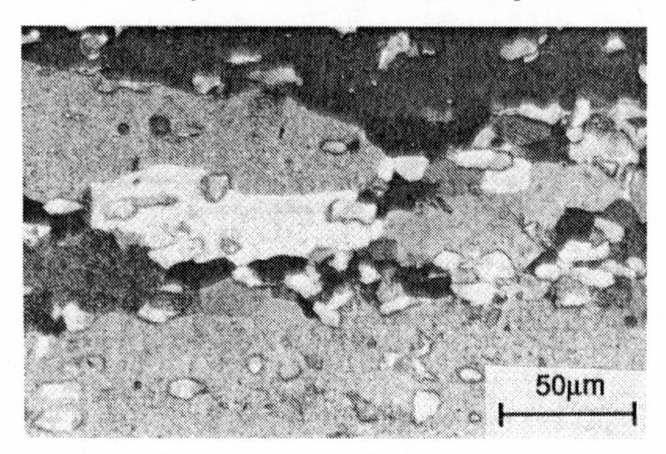

Fig. 11.5.2 An Al–10 vol% Al_2O_{3P} composite exhibiting a recrystallised grain size. Anodising was carried out for one minute at 20 V using Barker's reagent (2% HBF_4).

to about 15 V and this breaks up as it is further increased to 20 V. The current density is typically 5 mA mm^{-2} at this point. The specimen is shaken to detach the film, removed and washed.

A similar arrangement can be used to produce a decorative anodised finish on the surface of the specimen. For example, this can be particularly useful in producing grain contrast with aluminium alloys[1]. This can be done at room temperature, with Barker's reagent (2% HBF_4) as the electrolyte, a voltage of 20–30 V and a current density of 10 mA mm^{-2}. The prior mechanical polishing procedure is critical for anodising, as relief will obscure the grain structure severely. It may be noted that for aluminium anodised in this way, an amorphous or very fine-grained surface film of fluorine-containing alumina is produced[2,3], in the form of aligned furrows. The alignment direction depends on the orientation of the underlying grain and, as the film itself is optically anisotropic, contrast is produced in the polarising microscope which reveals the underlying grain structure. The observed colours are hence independent of film thickness and they can be used to make qualitative deductions about the nature of any crystallographic texture in the matrix[4,5]. It is also possible to observe contrast in the matrix giving information on the subgrain structure and distribution of plastic strain. For example, Fig. 11.5.2 shows an optical micrograph of an anodised Al-based composite after extrusion, revealing the deformation pattern around the reinforcing particles. Grain contrast can also be seen in the scanning electron microscope, notably from channeling patterns[6] and variations in back-scattered intensities (preferably with a low threshold voltage detector). The application of these techniques to Al MMCs has been described[1].

Pros	Cons
√ potentially very informative	× painstaking work may be necessary
√ equipment needed is relatively cheap	× appearances can be misleading
√ specimens may be very small	

References

1. R. A. Shahani (1991) *Microstructural Development during Thermomechanical Processing of Aluminium-Based Composites*, PhD Thesis, Univ. of Cambridge.
2. H. Gahm, F. Jeglitsch and E. M. Horl (1982) Investigation of the Structure of Chemically Deposited Films Produced by Precipitation Etching, *Prac. Met.*, **19**, pp. 369–90.
3. R. J. Gray, R. S. Crouse and B. C. Leslie (1976) Decorative Etching, in *Metallographic Specimen Preparation*, J. L. McCall and W. M. Mueller (eds.), Plenum, New York, pp. 192–8.
4. S. Kroger, E. Dahlem-Klein, H. Weiland and H. J. Bunge (1988) Texture Determination in Aluminium Alloys Using Colour Metallography, *Textures and Microstructures*, **10**, pp. 41–8.
5. T. O. Saertre, J. K. Solberg and N. Ryum (1986) Variation in Polarized Light Intensity with Grain Orientation in Anodized Aluminium, *Metallog.*, **19**, pp. 345–57.
6. H. J. Lee, R. D. Doherty, E. A. Feest and J. M. Titchmarsh (1983) Structure and Segregation of Stir Cast Al Alloys, in *Solidification Tech. in the Foundry and Casthouse*, The Metals Soc., London.

11.6 TEM specimen preparation

11.6.1 Basic principle and capability

The aim of transmission electron microscopy specimen preparation is to produce large areas of uniformly electron transparent material (thickness $<0.5\ \mu m$*) without introducing artefacts into the composite microstructures. Good specimen preparation is something of a black art, but normally it can be broken down into three distinct steps:*

- *forming a disc*
- *polishing and thinning*
- *final thinning and perforation.*

Throughout each of these steps it is important to bear in mind the potential sources of microstructural damage which may be caused by the preparation process.

11.6.2 Application to MMCs

The main difficulties concerning specimen preparation are associated with obtaining a good extent of thin area, retaining the original dislocation microstructure and, except for composites comprising very fine scale particles or fibres <0.5 μm, in thinning the reinforcement sufficiently so as to be able to examine the reinforcement and interface regions.

Disc forming

This step is usually carried out prior to polishing and thinning, but if very thin specimens are required it is sometimes best carried out afterwards[1]. Electric discharge machining is a good method for producing a 3 mm or 2.3 mm disc as surface damage is restricted to <200 μm beneath the surface[2] (50–100 V). Punching or, in the case of very thin specimens, cutting with a razor blade can also be used, but the damage zone is somewhat larger and these methods carry the risk of bending the specimen.

Polishing and thinning

The aim of this step is to produce a specimen suitable for final perforation, which in most cases requires a specimen thickness of around 0.5 mm. This is usually carried out by wet grinding. For SiC reinforced systems care must be taken not to contaminate the specimen with SiC from the grinding wheels. In situations where shadowing of the electron beam is a problem because of the large scale of the reinforcement, but where dislocation microstructure is not of interest, good results can be achieved[1] by grinding to a thickness of ~ 25 μm and then final polishing using a napless cloth to minimise surface relief (§11.5.2).

Final thinning and perforation

There are basically two methods available for specimen perforation: ion beam milling and electropolishing. ***Ion beam milling*** causes little surface damage (less than 20 nm with 5 kV, 5 mA current and an incidence angle[3] of 15°) and, compared with electropolishing, has the advantage that the reinforcement is thinned, if at a somewhat slower rate than most metals. Typically, metals thin at approximately 1 μm/hr per ion beam. For the

Al–SiC system, thinning rates can be made more or less equal by bleeding in oxygen which reduces the milling rate of the aluminium[4]. In the light of this, it is perhaps not surprising that the milling rate of Al_2O_3 fibres and particles is notoriously slow. In all cases the specimen should be cooled by liquid nitrogen throughout the milling process in order to limit damage by the beam. This is especially important for Al matrix systems, because of the ease of dislocation recovery. However, because of the large disparity in expansion coefficients, Al-based composites are especially vulnerable to both low and elevated temperature excursions during preparation. Additional care must be exercised with Al–Li for which thinning can denude Li from the surface as well as damage δ' precipitates[1].

Because of the incidence angle of the beam, a well polished specimen free of surface relief is needed prior to milling. The grinding of a 'dimple' on one surface of the specimen before milling is often beneficial in terms of increasing the extent of thin area, especially for high reinforcement content specimens. ***Electropolishing*** has the advantage that it is an especially quick means of specimen preparation, the thinning process taking no more than a couple of minutes to complete. Unfortunately, the ceramic is not attacked by this technique so that apart from very fine whisker and dispersoid containing systems electron microscope observation is restricted to the matrix. However, large thin areas can be produced using the conditions outlined in Table 11.6.1. For the preparation of difficult specimens a twin jet electropolisher has been found to give better results than conventional electropolishing[5]. A problem for Al–Li systems is that the matrix reacts rapidly with moisture to form the hydroxide so that after electropolishing an immediate wash in dry

Table 11.6.1 *Polishing conditions for various composites and dispersion hardened systems using either* (1) 20% *perchloric acid in ethanol,* (2) 25% *nitric acid with methanol or* (3) 25% *phosphoric acid,* 25% *ethylene glycol,* 50% *distilled water*

System	Electrolyte	Voltage (V)	Current (A)	Temperature (°C)
Al/Al_2O_3 or SiC[5]	1	20	0.1	−20
Al–Li/SiC[1]	2	20		−30
Stainless steel/Al_2O_3[5]	1	18	0.2	−20
Zircalloy/Y_2O_3[5]	1	14	0.1	−20
Ferritic iron/Al_2O_3[5]	1	18	0.05	−20
Cu/Al_2O_3 or ZrO_2[6]	3	10		R.T.

methanol is required, making sure that the electrolyte does not get trapped behind the reinforcement[1].

Examples: (refer to table 11.6.1)

Pros	Cons
Ion beam milling	
√ thins both matrix and reinforcement	× time consuming
√ good for interface studies	
Electropolishing	
√ relatively cheap and easy to undertake	× only suitable for viewing matrix region
√ quick	

References

1. P. B. Prangnell (1991) *Development of Matrix and Interface Microstructures and Effect on Mechanical Behaviour of SiC_P/Al MMCs*, PhD Thesis, Univ. of Cambridge.
2. L. E. Samuels (1971) *Metallographic Polishing Methods*, Pitman, Melbourne, p. 79.
3. D. G. Howitt and R. H. Geis (eds.) (1981) *Analytical Electron Microscopy*, San Francisco Press, CA, p. 252.
4. P. B. Prangnell (1991) Private Communication.
5. J. Lindbo and T. Leffers (1972) Preparing Dispersion Hardened Materials for TEM, *Metallography*, **5**, pp. 473–7.
6. J. Lindbo (1991) Private Communication.

11.7 Characterisation of reinforcement parameters

11.7.1 Basic principle and capability

As is clear from Chapter 9, there are a great many process routes by which MMCs can be made. For each route the precise choice of process variables will control the resulting composite microstructure, which in turn will control composite performance. Reinforcement parameters are very sensitive to fabrication route, and thus it is essential that they can be accurately evaluated. Perhaps the most important reinforcement parameters in terms of dictating performance are the following:

- *inclusion aspect ratio*
- *inclusion orientation*
- *inclusion distribution.*

These are discussed in the following sections.

11.7.2 Inclusion aspect ratio

Fibre aspect ratio has a marked influence on the properties of aligned composites. Consequently, there is considerable interest in its measurement, especially because of the high degree of fibre fracture which can occur during fabrication (aspect ratios typically falling from >100 to ~3–20). Given that fibre breakage can be extensive, care must be exercised during aspect ratio analysis to retain the smaller fragments, so as not to overestimate the aspect ratio. Probably the simplest method of measurement is that described by Arsenault[1]. The reinforcement is obtained in suspension by dissolving away the matrix using a suitable acid. This is followed by sedimentation onto a glass or a metallic substrate upon which the fibres can be observed easily under a scanning electron or optical microscope. Analysis can then be undertaken either manually or by image analysis. The results in Fig. 11.7.1 were obtained in this manner. A theoretical study[3] has shown that for modelling all but the most grossly skewed of aspect ratio distributions, it is not necessary to take into account the precise distribution, but that the use of the *volume*-averaged aspect ratio is usually sufficient. For almost equiaxed particles, the effective aspect ratio distribution can be measured adequately on polished sections, because the requirement of identifying the direction of the greatest dimension is less critical when the aspect ratio is low.

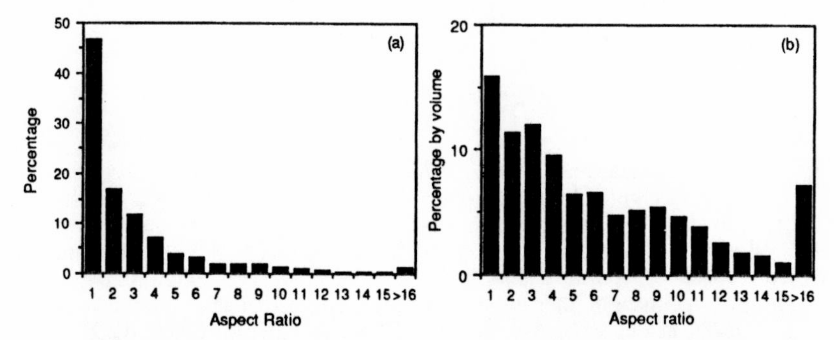

Fig. 11.7.1 (a) Number- and (b) volume-averaged aspect ratio distributions for extruded Al–5 vol% SiC_W composite[2]. The very high proportion of short fibres and fragments present in most MMCs results in a volume-averaged aspect ratio which is significantly larger than the simple frequency-averaged value. In the present case the mean values were 2.4 and 5.5 for the frequency- and volume-averaged distributions respectively.

11.7.3 Inclusion orientation

Composite properties are naturally very sensitive to reinforcement alignment. Three distinct methods for the assessment of the degree of alignment have been proposed:

- *microscopical observation*
- *Fourier transform analysis*
- *crystallographic diffraction.*

Microscopical observation[4,5]

Orientations of individual fibres are calculated on the basis of their elliptical intersection with a polished surface (Fig. 11.7.2(a)). This method

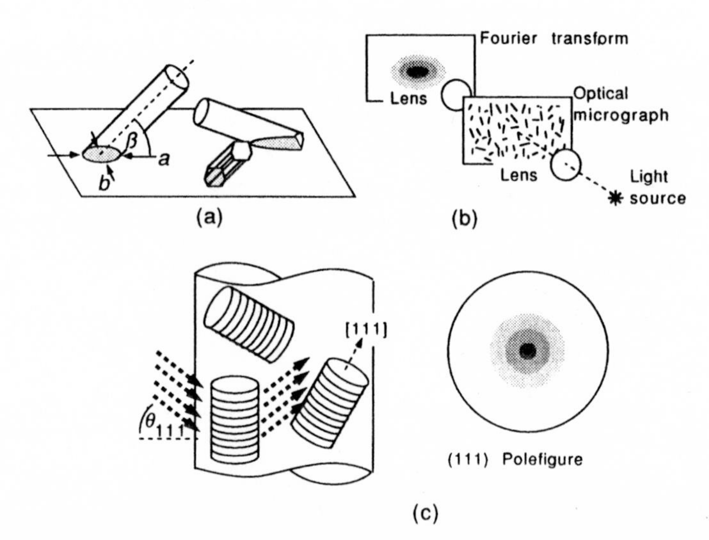

Fig. 11.7.2 (a) Calculation of fibre misalignment ($\beta°$) from the major, a, and minor, b, axes of elliptical fibre cross-sections using the relation $\beta = \sin^{-1}(b/a)$. Note that this method only really works for long accurately circular fibres, and that the greater probability of intersecting fibres almost normal to the plane means that correction for counting bias must be made[8], by (b) taking the Fourier transform of the image (either practically on an optical bench or mathematically by computer). The degree of alignment as well as other reinforcement distribution parameters can then be deduced from deviations from circular symmetry; (c) provided the fibre crystallography can be related to the fibre axis, the variation in diffracted peak intensity with specimen orientation can be interpreted in terms of the fibre alignment. Note that when the fibre axis is parallel to, for example, the $[0001]_{hcp}$ a unique solution is obtained, whereas this is not so for $[111]_{fcc}$ because of the four symmetry-related $\langle 111 \rangle$ directions[8].

works well for large (clearly resolvable) circular fibres of reasonable length, but is of dubious value for fine mis-shapen short whiskers.

Fourier transform analysis[6,7]

This is an elegant method, which involves the diffraction of light on passing through an optical micrograph (Fig. 11.7.2(b)). It does, however, require a number of polished sections, and relies on an adequate removal of the sectioning bias.

Crystallographic analysis

Neutron (or possibly X-ray) diffraction provides a quick, and, for fairly sharp orientation distributions, accurate method of determining the orientation distribution of the reinforcing inclusions[8], provided the crystallographic orientation of the reinforcement can be related to its shape. (For instance it is suited to SiC_W which grows parallel to (111), but not $SiC_{monofilament}$ (radial columnar texture), Al_2O_{3F} (random polycrystalline) or C (amorphous)). In this method, texture information expressed as polefigures or orientation distribution functions is interpreted in terms of fibre alignment (Fig. 11.7.2(c)).

Pros	Cons
Microscopical observations	
√ simple to carry out	× time consuming
√ cheap	× requires many polished sections
	× yields only a number average distribution
	× poor for misshapen fibres
	× difficult to be sure surface is flat
Fourier transform analysis	
√ cheap	× requires many polished sections
√ Fourier transform also includes distribution information	× difficult to separate and quantify the influence of various factors
Crystallographic analysis	
√ fast	× requires sophisticated equipment
√ gives a volume average	× rarely a unique interpretation
√ samples a large population	× ambiguous for poorly aligned systems if no unique crystal direction
√ can use with fine fibres	× not applicable to many types of fibre

Once the distribution of fibre orientations is known, there are a variety of methods by which misorientation can be taken into account[10–12]. The variation in the stiffness of an aligned Al/SiC_W composite with the angle at which the load is applied, has recently been measured[13], and has been shown to be in agreement with the predictions of the Eshelby model.

Examples:

- microscopical observation – Al/SiC_W[8], Al/Al_2O_{3F}[9]
- Fourier transform analysis – Al/SiC_W[7]
- crystallographic analysis – Al/SiC_W[8].

11.7.4 Inclusion distribution

Although volume average-related properties, such as Young's modulus, are insensitive to local volume fraction variations and nearest neighbour distances, failure processes are dependent on the local environment (§7.4). Consequently, the spatial distribution of the reinforcement is of considerable interest. Most methods of quantification are based on the use of tessellations, the most common being the ***Dirichlet tessellation***[14]. This is simply a network of polygons generated around the particles such that all points within a polygon lie closer to the centre of the enclosed particle than to any other (Fig. 11.7.3(a)).

Deviations from a random distribution are best identified by comparison of various parameters with those for computer-generated random distributions of non-overlapping idealised shapes†. In this context, two parameters of interest[14] are the ratio of observed to predicted nearest neighbour distances (Q), and the ratio of observed to expected variances of this distance (R). The following interpretations have been proposed[15]:

$Q < 1$	$R < 1$	clustered sets
$Q > 1$	$R < 1$	short-range ordered sets
$Q < 1$	$R > 1$	sets of clusters with superimposed background of ordered points.

The tessellations have also been used to quantify local fluctuations in volume fraction[16] (although it is important to consider what is meant by local), as have hardness measurements (200 μm square impressions –

† At high volume fractions care must be exercised to ensure that the shapes are representative.

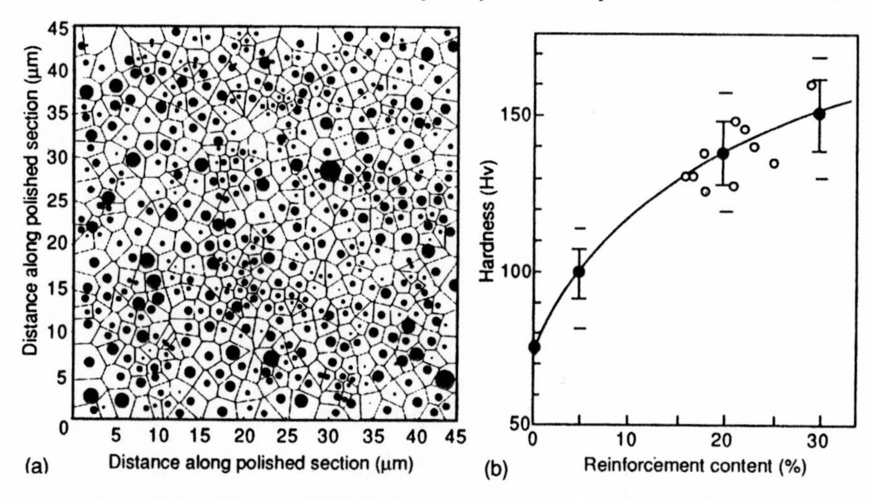

Fig. 11.7.3 (a) A Dirichlet tessellation for Al/SiC_p[16]; (b) the correlation between hardness measurements (200 μm square) and local volume fraction[9], the extreme bars show the range of hardness measurements, and the error bars the standard deviation. The open circles about the 20% mark illustrate the scatter recorded for a nominally 20% volume fraction system.

Fig. 11.7.3(b))[9]. Banding and other distribution anisotropies on the other hand, have been identified by plotting histograms of the angle of lines drawn between nearest neighbours, or by measuring the mean intercept length of the Dirichlet cells in different directions[14].

A related problem is the quantification of the preference or disinclination of a growing crack for locally high volume fraction regions. With this objective in mind, a crack path preference parameter Γ has been defined[17] in terms of the fraction of the total crack length L_{tot} taken up by each fractured particle (l_i):

$$\Gamma = \left(\frac{\sum_i l_i}{L_{tot}}\right) \Big/ f_{loc} \qquad (11.7.1)$$

The local volume fraction term (f_{loc}) accounts for local fluctuations in particle volume fraction. Consequently, a value of unity indicates a random path, whilst a value greater (or less) than unity indicates a preference (disinclination) for particle cracking/interfacial debonding as compared with matrix failure.

Examples:

- Dirichlet tessellations (Al–Gr[14])

- Hardness measurements (Al–SiC[16])
- Fracture path preference (Al–SiC[18]).

References

1. R. J. Arsenault (1984) The Strengthening of Al Alloy 6061 by Fibre and Platelet SiC, *Mat. Sci. & Eng.*, **64**, pp. 171–81.
2. P. J. Withers (1988) *The Application of the Eshelby Method of Internal Stress Determination for Short Fibre Metal Matrix Composites*, PhD Thesis, Univ. of Cambridge.
3. Y. Takao and M. Taya (1987) The Effect of Variable Fibre Aspect Ratio on the Stiffness and Thermal Expansion Coefficients of a Short Fibre Composite, *J. Comp. Mat.*, **21**, pp. 140–56.
4. D. Hull (1981) *An Introduction to Composite Materials*, Cambridge Univ. Press, Cambridge.
5. M. Vincent and J. F. Agassant (1986) Experimental Study and Calculation of Short Glass Fibre Orientation in a Center Gated Molded Disc, *Polymer Comp.*, **7**, pp. 76–83.
6. T. W. Chou, R. L. McCullough and R. Byron Pipes (1986) Composites, *Sci. Am.*, **255**, pp. 166–77.
7. S. Øvland and K. Kristiansen (1988) Characterisation of Homogeneity and Isotropy of Composites by Optical Diffraction and Imaging, in *Mech. & Phys. Behav. of Met. and Ceramic Mat. Composites, Proc. of 9th Risø Symp.*, S. I. Andersen, H. Lilholt, and O. B. Pedersen (eds.), pp. 527–32.
8. D. Juul Jensen, H. Lilholt and P. J. Withers (1988) Determination of Fibre Orientations in Composites with Short Fibres, *ibid.*, pp. 413–20.
9. D. van Hille, S. Bengtsson and R. Warren (1989) Quantitative Metallographic Study of Fibre Morphology in a Short Fibre Al_2O_3 Fibre Reinforced Al Alloy Matrix, *Comp. Sci. & Tech.*, **35**, pp. 195–206.
10. C. C. Chamis and G. P. Sendeckyj (1968) Critique on Theories Predicting Thermoelastic Properties of Fibre Composites, *J. Comp. Mat.*, **2**, pp. 332–58.
11. T. W. Chou and S. Nomura (1981) Fibre Orientation Effects on the Thermoelastic Properties of Short Fibre Composites, *Fibre Sci. and Tech.*, **14**, pp. 279–91.
12. L. M. Brown and D. R. Clarke (1975) Work Hardening due to Internal Stresses in Composite Materials, *Acta Metall.*, **23**, pp. 821–30.
13. T. J. Warner and W. M. Stobbs (1989) Yield Stress Anisotropy of Short Fibre MMCs, *Acta Metall.*, **37**, pp. 2873–81.
14. W. A. Spitzig, J. F. Kelly and O. Richmond (1985) Quantitative Characterisation of Second Phase Populations, *Metallog.*, **18**, pp. 235–61.
15. H. Schwartz and H. E. Exner (1983) The Characterisation of the Arrangement of Feature Centroids in Planes and Volumes, *J. Micros.*, **129**, pp. 155–70.
16. W. H. Hunt Jr., O. Richmond and R. D. Young (1987) Fracture Initiation in Particle Hardened Materials with High Volume Fraction, in *Proc. ICCM VI/ECCM2*, London, F. L. Matthews *et al.* (eds.), **2**, pp. 209–23.
17. E. E. Underwood (1989) Recent Advances in Quantitative Fractography (1989) in *Fracture Mechanics: Microstructure and Micromechanics*, S. V. Nair *et al.* (eds.), ASM, pp. 87–109.
18. I. Sinclair and J. F. Knott (1990) Fatigue Crack Growth in Al–Li/SiC_P MMCs, in *Proc. ECF8*, Turin, D. Firrao (ed.), pp. 303–9.

11.8 Archimedean densitometry

11.8.1 Basic principle and capability

Accurate measurement of the density of a sample can be used to evaluate the pore content, provided the density of corresponding fully sound material is known. Measurements of porosity level are possible to within 0.5% (absolute value), while successive measurement during loading facilitates the monitoring of cavity formation (0.1% changes). On the other hand if the composite is fully dense, reinforcement content can be evaluated (~ ±1%, depending on the relative particle/matrix densities); this is especially useful for sprayed material (see §9.1.2), because the volume fraction of reinforcement actually incorporated into the composite is often unknown.

The most reliable method of density measurement[1-3] simply involves weighing the sample in air and in another fluid of known density. Application of Archimedes' principle leads to the following expression for the density (ρ) of the sample in terms of measured weights (W)

$$\rho = \left(\frac{W_a \rho_L - W_L \rho_a}{W_a - W_L} \right) \tag{11.8.1}$$

where the subscripts a and L refer to air and the second fluid (normally a liquid). In order to improve the precision, this liquid should have a high density. It must also be chemically stable, have a low vapour pressure and a low well-defined surface tension. This latter property ensures that the liquid will flow up the sides of the suspension wire at the air–liquid interface and give a zero contact angle. Di-ethyl phthalate has been used in the past, but the liquid offering the best combination of properties for this application[4,5] is probably perfluoro-1-methyl decalin. A controlled temperature weighing environment is desirable, to minimise fluctuations in buoyancy forces. A balance with a sensitivity of ±10 μg or better is normally required. For a typical specimen weighing about a gram, this will in practice allow porosity levels down to about 0.3–0.5 vol% to be measured routinely.

A problem can arise with use of this method to measure porosity if any of the pores are surface-connected. The liquid may in this case progressively enter the porosity network during weighing, giving variable readings. A possible solution is to coat the sample with a thin layer of a lacquer which is impervious to the liquid. A weighing system with two scale pans in series[5] is then used, one above the liquid and one immersed in it. This assembly is weighed (a) with no sample, (b) with the uncoated sample on the upper scale pan, (c) with the coated sample on the upper scale pan and (d) with the coated sample immersed in the liquid on the lower scale pan. The density ρ of the specimen is given[5] from the recorded sample weights W by

$$\rho = W_{\text{ua}}\left[\left(\frac{W_{\text{ca}} - W_{\text{cL}}}{\rho_{\text{L}} - \rho_{\text{a}}}\right) - \left(\frac{W_{\text{ca}} - W_{\text{ua}}}{\rho_{\text{c}} - \rho_{\text{a}}}\right)\right]^{-1} + \rho_{\text{a}} \qquad (11.8.2)$$

where the subscripts a, L, u and c refer respectively to air, liquid, uncoated and coated respectively. The density of the coating, ρ_{c}, must be known, but a very high precision is not essential for this, provided the coating is relatively thin.

11.8.2 Application to MMCs

The method can only be used to reveal porosity content when the density of the corresponding sound material is accuratey known. For a composite, the content and density of the reinforcement are therefore required. Although this is often known, there may be cases when there is some uncertainty about both the precise ceramic content and the level of porosity. A possible solution is to estimate the ceramic content metallographically. While this is often unreliable, it is preferable to any attempt to measure porosity in this way; it is well-established that smearing of matrix over pores during mechanical polishing makes this procedure very inaccurate and even chemical polishing does not reliably reveal all pores, particularly if they are relatively small. For cases where porosity and reinforcement content are both unknown, it may be preferable to measure the specimen density before and after a consolidation procedure, such as hot isostatic pressing, designed to eliminate all porosity[6]. This will allow both reinforcement and porosity contents to be identified. Evidently, the density of the matrix and of the reinforcement when fully sound must be accurately known. There are some cases where particular attention should be paid to the reinforcement in this regard; for example, carbon fibres are available with a relatively wide range of density, while CVD monofilaments may have different types of core and some reinforcements, such as alumina, can appear in forms having different phase constitutions and hence different densities.

Pros	Cons
√ non-destructive	× no size, shape or distribution information
√ can give porosity and ceramic contents	× difficult to apply to small volumes
√ can follow changes in a given sample	× sensitive balance needed
√ fairly quick and simple to carry out	× very precise true densities needed
√ can study surface-connected porosity	

References

1. N. A. Pratten (1981) Review: The Precise Measurement of the Density of Small Samples, *J. Mat. Sci.*, **16**, pp. 1737–47.
2. H. A. Bowman and R. M. Schoonover (1967) Procedure for High Precision Density Determination by Hydrostatic Weighing, *J. Res. Nat. Bur. Stand.*, **71C**, pp. 179–98.
3. D. Cawthorne and W. D. J. Sinclair (1972) An Apparatus for Density Determination on Very Small Solid Samples, *J. Phys. E.*, **5**, pp. 531–3.
4. W. A. Oddy and M. J. Hughes (1971) The Specific Gravity Method for the Analysis of Gold Coins, in *Methods of Chemical and Metallurgical Investigation of Ancient Coinage, Symp., 9–11 Dec.*, E. T. Hall and D. M. Metcalfe (eds.), Royal Numismatic Soc.
5. A. F. Whitehouse, C. M. Warwick and T. W. Clyne (1991) The Electrical Resistivity of Copper Reinforced with Short Carbon Fibres, *J. Mat. Sci.*, **26**, pp. 6176–82.
6. C. A. Lewis, W. M. Stobbs and P. J. Withers (1991) Transformation-Induced Stress Relief in Metal Matrix Composites, *Metal Matrix Composites – Processing, Microstructure and Properties, Risø 12th Int. Symp.*, N. Hansen, D. J. Jensen, T. Leffers, H. Lilholt, T. Lorentzen, A. S. Pedersen, O. B. Pedersen and B. Ralph (eds.), pp. 483–8.

11.9 Thermal and electrical conductivity

11.9.1 Basic principle and capability

The aim is to measure the thermal or electrical conductivity to high accuracy. In practice, this usually involves burst or pulsed flow in the case of heat conduction, while for electrical conductivity steady state current flow is employed. Typical measurements range from <10 *to* >400 $W\,m^{-1}\,K^{-1}$ *to* $\pm 1\%$ *precision for thermal conductivity, and from* <2 *to* >200 $\mu\Omega$ *cm to* $\pm 1\%$ *precision for electrical conductivity.*

The thermal conductivity of a material is the proportionality constant (K_{ij}) between a thermal gradient and the associated heat flux (§8.1)

$$q_i = -K_{ij} T'_j \tag{11.9.1}$$

Direct measurement requires the establishment of a well-defined, constant gradient; it is difficult to establish such a steady state quickly and without lateral heat exchange leading to variations in this gradient with distance. It is therefore common to use unsteady methods to measure the thermal diffusivity (given by the conductivity divided by the volume specific heat), which is a measure of the rate at which a material will come into thermal equilibrium with its surroundings. The most convenient method is the laser flash technique[1,2], in which one face of a relatively thin, parallel-sided disc is subjected to a short duration (~1 ms) heat pulse and the subsequent thermal history of the other face is carefully recorded – usually with an infra-red pyrometer. The thermal diffusivity can be inferred from this history. The temperature changes are relatively small, so that variations in the thermal properties with temperature are not important. The sample is usually of relatively large cross-section, ensuring that heat flow is essentially unidirectional. The specific heat of the material must be obtained separately, but this is not normally a problem as this property is a scalar (non-directional) and is not usually sensitive to microstructure. If there are a number of phases present, then the effective specific heat should be given by a weighted mean of those of the individual phases. In many cases, the data required can be obtained from standard handbooks such as Touloukian and Ho[3].

Electrical conduction conforms to the same mathematical laws as thermal conduction, with the analogous equation to eqn (11.9.1) usuallly written as

$$j_i = -\sigma_{ij} E_j \tag{11.9.2}$$

in which j is the current density, σ is the conductivity and E is the electric field strength. An advantage in measuring electrical, as opposed to thermal, conductivity is that a selected electric field strength is very rapidly established in the specimen and, because materials having effectively zero electrical conductivity are readily available, lateral current leakage is easily avoided. The conductivity (which is often expressed as its reciprocal, the resistivity) can therefore be directly measured in a steady state more conveniently than in the thermal case. Most techniques involve measurement of the potential difference across two planes, between which a known current density is flowing. The main difficulties arise from potential drops within the circuit designed to measure this difference. These can be overcome[4,5] by passing a fixed frequency AC current through the specimen and identifying the component of the total potential difference within the sensing circuit which is modulated at this frequency, thus eliminating the effects of contact resistance and other extraneous resistances. Measurements are quick and easy to make, but problems of electrical noise and thermal drift become more troublesome as the specimen resistance falls; for low resistivity materials, it may therefore be desirable to keep the cross section small and/or increase the length of the specimen. Higher temperatures often cause problems of increased thermal noise in the electrical circuit and difficulties in finding suitable materials for the insulating sample support arrangement.

11.9.2 Application to MMCs

Thermal measurements on MMCs are readily made using the laser flash method and specimen preparation is relatively undemanding. For very high conductivity materials, thicker specimens are desirable, but even measurements on copper-based MMCs do not require inconveniently thick samples. In order to analyse the behaviour, conductivity data are required for both matrix and reinforcement. Matrix values can be quite sensitive to the exact composition[6]. Complications can arise for some reinforcements, particularly ceramics having relatively high conductivities as a result of heat transfer by phonons. Phonon scattering at grain or phase boundaries may affect the conductivity, so that errors may arise from differences in the constitution and size of the ceramic within the composite compared with that in a bulk conductivity sample. For discontinuous reinforcements carrying a significant heat flux, the nature of the interface will be important.

Electrical measurements on MMCs usually present few difficulties, although there may be problems with transverse measurements on high conductivity materials in obtaining samples long enough to give a sufficiently high resistance. In making comparisons with data for the unreinforced matrix, there is a dependence on microstructural features such as dislocation density, vacancy concentration and grain size, but the associated changes in resistivity are relatively small[4,7,8]. In the vast majority of cases, the reinforcement will carry no current density and so will act as holes. Electrical properties of the interface should therefore have no effect, although it has been suggested[5] that electron scattering at the interfaces will raise the resistivity; this would probably only be significant with exceptionally fine scale reinforcement.

Examples:

- Thermal conductivity of long fibre MMCs[9]
- Thermal conductivity of particulate MMCs and interfacial effects[10,12]
- Electrical conductivity of long fibre MMCs[5,9,11]
- Electrical conductivity of short fibre MMCs[4].

Pros	Cons
Thermal	
√ quick to carry out	× equipment is fairly complex
√ easy to measure over wide range of T	× need to obtain specific heat data
√ non-destructive	× sample must be parallel-sided
√ can measure in different directions	
√ sample can be very short in the measurement direction	
Electrical	
√ very quick to carry out	× difficult for elevated temperature
√ equipment is relatively simple	× sample cannot be very short
√ non-destructive	× sample must be parallel-sided
√ can measure in different directions	

References

1. W. J. Parker, R. J. Jenkins, C. P. Butler and G. L. Abbott (1961) Flash Method of Determining Thermal Diffusivity, Heat Capacity and Thermal Conductivity, *J. Appl. Phys.*, **32**, pp. 1679–84.
2. R. Taylor (1980) Construction of Apparatus for Heat Pulse Thermal Diffusivity Measurements from 300–3000 K, *J. Phys. E. Sci. Instr.*, **13**, p. 1193.
3. Y. S. Touloukian and C. Y. Ho (1973) *Thermophysical Properties of Solid Materials*, vol. **1–16**, Plenum, New York.
4. A. F. Whitehouse, C. M. Warwick and T. W. Clyne (1991) The Electrical Resistivity of Copper Reinforced with Short Carbon Fibres, *J. Mat. Sci.*, **26**, pp. 6176–82.
5. D. Abukay, K. V. Rao and S. Arajs (1977) Electrical Resistivity of Aluminium–Boron Composites between 78 K and 400 K, *Fibre Science & Technology*, **10**, pp. 313–18.
6. P. G. Klemens and R. K. Williams (1986) Thermal Conductivity of Metals and Alloys, *Int. Met. Reviews*, **31**, pp. 197–215.
7. H. G. van Bueren (1955) Electrical Resistance and Plastic Deformation of Metals, *Z. Mattalk.*, **46**, pp. 272–81.
8. D. K. Crampton, H. L. Burghoff and G. T. Stacey (1941) Effect of Cold Work upon Electrical Resistivity of Copper Alloys, *Trans AIME*, **143**, pp. 228–45.

9. K. Kuniya, H. Arakawa, T. Kanani and A. Chiba (1987) Thermal Conductivity, Electrical Conductivity and Specific Heat of Copper–Carbon Fibre Composite, *Trans. Jap. Inst. Met.*, **28**, pp. 819–26.
10. A. J. Reeves, R. Taylor and T. W. Clyne (1991) The Effect of Interfacial Reaction on Thermal Properties of Titanium Reinforced with Particulate SiC, *Mat. Sci. and Eng.*, **A141**, pp. 129–38.
11. S. DeBondt, L. Froyen and A. Deruyttere (1988) Electrical Conductivity of Metal Matrix Composites, in *Mechanical and Physical Behaviour of Metallic and Ceramic Composites, 9th Risø Int. Symp.*, S. I. Anderson, H. Lilholt and O. B. Pedersen (eds.), Risø Nat. Lab., pp. 345–8.
12. F. H. Gordon and T. W. Clyne (1991) Transport Properties of Short Fibre SiC-reinforced Ti, in *Metal Matrix Composites – Processing Microstructure and Properties, 12th Risø Int. Symp.*, N. Hansen, D. Juul-Jensen, T. Leffers, H. Lilholt, T. Lorentzen, A. S. Pedersen, O. B. Pedersen and B. Ralph (eds.), Risø Nat. Lab., pp. 361–6.

11.10 High temperature extensometry

11.10.1 Basic principle and capability

The aim is very straightforward; to measure displacement (extension) to ± 1 *μm precision at temperatures up to at least 1500°C. There are a number of techniques available for the measurement of strain over a range of temperature. These may be divided into contacting and non-contacting methods. Non-contacting methods, often utilising optical devices, are attractive when high temperatures are involved and these are becoming increasingly sophisticated and popular.*

The main contacting methods are strain gauges, clip gauges and Linear Variable Displacement Transducers (LVDTs). Of these, the first is not applicable to high temperature and the second can give problems as a result of high thermal gradients in the clip arms and specimen attachment difficulties. The use of LVDTs is common in dilatometers, the main problem being compensation for expansion of the rod linking specimen and transducer. This is usually tackled by using low expansivity material, such as silica, for this rod and compensating for the effect by measuring length changes relative to a tube of the same material supporting the specimen. Contacting methods are generally problematic when the specimen is being loaded at elevated temperature. High precision sensing through the grips and loading train can be very difficult when these are subject to significant thermal expansion.

Optical methods offer potential for direct sensing of the specimen strain. These methods may be sub-divided into interferometric and scanning devices. The former are based on beam splitting and interference after

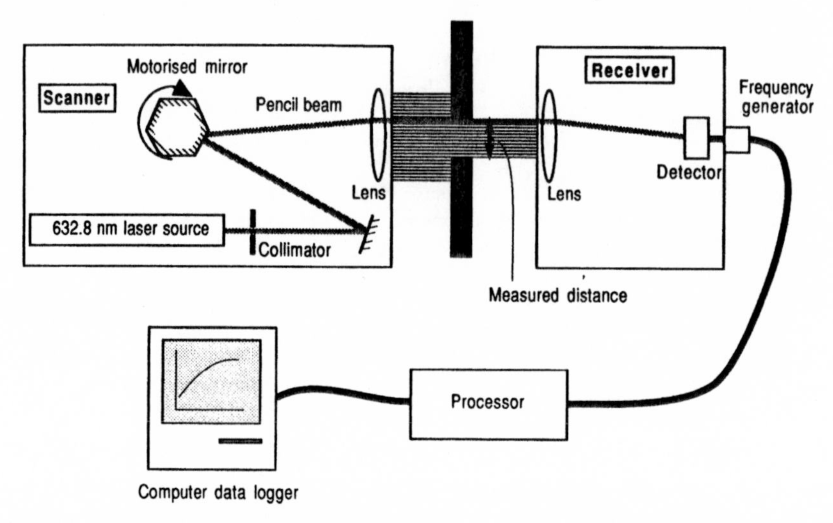

Fig. 11.10.1 A scanning laser extensometer obviates the need for contact with the specimen under observation. Instead, the distance between two beam obstructing tags, which are either part of, or are directly attached to, the specimen, is continuously monitored by the laser beam.

recombination. Various techniques are used to resolve and count fringe patterns. Four interferometric designs applied at high temperature are those of Moiré[1], Fabry-Perot[2], Michelson[3] and Fizeau[4]. These techniques have a basic resolution of about $\lambda/2$, where λ is the wavelength of the light used to form the interference pattern; the Moiré method is potentially more sensitive, but it is not well-suited to elevated temperature operation. The need for mirrors bonded to the specimens tends to make the setting up and alignment operation rather delicate for all of these methods. In addition, thermal cycling conditions can lead to stability problems with the measuring system.

Scanning optical techniques tend to be more robust, but with comparable accuracy to interferometric devices. A typical arrangement[5] is shown in Fig. 11.10.1. A laser pencil beam scans at a well-defined velocity and the elapsed time between this beam passing two protrusions on the gauge length of the specimen is measured to high precision. The specimen can also be scanned transversely, eliminating the need for any protrusions. This technique is particularly well suited to thermal cycling creep measurements. Some general points about temperature gradients within specimens and optimum methods of heating and cooling under such conditions have also been identified[5,6].

Pros	Cons
Contacting	
√ equipment can be relatively cheap	× often unsuited to high temperature × errors from component expansion × sensitive to ambient fluctuations
Non-contacting interferometric	
√ high precision possible	× equipment is relatively expensive × delicate to set up × sensitive to ambient fluctuations
Non-contacting scanning	
√ high precision possible √ fairly easy to set up √ not sensitive to ambient fluctuations √ transverse measurements easy to make	× equipment is relatively expensive

References

1. D. E. Bowes, D. Post, C. T. Herakovich and D. R. Tenney (1981) Moiré Interferometry for Thermal Expansion of Composites, *Experimental Mechanics*, **21**, pp. 441–8.
2. V. E. Bottom (1964) Fabry-Perot Dilatometer, *Rev. Sci. Instrum.*, **35**, pp. 364–6.
3. E. G. Wolff and S. A. Eselun (1979) Double Michelson Interferometer for Contactless Thermal Expansion Measurement, in *Proc. of SPIE, 192, Interferometry*, pp. 204–8.
4. S. S. Tompkins (1989) Techniques for Measurement of the Thermal Expansion of Advanced Composite Materials, in *Metal Matrix Composites: Testing, Analysis and Failure Modes*, ASTM STP 1032, W. S. Johnson (ed.), pp. 54–67.
5. J. A. G. Furness and T. W. Clyne (1991) The Application of Scanning Laser Extensometry to Explore Thermal Cycling Creep of Metal Matrix Composites, *Mat. Sci. & Eng.*, **A141**, pp. 199–207.
6. G. A. Hartman and S. M. Russ (1989) Techniques for Mechanical and Thermal Testing of Ti_3Al/SCS-6 Metal Matrix Composites, in *Metal Matrix Composites: Testing, Analysis and Failure Modes*, ASTM STP 1032, W. S. Johnson (ed.), pp. 43–53.

11.11 Damage event detection by acoustic emission

11.11.1 Basic principle and capability

The principle is simple; to 'listen' to damage events taking place upon loading. Interpretation, in terms of the different damage mechanisms, is more difficult. However, the acoustic emission technique offers the potential to monitor damage accumulation in real time, and to distinguish between different micro-damage mechanisms and to establish their spatial and temporal coordinates.

Equipment for acoustic emission (AE) measurements is commercially available, and has long been used for polymer matrix composites[1]. The spatial location of the event can be identified approximately by the time delay at each of two or more sensors. Post-test analyses of event amplitude, frequency, and duration facilitate the identification of different sources of AE, while spatial filtering enables one to monitor a preselected gauge volume[2] free from gripping noise. A highly sensitive resonance type transducer (~200–400 kHz) is recommended for amplitude and energy analysis, whilst a wide-band transducer must be used for frequency analysis. Waveguides can be used to improve transducer contact with curved surfaces.

11.11.2 Application to MMCs

The objective is concisely stated by Awerbuch and Bakuckas[2]; 'A major challenge in the application of the AE technique has been to identify the specific failure mechanisms and to monitor the failure processes in composites subjected to external stimulation.' The aim is to seek a correspondence between the AE characteristics of the signal [amplitude, duration, energy (essentially amplitude times pulse duration), frequency spectra and counts per event] and the different modes of failure that are known to occur in a composite. The recorded amplitude is the most commonly used characteristic, but it should be remembered that this is related to the test geometry and the propagating medium, as well as to the position of initial event, so that its value has no absolute significance[3]. For plane wave propagation (e.g. through a parallel-sided waveguide or, to a good approximation, with a set-up similar to that in Fig. 11.11.1), attenuation of the signal amplitude with distance is small (~few dB/m in metals). This means that similar events will give similar amplitudes whatever their location within the gauge volume. (Note that when precise measurement of amplitude or energy is required, attenuation must be taken into account.) Frequency-based analysis has the advantage that it is insensitive to the sample sensor distance or the thickness of the coupling medium, but since each mechanism tends to have a complex frequency spectrum signature which is dependent on specimen

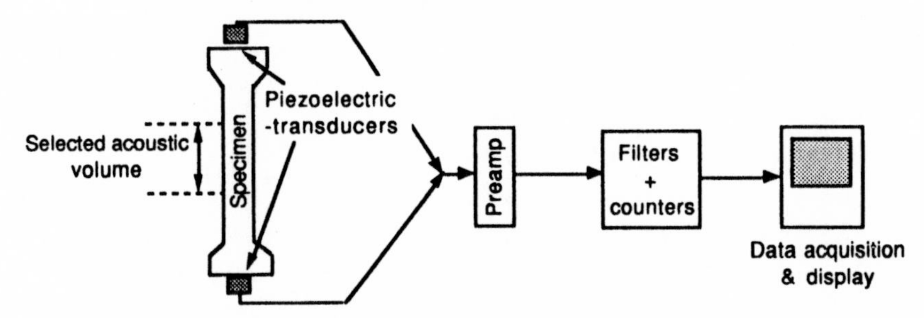

Fig. 11.11.1 Schematic diagram of a typical acoustic emission system.

geometry, the identification of specific mechanisms in this way is more difficult.

Application to long fibre composites

Interpretation of the acoustic emission can be made on an ***amplitude*** basis†; for example, for Al–Al_3Ni eutectic (Al_3Ni fibres), characteristic amplitudes were found[2] to be associated with the following events:

Friction generated interfacial failure, fracture surface fretting, shear deformation	Plastic deformation and cracking of matrix	Fibre breakage
40–65 dB	65–90 dB	>90 dB

Another important aspect is the sequence of the various damage events; for example, in the above work, mid-range events were found to occur initially, followed by (localised) high amplitude fibre fracture events at higher strains[2]. ***Frequency analysis*** has also been used to characterise damage, for example, Nakanishi *et al.*[3] found that for resin–$CaCO_3$ filler–glass fibre polymer matrix systems, the frequency ranges 80–180 kHz, 180–250 kHz, 240–450 kHz and 250–400 kHz were associated with resin cracking, fibre debonding of $CaCO_3$ particles and their cracking and fibre breakage, respectively. A similar study of the loading of Al–60 vol% SiC (Nicalon®) unidirectional composites[4] indicated that

† Characteristic amplitudes cannot be ascribed to specific events universally, but must be identified separately in every case. The measured amplitude is often given in dB with the 10 μV (say) peak amplitude at the pre-amp input arbitrarily set to 0 dB.

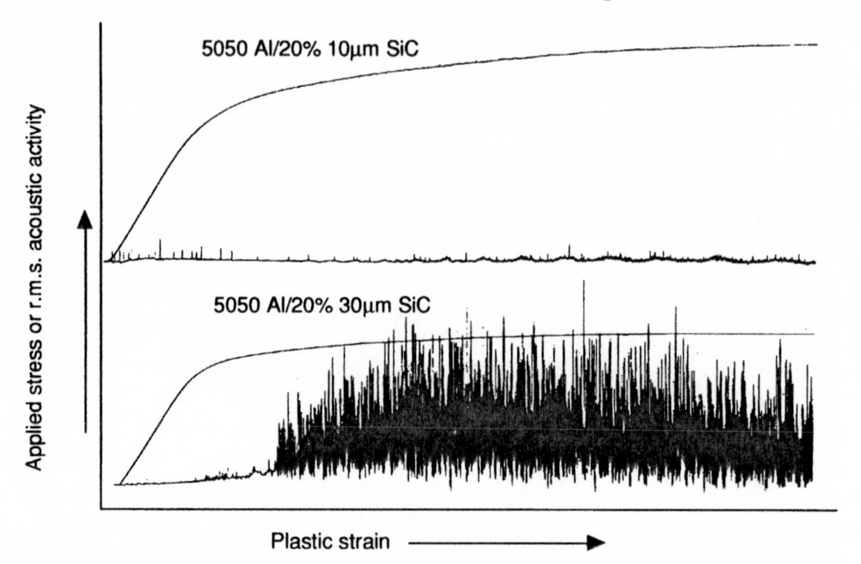

Fig. 11.11.2 Root mean square acoustic activity and the corresponding stress–strain curves for 20 vol% SiC particle (10 μm (upper) and 30 μm (lower))/5050 Al composites[6]. Note that emissions were detected throughout loading and correspond to both interfacial decohesion and particle cracking: no emissions above threshold were detected for similar 3 μm containing material.

in this case fibre breakage and interface debonding were associated with frequencies of 468, 300, 89 and 491, 141, 100 kHz respectively.

Application to discontinuous composites

As with long fibre composites, interpretation is usually based on the idea that high amplitude events correspond to reinforcement failure, while interface failure and plastic deformation correspond to lower amplitude events[5] (Fig. 11.11.2).

Examples:

- unreinforced Al: plastic deformation[7]
- long fibre; fracture (B–Al, SiC–Al, SiC–Ti–6Al–4V[2], SiC–Al[4], Al–Al_3Ni)[8], stress corrosion cracking[9] (successful operating conditions[5]: threshold 1 V, gain 40 dB pre-amp, post-amp gain 40 dB, dead time 3 ms, envelope 10 ms)
- discontinuous; fracture (Al–SiC_P[6,10], Al–SiC_W[5,10]).

Pros	Cons
√ relatively cheap and easy to undertake	× restrictions on specimen size and shape
√ facilitates detection of damage accumulation within bulk specimens	× characteristic frequencies, amplitudes, etc., not universal and must be identified in individual cases
	× a number of mechanisms have similar acoustic signatures

References

1. R. G. Liptai (1972) Acoustic Emission from Composite Materials, in *Composite Materials: Testing and Design*, ASTM STP 497, pp. 285–98.
2. J. Awerbuch and J. G. Bakuckas (1989) On the Applicability of Acoustic Emission for Monitoring Damage Progression in MMCs, in *Metal Matrix Composites: Testing, Analysis and Failure Modes*, Philadelphia, ASTM STP 1032, W.S. Johnson (ed.), pp. 68–99.
3. H. Nakanishi, M. Suzuki, M. Iwamoto, E. Jinen, Z. Maekawa and K. Koike (1987) A Study on Fracture Mechanisms of Class-A SMC by AE Method, in *Proc. ICCM VI/ECCM2*, London, F. L. Matthews *et al.* (eds.), Elsevier, **1**, pp. 395–404.
4. S. De Bondt, L. Froyen and A. Deruyttere (1988) Monitoring Failure Processes in Unidirectional Al–SiC Composite by AE, *Mech. & Phys. Behav. of Met. & Ceramic Mat. Composites, 9th Risø Symp.*, S. I. Andersen, H. Lilholt and O. B. Pedersen (eds), pp. 339–43.
5. J. Awerbuch, J. Goering and K. Buesking (1988) Minimechanics Analysis and Testing of Short Fibre Composites: Experimental Methods and Results, in *Testing Tech. of MMCs*, Philadelphia, P. R. DiGiovanni and N. R. Adsit (eds.), pp. 121–42.
6. P. M. Mummery, B. Derby, D. Buttle and C. B. Scruby (1991) Micromechanisms of Fracture in Particle-Reinforced MMCs: Acoustic Emission and Modulus Reduction, in *Proc. Euromat '91*, T. W. Clyne and P. J. Withers (eds.), Cambridge, UK, **2**, pp. 441–7.
7. J. R. Frederick and R. K. Felbeck (1972) Dislocation Motion as a Source of AE, in *Acoustic Emission*, ASTM STP 505, pp. 129–39.
8. D. O. Harris, A. S. Tatelman and F. A. Darvish (1972) Detection of Fibre Cracking by AE, in *Acoustic Emission*, ASTM STP 505, pp. 238–49.
9. M. Kumosa (1987) Acoustic Emission Monitoring of Stress Corrosion Cracks in Aligned GRP, *J. Phys. D: Appl. Phys.*, **20**, pp. 69–74.
10. C. Johnson, K. Ono and D. Chellman (1985) Acoustic Emission Behaviour of MMCs, *2nd Int. Conf. on AE*, Nevada USA, 28 Oct.–2 Nov., pp. S263–S268.

12
Applications

The commercial exploitation of MMCs is now becoming significant. A number of applications have emerged in which their advantages over traditional materials are such that full-scale substitution is now taking place. In other instances more detailed engineering data are needed and/or manufacturing or design problems need to be resolved. In surveying the scope for commercial usage, it should be recognised that in many cases it is the potential for achieving a desirable suite of properties which makes MMCs so attractive. Nevertheless, it is helpful to examine the potential advantages in turn, and this is done in the first part of the current chapter, with illustrative examples of usage mentioned where appropriate. In the second part, attention is concentrated on several specific applications for which MMCs are either already in use or are undergoing commercialisation.

12.1 Engineering properties of MMCs

12.1.1 Stiffness enhancement

Potential for the enhancement of stiffness, and specific stiffness, is one of the most attractive features of MMCs. Stiffness is a critical design parameter for many engineering components, as the avoidance of excessive elastic deflection in service is commonly the overriding consideration, and the incentive to achieve even a modest increase is often very high indeed. This is the case for many rotating parts, support members, structural bodywork, etc., for which metals offer essential combinations of toughness, formability, environmental stability and strength. However, with very few exceptions (Al–Li being the prime example), there is no scope for increasing the stiffness of a metal by minor additives or microstructural control. In some instances, recourse has been necessary

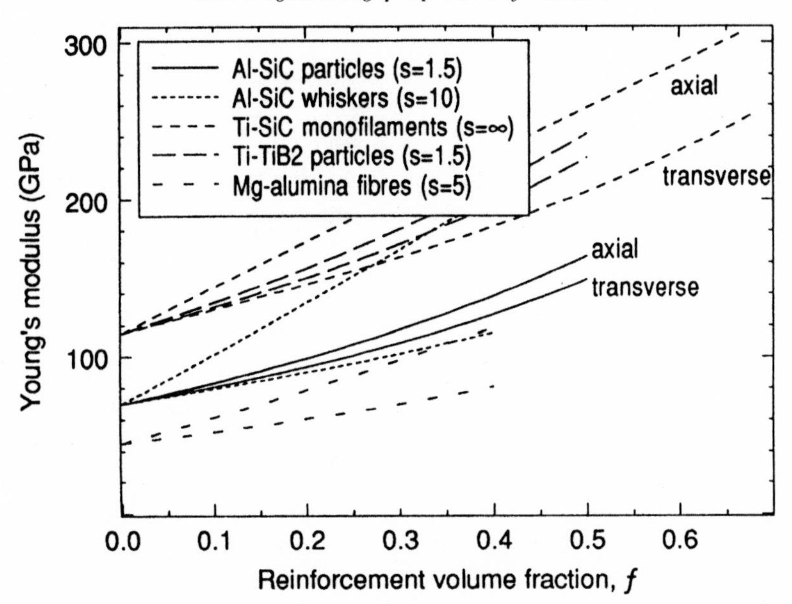

Fig. 12.1 Predicted Young's modulus as a function of reinforcement content for several types of MMC.

to expensive or otherwise difficult metals, such as beryllium – now being used, for example, for rocket and navigational components for spacecraft[1,2], despite its problems of brittleness and toxicity. That significant enhancement of stiffness is achieved in commercially available MMCs can be clearly seen from the data shown in Fig. 12.1.

Examples of applications dependent primarily on stiffness include drive shafts (§12.2.2), instrument racks (§12.2.3), bicycle frames (§12.2.5), inertial guidance spheres for Trident missiles (made of 40% SiC_P/Al, replacing beryllium) and the cross-boom of the 'Sail America' catamaran (entered for the America's Cup), a large component (~10 m long and 45 cm in diameter) requiring high specific stiffness to resist independent flexing of the two sections. This was fabricated from Al–SiC_P by extrusion and is believed to be the largest single component made of MMC material to date.

One of the earliest specific stiffness-critical applications of MMCs occurred in the US Space Shuttle, in which a structural component in the cargo bay section framework was made from a 60% boron monofilament reinforced Al-based composite[3]. This was a small tubular strut, in which the fibres were aligned parallel to the applied load. In fact, the load-bearing skeleton of the Shuttle cargo bay was made up of about 300 of these struts, to give an exceptionally lightweight and stiff structure[4]. Other, less specialised, applications may yet emerge. For example, specific stiffness is reported to be the critical limiting factor in designing the latest high speed railway coaches[5], which are longer than their predecessors and have more stringent dynamic stability requirements.

12.1.2 Strength enhancement

The enhancement of strength by addition of reinforcement, expressed in terms of yield (0.2% proof) stress or failure stress, can be quite substantial (see Chapter 4). In addition, fatigue resistance at low ΔK can be enhanced, although the fracture toughness and ductility are usually reduced (Chapter 7). In general, however, there are relatively few applications where the main attraction of using MMCs lies in the greater strength offered (at least at room temperature). Nevertheless, some applications, such as aircraft landing gear[4], have been identified as requiring light materials with enhanced strength and resistance to low cycle mechanical fatigue (Ti–6Al–4V is commonly used at present) and there may be scope for development of MMCs in such areas. A type of MMC which is potentially quite tough is the so-called ARALL® laminated material[6], made up of alternate layers of Al and polymeric long fibre composite. This is, however, rather limited in terms of component shapes and types of possible application. Furthermore, the strength and, particularly, the toughness are often important in situations where the enhanced stiffness is being exploited by the use of thinner sections, which therefore carry higher stresses.

12.1.3 Increased creep resistance

The enhancement of creep resistance is potentially one of the most significant areas of exploitation of MMCs in that very substantial improvements can result from the addition of fibres, particularly long fibres (§5.2.3). One potential application is in the development of the gas turbine jet engine, where the aim is to replace Ni-based rotor blades and discs towards the back (hot) part of the engine with components made of some lighter substitute material. An obvious candidate would be alloys based on Ti, but their poor creep resistance is a serious handicap. The incorporation of long ceramic fibres into Ti is therefore very attractive, as it would certainly raise the creep resistance substantially and would also give a highly beneficial increase in stiffness (§12.2.9).

12.1.4 Increased wear resistance

While different wear applications[7] require different reinforcement types to achieve optimal wear rate reduction (§8.2.2), it is common for wear rates to be reduced by factors of up to ten by the introduction of the reinforcement. Furthermore, it is often advantageous to control the distribution of reinforcement, so as to provide material of high wear resistance in selected surface areas while other regions are suitably tough, strong, thermally conducting, etc. This can be done by selective reinforcement of critical areas, for example by placement of fibre preforms within a die prior to casting (see §12.2.1) or by the use of composite or graded composition coatings (§9.1.2) or by surface treatment (e.g. friction surfacing[7]). Commonly, it is important for wear resistance to be combined with other properties, such as a high thermal conductivity (to dissipate frictional heat) and a high stiffness (to avoid wear from excessive deflections).

12.1.5 Density reduction

A relatively low density is an attractive feature of many MMC materials. In many cases of interest, addition of the reinforcement raises the density slightly (see Appendix II), but the increase is usually more than offset by enhancement of stiffness, strength, etc. However, a number of instances can be identified where the addition of reinforcing particles or fibres would lower the density markedly. For example, there is an enormous incentive to modify nickel-based superalloys for use in gas turbines so as to reduce their density, provided this can be done without unduly impairing properties such as creep resistance, corrosion resistance and toughness. One of the main problems in doing this by the incorporation of ceramic fibres or particles in Ni, for use at high temperature, is that interfacial attack is difficult to avoid. Furthermore, even without reaction products, the presence of large brittle constituents tends to disturb the complex and highly efficient strengthening mechanisms operative in these materials and to cause significant embrittlement.

A further example is provided by the reinforcement of lead electrodes in rechargeable electrical batteries[8], in which their high density is otherwise a considerable handicap. In this case, care must be exercised so as not to disturb the electrochemical operation of the cell or to risk the release of contaminants into the electrolyte. Finally, there is interest in increasing the unsupported span of electrical transmission lines. While steel-cored lines have been tried, another approach is to use low density carbon fibre-reinforced copper cables, but substantial problems of manufacture, etc., would need to be overcome. While high performance Cu–C fibre composites[2] can be made by electroplating the fibres (§6.4.1), this is unlikely to be viable for large volume production and another method would have to be found. In all of the above examples, considerable research and development is needed before commercial usage is likely to occur.

12.1.6 Thermal expansion control

Just as the differences in stiffness between metals and ceramics can be exploited in the enhancement of stiffness, the low thermal expansivity of ceramics (see Appendix II and Fig. 5.1 in §5.1.1) can be used to tailor the composite expansivity to match that of many different materials (Fig. 12.2). This may be useful for a wide range of applications, from

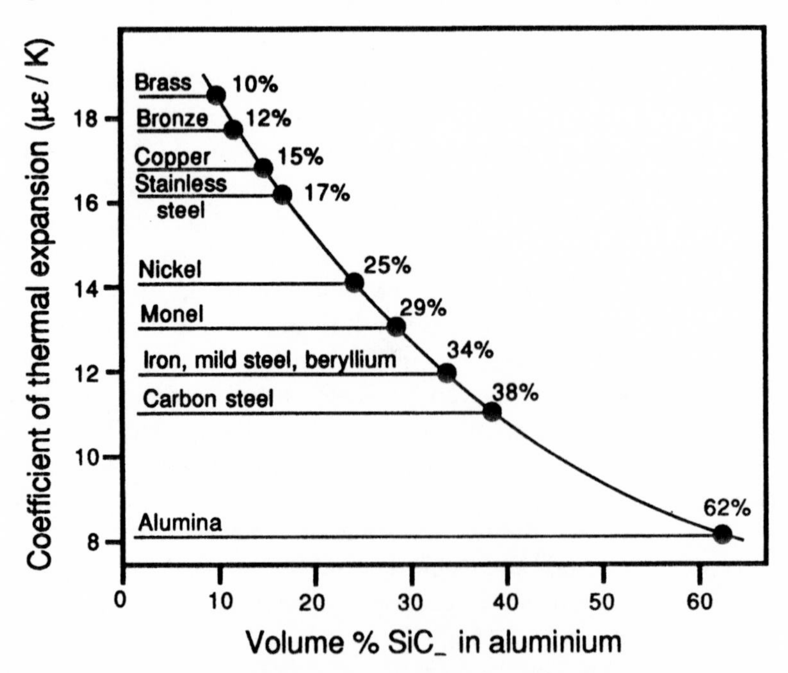

Fig. 12.2 Predicted dependence of the coefficient of thermal expansion on reinforcement content for Al–SiC$_P$ composites, showing how the value can be matched to those of a wide range of materials.

substrates for microelectronic devices (see §12.2.8), where close matching to a ceramic component is desirable, to components in optical platforms and laser mirrors, where distortions must be kept to a minimum while the assembly is exposed to relatively large changes in temperature. Examples of the latter type of application are provided by the satellite boom/waveguide (§12.2.7) and the Al–C fibre antenna support boom structure used on the Hubble Space Telescope[1]. These components require very high axial stiffness whilst exhibiting ultra low (zero) thermal expansivity, so as to retain dimensional stability on passing into and out of direct sunlight.

12.2 Case studies

The potential advantages of MMCs, and hence the types of application for which they are likely to be attractive, have been well documented. For example, several reviews have been published covering general features of their use for applications in the automobile[9–13], aeronautical[14,15], astronautical[1,2,16] and general engineering[4,8,12] fields. While commercial applications are now beginning to emerge, two factors have conspired to keep volume production well below early expectations. Firstly the aerospace industry requires large safety margins, which are difficult to satisfy in view of the relatively low toughness of MMCs, and long materials testing times. This has meant that, at the time of writing, no MMC components are flying in commercial aircraft. Secondly, the automotive industry is highly cost sensitive. This has ruled out continuous fibre systems, and has stimulated considerable advances in cost effective processing of cast Al/particulate and cast Al/fibre preform technology.

In the following sections, attention is concentrated on applications which are currently marketed, or are in the process of becoming commercialised, and serve to illustrate various features concerning the exploitation of MMCs. At the beginning of each section, the reasons for preferring an MMC over more conventional engineering materials, and the composite system being used, are listed.

12.2.1 Diesel engine piston

Al alloy/5% Al_2O_3 short fibre

- good wear resistance
- high temperature strength

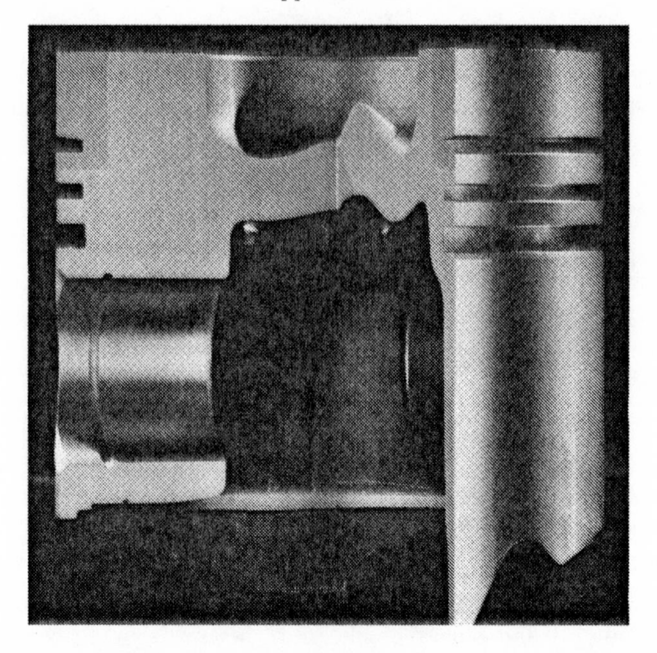

Fig. 12.3 Photograph[21] of a diesel engine piston, showing the region of fibre reinforcement (darker area) in the land (support region) of the groove for the piston ring.

- scope for selective reinforcement
- high thermal stability
- good thermal conductivity

This application represents one of the major early successes in the industrial use of MMCs. Production of these pistons in Japan has been increasing steadily over the past several years and now runs to millions of units annually. Originally, a Ni cast iron (Ni-resist®) insert was used in the piston ring area in order to prevent seizure of the piston ring with the top ring groove and bore. Unfortunately, this impaired heat flow and increased weight and wear rate[17]. In 1983 Toyota Motor Co./Art. Metal Manufact. Co. replaced this with a 5% Al_2O_{3F} short fibre insert, thus reducing the weight by 5–10%. This was achieved by squeeze casting[10,18–20] into an alumina preform to produce a selectively reinforced component (Fig. 12.3). In standard tests, wear was reduced by over four times[17] and seizure stress doubled, relative to the unreinforced Al alloy. This was combined with four times the thermal conductivity of the Ni-resist® insert.

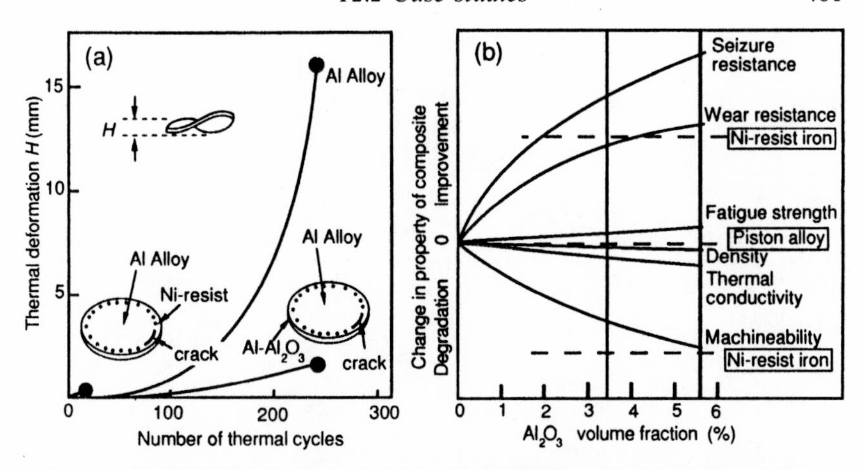

Fig. 12.4 (a) Deformation and cycle life characteristics of the three candidate piston material combinations[17]. (b) The relation between Al_2O_3 fibre volume fraction and piston performance indicators. For reference, the approximate machineability and wear resistance values for the equivalent Ni-resist® insert are shown by the dashed lines.

Another important factor is the thermal fatigue life[22], which is limited by cracking between the ring groove and the piston itself, or by dimensional instability. As illustrated in Fig. 12.4(a), the Al_2O_{3F} insert out-performs both the Ni cast insert and the base alloy. The fibre content selected represents a compromise between improved wear and seizure resistance, combined with good machinability relative to the Ni-resist®, and acceptably small deterioration in fatigue strength and thermal conductivity relative to the unreinforced alloy (Fig. 12.4(b)). Finally, it may be noted that homogeneously reinforced pistons cast from Al/20% SiC_P are also being developed[10].

12.2.2 Automotive drive shaft

Al alloy/20% Al_2O_3 particulate

- high stiffness
- low density
- acceptable toughness

An automobile drive shaft is simply a tube, which transmits the power to the differential, so that it can be distributed to the wheels (Fig. 12.5).

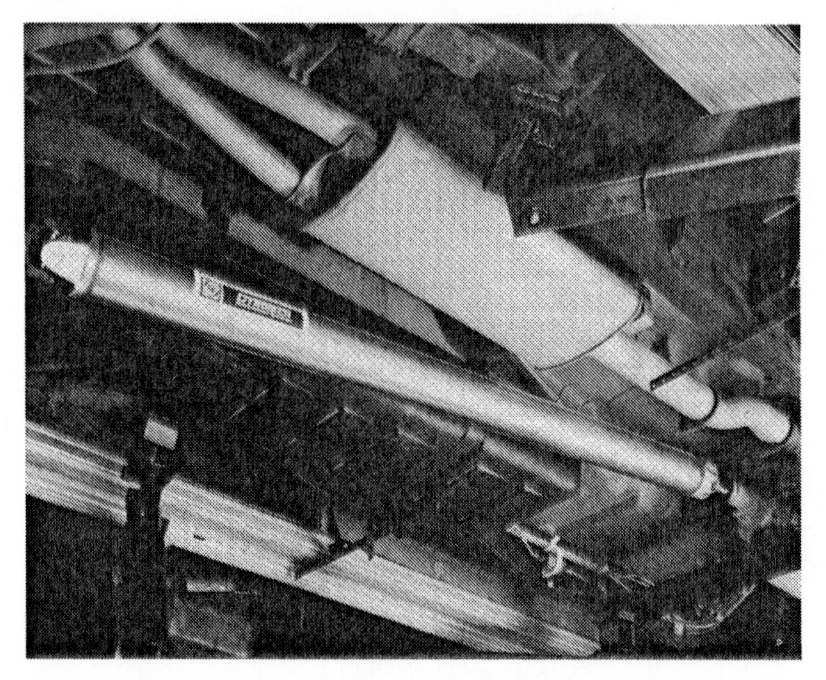

Fig. 12.5 Photograph[23] of a Duralcan® 6061 Al–20 vol% Al_2O_{3P} drive shaft.

This places onerous dynamic stability requirements on the shaft, as well as a need for high torque resistance. The critical velocity at which dynamic instability occurs (ω_*) depends on the shaft length (L), the inner and outer radii (R_i, R_o) of the tube and the stiffness and density of the tube material[23]

$$\omega_* = \frac{15\pi}{L^2}\sqrt{(R_o^2 + R_i^2)}\sqrt{\left(\frac{E}{\rho}\right)} \tag{12.1}$$

Consequently, there is no advantage to be gained in changing from steel to aluminium, since their specific stiffnesses (26.6 $km^2\ s^{-2}$ and 25.9 $km^2\ s^{-2}$ respectively) are very similar (and close to those of Ti and Mg). However, substitution[24] with an MMC such as Duralcan® 6061 Al–20% Al_2O_{3P} leads to a substantial improvement in this value [$(E/\rho) \sim 34.7\ km^2\ s^{-2}$]. From eqn (12.1), this would allow a significant increase (~14%) in the maximum shaft rotation speed. Such a choice has cost and other advantages over an identified alternative of graphite–epoxy wrapped aluminium shafts[23]. The MMC shaft is at present direct extruded by the seamless technique, which involves the penetration of the billet by the mandrel prior to extrusion[24].

12.2.3 Equipment rack

Al alloy/25% SiC particulate

- high stiffness
- low density
- good electrical conductivity

Racks used to house electrical equipment on Lockheed production aircraft provide an example of an application where a high specific stiffness is required, combined with good electrical conductivity. The racks are quite sizeable constructions, as can be seen[25] in Fig. 12.6. The 6061 Al–25% SiC_P composite racks supplied by DWA Composite Specialties Inc. are ~65% stiffer than the 7075 Al racks they replaced[26], which were flexing too much under the gravitational forces caused by the twisting and turning of the aircraft. Carbon–epoxy was tried, but gave earthing problems because their aluminium foil coatings tended to flake off as the components were inserted and removed. The good electrical conductivity of Al-based composites meant that this problem was avoided.

12.2.4 Brake rotor disc

Al–Si alloy/20% SiC particulate

- good wear resistance
- low density
- high thermal conductivity

Traditionally, brake rotor discs are made from cast iron, which has good wear resistance and thermal stability. However, a low density is particularly desirable in rotating components of this type, making replacement with Al MMCs highly attractive. Obviously, excellent wear resistance is required. In October 1991, decisions were made by both Ford and Toyota to adopt Duralcan® cast Al–10% Si/20% SiC_P for brake rotor discs, although various trials will delay their introduction until after 1995. A photograph[27] showing a disc made from MMC is shown in Fig. 12.7. An inclusion content of 20% was chosen to combine good wear resistance (increases with SiC content – see §8.2.2) with a thermally

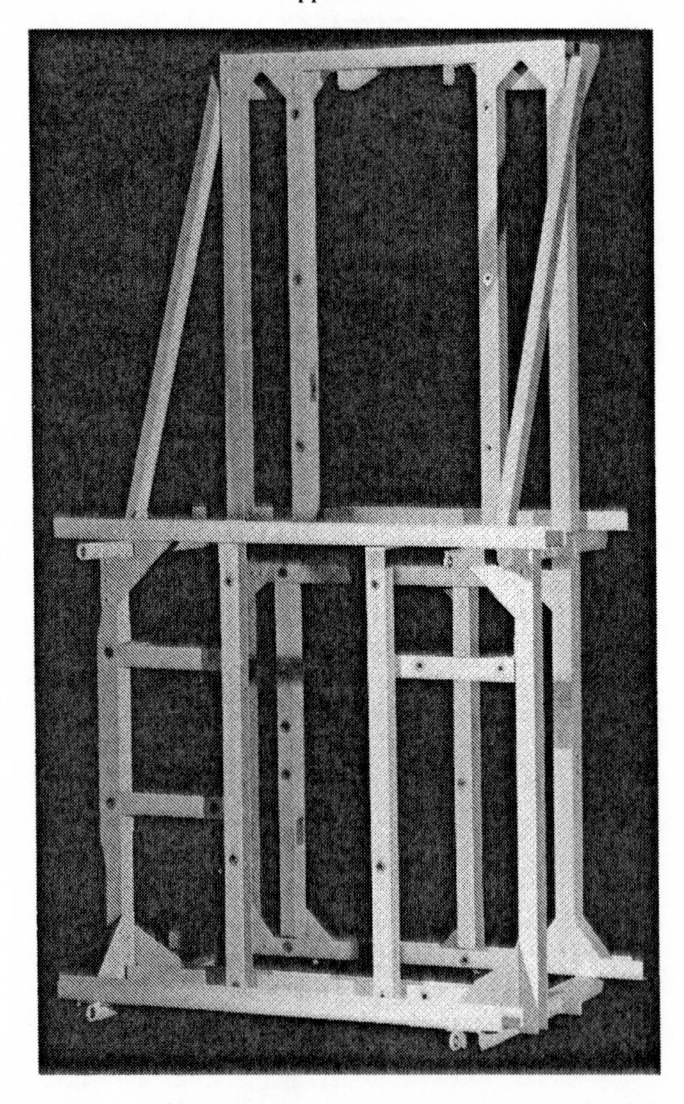

Fig. 12.6 Photograph[25] of an instrument rack (~2 m in length) fabricated from 6061 Al–25 vol% SiC_P for use in Lockheed production aircraft.

and mechanically stable transfer/contact layer. The matrix is over-aged so as to prevent property degradation in-service. The disc is normally cast. However, the air cooling channels complicate the casting process, requiring disposable inserts or the use of lost foam. Asbestos-free inorganic fibres have been used as the friction materials.

Fig. 12.7 Photograph[27] of a rotor brake disc, made from Duralcan® cast Al–10 wt% Si/20 vol% SiC_P.

12.2.5 Bicycle frame

Al alloy/10% Al_2O_3 or 20% SiC particulate

- high stiffness
- low density
- good fatigue resistance

Both Duralcan® and BP have developed bicycle frames for commercial sale. The Duralcan® 6061 Al–10% Al_2O_{3P} composite material is used[24] in the 'Stumpjumper M2' mountain bike manufactured by Specialised Bicycle Components Inc., while BP 2124 Al–20% SiC_P material is used for the frame of Raleigh's racing bikes[25]. Both models have been successfully tested in extensive sports trials. In the former case, tubing of ~1.5 mm wall thickness is extruded via the porthole-bridge-tooling technique, which involves rejoining under pressure around the tube mandrel. The tube sections are then MIG fusion welded, using conventional techniques. In the case of the BP frame, the material is made by a

Fig. 12.8 Photograph[24] of the Stumpjumper M2 bicycle in action. The frame is made of Duralcan® 6061 Al–10 vol% Al_2O_{3P}.

powder route and then the tubing is adhesively joined. As well as improved specific stiffness, both bikes have proven to have exceptionally good fatigue endurance, presumably because of the enhanced value of ΔK_{th} (see §7.3.2) compared with unreinforced material. A photograph[24] of the Stumpjumper bicycle in action is shown in Fig. 12.8.

12.2.6 Engine cylinder block

Al–Si alloy/12% Al_2O_3 short fibre + 9% carbon short fibre

- good wear resistance
- good thermal fatigue resistance
- low density
- good high temperature stability and strength
- strong damping
- thin sections castable

The cylinder block of the internal combustion engine is of major technological significance, being the most substantial component in the product with the greatest annual sales value in the world. A breakthrough in the use of MMCs for this component would signify their arrival as materials of prime industrial importance. Developments are currently taking place which may well lead to such a breakthrough. Cylinder blocks have traditionally been made of cast iron. Aluminium is very attractive as a replacement material, but it has inadequate resistance to fretting, abrasive and erosive wear and to thermal fatigue. Improvements in power/weight ratios have resulted from the use of Al-based blocks with cylinder liners of cast iron. In a logical development of this trend, Honda have built and tested[28,29] Al blocks containing Al–MMC liners. Honda 'Prelude' engines, with a 16-value, 2-litre block, have been cast from a (hyper-eutectic) Al–Si alloy, incorporating hybrid preforms of a mixture of carbon and alumina fibres. Tests[29] have shown that the performance of these engines represents a further significant improvement over that achieved with cast iron liners – see Fig. 12.9. In order to bring this about, a new pressure die casting process, using pressures of about 25 MPa, has been developed for the application[29].

12.2.7 Boom/waveguide for space telescope

Al alloy/60% carbon long fibre

- high axial stiffness
- low density
- ultra low axial thermal expansivity
- good electrical conductivity

The square section tubular component[30] shown in Fig. 12.10 doubles as a structural boom and as a waveguide, for service on a NASA Space

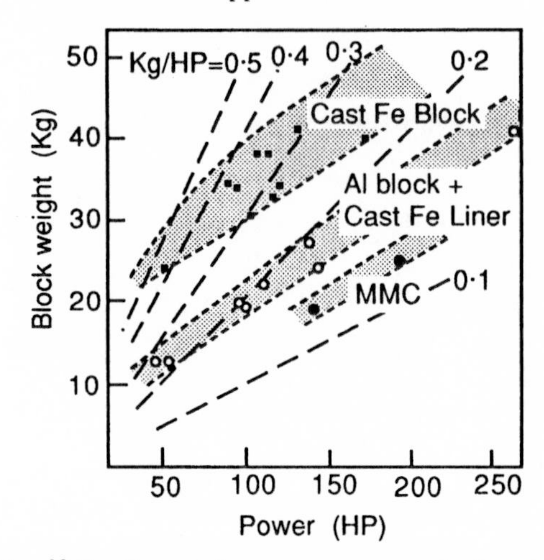

Fig. 12.9 Test data[29] for the maximum power output of various engine blocks, plotted against the mass of the block, for conventional cast iron blocks, for Al alloy with cast iron liners and for Al alloy with integral carbon/alumina fibre reinforced liners.

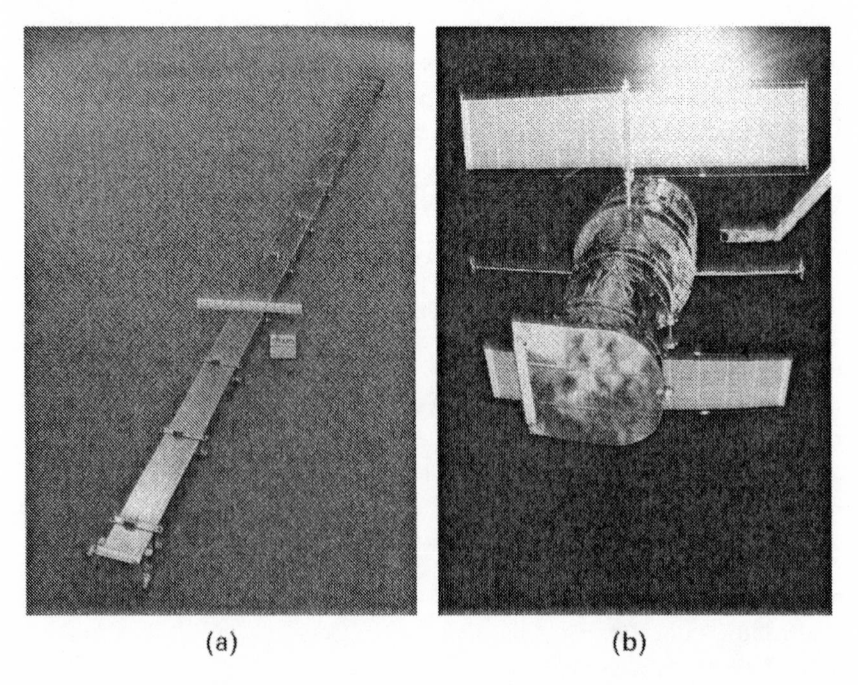

Fig. 12.10 Photograph[30] of a satellite boom/waveguide structure fabricated from an Al–carbon long fibre MMC. The fibres are aligned parallel to the axis of the boom. (b) Satellite assembly.

Telescope satellite. A very high dimensional accuracy and stability is required, under quite substantial temperature changes. The Al–C fibre MMC met all the property goals, while achieving a 30% weight saving compared with a previous design based on an assembly of Al and carbon–epoxy composite materials. One important advantage of metallic, as opposed to polymeric, composites for use in space is their greater environmental stability. This is particularly so under ionising radiations, which tend to cause chemical degradation in polymers.

12.2.8 Substrate/housing for microelectronics package

Al alloy/20–65% SiC particulate

- thermal expansion matched (8 $\mu\varepsilon\ K^{-1}$)
- high thermal conductivity ($>120\ W\ m^{-1}\ K^{-1}$)
- suitable for brazing
- low density
- electrically conducting
- dimensional stability

Many microelectronic devices require highly stable environments. For example, microwave radar and communication systems need to be housed so that they are shielded from stray fields and are mechanically, thermally and electrically stable. Kovar® (high Ni steel) or brazed steel/molybdenum housings have been used[31] to support ceramic (Al_2O_3) substrate electronic packages. These materials have been chosen largely because they give fairly close matching of thermal expansivity with alumina (8 $\mu\varepsilon\ K^{-1}$). However, they are dense and have low thermal conductivities. This latter point has become a major drawback as progressive increases in component density have led to a greater need for effective heat dissipation[32].

The solution has been to use 6061 Al–20% SiC_P (DWA Composite Specialties Inc.) or Al(356)–65% SiC_P (Cercast)[31] composite housings. The latter has a thermal expansivity of 8.2 $\mu\varepsilon\ K^{-1}$ and a thermal conductivity of 150 $W\ m^{-1}\ K^{-1}$. Packaging weight is also reduced[32] significantly ($>65\%$), and machining and brazing distortion minimised ($<\pm 50\ \mu m$). In addition, if cast it is possible to leave an unreinforced region on top of the side-walls to aid welding of the cover onto the housing[31]. Components such as these, including those with thin section walls, can be successfully die cast in these particulate MMCs[27]. There are many similar electronic device substrate applications, with power dissipation requirements, for which the use of MMCs is being investigated. Other types of MMC, such as Cu–carbon fibre[33] are also being explored for these applications.

12.2.9 Aero-engine components

Ti alloy/40% SiC monofilaments

- high temperature performance
- increased strength
- low density
- component simplification
- increased stiffness

Initially (1970s), attention was concentrated on the improvement of creep resistance of rotor blades through the reinforcement of Al with B fibres[34], but poor foreign object damage tolerance proved a serious obstacle. Recently, interest has focused on axisymmetric aero-engine components, many of which are ideally suited to the excellent unidirectional properties that can be achieved with hoop-wound titanium-based long fibre reinforced composites. Research and development programmes aimed at specific components are currently active in several countries, although little technical information is available in the open literature. While firmly at the developmental stage, this case study merits inclusion because of the radical implications for design and the potential for drastic improvements in engine weight and performance.

In a conventional engine, much of the weight consists of mechanical fixings and spacers, which are independently supported and direct flow through the engine. Fig. 12.11 illustrates a series of modifications which could be made so as to reduce weight as a result of the exploitation of the high performance of Ti composites[35]. It should be noted that, as this is a rotating system, weight savings at the periphery are especially helpful because they reduce the necessary radial support. One-piece bladed discs (bliscs) reduce the parasitic weight, but much greater savings are possible if composite hoop-wound ring structures are employed. These eliminate the need for weighty support structures through their high circumferential specific strength and stiffness, combined with acceptable radial properties (~50% lower). The main problems lie in developing technically and economically viable fabrication routes (Chapter 9) and in avoiding impairment of strength and toughness through microstructural degradation. For example, protective fibre coating procedures must be

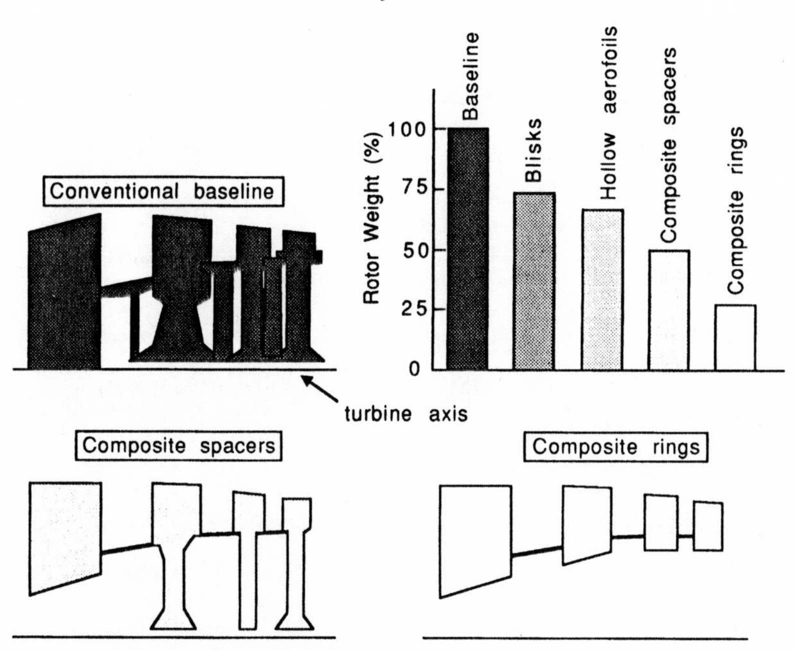

Fig. 12.11 Upon incorporation of around 40% of SiC fibres, the stiffness of Ti can be doubled and its strength improved by 50%. In addition the density is markedly reduced (10%). These increases in performance offer exciting possibilities for engine design and weight reduction. This diagram illustrates that if composite rings and spacers can be developed, weight savings of nearly 75% are possible[35].

devised which offer satisfactory resistance to interfacial degradation in service (§6.3.2).

References

1. J. Dauphin, B. D. Dunn, M. D. Judd and F. Levadou (1991) Materials for Space Application, *Metals & Materials*, **6**, pp. 422–30.
2. T. M. F. Ronald (1989) Advanced Materials to Fly High in NASP, *Adv. Mats. & Procs.*, **6**, pp. 29–37.
3. R. R. Irving (1983) Metal Matrix Composites Pose a Big Challenge to Conventional Alloys, *Iron Age*, **226**, pp. 35–9.
4. R. L. Trumper (1987) Metal Matrix Composites – Applications and Prospects, *Metals & Materials*, **2**, pp. 662–7.
5. R. Ford (1990) Can BR Afford IC250?, *Modern Railways*, **11**, pp. 564–5.
6. L. B. Vogelesang and J. W. Gunnink (1983) ARRALL®, a Material for the Next Generation of Aircraft – a State of the Art, in *High Performance Composite Materials – 4th Int. SAMPE Conf.*, Bordeaux, France, G. Jube, A. Massiah, R. Naslain and M. Popot (eds.), SAMPE, pp. 81–92.

7. P. Chesney (1990) Steel/Ceramic MMCs for Wear Applications, *Metals & Materials*, **6**, pp. 373–6.
8. P. Rohatgi (1990) Advances in Cast MMCs, *Adv. Mats. & Procs.*, **137**, pp. 39–44.
9. F. Folgar, W. H. Kreuger and J. G. Goree (1984) Fiber FP Metal Matrix Composites in Reciprocating Engines, *Ceram. Eng. Proc.*, **5**, pp. 643–8.
10. P. Rohatgi (1991) Cast Aluminium Matrix Composites for Automotive Applications, *J. Metals*, **43**, pp. 10–15.
11. M. R. Jolly (1990) Opportunity for Aluminium Based Fibre Reinforced Metal Matrix Composites in Automotive Castings, *The Foundryman*, **11**, pp. 509.
12. E. A. Feest (1986) Metal Matrix Composites for Industrial Application, *Mater. Design*, **7**, pp. 58–64.
13. M. Hunt (1989) Automotive MMCs: Better and Cheaper, *Materials Engineering*, **10**, pp. 45–50.
14. D. Driver (1985) Developments in Aero Engine Materials, *Metals & Materials*, **1**, pp. 345–54.
15. D. Charles (1990) Ready for Take Off, *Metals & Materials*, **6**, pp. 59–62.
16. K. Marsden (1985) Commercial Potentials for Composites, *J. Metals*, **37**, pp. 59–62.
17. T. Donomoto, K. Funatani, N. Miura and N. Miyake (1983) *Ceramic Fibre Reinforced Piston for High Performance Diesel Engines*, SAE paper 830252.
18. M. W. Toaz, P. R. Bowles and D. L. Mancini (1987) Squeeze Casting Composite Components for Diesel Engines, *Ind. Heat.*, **54**, pp. 17–19.
19. R. R. Bowles, D. L. Mancini and M. W. Toaz (1987) Metal Matrix Composites Aid Piston Manufacture, *Manuf. Eng.*, **98**, pp. 61–2.
20. J. Dinwoodie (1987) *Applications for MMCs Based on Short Staple Alumina Fibres*, SAE paper 870437.
21. E. A. Feest (1988) Exploitation of the Metal Matrix Composites Concept, *Metals & Materials*, **4**, pp. 273–8.
22. M. R. Myers and F. Chi (1991) *Factors Affecting the Fatigue Performance of Metal Matrix Composites for Diesel Pistons*, SAE paper 910833.
23. W. R. Hoover (1991) *DURALCAN Composite Driveshafts*, Duralcan USA, San Diego, California.
24. W. Dixon (1991) Processing, Properties and Applications of DURALCAN Composite Tubes, in *ITA Seminar*.
25. A. Tarrant (1991) Particle Reinforced Aluminium Alloys. Maximising the Benefits – Minimising the Risks, in *Metal Matrix Composites – Exploiting the Investment*, London, Inst. of Metals, paper 6.
26. Unattributed (1989) Materials in Action, *Adv. Mats. & Procs.*, **6**, pp. 22.
27. W. R. Hoover (1991) Die Casting of Duralcan Composites, in *Metal Matrix Composites – Processing, Microstructure and Properties*, Risø, N. Hansen, D. J. Jensen, T. Leffers, H. Lilholt, T. Lorentzen, A. S. Pedersen, O. B. Pedersen and B. Ralph (eds.), Risø Nat. Lab., Denmark, pp. 387–92.
28. T. Hayashi, H. Ushio and M. Ebisawa (1989) *The Properties of Hybrid Fiber Reinforced Metal and its Application for Engine Block*, SAE paper 890557.
29. M. Ebisawa, T. Hava, T. Hayashi and H. Ushio (1991) *Production Process of Metal Matrix Composite (MMC) Engine Block*, SAE paper 910835.
30. J. F. Dolowy, private communication (January 1992).
31. X. Dumant, S. Kennerknecht and R. Tombari (1990) *Investment Cast MMCs*, Cercast, Montreal, EM90-441.

32. C. Thaw, R. Minet, J. Zemany and C. Zweben (1987) MMC Microwave Packaging Components, *J. Metals*, **35** (May), p. 55.
33. K. Kuniya, H. Arakawa, T. Sakaue, H. Minorikawa, K. Akeyama and T. Sakamoto (1986) Application of Copper–Carbon Fibre Composites to Power Semiconductor Devices, *J. Jap. Inst. Metals*, **50**, pp. 583–9.
34. J. F. Garibotti (1978) Trends in Aerospace Materials, *Astronautics and Aeronautics*, **16**, pp. 70–81.
35. D. Driver (1989) Towards 2000 – The Composite Engine, in *Meeting of Australian Aeronautical Society*.

Appendix I

Nomenclature

Parameters

a	(m)	crack length
a	($m^2\ s^{-1}$)	thermal diffusivity
A	(m^2)	cross-sectional area
A	(m^2)	surface area
b	(m)	Burgers vector
B	(–)	wear coefficient
c	($J\ m^{-3}\ K^{-1}$)	volume specific heat
C	(Pa)	stiffness (tensor of 4th rank)
C^{-1}	(Pa^{-1})	compliance (tensor of 4th rank)
d	(m)	fibre/particle diameter
D	(m)	diameter of abrading particle
D	($m^2\ s^{-1}$)	diffusion coefficient
$\bar{D}$	($m^2\ s^{-1}$)	'diffusional mean' diffusion coefficient (eqn (5.20))
E	(Pa)	Young's modulus
E	($V\ m^{-1}$)	electric field strength
f	(–)	reinforcement volume fraction
F	(N)	point load
$\mathcal{G}$	($J\ m^{-2}$)	strain energy release rate
G	($J\ mole^{-1}$)	free energy
G	(Pa)	shear modulus
h	(m)	height
h	($W\ m^{-2}\ K^{-1}$)	heat transfer coefficient
H	(Pa)	hardness
I	(–)	identity matrix
j	($A\ m^{-2}$)	electrical current density
k	($J\ K^{-1}$)	Boltzmann's constant

K	(Pa)	bulk modulus
K	(Pa$\sqrt{m}$)	stress intensity
K_c	(Pa$\sqrt{m}$)	critical stress intensity under mode I loading
K	(W m^{-1} K^{-1})	thermal conductivity
L	(m)	fibre half-length
m	(–)	Weibull modulus
n	(–)	stress exponent
N	(–)	number of loading cycles
N	(m^{-3})	spatial number density
P	(–)	deviatoric component of the mean matrix stress per unit applied stress
P	(Pa)	pressure
q	(W m^{-2})	heat flux
Q	(J $mole^{-1}$)	activation energy
Q	(m^2)	volumetric wear rate
Q	(Pa K^{-1})	deviatoric component of the mean matrix stress per unit temperature change
Q^{-1}	(–)	quality factor
r	(m)	radial distance from fibre axis
r_0	(m)	fibre radius
r_Y	(m)	plastic zone radius
R	(J K^{-1} $mole^{-1}$)	universal gas constant
R	(m)	far field radial distance from fibre axis
R	(W $m^{-1/2}$)	thermal shock resistance figure of merit
s	(–)	fibre aspect ratio ($2L/d = L/r_0$)
S	(–)	Eshelby tensor (tensor of 4th or 2nd rank)
t	(s)	time
T	(K)	absolute temperature
$\bar{T}$	(K)	'diffusional mean' temperature (eqn. (5.20))
T'	(K m^{-1})	thermal gradient
u	(m)	displacement in z direction (fibre axis)
U	(J m^{-3})	energy density
U	(J)	energy absorbed at interface
V	(m^3)	volume
w	(m)	width
W	(N)	load
W	(Pa)	deviatoric component of the mean matrix stress per unit plastic strain
x, y	(m)	distance (transverse)
z	(m)	distance (axial)

α	(K^{-1})	thermal expansion coefficient (tensor of 2nd rank)
δ	(m)	thickness of grain boundary (eqn (5.11))
δ	(m)	crack opening displacement
ε	(–)	strain (tensor of 2nd rank)
ε_g	(–)	strain for growth of voids
ε_n	(–)	strain for nucleation of voids
$\dot{\varepsilon}$	(s^{-1})	elongational strain rate
ϕ	(–)	ratio of 'interplanar spacing' to length of 'crosslinking' fibre (eqns (9.1–9.4))
γ	(–)	shear strain
γ	($J\,m^{-2}$)	surface energy
$\dot{\gamma}$	(s^{-1})	shear strain rate
λ_v	(m)	spacing of voids
μ	(–)	coefficient of sliding friction
κ	(–)	ratio of thermal conductivities
ν	(–)	Poisson's ratio
ρ	(m^{-2})	dislocation density
ρ	($\Omega\,m$)	electrical resistivity
σ	(Pa)	stress (tensor of 2nd rank)
σ	($\Omega^{-1}\,m^{-1}$)	electrical conductivity
τ	(Pa)	shear stress
ξ	(–)	ratio of 'interplanar spacing' to interfibre spacing (eqns (9.1–9.4))

Subscripts

These generally qualify the type or nature of the parameter – e.g. σ_{YM} is the matrix yield stress

0	initial
1	*x* direction
2	*y* direction
3	*z* direction (along fibre axis)
app	apparent
b	boundary
C	composite
CL	crosslinking
com	compression
e	fibre end

eff	effective
esf	effectively stress-free
fr	frictional sliding
F	fibre
G	ghost
H	hydrostatic
I	inclusion (particle, fibre or other reinforcement)
i	interfacial
i	infiltration
loc	local
M	matrix
mp	melting point
p	pipe
p	preform
P	particle
r	radial
RM	Rule of Mixtures
t	stress transfer
ten	tension
th	threshold
trans	transverse
U	failure (ultimate tensile)
v	volume
v	voids
W	whisker
Y	yield (0.2% proof stress often taken)
θ	hoop
*	critical (e.g. debonding or fracture)

Superscripts

These generally refer to the origin of the phenomena – e.g. σ^A is the applied stress

A	applied loading
C	constrained (strain)
$\Delta\nu$	differential Poisson contraction
ΔT	thermal loading (a temperature change)
fh	forest hardening

Orow	Orowan strengthening
P	plastic (strain)
P*	effective plastic (strain)
R	relaxation (strain)
sg	strain gauge
ss	source shortening
T	transformation (strain) for homogeneous (ghost) inclusion
T*	misfit (strain) for real inclusion
XR	X-ray

Prefices and other qualifying symbols

ΔX	finite change in parameter X
X'	differential of X with respect to distance
X''	second differential of X with respect to distance
$\dot{X}$	differential of X with respect to time
$\bar{X}$	volume average of parameter X
$\langle X \rangle$	volume average of parameter X, over and above that in homogeneous material. (For example, the stress in a fibre within a composite under an applied load σ^A is equal to $\sigma^A + \langle\sigma\rangle_I^A$.)

Appendix II

Matrices and reinforcements – selected thermophysical properties

Material	Diameter (μm)	Young's modulus (GPa)	Poisson ratio	Tensile strength (GPa)	Density (Mg/m^3)	Melting point (K)	CTE (με/K)	Thermal conductivity (W/mK)
Matrices								
Al (1100)	–	70	0.33	0.17	2.7	933	23.6	230
Al–Cu (2024)	–	73	0.33	0.47	2.8	915	23.6	190
Al–Mg–Si (6061)	–	70	0.33	0.38	2.7	925	23.6	180
Al–Zn (7075)	–	72	0.33	0.57	2.8	925	22.0	130
Al–Li (8090)	–	80	0.33		2.55	930		
Cu (C10100)	–	117	0.34	0.22	8.9	1356	17.0	391
Mg	–	45	0.35	0.13	1.8	922	26.6	107
Fe (mild steel)	–	208	0.28	0.43	7.8	1810	17.2	60
Ti	–	110	0.36	0.24	4.5	1940	9.5	22
Ti–6Al–4V	–	115	0.36	0.95	4.4	1920	9.0	6
Zn	–	105	0.35	0.12	7.1	693	31.0	120
Ni	–	214	0.31	0.30	8.9	1728	13.0	89
Ni superalloy	–	214	0.31	1.30	7.9	1550	12.0	11
Pb	–	14	0.36	0.02	11.3	600	30.0	25
Reinforcements								
Single crystals (particles and whiskers)								
Al_2O_3	1–50	430	0.24		3.8	2313	7	~100
B_4C	1–50	480		2.1	2.5	2623		
Diamond		1000	0.07–0.25		3.2		0.5	600
SiC	1–50	450	0.17		3.2		4.0	~100
TiB_2		350–570	0.13–0.19		4.5		8.1	~100
TiC		230–400	0.19		4.9			
WC		>530			15.7			
SiC_W	0.1–1	450, <700	0.17	3.1, <21.0	3.2		4.0	>16
Si_3Ni_{4W}	0.1–10	350–385		2–4, <14	3.1	1600*		

Material	Diameter (μm)	Young's modulus (GPa)	Poisson ratio	Tensile strength (GPa)	Density (Mg/m^3)	Melting point (K)	CTE (με/K)	Thermal conductivity (W/mK)
Reinforcements (*cont.*)								
Metallic wires								
Patented steel	100	220	0.28	4.2, <13.0	7.9	1800	13.3	29
W wire	25–380	405	0.29	1.7–3.9	19.3	3660	4.5	168
Spun fibres (multifilaments and staple fibres)								
Al_2O_3 (Saffil®)	3	285		1.5	3.5	1600*/ 2313	7.7	
Al_2O_3 (DuPont FP)	20	380		1.3–2.1	4.0	1650*/ 2313	8.3	
C pitch	5–13	380–690		2.0	2.0	700*/ 3950	−1.4 (axial) 10 (trans)	355 (axial)
C PAN (high strength)	7	230		4.8	1.9	700*/ 3950	−1.2 (axial)	
C PAN (high modulus)		412		2.4			10 (trans)	
SiC (Nicalon)	15	180		>2.4, 8.3	2.55	1095*/ 2700†	3.0	
Monofilaments (CVD fibres)								
B (CVD W core)	50–100	400		2.7–7.0	2.6	2300	5.0	~38
Borsic (CVD)	100	415		2.93	2.77	2600	5.0	38
SiC (C core CVD)	150	345		3.8	3.29	2700†	4.0	16
SiC (W core CVD)	100	415		3.8	3.29	900*/ 2700†	4.0	16

* Denotes a maximum use temperature due to oxidation or other chemical reaction.
† Sublimation temperature.

These data are taken from various sources, including the following compilations[1–10]

1. J. W. Weeton, D. M. Peters and K. L. Thomas (1987) *An Engineers Guide to Composite Materials*, ASM, Ohio.
2. A. Kelly (1966) *Strong Solids*, Oxford University Press.
3. M. F. Ashby and D. R. H. Jones (1980) *Engineering Materials 2 – An Introduction to their Properties and Applications*, Pergamon, Oxford.
4. M. F. Ashby and D. R. H. Jones (1988) *Engineering Materials 1 – An Introduction to Microstructures, Processing and Design*, Pergamon, Oxford.
5. H. J. Frost and M. F. Ashby (1982) *Deformation Mechanism Maps – The Plasticity and Creep of Metals and Ceramics*, Pergamon, Oxford.
6. E. A. Brandes (ed.) (1983) *Metals Reference Book*, Butterworth, London.
7. G. H. Geiger and D. R. Poirier (1973) *Transport Phenomena in Metallurgy*, Addison-Wesley, New York.
8. A. P. Levitt (1970) *Whisker Technology*, Wiley and Sons, New York.
9. A. Kelly (ed.) (1989) *Concise Encyclopedia of Composite Materials*, Pergamon, Oxford.
10. J. M. Gere and S. P. Timoshenko (1987) *Mechanics of Materials*, Van Nostrand Reinhold, Wokingham.

Appendix III

The basic Eshelby S tensors

Following the notation adopted by Brown and Clarke[1] (omitting the redundant π's), the S tensor has the following form:

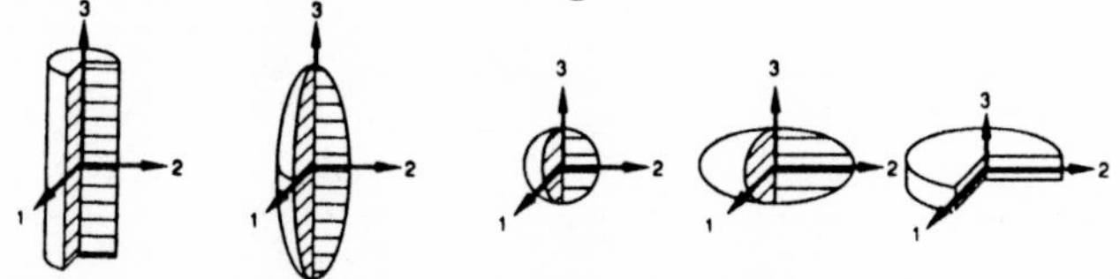

S_{ijkl}	Fibres	Prolate spheroids	Spheres	Oblate spheroids	Plates
S_{3333}	0	$\frac{4Q}{3} + RI_3 + 2s^2T$	$\frac{7-5\nu}{15(1-\nu)}$	$\frac{4Q}{3} + RI_3 + 2s^2T$	1
$S_{1111} = S_{2222}$	$\frac{5-\nu}{8(1-\nu)}$	$Q + RI_1 + \frac{3T}{4}$	$\frac{7-5\nu}{15(1-\nu)}$	$Q + RI_1 + \frac{3T}{4}$	0
$S_{1122} = S_{2211}$	$\frac{-1+4\nu}{8(1-\nu)}$	$\frac{Q}{3} - RI_1 + \frac{4T}{3}$	$\frac{-1+5\nu}{15(1-\nu)}$	$\frac{Q}{3} - RI_1 + \frac{4T}{3}$	0
$S_{1133} = S_{2233}$	$\frac{\nu}{2(1-\nu)}$	$-RI_1 - s^2T$	$\frac{-1+5\nu}{15(1-\nu)}$	$-RI_1 - s^2T$	0
$S_{3311} = S_{3322}$	0	$-RI_3 - T$	$\frac{-1+5\nu}{15(1-\nu)}$	$-RI_3 - T$	$\frac{\nu}{1-\nu}$
$S_{1212} = S_{1221} = S_{2112}$ $= S_{2121}$	$\frac{3-4\nu}{8(1-\nu)}$	$\frac{Q}{3} + RI_1 + \frac{T}{4}$	$\frac{4-5\nu}{15(1-\nu)}$	$\frac{Q}{3} + RI_1 + \frac{T}{4}$	0
$S_{1313} = S_{1331} = S_{3113}$ $= S_{3131} = S_{3232}$ $= S_{3223} = S_{2332}$ $= S_{2323}$	$\frac{1}{4}$	$2R - \frac{I_1R}{2} - \frac{1+s^2}{2}T$	$\frac{4-5\nu}{15(1-\nu)}$	$2R - \frac{I_1R}{2} - \frac{1+s^2}{2}T$	$\frac{1}{2}$
For all other S_{ijkl}	0	0	0	0	0

In this Table, the constants are given by

$$I_1 = \frac{2s}{(s^2-1)^{3/2}}(s(s^2-1)^{1/2} - \cosh^{-1} s) \quad \text{for a prolate ellipsoid}^\dagger \ (s > 1)$$

$$I_1 = \frac{2s}{(1-s^2)^{3/2}}(\cos^{-1} s - s(1-s^2)^{1/2}) \quad \text{for an oblate spheroid } (s < 1)$$

$$Q = \frac{3}{8(1-\nu)} \qquad R = \frac{1-2\nu}{8(1-\nu)} \qquad T = Q\frac{(4-3I_1)}{3(s^2-1)} \qquad I_3 = 4 - 2I_1$$

Note that in the contracted notation (no summation over subscripts):

$$S_{iiii} = S_{ii}; \quad S_{iijj} = S_{ij}; \quad S_{ijij} = \tfrac{1}{2}S_{3+k,3+k} \quad (i, j, k = 1 \text{ to } 3, i \neq j \neq k)$$

Details of other S tensors, including the case of materials which are only transversely isotropic, are given elsewhere[2,3].

For heat and electrical conduction, the S tensor is of second rank:

S_{ij}	Long fibre	Prolate spheroid	Sphere	Oblate spheroid	Plate
S_{33}	0	$1 - 2S_{11}$	$\frac{1}{3}$	$1 - 2S_{11}$	1
$S_{11} = S_{22}$	$\frac{1}{2}$	$\frac{1}{4}I_1$	$\frac{1}{3}$	$\frac{1}{4}I_1$	0
For all other S_{ij}	0	0	0	0	0

References

1. L. M. Brown and D. R. Clarke (1975) Work Hardening Due to Internal Stresses in Composite Materials, *Acta Metall.*, **23**, pp. 821–30.
2. O. B. Pedersen and P. J. Withers (1992) Iterative Estimates of Internal Stresses in Short Fibres MMCs, *Phil. Mag.*, **65A**, pp. 1217–33.
3. T. Mura (1987) *Micromechanics of Defects in Solids*, Nijhoff, The Hague.

† Note that $\cosh^{-1} x = \ln[x + \sqrt{(x^2-1)}]$.

Appendix IV

Listing of a program for an Eshelby calculation

```
PROGRAM Youngs_moduli;          {Calculates axial & transverse Youngs moduli of composite using eqn(3.32)}

    USES
        SANE;                                              {Pascal numerical algorithm library}

    TYPE
        Col_Matrix = ARRAY[1..6] OF extended;
        Col_Matrix_Int = ARRAY[1..6] OF integer;
        Sq_Matrix = ARRAY[1..6, 1..6] OF extended;

    VAR                                                    {Global variable names}
        i, steps: integer;
        Cm, Cf, Ccomp, Scomp, ident, S_tens: Sq_Matrix;    {C_M, C_I, C_comp, C^-1_comp, I, S}
        f, s, nu, nuf, Em, Ef, E3C, E1C, flower, fupper, span: extended;   {f, s, ν, ν_I, E_M, E_I, E_3Comp, f_min, f_max, span}
        fName: STRING;
        file_axial, file_trans: text;                      {data files for axial and transverse data}

    PROCEDURE Open_Text_Window;                            {Routine for preparing input data screen}
        VAR
            textrctngl: Rect;
    BEGIN
        HideAll;
        SetRect(textrctngl, 10, 50, 500, 330);
        SetTextRect(textrctngl);
        ShowText;
        writeln('Calculation of Composite Young''s moduli');
        writeln;
    END;

    PROCEDURE Input_Data;                                  {inputting of composite paramaters}
    BEGIN
        write('Matrix Young''s modulus (GPa): ');          {System data}
        readln(Em);                                        {matrix modulus}
        write('Fibre Young''s modulus (GPa):');
        readln(Ef);                                        {fibre modulus}
        write('Matrix Poisson''s ratio : ');
        readln(nu);                                        {matrix Poissons's ratio}
        write('Fibre Poisson''s ratio : ');
        readln(nuf);                                       {fibre Poisson's ratio}
        write('Fibre aspect ratio :');
        readln(s);                                         {fibre aspect ratio}
        write('f lower limit =');                          {Range of fibre volume fractions}
        readln(flower);                                    {fmin}
        write('f upper limit =');
        readln(fupper);                                    {fmax}
        span := fupper - flower;
        write('No. of points =');
        readln(steps);                                     {No of data points}
        write('Prefix for Eax & Etr files: ');             {Files for storage of results}
        readln(fName);                                     {filename prefix}
        rewrite(file_axial, concat(fName, '-Eax(f)'));
        rewrite(file_trans, concat(fName, '-Etr(f)'));
    END;
```

```
PROCEDURE Fibre_and_Matrix_Stiffness_Tensors;                    {calculating the stiffness tensors}
   VAR
      i, j: integer;
      mii, mij, m44, fii, fij, f44: extended;
BEGIN
   FOR i := 1 TO 6 DO
      BEGIN
         FOR j := 1 TO 6 DO
            ident[i, j] := 0;                                          {identity matrix set up}
         ident[i, i] := 1;
      END;
   Cm := ident;
   Cf := ident;
   Ccomp := ident;                          {Comp. stiffness tensor initialised as identity matrix}

   mii := Em * (1 - nu ) / ((1 - 2 * nu ) * (1 + nu));              {CMii=EM(1-v)/(1-2v)(1+v)}
   mij := Em * nu  / ((1 - nu - 2 * nu * nu));                          {CMij=EMv/(1-2v)(1+v)}
   m44 := Em / (2 * (1 + nu));                                               {CM44=EM/2(1+v)}
   fii := Ef * (1 - nuf) / ((1 - 2 * nuf) * (1 + nuf));           {CIii=EI(1-vI)/(1-2vI)(1+vI)}
   fij := Ef * nuf  / ((1 - nuf - 2 * nuf * nuf));                    {CIij=EIvI/(1-2vI)(1+vI)}
   f44 := Ef / (2 * (1 + nuf));                                            {CI44=EI/2(1+vI)}

   FOR i := 1 TO 3 DO
      BEGIN
         FOR j := 1 TO 3 DO
            BEGIN
               Cm[i, j] := mij;
               Cf[i, j] := fij
            END;
         Cm[i, i] := mii;
         Cf[i, i] := fii;
      END;
   FOR i := 4 TO 6 DO
      BEGIN
         Cm[i, i] := m44;
         Cf[i, i] := f44
      END;                          {Matrix & fibre isotropic, stiffness tensors completed}
END;

PROCEDURE Eshelby_S_Tensor;   {Evaluates S_tensor for given incl aspect ratio, s, & matrix Poissons ratio, nu}
                                         {Only the prolate ellipsoid is dealt with in this version}
   VAR
      A, B,I1,Q,R,T: extended;
BEGIN                                {see Appendix III  for complete list of S tensor components}
   S_tens:=ident;
   A:=ln(s+sqrt(s*s-1));
   I1:=2*s*(s*sqrt(s*s-1)-A)/((s*s-1)*sqrt(s*s-1));
   Q:=3/(8*(1-nu))
   R:=(1-2*nu)/(8(1-nu));
   T:=Q*(4-3I1)/(3*(s*s-1))
   S_tens[1,1]:=Q+R*I1+3*T/4;
   S_tens[2,2]:=S_tens[1,1];
   S_tens[3,3]:=4*Q/3+R*(4-2I1)+2*s*s*T;
   S_tens[1,2]:=Q/3-R*I1+4T/3;
   S_tens[2,1]:=S_tens[1,2];
   S_tens[1,3]:=-R*I1-s*s*T;
   S_tens[2,3]:=S_tens[1,3];
   S_tens[3,1]:=-R*(4-2*I1)-T;
   S_tens[3,2]:=S_tens[3,1];
   S_tens[4,4]:=4*R-I1*R-(1+s*s)*T;                                               {S44=2S1313}
   S_tens[5,5]:=S_tens[4,4];
   S_tens[6,6]:=2*Q/3+2*I1*R+T/2;                                                 {S66=2S1212}
END;

PROCEDURE Second_by_Second (a, b: Sq_Matrix; VAR r: Sq_Matrix);      {product of two 6x6 matrices}
   VAR
      i, j, k: integer;
BEGIN
   FOR i := 1 TO 6 DO
      FOR j := 1 TO 6 DO
         BEGIN
            r[i, j] := 0;
            FOR k := 1 TO 6 DO
               r[i, j] := r[i, j] + a[i, k] * b[k, j];
         END;
END;
```

```
PROCEDURE Second_by_Scalar (a: Sq_Matrix; b: extended; VAR r: Sq_Matrix);  {product of scalar & 6x6 matrix}
   VAR
      i, j: integer;
BEGIN
   FOR i := 1 TO 6 DO
      FOR j := 1 TO 6 DO
         r[i, j] := a[i, j] * b;
END;

PROCEDURE Second_Add (a, b: Sq_Matrix; VAR r: Sq_Matrix; sign: integer);      {addition of two 6x6 matrixces}
   VAR
      i, j: integer;
BEGIN
   FOR i := 1 TO 6 DO
      FOR j := 1 TO 6 DO
         r[i, j] := a[i, j] + sign * b[i, j];
END;

PROCEDURE Core_Calculation_of_Ccomp;                                {Implementation of equation (3.32)}
   VAR
      m1, m2, m3, m4, m5, m6, m7, m8, m9, m10, m11, m12: Sq_Matrix;              {dummy matrices}
BEGIN
   Second_Add(Cf, Cm, m1, -1);                                                    {m1=(C_I-C_M)}
   Second_Add(S_tens, ident, m2, -1);                                                {m2=(S-I)}
   Invert(Cm, m3);                                     {library inversion routine} - {m3=(C_M)^-1}
   Second_by_Scalar(m2, f, m4);                                                     {m4=f(S-I)}
   Second_Add(S_tens, m4, m5, -1);                                                {m5=S-f(S-I)}
   Second_by_Second(m1, m5, m6);                                        {m6=(C_I-C_M)[S-f(S-I)]}
   Second_Add(m6, Cm, m7, +1)                                       {m7=(C_I-C_M)[S-f(S-I)]+C_M}
   Invert(m7, m8);                                            {m8=[(C_I-C_M)[S-f(S-I)]+C_M]^-1}
   Second_by_Second(m8, m1, m9);                       {m9=[(C_I-C_M)[S-f(S-I)]+C_M]^-1(C_I-C_M)}
   Second_by_Second(m9, m3, m10);            {m10=[(C_I-C_M)[S-f(S-I)]+C_M]^-1(C_I-C_M)(C_M)^-1}
   Second_by_Scalar(m10, f, m11);           {m11=f[(C_I-C_M)[S-f(S-I)]+C_M]^-1(C_I-C_M)(C_M)^-1}
   Second_Add(m3, m11, m12, -1)    {m12=(C_M)^-1-f[(C_I-C_M)[S-f(S-I)]+C_M]^-1(C_I-C_M)(C_M)^-1}
   Invert(m12, Ccomp);       {C_comp=[(C_M)^-1-f[(C_I-C_M)[S-f(S-I)]+C_M]^-1(C_I-C_M)(C_M)^-1]^-1}
END;

PROCEDURE Find_Composite_Moduli;
BEGIN
   Core_Calculation_of_Ccomp;
                          {Now find the engineering constants from the composite stiffness tensor}
                           {This is done after first obtaining the compliance tensor by inversion}
   Invert(Ccomp, Scomp);
   E3C := 1 / Scomp[3, 3];          {Only axial and transverse Youngs moduli are evaluated in this version}
   E1C := 1 / Scomp[1, 1];
END;


BEGIN
   Open_Text_Window;                                                        {Start of MAIN program}
   Input_Data;
   Fibre_and_Matrix_Stiffness_Tensors;
   Eshelby_S_Tensor;
   FOR i := 0 TO steps DO                   {start stepping volume fraction between fmin and fmax}
      BEGIN
         f := flower + i * span / steps;
         Find_Composite_Moduli;
         writeln(file_axial, f : 30, E3C : 30);          {write fib. vol. frac. / axial Young's modulus}
         writeln(file_trans, f : 30, E1C : 30)       {write fib. vol. frac. / transverse Young's modulus}
      END;
   close(file_axial);
   close(file_trans);
END.
```

Author index

The work of the author mentioned is cited on the page(s) indicated

Subject index